T0176888

Integrating Biological Control into Conservation Practice

Integrating Biological Control into Conservation Practice

EDITED BY

Roy G. Van Driesche

Department of Environmental Conservation, University of Massachusetts, USA

Daniel Simberloff

Department of Ecology & Evolutionary Biology, University of Tennessee, USA

Bernd Blossey

Department of Natural Resources, Cornell University, USA

Charlotte Causton

Charles Darwin Foundation, Galápagos, Ecuador

Mark S. Hoddle

Department of Entomology, University of California, USA

David L. Wagner

Department of Ecology & Evolutionary Biology, University of Connecticut, USA

Christian O. Marks

The Nature Conservancy, Connecticut River Program, USA

Kevin M. Heinz

Department of Entomology, Texas A & M University, USA

Keith D. Warner

Center for Science, Technology, and Society, Santa Clara University, USA

WILEY Blackwell

This edition first published 2016 © 2016 by John Wiley & Sons, Ltd.

Registered Office
John Wiley & Sons, Ltd, The Atrium, Southern Gate, Chichester, West Sussex, PO19 8SQ, UK

Editorial Offices
9600 Garsington Road, Oxford, OX4 2DQ, UK
The Atrium, Southern Gate, Chichester, West Sussex, PO19 8SQ, UK
111 River Street, Hoboken, NJ 07030-5774, USA

For details of our global editorial offices, for customer services and for information about how
to apply for permission to reuse the copyright material in this book please see our website at
www.wiley.com/wiley-blackwell.

The right of the author to be identified as the author of this work has been asserted in accordance with
the UK Copyright, Designs and Patents Act 1988.

All rights reserved. No part of this publication may be reproduced, stored in a retrieval system, or
transmitted, in any form or by any means, electronic, mechanical, photocopying, recording or otherwise,
except as permitted by the UK Copyright, Designs and Patents Act 1988, without the prior permission
of the publisher.

Designations used by companies to distinguish their products are often claimed as trademarks. All brand
names and product names used in this book are trade names, service marks, trademarks or registered
trademarks of their respective owners. The publisher is not associated with any product or vendor
mentioned in this book.

Limit of Liability/Disclaimer of Warranty: While the publisher and author(s) have used their best efforts
in preparing this book, they make no representations or warranties with respect to the accuracy or
completeness of the contents of this book and specifically disclaim any implied warranties of
merchantability or fitness for a particular purpose. It is sold on the understanding that the publisher
is not engaged in rendering professional services and neither the publisher nor the author shall be liable
for damages arising herefrom. If professional advice or other expert assistance is required, the services
of a competent professional should be sought.

Library of Congress Cataloging-in-Publication data applied for

ISBN: 9781118392591

A catalogue record for this book is available from the British Library.

Wiley also publishes its books in a variety of electronic formats. Some content that appears in print may not
be available in electronic books.

Set in 8.5/12pt Meridien by SPi Global, Pondicherry, India

Printed in Singapore by C.O.S. Printers Pte Ltd

1   2016

Contents

List of contributors

Andersen, Jeremy C
Department of Environmental Science, Policy, & Management, University of California, Berkeley, USA, jandersen@berkeley.edu

Blossey, Bernd
Department of Natural Resources, Cornell University, Ithaca, New York, USA, bb22@cornell.edu

Causton, Charlotte
Charles Darwin Foundation, Puerta Ayora, Santa Cruz, Galápagos Islands, Ecuador, CAUSTON@rockbug.net

Center, Ted D
USDA ARS Invasive Species Laboratory (retired), Ft. Lauderdale, Florida, USA, tdcenter@comcast.net

Duan, Jian J
USDA ARS Beneficial Insects Introduction Research Unit, Newark, Delaware, USA, Jian.Duan@ARS.USDA.GOV

Fowler, Simon
Landcare Research, Manaaki Whenua, New Zealand, FowlerS@landcareresearch.co.nz

Heinz, Kevin M
Department of Entomology, Texas A & M University, College Station, TX, USA, kmheinz@tamu.edu

Hoddle, Mark S
Department of Entomology, University of California, Riverside, California, USA, mark.hoddle@ucr.edu

Johnson, M. Tracy
USDA Forest Service, Pacific Southwest Research Station, Institute of Pacific Islands Forestry, Volcano, Hawaii, USA, tracyjohnson@fs.fed.us

Kaufman, Leyla
Department of Entomology, University of Hawaii, Manoa, Hawaii, USA, leyla@hawaii.edu

Marks, Christian O
Connecticut River Program, The Nature Conservancy, Northampton, MA, USA, cmarks@TNC.ORG

Messing, Russell H
Department of Entomology, University of Hawaii, Manoa, Hawaii, USA, messing@hawaii.edu

Meyer, Jean-Yves
Délégation à la Recherche, Tahiti, French Polynesia, yves.meyer@recherche.gov.pf

Montgomery, Michael E
Northern Research Station, USDA Forest Service (retired), Hamden, Connecticut, USA, michaelemontgomery@fs.fed.us

Pratt, Paul D
USDA ARS Invasive Species Laboratory, Ft. Lauderdale, Florida, USA, Paul.Pratt@ARS.USDA.GOV

Purcell, Mary
USDA ARS Invasive Species Laboratory, Ft. Lauderdale, Florida, USA, mpurcell@nifa.usda.gov

Rayamajhi, Min B
USDA ARS Invasive Species Laboratory, Ft. Lauderdale, Florida, USA, Min.Rayamajhi@ARS.USDA.GOV

Sheppard, Andy W
Commonwealth Scientific and Industrial Research Organisation (CSIRO), ACT, Australia, Andy.Sheppard@csiro.au

Simberloff, Daniel
Department of Ecology & Evolutionary Biology, University of Tennessee, Knoxville, TN, USA, tebo@utk.edu

Tipping, Phil W
USDA ARS Invasive Species Laboratory, Ft. Lauderdale, Florida, USA, Philip.Tipping@ars.usda.gov

Van Driesche, Roy G
Department of Environmental Conservation,
University of Massachusetts,
Amherst, MA, USA, vandries@cns.umass.edu

van Klinken, Rieks
Commonwealth Scientific and Industrial Research
Organisation (CSIRO), Brisbane, Queensland, Australia, Rieks.
VanKlinken@csiro.au

Wagner, David L
Department of Ecology & Evolutionary Biology,
University of Connecticut, Storrs, Connecticut, USA,
david.wagner@uconn.edu

Warner, Keith D
Center for Science, Technology, and Society,
Santa Clara University, California, USA,
kdwarner@gmail.com

Preface

The magnitude of threat posed to native ecosystem function and biodiversity by some invasive vertebrates, insects, pathogens, and plants is enormous and growing. At the landscape level, after damaging invaders are beyond eradication, a variety of habitats and ecosystems, on islands and continents, in all parts of the world may be affected and require some form of restoration. Biological control offers substantial opportunity to reduce the damage from invasive insects and plants, two of the most frequent and damaging groups of invasive species.

The purpose of this book is to address a nearly 25-year-old rift (from the seminal article by Howarth [1991]) that opened between conservation/restoration biologists and biological control scientists, particularly in the United States, so that in the future conservation biologists and biological control scientists might work together better to restore native ecosystems damaged by invasive species. The planning for this book originated in an informal meeting of conservation biologists, invasion biologists, and biological control scientists in October 2009, in Sunapee, New Hampshire, following a meeting that year on biological control for the protection of natural areas, held in Northampton, Massachusetts.

The tension between biological control and conservation biology had two causes. The first was that by the 1960s biological control agents introduced earlier to protect grazing or agricultural interests were found attacking native plants and insects in natural areas. More extensive search found other cases of such non-target impacts (Johnson and Stiling, 1996; Louda et al., 1997; Strong, 1997; Boettner et al., 2000; Kuris, 2003), tarnishing the use of biological control for a generation of conservation biologists and restoration ecologists. Any discussion of potential use of biological control agent to mitigate pest problems prompted the question: "What will it eat next if it controls the target?" This question is today routinely asked by undergraduates, graduate students, and the general public, but fails to recognize the dietary restrictions of many biological control agents. Mechanisms of population dynamics exist that cause insects with specialized diets, unlike vertebrates, to lose host-finding efficiency when the density of their prey or host plant declines, resulting in lower realized fecundity and a decrease in population size. Therefore, for specialized biological control agents, the answer to "what will they eat next" is "the same, just less of it as it becomes harder to find." Others were concerned that agents would attack non-target species due to evolutionary expansion of their host ranges. However, while host shifts do frequently occur over evolutionary time (Stireman, 2005; Barrett and Heil, 2012), such changes have rarely been documented among insects introduced for biological control.

The second reason for the lack of understanding that developed between biological control and conservation/restoration scientists was research compartmentalization, with each group defining itself into its own sub-disciplines, attending different meetings and publishing in different journals. This is true both for conservation/restoration biologists (who publish in *Conservation Biology*, *Restoration Ecology*, *Biological Invasions*, etc.) and biological control scientists (*BioControl*, *Biological Control*, *Biological Control Science and Technology*, etc.). Opportunities to talk at length between these groups were, therefore, rare.

If invasive species were not one of the most important drivers of ecological degradation across natural ecosystems, the status quo could continue indefinitely. But they are and we must confront them as efficiently as possible. Conservation biologists should no longer leave a good tool unused and biological control scientists should no longer work in isolation from conservation biologists with special knowledge of the invaded ecosystems. The goal of this book is to discuss these issues in ways that make sense to both groups and find ways to work together better.

References

Barrett, L. G. and M. Heil. 2012. Unifying concepts and mechanisms in the specificity of plant-enemy interactions. *Trends in Plant Science* **17**: 282–292.

Boettner, G. H., J. S. Elkinton, and C. J. Boettner. 2000. Effects of a biological control introduction on three nontarget native species of saturniid moths. *Conservation Biology* **14**: 1798–1806.

Howarth, F. G. 1991. Environmental impacts of classical biological control. *Annual Review of Entomology* **36**: 485–509.

Johnson, D. M. and P. D. Stiling. 1996. Host specificity of *Cactoblastis cactorum* (Lepidoptera: Pyralidae), an exotic *Opuntia*-feeding moth, in Florida. *Environmental Entomology* **25**: 743–748.

Kuris, A. M. 2003. Did biological control cause extinction of the coconut moth, *Levuana iridescens*, in Fiji? *Biological Invasions* **5**: 133–141.

Louda, S. M., D. Kendall, J. Connor, and D. Simberloff. 1997. Ecological effects of an insect introduced for biological control of weeds. *Science* **277** (5329): 1088–1090.

Stireman, J. O. 2005. The evolution of generalization? Parasitoid flies and the perils of inferring host range evolution from phylogenies. *Journal of Evolutionary Biology* **18**: 325–336.

Strong, D. R. 1997. Fear no weevil? *Science* (Washington) **277** (5329): 1058–1059.

CHAPTER 1

Integrating biological control into a conservation context: why is it necessary?

Kevin M. Heinz[1], Roy G. Van Driesche[2], and Daniel Simberloff[3]

[1] Department of Entomology, Texas A & M University, USA

[2] Department of Environmental Conservation, University of Massachusetts, USA

[3] Department of Ecology & Evolutionary Biology, University of Tennessee, USA

Potential problems if integration is lacking

The basic argument of this book is that, for pests of wildlands[1], biological control should be one of the tools considered for use. Not to do so would lead to inadequate restoration for many pests because, while they might be controlled in small areas, they would remain uncontrolled over much of the landscape. We further argue that biological control will be done better if integrated into conservation biology because that will force greater consideration of the role of the invader as the true source, or not, of ecosystem degradation (see Chapter 2) and would incorporate into the control program more detailed knowledge of the invaded community's ecology, which may exist best within the conservation biology community. Finally, we argue that biological control in areas of conservation importance can be done safely with modern methods of evaluation for assessing pest impact and natural enemy host range.

When conservation biologists seek to restore natural communities damaged by invasive species, if they give no thought to biological control, their efforts may be far less successful. Without biological control in the mix of potential tools, restoration efforts move toward eradication if possible, suppression over large areas by changing processes (e.g., fire, flood, or grazing regimes) at the landscape level if relevant, or suppressing the invader on small patches with chemical or mechanical tools if these methods work and money can be found for long-term management. Many invaders, however, cannot be eradicated if they are widespread, or their biology may not be appropriate to control over the long term with pesticides or mechanical tools. Similarly, while some plants or insects may have become highly invasive because people have altered historical landscape processes (MacDougall and Turkington, 2005), this factor surely does not account for the damage caused by some invaders. Certainly, it applies to few if any invasive insects: virtually none of the invasive insects that have so damaged North American forests (Campbell and Schlarbaum, 1994; Van Driesche and Reardon, 2014) could be said to have such factors driving their destructive effects. In contrast, some invasive plants quite likely are augmented in their densities by such forces, but clearly not all are. This leaves many highly damaging insects and plants for which restoration of ecological processes toward historical norms will not lead to restoration of the ecosystem. In such cases, then, restoration efforts are limited to saving fragments through intensive efforts at the preserve rather than the landscape level. While these efforts may protect rare species with small, threatened ranges, they do nothing to preserve average habitat conditions for the bulk of species across the broader landscape. Working with biological control scientists can sometimes provide a solution that can safely (if well conceived and executed) protect the landscape rather than just a few isolated preserves.

---

[1] For purposes of this book, the term "wildlands" does not equal wilderness nor does the term "natural" mean "pristine." Rather the term wildlands is taken to mean places, both land and water, that are not intensively managed.

Integrating Biological Control into Conservation Practice, First Edition. Edited by Roy G. Van Driesche, Daniel Simberloff, Bernd Blossey, Charlotte Causton, Mark S. Hoddle, David L. Wagner, Christian O. Marks, Kevin M. Heinz, and Keith D. Warner.
© 2016 John Wiley & Sons, Ltd. Published 2016 by John Wiley & Sons, Ltd.

To succeed at biological control is not easy and requires cross-disciplinary collaborations to understand fully the implications of releasing natural enemies of the invader. If such collaborations with conservation biologists are lacking, decisions may be taken that undervalue certain native species, miss important ways in which these species are interacting, or fail to consider fully the potential impacts of the introduced biological control agents on the native ecosystem or what other forces may be at work driving ecosystem change. If biological control scientists work within a broader restoration team that includes conservation biologists, these potential pitfalls are more likely to be recognized and avoided.

Carrying out a biological control program typically requires a commitment to travel to the invader's native range and determine what natural enemies affect the invader's population dynamics there and which of these are plausibly sufficiently specialized that they might be safe for release in the invaded region. These demands require training in natural enemy biology and population dynamics, as well as knowledge of foreign cultures and geography. If the targeted invader is a plant, the biological control scientist must also have extensive understanding of plant taxonomy, physiology, and how both biotic and abiotic factors affect plant demography. If the invader is an insect, the practitioner must also be familiar with the taxonomy and biology of parasitoids or predators, how to rear them, and how they overcome host defenses. Training in these diverse subjects may leave little time to develop a deep appreciation for the community ecology and details of the particular ecosystems invaded by the pest. This leaves the biological control scientist vulnerable to making decisions that fail to take such information fully into account, and hence underscores the value of collaborative projects within a conservation biology framework, working with specialists on the ecology of the invaded communities.

Book organization

The practices of biological control and ecological restoration can be viewed as large-scale field experiments that unintentionally test many fundamental principles in ecology, as noted previously for both biological control (e.g., Hawkins and Cornell, 1999; Wajnberg et al., 2001; Roderick et al., 2012) and species conservation

and habitat restoration (e.g., Young, 2000; Groom et al., 2005). Several issues need addressing when one attempts to integrate biological control of pests of wildlands into the larger framework of conservation biology. In the chapters that follow, experts illustrate some of the problems that can arise when such integration is lacking and provide insights for avoiding problems that may affect the management program or conservation interests.

In Chapter 2, readers are presented with a conceptual framework for confirming whether an invasive species is the primary cause of environmental change and for deciding how to minimize its impacts, potentially as part of a larger package of restoration activities. Approaches potentially able to generate the desired outcomes are discussed and illustrated with the example of conservation threats to floodplain forests in New England. Chapter 3 subsequently addresses the means (tools) available to control invasive species. Depending on circumstances, control goals may be eradication, human-sustained invader suppression with periodic mechanical or chemical control plus monitoring, or permanent area-wide invader suppression through alteration of ecosystem processes or programs of biological control. Once goals are set, a variety of tools may be relevant and are discussed (mechanical, chemical, biological, combinations) in terms of the system or pest attributes that affect efficacy, control cost, and effects on the environment. Chapter 4 examines tradeoffs among risks posed by major control methods using case histories of particular projects. Chapter 5 continues this discussion through an examination of how the risks and benefits of biological control projects against wildland pests can best be recognized and compared, through the planned interaction of biological control scientists and conservation biologists. At the end of these chapters, readers should have a better understanding of when biological control may be the right or wrong option.

The next block of chapters shifts to the practice of biological control within the context of environmental restoration projects. Chapter 6 discusses the importance of systematics and accurate taxonomic identification, both of pests and natural enemies, for biological control programs. The discussion includes recent developments in molecular techniques applicable to modern biological control programs. Chapter 7 addresses our ability to forecast unwanted impacts of biological control, describing the nature of the concern, reviewing the historical record, and ending with a discussion of unresolved issues. Chapters 8 and 9 discuss how to measure and evaluate outcomes of biological control projects. Because biological control is costly in terms of financial and human resources, there is an increasing demand for accountability as to efficacy when biological control is used to restore or protect native ecosystems or species. Addressed directly in these chapters are the difficult tasks associated with delineating the damaged system's starting conditions and measuring the progress toward achieving restoration goals. Chapter 8 takes a broad conceptual view of the task, while Chapter 9 reviews techniques used for such assessments and their limits and requirements for application. Chapter 10 discusses a series of biological control projects conducted in wildland ecosystems. These cases provide concrete examples of the kinds of damage that can be corrected with biological control, and the discussions of project details highlight the variety of issues that can affect such work.

Concluding chapters address societal and economic matters. Chapter 11 discusses laws and regulations that affect biological control. The evolution of regulations and regulatory agencies from several parts of the world are reviewed, which provides the context for recommendations for improvements in biological control regulations. Chapter 12 describes how conflicts among groups may arise during a biological control project. The focus of the chapter is on methods for setting goals and resolving disagreements that are either initially present or arise during the conduct of the project. Chapter 13 discusses ethical principles related to the introduction of non-native species, focusing on processes and goals that can help resolve disagreements among parties in conflict. In Chapter 14, we discuss economic issues associated with species invasions and their biological control in wildlands. Chapter 15 describes steps to reform the practice of biological control and integrate its use against pests of wildlands into a conservation framework. It also makes recommendations for changes needed to make biological control of agricultural and ornamental pests at least environmentally neutral.

We end by returning to the central message of the book, looking to the future and describing activities likely to further the integration between biological control activities and those of conservation biologists and restoration ecologists.

Acknowledgments

We thank Bernd Blossey, Charlotte Causton, and David Wagner for reviewing Chapter 1.

References

Campbell, F. and S. E. Schlarbaum. 1994. *Fading Forests I.* Natural Resource Defense Council. New York.

Groom, M. J., G. K. Meffe, and C. R. Carroll. 2005. *Principles of Conservation Biology*, 3rd edn. Sinauer Associates. Amherst, Massachusetts, USA.

Hawkins, B. A. and H. V. Cornell. 1999. *Theoretical Approaches to Biological Control*. Cambridge University Press. New York. 424 pp.

MacDougall, A. S. and R. Turkington. 2005. Are invasive species the drivers or passengers of change in degraded ecosystems? *Ecology* **86**: 42–55.

Roderick, G. K., R. Hufbauer, and M. Navajas. 2012. Evolution and biological control. *Evolutionary Applications* **5**: 419–423.

Van Driesche, R. G. and R. Reardon (eds.). 2014. *The Use of Classical Biological Control to Preserve Forests in North America.* FHTET 2013-2 September 2014. USDA Forest Service. Morgantown, West Virginia, USA.

Wajnberg, E., J. K. Scott, and P. C. Quimby (eds.). 2001. *Evaluating Indirect Ecological Effects of Biological Control.* CABI Publishing. New York. 261 pp.

Young, T. P. 2000. Restoration ecology and conservation biology. *Biological Conservation* **92**: 73–83.

CHAPTER 2

Designing restoration programs based on understanding the drivers of ecological change

Christian O. Marks[1] and Roy G. Van Driesche[2]

[1] *Connecticut River Program, The Nature Conservancy, USA*

[2] *Department of Environmental Conservation, University of Massachusetts, USA*

Overview of concepts

Introduction

The activities of conservation planning and biological control of invasive species are both continuing to evolve, requiring greater collaboration between these disciplines to achieve mutual goals pertaining to invasive species management (Chapter 1). Invasive species can be a factor contributing to ecological degradation (Simberloff, 2011; Kumschick et al., 2015). Even reserves in relatively intact ecosystems in remote regions can be threatened by exotic species invasions. Often this impact is not recognized until after the invasive species has become too abundant and widespread for eradication or even containment (e.g., Herms and McCullough, 2014). Long term, such pervasive invader populations are usually prohibitively expensive to suppress using conventional chemical and mechanical methods, especially as the infested area increases to tens or hundreds of thousands of hectares. Development of an effective biological control program is a potential alternative for managing an invasive pest, but biological control frequently must be integrated into the broader conservation plans of the local ecosystem because invasive species, particularly invasive plants, are rarely the only factor contributing to ecological degradation, as we will illustrate. Even where an invasive species is the leading cause of ecological degradation, its control alone may not accomplish restoration goals, and additional measures may be necessary (Chapter 3). Moreover, funding for conservation is limited, necessitating a strategic approach and a clear vision of what the intended end goal will be for the restoration.

In this chapter, we briefly review the conservation planning process, focusing on the roles invasive species play in ecological change. We pay particular attention to how to determine if an invasive species rises to the level of threat that warrants development of a biological control program, which we illustrate with a representative case study – the restoration of Connecticut River floodplain forests in the northeastern United States. A lack of integration into a wider restoration planning process has sometimes resulted in criticism of past biological control programs. For example, biological control of purple loosestrife (*Lythrum salicaria* L.) is one of the most widespread biological control programs for weeds in North America (Wilson et al., 2009), yet the necessity of controlling this invader has been questioned by some ecologists (Anderson, 1995) – although some of these concerns have since been rebutted (Blossey et al., 2001). More notably, in another case, a lack of integration of the biological control of saltcedar (*Tamarix* species) into a wider plan for the ecological restoration of riparian communities in the southwestern United States has resulted in controversy among various interest groups (see Chapter 4; or Dudley and Bean, 2012). Saltcedar is a widespread invader of riparian areas along southwestern rivers with well-known, large negative ecological impacts, but on some rivers it has also become one of the few remaining riparian tree species (Tracy and DeLoach, 1999; Sher and Quigley, 2013). The release of a highly effective biological control agent for

Integrating Biological Control into Conservation Practice, First Edition. Edited by Roy G. Van Driesche, Daniel Simberloff, Bernd Blossey, Charlotte Causton, Mark S. Hoddle, David L. Wagner, Christian O. Marks, Kevin M. Heinz, and Keith D. Warner.
© 2016 John Wiley & Sons, Ltd. Published 2016 by John Wiley & Sons, Ltd.

saltcedar, without also taking action to increase recruitment of native floodplain tree species like willows (*Salix*) and cottonwoods (*Populus*), may have resulted in a loss of some marginal nesting habitat for the federally listed endangered southwestern willow flycatcher (*Empidonax traillii extimus* Phillips) (Finch et al., 2002; Smith and Finch, 2014). On some southwestern rivers, modifying operations at dams to restore a more natural flood regime downstream, alone or in combination with saltcedar biological control, may be more effective at restoring floodplain function, including natural recruitment of the native riparian trees that the flycatcher prefers for nesting (Cooper et al., 2003; Richard and Julien, 2003; Shafroth et al., 2005; Ahlers and Moore, 2009; Hultine et al., 2009; Merritt and Poff, 2010; Dudley and Bean, 2012). These examples show how important it is to evaluate the factors that are influencing ecosystem function and degradation before irreversible actions are taken. The mere high dominance by an invasive species is not necessarily equivalent to degradation of ecological function. Therefore, it is necessary to rank invasive species not just against each other for control priority, but also to rank their control against other conservation actions that may have a greater positive impact. It is critical to think holistically about how the system functions before designing a plan of action.

Ecological restoration planning process

The motivations for carrying out ecological restoration are diverse and depend on the stakeholders' values. These motivations can include anything from landscape aesthetics and protection of endangered species to conservation of ecosystem services. The first step in the planning process is to achieve a consensus among stakeholders on what aspects of the ecosystem are valued, as well as what outcomes are desired for the restoration activity. This goal-setting process is subjective, and it is important to achieve a consensus among stakeholders early to avoid conflicts later, when program momentum may be significant, making change difficult or costly (Chapter 12). Next, one needs to understand the threats that have led to past declines in the aspects of the ecosystem where restoration is desired. Specifically, one needs to develop an understanding of system change with the best science available at the time, being aware that our knowledge of the system is usually incomplete. Consequently, it is important to be explicit about one's assumptions of what is driving change in the system

because they could be incorrect (Wilkinson et al., 2005), and scientists should seek to test such assumptions to guide restoration in an adaptive management framework (Westgate et al., 2013).

Invasive species and system change

High abundance of invasive species in wildlands is often associated with dramatic ecosystem alterations, such as eutrophication of soil or water bodies (Green and Galatowitsch, 2002; Perry et al., 2004; Silliman and Bertness, 2004; Kercher et al., 2007), overgrazing (Knight et al., 2009; HilleRisLambers et al., 2010; Dornbush and Hahn, 2013), and altered disturbance regimes such as fire and flooding (Cooper et al., 2003; Katz and Shafroth, 2003; Keeley, 2006; MacDougall and Turkington, 2007; Stromberg et al., 2007; Merritt and Poff, 2010; Metz et al., 2013; Greet et al., 2013; Schmiedel and Tackenberg, 2013; Terwei et al., 2013; Reynolds et al., 2014). However, it is not always immediately obvious to what degree non-native species invasions are the cause or the consequence of the ecological change, or both. Determining the answer to this question is crucial to deciding if the most effective strategy is more likely to be restoring the physical environment and key ecological processes or starting a biological control program, or if both may be necessary.

MacDougal and Turkington (2005) defined invasive species that thrive on ecological change, such as altered ecosystem properties or a shift in disturbance regimes, as passengers (see Figure 2.1). Owing to their high density in degraded ecosystems, passengers appear more damaging than they actually are. If the ecosystem

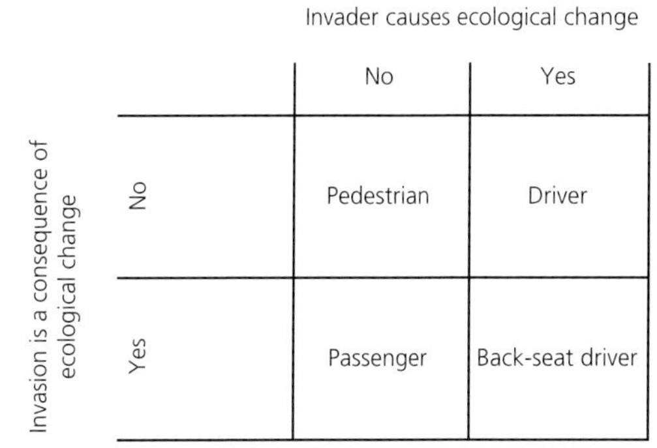

Figure 2.1 A chart to classify the ecological role of an invasive species on the spectrum from invasion being a consequence of ecological change to invasion being the cause of ecological change.

stressor that has allowed the passenger to proliferate is removed, one would expect passenger populations to decline. MacDougal and Turkington (2007) argued, for example, that the *Poa pratensis* L. invasion of Garry oak (*Quercus garryana* Douglas ex Hook.) savannas in British Columbia was a consequence of fire suppression. The failure of native vegetation to respond to *Poa* removal indicated that *Poa* was not the cause of change, only associated with it. Follow-up experiments found that restoration of fire to these ecosystems reduced invader abundance and promoted native species' recovery (MacDougall and Turkington, 2007).

Exceptions to the autogenous recovery of native populations following removal of the ecosystem stressor include situations where there are strong feedbacks between biotic factors and the physical environment (Suding et al., 2004). Specifically, once an invasive species is dominant, it might change the environment in ways that would favor its continued dominance even after the factor promoting its initial establishment was removed. For example, marsh disturbances such as ditching create microsites with better soil aeration where invasive common reed (*Phragmites australis* [Cav.] Trin. ex Steud.) can establish (Bart and Hartman, 2003; Chambers et al., 2003; Lathrop et al., 2003; Silliman and Bertness, 2004). Once established, *Phragmites* can transfer air within a clone via its hollow stalks, enabling it to spread to the rest of the marsh, forming large monospecific patches (Bart and Hartman, 2000; Lathrop et al., 2003). In another example, native deer herbivory was shown to accelerate forest invasion of garlic mustard (*Alliaria petiolata* [M. Bieb.] Cavara & Grande), Japanese barberry (*Berberis thunbergii* DC), and Japanese stiltgrass (*Microstegium vimineum* [Trin.] A. Camus), but was not as important as canopy disturbance or propagule pressure in explaining different levels of invasive weed abundance (Eschtruth and Battles, 2009). Once these invasive, non-native forest understory plants became abundant, propagule pressure would remain high even if canopy disturbance and deer herbivory were reduced. In such cases, restoration success would require both reducing the ecosystem stressor that had led to ecological degradation and suppressing the invasive species to reduce propagule pressure. Similarly, native plant propagules may be too scarce for native plants to recolonize on their own even after deer and invasive plant populations have been reduced, thus necessitating native plant seed addition or planting

(Tanentzap et al., 2009, 2011, 2013; Collard et al., 2010; Royo et al., 2010; Dornbush and Hahn, 2013). Holistic restoration approaches are especially important in urban and suburban areas, where there are usually multiple interacting stressors including invasive plants (Sauer, 1998).

In contrast to ecological passengers, MacDougal and Turkington (2005) defined drivers as invasive species that are both able to proliferate unaided by external ecological change and cause considerable damage. An example of an invasive driver is the fungal pathogen *Cryphonectria parasitica* (Murrill) Barr, the causal agent of chestnut blight. This fungus was accidentally introduced from Asia into North America, where it killed virtually all mature American chestnut (*Castanea dentata* [Marshall] Borkh.), the tree that once dominated many eastern North American forests (Braun, 1950). Attempts at biological control of the chestnut blight fungal pathogen with viruses were successful in Europe but not in eastern North America (Anagnostakis, 2001; Milgroom and Cortesi, 2004). Current efforts at restoring American chestnut are instead focused on breeding blight-resistant hybrids (Jacobs, 2007; Anagnostakis, 2012). Other examples of pure drivers of ecological change are the cottony cushion scale (*Icerya purchasi* Maskell), a phloem-sucking insect that caused many native plant populations in the Galápagos Islands to decline (Chapter 10), and laurel wilt, a disease caused by an invasive fungus vectored by the non-native redbay ambrosia beetle (*Xyleborus glabratus* Eichhoff), which is causing extensive mortality of redbay (*Persea borbonia* [L.] Spreng.) in the southeastern United States (Spiegel and Leege, 2013). Clearly, drivers are the most threatening invasive species and thus should receive a high priority on lists of candidate invaders for developing control programs.

Although originally set up as a dichotomy, the distinction between drivers and passengers is more accurately thought of as a spectrum, with many invasive species being intermediate cases where their proliferation has benefited from wider ecosystem change, but their high abundance also affects the ecosystem. Bauer (2012) has called these intermediate cases back-seat drivers, and his review suggests that most invasive plant species are back-seat drivers. Berman et al. (2013) proposed that invasive non-native ants in New Caledonia are back-seat drivers whose initial invasion is associated with disturbance, such as forest clearing, but which subsequently also harms native ant communities. Similarly, experimental manipulations have shown that invasion by the red imported fire ant (*Solenopsis invicta* Buren.) in the southeastern United States is driven by disturbance (King and Tschinkel, 2008). Many studies have documented large impacts by non-native fire ants on native ants and other native arthropods through competition and predation (Porter and Savignano, 1990; Gotelli and Arnett, 2000; Wojcik et al., 2001; Sanders et al., 2003). Decapitating flies in the genus *Pseudacteon* (e.g., *P. tricuspis* Borgmeier) were imported from Argentina and released as biological control agents of the red imported fire ant because the type of disturbance that promotes fire ant invasion has become unavoidable in much of the landscape, resulting in substantial damage to crops, livestock, human health, electrical equipment, and wildlife (Porter et al., 2004). Thus, where system changes that have enabled invasion by a back-seat driver are irreversible, there may be a sufficiently compelling argument for developing a biological control program.

Finally, there are non-native species whose establishment is not associated with significant ecological change either as a cause or consequence. We have labeled these species as pedestrians in Figure 2.1 to highlight the difference in pace of change. It is important to remember that the categories in Figure 2.1 are not immutable; many of today's invasive driver species were pedestrians receiving little notice during the first century of colonization in their new range (Kowarik et al., 1995; Crooks, 2005). With the right ecological or evolutionary changes, species can quickly switch between these categories. Moreover, local context matters; an invasive species that acts like a back-seat driver or passenger in one area may act like a driver in another part of its invaded range or in a different habitat (Wilson and Pinno, 2013). Therefore, in cases where there are no obvious large impacts by an invader in a particular ecosystem, further study elsewhere may be necessary to make a well-informed assessment of their overall impact in the invaded range.

Ranking invasive species for classical biological control

Central to ranking ecological threats for remediation is a consensus on what level of impact is sufficient to require conservation action. For example, The Nature Conservancy's conservation planning process ranks threats (both biotic and abiotic) according to scope, severity, and irreversibility (also referred to as permanence). With

respect to an invasive species, scope could be the area or percentage of a habitat likely to become threatened by the invader over the coming decade. Severity could be thought of as the level of damage to native biota in the invaded area that can reasonably be expected from the threat given the continuation of current circumstances and trends. Severity is the seriousness of the impact. For example, an insect pest invasion that causes high mortality of its tree host would be considered a more severe threat than one that only reduced the tree's growth rate. Irreversibility (or permanence) is the degree to which the effects of a threat cannot be reversed by restoration. For instance, the effects of the most damaging non-native species, once they become widespread, are difficult to reverse. Therefore preventing invaders from establishing, through early detection and elimination of incipient populations, generally receives high priority in conservation planning.

To help answer the question of how severe the threat posed by an invasive species needs to be to warrant the development of a biological control agent, given the costs and risks involved, we suggest using the following ranking, keeping in mind that ranking will vary depending on conservation goals and context, particularly stakeholder values. Invasive species that change community composition by taking up space and resources but do not destroy native biota should receive a low rank, especially if they are largely passengers of other ecological changes. Many non-native plants fall into this category. A more severe threat is posed by invasive species that cause a high rate of mortality in an important native species such as one of the following types: a community dominant, an endangered species, an ecosystem engineer, or an economically important species. Perhaps the most severe threat is posed by invasive species that have large undesirable impacts on ecosystem function. Such changes in ecosystem function include altered disturbance regimes, such as increases in fire intensity and frequency, large persistent changes in ecosystem properties like soil chemistry through salinization or nitrogen fixation, reductions in ecosystem services like drinking water supply through dramatic increases in transpiration, and qualitative changes in vegetation structure like conversion of forest to scrubland or grassland or vice versa.

Ranking of invasive species for control can be complicated if a species has both positive and negative ecological impacts. Consider the case of common reed (*P. australis*) invading North American marshes. This reed results in an almost two-thirds decrease in native plant species richness (Silliman and Bertness, 2004); however, its presence has also been shown to increase tidal marsh soil accretion rates, increasing resilience to sea level rise and storm surges (Rooth and Stevenson, 2000; Rooth et al., 2003). Imposing a hierarchy on different types of impacts can help in making decisions regarding invasive species whose effects are both positive and negative. In the case of common reed (*Phragmites*) it was decided that loss of native plant and bird diversity was sufficient reason to embark on a biological control program (Tewksbury et al., 2002; Blossey, 2003). Crucial to a well-informed decision-making process is quantifying the ecological impacts of an invader and understanding the causal mechanisms driving invasions before embarking on expensive control measures, biological or otherwise. Such an approach has not yet been widely adopted owing to a lack of relevant research results available to conservation managers.

To help guide the assessment of potential targets for developing a classical biological control program, we developed a decision tree (Figure 2.2). The first step in the process is to determine if the invasion is caused by some independent ecosystem change, such as eutrophication, overgrazing, or altered disturbance regime. If so, addressing this other stressor may be more important than attempting to control the invasive species directly. Next, one should assess the level of impact the invader is having. Given limited resources for conservation, control efforts should focus on the invaders with the most severe impacts. These first two steps need not necessarily involve lengthy scientific investigations. For example, in the case of the invasion of the emerald ash borer (*Agrilus planipennis* Fairmaire) in Michigan, it was immediately obvious that the pest was able to invade relatively unaltered ash forests and had severe impacts through causing high rates of mortality of a commercially and ecologically

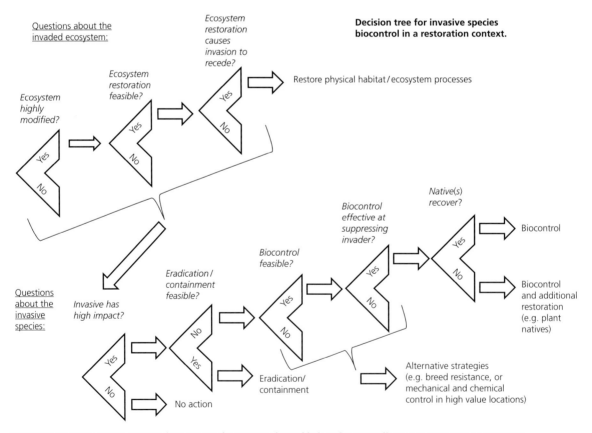

Figure 2.2 Decision tree to assist in determining the strategy that is likely to be most effective at restoring an ecosystem.

important tree species. Given those facts, an eradication/ containment program was implemented. After this effort failed due to the sometimes large dispersal distances by adult emerald ash borers, a biological control program was initiated (Herms and McCullough, 2014). Biological control, however, is not always feasible or successful. For instance, control of an insect vector of a systemic disease is unlikely to reduce the vector's abundance sufficiently to prevent disease transmission (e.g., Fuester et al., 2014). In such cases an alternative conservation strategy, for example breeding disease-resistant varieties, may be more successful. Even where biological control is successful at suppressing invader populations, populations of native species may not recover satisfactorily and additional restoration actions such as planting native plants or reducing herbivory may be necessary (Lake et al., 2014).

Despite the potential challenges, conservation managers are increasingly looking to biological control as a more sustainable solution to invasive species management because conventional mechanical and chemical control needs a high level of investment indefinitely. A further advantage of biological over chemical and mechanical control is that it does not stop at the property line of participating landowners, thus avoiding a major source of re-invasion. A growing trend is to focus mechanical and chemical invasive species control on sites of exceptionally high conservation value (e.g., eagle nest tree threatened by invasive vines) or preserve locations with high public visibility where the desire is to keep them as a natural history museum (e.g., nature centers with an environmental education mission) to reduce costs. Thus, for invasive species that are already widespread, developing an effective biological control program may be the only potentially feasible way to suppress the invader's population and its damage over the long term and at geographic scales larger than a few hundred hectares. Given this realization, we would argue that conservation organizations and especially state and federal agencies should increasingly shift conservation resources used for mechanical and chemical invasive species control in the past towards developing more classical biological control programs for the most serious invasive insect and plant species in the future (Van Driesche et al., 2010; Martin and Blossey, 2013).

Designing a restoration plan using Connecticut River floodplain forests as a model

Aquatic habitats, riparian areas, and wetlands are the focus of much restoration work because of their disproportionate importance to ecosystem services (Costanza et al., 1997; Zedler and Kercher, 2005; Moreno-Mateos et al., 2012). These habitats are disturbed by floods and human activities, and they accumulate water, energy, sediments, nutrients, pollutants, seeds, and other propagules from the rest of the watershed, making them prone to invasion (Zedler and Kercher, 2004; Richardson et al., 2007; Catford and Jansson, 2014). A high abundance of invasive plants is, for example, a common cause of failure in wetland mitigation (Minkin and Ladd, 2003).

The Connecticut River was identified as a conservation priority through a regional "conservation action planning" (CAP) initiative convened by The Nature Conservancy (TNC) in New England in the 1990s. The Connecticut is New England's longest river, supporting extensive biodiversity and acting as a migration corridor for diadromous fish and birds. Its water constitutes 70% of the freshwater inflow to Long Island Sound and its watershed provides the drinking water supply for Boston, Springfield, Hartford, and many smaller cities and towns. The Connecticut River tidal wetlands and estuary were recognized at the 1994 Ramsar Convention as a wetland area of international importance. Its floodplains contain some of the region's most fertile agricultural soils and provide natural flood protection for downstream cities. Instead of addressing the needs of individual species, CAP recommend focusing on the processes and habitats that maintain the health of this critical ecosystem (Nislow et al., 2010).

Floodplain forests, low-lying forested areas along rivers that flood periodically, are considered one of the rarest and most threatened natural community types in New England. They provide valuable habitat for wildlife as well as absorbing flood waters, sediments, and pollutants. Many invasive species thrive in such nutrient-rich sites with a history of agricultural use, making the restoration planning for Connecticut River floodplain forests a good example for discussion of issues relevant to invasive species management. In this case study, we will examine the drivers of ecosystem change in New England's floodplain forests, focusing on the role of invasive non-native species and what are likely to be effective restoration strategies. Rather than taking the perspective of biological control scientists or even invasive species biologists, we are taking the perspective of the conservation planner where invasive species control may or may not be a priority conservation strategy, depending on the relative importance of other threats to this ecosystem. Application of the decision tree (Figure 2.2) led to contrasting strategies for restoration and invasive species management for different invasive species groups and types of floodplain forest habitats, as described below.

Restoring physical processes to suppress invasive plants

The most important process affecting floodplain forests is flooding (Junk et al., 1989). Flooding and associated sediment movement not only govern species composition but also shape the morphology of the river channel and floodplain (Hupp, 2000). A study of vegetation composition in relation to flood regime at 103 floodplain forest sites located throughout the Connecticut River watershed found that the abundance of both native upland trees and exotic invasive shrubs declined with increasing flooding, whereas the abundance of native floodplain tree species increased (Marks et al., 2014). Flood-intolerant invasive plants that were increasingly suppressed with increasing flood duration include *Acer platanoides* L., *Aegopodium podagraria* L., *Ailanthus altissima* (Mill.) Swingle, *A. petiolata*, *B. thunbergii*, *Celastrus orbiculatus* Thunb., *Cynanchum louiseae* Kartesz & Gandhi, *Elaeagnus umbellata* Thunb., *Euonymus alatus* [Thunb.] Siebold, *Frangula alnus* Mill., *Lonicera morrowii* A. Gray, *Rhamnus cathartica* L., and *Rosa multiflora* Thunb.

In contrast to the invasive shrubs and trees, a few non-native herbaceous species were found to be able to tolerate extended flooding. These included *Fallopia japonica* [Houtt.] Ronse Decr., *Fallopia* x *bohemica*, *Lysimachia nummularia* L., *L. salicaria*, *Microstegium vimineum* [Trin.] A. Camus., *Phalaris arundinacea* L., and *P. australis*. Three of these species (*L. salicaria*, *P. arundinacea*, and *P. australis*) are shade-intolerant marsh species and thus not a threat to floodplain forests. At floodplain forest sites where restoration of extended flooding can be accomplished, it is likely that problems with invasive plants can be effectively reduced.

Succession is the process of change in ecological communities after disturbance and as such provides a useful guide for restoration (Whisenant, 2005). In floodplains,

succession is initiated by the formation of new bars, which frequently happens during a large flood event. Pioneer species such as willows (*Salix*) colonize these bars. Growing pioneer trees and shrubs stabilize the bars and promote the accretion of more sediment, thereby improving conditions for colonization by late successional floodplain forest species that are both less flood tolerant and more shade tolerant (Dietz, 1952; Shelford, 1954; Lindsey et al., 1961; Hosner and Minckler, 1963; Johnson et al., 1976; Nanson and Beach, 1977; Bertoldi et al., 2009; Meitzen, 2009; Gurnell et al., 2012). Relatively few invasive species currently occur in these pioneer bar habitats on the Connecticut River because flooding is typically too severe and sediment accretion rates are high (Marks et al., 2014). Another factor promoting native dominance on bars is that native floodplain pioneers like *Salix nigra* Marshall, *Populus deltoides* Bartram ex Marshall, and *Acer saccharinum* L. produce their seeds in spring and are wind and water dispersed, an ideal strategy to reach fresh sediment seed beds as flood waters from the spring freshet recede (Mahoney and Rood, 1998), in contrast to most invasive shrubs and woody vines, whose bird-dispersed seeds occur in the fall. Thus, in un-channelized rivers the dynamics of lateral channel migration and bar formation create habitats that are relatively resistant to invasion by the existing suite of bird-dispersed, invasive shrubs in northeastern North America. It is crucial to maintain these physical processes to protect this habitat. Specifically, it is critical to avoid bank hardening and to sustain natural flood and sediment-transportation regimes as much as possible (Shankman, 1993; Schnitzler, 1995; Fierke and Kauffman, 2005; Leyer, 2006).

The exception to this natural invasion resistance of river bars is Japanese knotweed (collectively, *F. japonica*, *F. sachalinensis* [F. Schmidt ex Maxim.] Ronse Decr. and their hybrid, *F.* × *bohemica*), which readily colonizes bars and riverbanks especially on high-gradient rivers where flooding is naturally brief and scour from high flows moves knotweed rhizomes, which can re-sprout after being deposited on downstream bars. Japanese stiltgrass (*M. vimineum*) invasion may similarly benefit from water dispersal of seeds in riparian areas (Eschtruth and Battles, 2011). Japanese knotweed can reach a very high level of dominance in this habitat, which interferes with recruitment of trees and other native plants (Figure 2.3) (Urgenson et al., 2012). Thus, Japanese

Figure 2.3 Forlorn TNC intern standing surrounded by Japanese knotweed in a high-gradient river floodplain forest on the Green River in Massachusetts, June 23, 2009. Photo credit, Christian Marks.

knotweed not only transforms the understory of flood-plain forests on high-gradient rivers but can eventually also reduce riparian forest cover by preventing tree seedling recruitment. A lack of riparian trees, with their extensive root systems, increases bank erosion (Secor et al., 2013). This ecological impact of Japanese knotweed was one of the motivations behind an international program (USA, Canada, and the UK) to attempt to develop an effective biological control project against Japanese knotweeds (Shaw et al., 2009; Grevstad et al., 2013). The first agent, the psyllid *Aphalara itadori* Shinji, in this project is currently under review in North America and being field-tested in England.

Assessing ecological impact of invasive species

Selection of species for control should ideally be based on quantitative evidence of their impact. While failure of native species to reproduce (as discussed above for Japanese knotweed's effect on riparian tree seedlings) is critical, so are higher mortality rates caused by effects of invasive species. Invasive vines and lianas are able to directly cause mortality of native plants and are thus of particular concern (Forseth and Innis, 2004; Hough-Goldstein et al., 2012; Center et al., 2013). The inva-sive liana oriental bittersweet (*C. orbiculatus*) causes

severe damage including mortality of mature trees in extensively invaded floodplain forests in Connecticut and Massachusetts among other states (Figure 2.4). However, severe impacts by *C. orbiculatus* are largely restricted to forest edges, canopy gaps, and heavily dis-turbed areas like old fields (McNab and Loftis, 2002; Kuhman et al., 2010; Pavlovic and Leicht-Young, 2011), which make up a relatively small part of the overall floodplain forest area and thus could be misleading as to the invader's true impact. However, a study of tree mortality in 103 Connecticut River floodplain forests estimated that 0.3% of floodplain forest trees were destroyed annually by oriental bittersweet (Marks and Canham, 2015). For mature trees where self-thinning is no longer an important cause of mortality (i.e., diameter at breast height [dbh] = 60 cm), lianas were second only to storms as a cause of floodplain tree mortality. Most of the liana-induced mortality of mature trees was due to invasive *C. orbiculatus* (43%) and native *Vitis riparia* Michx. and *V. labrusca* L. (35%), or a combination of *Celastrus* and *Vitis* (22%). By contrast, other abundant native lianas such as *Toxicodendron radicans* [L.] Kuntze and *Parthenocissus quinquefolia* [L.] Planch. do not appear to cause significant tree mortality.

Celastrus orbiculatus is dominant in the herb layer of Connecticut River floodplain forests about eight

Figure 2.4 Connecticut River floodplain forest breaking down under a heavy load of invasive oriental bittersweet and turning into a weedy vine thicket, West Springfield, Massachusetts, March 15, 2013. Photo credit, Christian Marks.

times more frequently than native *Vitis* species (Marks et al., 2014), and it is therefore going to affect tree recruitment and old field succession more often. *Celastrus orbiculatus* can become so dominant in the herb and shrub layer of forest openings that it prevents tree sapling recruitment. Similarly, researchers have observed that *C. orbiculatus* can arrest or even reverse succession in old fields (McNab and Meeker, 1987; Fike and Niering, 1999). Therefore the impact of *C. orbiculatus*-caused tree mortality may be cumulative, unlike other sources of mortality that result in only temporary forest canopy gaps. The Connecticut River floodplain forest mortality study estimated that floodplain forest basal area is currently destroyed by *C. orbiculatus* at a rate of 0.2% per year (Marks and Canham, 2015). If left unchecked for decades, the cumulative loss of forest area owing to *C. orbiculatus* could be comparable to the potential future impact of emerald ash borer. *Fraxinus* made up 7.4% of the floodplain forest in the study. At a constant rate of 0.2% per year, it would take just 39 years for *C. orbiculatus* to destroy a comparable 7.4% of the forest. Thus, although it moves more slowly than an insect pest or pathogen, the cumulative impact of this invasive liana may be just as severe.

While chemical/mechanical control programs against invasive vines can be mounted quickly, they are costly and difficult to sustain. Volunteers at the Silvio O. Conte National Wildlife Refuge along the Connecticut River helped us quantify the time needed for mechanical control of oriental bittersweet vines in two contexts. Along a heavily invaded floodplain forest edge, it took 115 man-hours/hectare (47 hours/acre) to cut bittersweet vines. In contrast, in a nearby old floodplain field dominated by cottonwood saplings that were starting to break down under smothering bittersweet vines, it took 435 man-hours/hectare (176 hours/acre) to cut bittersweet. The old field area was much harder to work in because the saplings were dense and there were many small vines, compared to fewer larger vines at the edge of the mature forest. These large labor costs would make it prohibitively expensive to control bittersweet at the landscape scale by conventional means and pose a financial burden even for restoration of individual floodplain forest sites if they are heavily infested. The development of an effective classical biological control agent is thus the only potentially financially viable means of reducing the impact of oriental bittersweet at the landscape scale. Trees falling into roads and onto power lines also have economic impacts, directly via cleanup costs and indirectly via power outages and blocked roads. Given the destructive potential of exotic vines and lianas (Forseth and Innis, 2004; Hough-Goldstein et al., 2012; Center et al., 2013), they should be ranked highly, not only in setting priorities for biological control but also in terms of the need for greater regulatory restrictions on the introduction of new plant species from overseas.

Eradiation and containment of a serious invader

Another potentially serious threat to floodplain forests is the Asian longhorned beetle (ALB) (*Anoplophora glabripennis* Motschulsky), which preferentially attacks species of *Acer* and *Populus*, common floodplain forest dominants on the Connecticut and many other northern US rivers. Repeated attack by ALB leads to tree mortality within a few years and the potential for damage in urban forests is large (Nowak et al., 2001). The threat to native forests, while still unknown, is potentially even larger. The eradication effort against ALB led by the USDA Forest Service has been focused mostly on urban areas, including one in Worcester, Massachusetts, at the edge of the Connecticut River watershed. This eradication program is an example of the early detection and rapid response approach to invasive species control that, if successful, would make a biological control program for ALB unnecessary. However, should ALB eradication fail in even one infested area and natural forests become extensively invaded, biological control would be a potentially feasible way to reduce the damage.

Biological control and breeding host resistance against pests and pathogens

Before the spread of Dutch elm disease (DED), American elm (*Ulmus americana* L.) was co-dominant with silver maple (*A. saccharinum*) in the canopies of floodplain forests on many northern rivers in the United States, including the Connecticut River (Nichols, 1916; Telford, 1926; Curtis, 1959). American elm was also the largest tree species in Massachusetts (Emerson, 1887). Even today, American elm is the most widespread and the second most abundant floodplain tree species in the Connecticut River watershed (Marks et al., 2014), but it now rarely lives long enough to reach the forest canopy (Figure 2.5). Green ash (*Fraxinus pennsylvanica* Marshall), which has a similar level of flood and shade tolerance as American elm, has to some degree replaced it in the southern part of the Connecticut River watershed, but green ash is now also threatened by emerald ash borer, which reached the Connecticut River watershed in Connecticut in 2013. Observations from formerly ash-dominated forest stands in the Great Lakes states where the emerald ash borer invasion began suggest that green ash might persist as an understory tree species, like American elm, because it reproduces early and seedlings are generally not attacked by emerald ash borer (Wagner and Todd, 2015). The reduction in floodplain forest

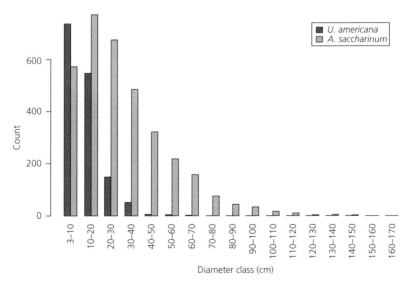

Figure 2.5 Tree size distribution for American elm (*U. americana*) and silver maple (*A. saccharinum*), the two most common tree species in Connecticut River floodplain forests when surveyed in 2008 to 2011. Note the rapid reduction of elms beyond 20 cm and absence over 60 cm dbh, in contrast to silver maple, likely due to Dutch elm disease. Unpublished data of Christian Marks.

stature associated with Dutch elm disease and emerald ash borer may have implications for bird species that prefer nesting or foraging in the upper canopy of riparian forests (Knutson and Klaas, 1998).

Given their destructiveness to ecologically and economically valuable trees, biological control has been considered for both Dutch elm disease and emerald ash borer. The more recently arrived emerald ash borer is the subject of an active biological control program and new parasitoids are still being released (Chapter 10). Although biological control of the bark beetles that spread DED and the deployment of hypovirulent strains of the fungus that causes DED were explored, neither of these biological control approaches were successful at reducing American elm mortality (Brasier, 2000; Fuester et al., 2014). Subsequent efforts at restoring American elm have instead focused on selecting surviving American elms for greater disease tolerance (Heybroek, 2000; Mittempergher and Santini, 2004; Townsend et al., 2005). While breeding of resistant elm varieties was initially done for use of elms in urban areas, programs are now underway to re-introduce highly disease-tolerant cultivars of native American elms into riparian habitats including Connecticut River floodplain forests (Slavicek et al., 2005; Slavicek and Marks, 2011; Knight et al., 2012). Although emerald ash borer causes high rates of mortality in the three native ashes of the region (*F. pennsylvanica, Fraxinus nigra* Marshall and *Fraxinus americana* L.) (Rebek et al., 2008; Knight et al., 2013), a few individuals appear to have some host resistance to the borer, and breeding native ash trees for greater host resistance is being explored as a restoration strategy to complement the biological control program (Koch et al., 2011).

Holistic ecological restoration and invasive species management

Rich, high-floodplain terraces typically have more invasive plants than other floodplain forest communities. High terraces are prone to non-native plant invasion, in part because soils are very fertile and floods are infrequent and of short duration, and because they have suffered from more intense disturbance by human activities, especially the historic clearance of native forest to create cropland. In our floodplain field work, we observed that many invasive woody plants (e.g., *R. multiflora*) had their greatest abundances in former agricultural land, as was also observed in other studies (McDonald et al., 2008; Mosher

et al., 2009; Kuhman et al., 2010, 2011). One exception to this general pattern that we observed in Connecticut River floodplain forests is Norway maple (*A. platanoides*), which does invade closed-canopy forests with deep shade in the understory, albeit slowly because of limited dispersal ability (Martin et al., 2010). These findings suggest that changes in land management could reduce the abundance of many but not all invasive plants species. In New England that would mean avoiding severe disturbances (like logging) in protected forests (Lee and Thompson, 2012), but this may conflict with other conservation goals, like creation of early successional habitat needed for certain declining animal species.

The invasive shrubs that we frequently encounter in abundance on high terraces in the Connecticut River floodplains include Japanese barberry (*B. thunbergii*), winged-euonymus or burning bush (*E. alatus*), Morrow's honeysuckle (*L. morrowii*), glossy buckthorn or alder buckthorn (*F. alnus*), multiflora rose (*R. multiflora*), and occasionally also autumn-olive (*E. umbellata*), Amur honeysuckle (*Lonicera maackii* [Rupr.] Herder), common buckthorn (*R. cathartica*), and Japanese wineberry (*Rubus phoenicolasius* Maxim.). It is not immediately obvious if these invasive shrubs have substantial ecological impacts on floodplain forests in New England beyond altering understory composition. For instance, our field data showed that there are still plenty of native tree seedlings to fill canopy gaps in floodplain forests even where there is a high abundance of non-native shrubs in the understory. The most concerning evidence of a negative impact by these invasive shrubs is on native insect herbivore communities and their predators. Native specialist as well as generalist Lepidoptera and arthropods fare poorly on non-native plants (Burghardt et al., 2010; Tallamy et al., 2010; van Hengstum et al., 2014). Thus, non-native plant invasions have consequences for higher trophic levels, such as birds that critically depend on insects as a source of protein for raising nestlings (Tallamy, 2004). This impact is substantial even where most trees are native and exotic invasive plants are largely restricted to the understory (Burghardt et al., 2009). While berries from invasive shrubs can provide an ample fall food source for migrating songbirds (Gleditsch and Carlo, 2011), that does not compensate for the decrease of insect prey during the nesting season. It seems fair to conclude from the evidence available that in New England forests the impacts of these invasive shrubs detract from the forests' ecological value but are not so disruptive as to cause transformative change, such as preventing forest regeneration.

Nutrient-rich forests with ample moisture like high-terrace floodplains, coves, and seeps are noted for their exceptional native plant species richness and as such are attractive for conservation in the region, but restoration of the native herb and shrub layer community may require more action than just control of invasive plants. For example, years of mechanical control of Japanese barberry on some TNC forest reserves in Connecticut resulted in relatively bare areas with some re-sprouting and re-invasion by barberry but little recruitment of native plants. The lack of recruitment of desired native plants may be due to a lack of native seed (Drayton and Primack, 2012), intense browsing by native deer (Knight et al., 2009; Collard et al., 2010; Royo et al., 2010; Tanentzap et al., 2011; Dávalos et al., 2014; Nuttle et al., 2014), or even effects of non-native earthworms on the leaf litter layer and the soil seed bank (Frelich et al., 2006; Hale et al., 2006; Nuzzo et al., 2009; Fisichelli et al., 2013). Successful restoration of rich high-terrace floodplain forests in particular locations will require determination of the causes of degradation so that those causes can be addressed, in conjunction with any potential invasive species control.

Biological control agents from other regions

Although some of the invasive plants discussed here may not reach the level of impact in Connecticut River floodplain forests to warrant a biological control program, they may have more transformative impacts in other ecosystem types or in other regions. For example, common buckthorn (*R. cathartica*) can transform the structure of forests in the midwestern United States and adjacent Canada where soils and climate are more suitable for buckthorn (Archibold et al., 1997; Knight et al., 2007; Mascaro and Schnitzer, 2007). Many insects from buckthorn's native range in Europe were tested as biological control insects for common (*R. cathartica*) or glossy buckthorn (*F. alnus*) by the Minnesota Department of Natural Resources and CABI (Gassmann et al., 2010), but none were host specific and sufficiently damaging to buckthorns, and the biological control program for buckthorns was discontinued after 11 years of research.

Sometimes biological control also happens on its own owing to biotic resistance from native organisms in the invader area. The invasive tree-of-heaven, *A. altissima*, is susceptible to a presumed-native North American soil-borne vascular wilt fungus *Verticillium nonalfalfae* Inderb., which is now being explored as a potential biological control agent for interstate movement within the United States (Schall and Davis, 2009a, b; Harris et al., 2013; Kasson et al., 2014; Snyder et al., 2014). Similarly, the invasive multiflora rose (*R. multiflora*) is highly damaged by rose rosette disease, a viral pathogen native to the Rocky Mountains that is gradually spreading eastward and may eventually suppress non-native roses along the Connecticut River, as has already happened in the midwestern states (Epstein et al., 1997; Epstein and Hill, 1999; Amrine, 2002; Jesse et al., 2006; Banasiak and Meiners, 2009; Jesse et al., 2010). In addition, a chalcid wasp (*Megastigmus aculeatus* var. *nigroflavus* Hoffmeyer) that is specialized to attack the seeds of *R. multiflora* was accidentally introduced into North America from its native range in Asia (Amrine, 2002). Thus it makes sense for limited conservation resources in a given region to focus on the most serious invaders, knowing that a few of the locally less serious invaders may eventually also come under biological control because of developments in other regions.

Conclusion for Connecticut River watershed case study

The first priority of a conservation strategy should be to restore, as much as is economically feasible, the physical processes that control community composition and ecosystem functions over the long term. Where physical processes like disturbance regimes and nutrient levels are relatively unaltered, generally fewer exotic plant species have invaded; where physical processes have been dramatically altered, invasive plant control alone is unlikely to achieve ecological restoration. Consistent with this view, TNC has made restoration of physical processes a primary focus for its Connecticut River Program. In particular TNC is collaborating with owners of large dams to modify operations for greater ecological benefit (Warner et al., 2014), as well as to protect and reforest floodplain land to give the river the space it needs for geomorphic processes to unfold in uninhibited fashion. However, some invasive species are able to invade and do substantial damage even where physical processes are intact. These species make compelling cases for classical biological control, particularly where invasion causes transformative ecological change. Specific examples where a biological control program could be justified by ecological impacts to Connecticut River floodplain forests include emerald ash borer, Japanese knotweed, and oriental bittersweet.

Acknowledgments

We thank Bernd Blossey, Charlotte Causton, David Wagner, Dan Simberloff, and Mark Hoddle for reviews of Chapter 2, as well as Mark Anderson, Jon Binhammer, Ellen Jacquart, Karen Lombard, Kim Lutz, and Rose Paul (six TNC colleagues that have extensive experience in conservation planning, ecological restoration, and invasive species management) for helpful discussions and their comments on earlier drafts. We also thank Cynthia Boettner and her summer interns at the Silvio O. Conte National Wildlife Refuge for their help in estimating the labor costs for mechanical control of Oriental bittersweet lianas.

References

Ahlers, D. and D. Moore. 2009. *A review of vegetation and hydrologic parameters associated with the Southwestern Willow Flycatcher, 2002 to 2008: Elephant Butte Reservoir Delta*, New Mexico, U.S.A. US Department of the Interior, Bureau of Reclamation.

Amrine, J. A. 2002. Multiflora rose, pp. 413–434. *In:* VanDriesche, R., S. Lyon, B. Blossey, M. S. Hoddle, and R. Reardon (eds.). *Biological Control of Invasive Plants in the Eastern United States*. USDA Forest Service FHTET. Morgantown, WV.

Anagnostakis, S. L. 2001. American chestnut sprout survival with biological control of the chestnut-blight fungus population. *Forest Ecology and Management* **152**: 225–233.

Anagnostakis, S. L. 2012. Chestnut breeding in the United States for disease and insect resistance. *Plant Disease* **96**: 1392–1403.

Anderson, M. G. 1995. Interactions between *Lythrum salicaria* and native organisms: a critical review. *Environmental Management* **19**: 225–231.

Archibold, O., D. Brooks, and L. Delanoy. 1997. An investigation of the invasive shrub European buckthorn, *Rhamnus cathartica* L., near Saskatoon, Saskatchewan. *Canadian Field Naturalist* **111**: 617—621.

Banasiak, S. E. and S. J. Meiners. 2009. Long term dynamics of *Rosa multiflora* in a successional system. *Biological Invasions* **11**: 215–224.

Bart, D. and J. M. Hartman. 2000. Environmental determinants of *Phragmites australis* expansion in a New Jersey salt marsh: an experimental approach. *Oikos* **89**: 59–69.

Bart, D. and J. M. Hartman. 2003. The role of large rhizome dispersal and low salinity windows in the establishment of common reed, *Phragmites australis*, in salt marshes: new links to human activities. *Estuaries* **26**: 436–443.

Bauer, J. T. 2012. Invasive species: "back-seat drivers" of ecosystem change? *Biological Invasions* **14**: 1295–1304.

Berman, M., A. Andersen, and T. Ibanez. 2013. Invasive ants as back-seat drivers of native ant diversity decline in New Caledonia. *Biological Invasions* **15**: 2311–2331.

Bertoldi, W., A. Gurnell, N. Surian, et al. 2009. Understanding reference processes: linkages between river flows, sediment dynamics and vegetated landforms along the Tagliamento River, Italy. *River Research and Applications* **25**: 501–516.

Blossey, B. 2003. A framework for evaluating potential ecological effects of implementing biological control of *Phragmites australis*. *Estuaries* **26**: 607–617.

Blossey, B., L. Skinner, and J. Taylor. 2001. Impact and management of purple loosestrife (*Lythrum salicaria*) in North America. *Biodiversity & Conservation* **10**: 1787–1807.

Brasier, C. M. 2000. Viruses as biological control agents of the Dutch elm disease fungus *Ophiostoma novo-ulmi*, pp. 201–212. *In:* Dunn, C. P. (ed.). *The Elms: Breeding, Conservation, and Disease Management*. Kluwer. Boston, Massachusetts, USA.

Braun, E. L. 1950. *Deciduous Forests of Eastern North America*. Blakiston. Philadelphia, Pennsylvania, USA.

Burghardt, K. T., D. W. Tallamy, and W. Gregory Shriver. 2009. Impact of native plants on bird and butterfly biodiversity in suburban landscapes. *Conservation Biology* **23**: 219–224.

Burghardt, K. T., D. W. Tallamy, C. Philips, and K. J. Shropshire. 2010. Non-native plants reduce abundance, richness, and host specialization in lepidopteran communities. *Ecosphere* **1**: 1–22.

Catford, J. A. and R. Jansson. 2014. Drowned, buried and carried away: effects of plant traits on the distribution of native and alien species in riparian ecosystems. *New Phytologist* **204**: 19–36.

Center, T. D., M. Rayamajhi, F. A. Dray, et al. 2013. Host range validation, molecular identification and release and establishment of a Chinese biotype of the Asian leaf beetle *Lilioceris cheni* (Coleoptera: Chrysomelidae: Criocerinae) for control of *Dioscorea bulbifera* L. in the southern United States. *Biological Control Science and Technology* **23**: 735–755.

Chambers, R. M., D. T. Osgood, D. J. Bart, and F. Montalto. 2003. *Phragmites australis* invasion and expansion in tidal wetlands: interactions among salinity, sulfide, and hydrology. *Estuaries* **26**: 398–406.

Collard, A., L. Lapointe, J.-P. Ouellet, et al. 2010. Slow responses of understory plants of maple-dominated forests to white-tailed deer experimental exclusion. *Forest Ecology and Management* **260**: 649–662.

Cooper, D. J., D. C. Andersen, and R. A. Chimner. 2003. Multiple pathways for woody plant establishment on floodplains at local to regional scales. *Journal of Ecology* **91**: 182–196.

Costanza, R., R. d'Arge, R. de Groot, et al. 1997. The value of the world's ecosystem services and natural capital. *Nature* **387**: 253–260.

Crooks, J. A. 2005. Lag times and exotic species: the ecology and management of biological invasions in slow-motion. *Ecoscience* **12**: 316–329.

Curtis, J. T. 1959. *The Vegetation of Wisconsin: An Ordination of Plant Communities*. University of Wisconsin Press. Madison, Wisconsin, USA.

Dávalos, A., V. Nuzzo, and B. Blossey. 2014. Demographic responses of rare forest plants to multiple stressors: the role of deer, invasive species and nutrients. *Journal of Ecology* **102**: 1222–1233.

Dietz, R. A. 1952. The evolution of a gravel bar. *Annals of the Missouri Botanical Garden* **39**: 249–254.

Dornbush, M. E. and P. G. Hahn. 2013. Consumers and establishment limitations contribute more than competitive interactions in sustaining dominance of the exotic herb garlic mustard in a Wisconsin, USA, forest. *Biological Invasions* **15**: 2691–2706.

Drayton, B. and R. B. Primack. 2012. Success rates for reintroductions of eight perennial plant species after 15 years. *Restoration Ecology* **20**: 299–303.

Dudley, T. L. and D. W. Bean. 2012. Tamarisk biological control, endangered species risk and resolution of conflict through riparian restoration. *Biological Control* **57**: 331–347.

Emerson, G. B. 1887. *A Report on the Trees and Shrubs Growing Naturally in the Forests of Massachusetts*, 4th edn. Little, Brown, and Co. Boston, Massachusetts, USA.

Epstein, A. H., J. H. Hill, and F. W. Nutter. 1997. Augmentation of rose rosette disease for biological control of multiflora rose (*Rosa multiflora*). *Weed Science* **45**: 172–178.

Epstein, A. H. and J. H. Hill. 1999. Status of rose rosette disease as a biological control for multiflora rose. *Plant Disease* **83**: 92–101.

Eschtruth, A. K. and J. J. Battles. 2009. Assessing the relative importance of disturbance, herbivory, diversity, and propagule pressure in exotic plant invasion. *Ecological Monographs* **79**: 265–280.

Eschtruth, A. K. and J. J. Battles. 2011. The importance of quantifying propagule pressure to understand invasion: an examination of riparian forest invasibility. *Ecology* **92**: 1314–1322.

Fierke, M. K. and J. B. Kauffman. 2005. Structural dynamics of riparian forests along a black cottonwood successional gradient. *Forest Ecology and Management* **215**: 149–162.

Fike, J. and W. A. Niering. 1999. Four decades of old field vegetation development and the role of *Celastrus orbiculatus* in the northeastern United States. *Journal of Vegetation Science* **10**: 483–492.

Finch, D. M., S. I. Rothstein, J. C. Boren, et al. 2002. Final recovery plan of the southwestern willow flycatcher (*Empidonax traillii extimus*). U.S. Fish and Wildlife Service, Region 2. Albuquerque, New Mexico, USA.

Fisichelli, N. A., L. E. Frelich, P. B. Reich, and N. Eisenhauer. 2013. Linking direct and indirect pathways mediating earthworms, deer, and understory composition in Great Lakes forests. *Biological Invasions* **15**: 1057–1066.

Forseth, I. N. and A. F. Innis. 2004. Kudzu (*Pueraria montana*): history, physiology, and ecology combine to make a major ecosystem threat. *Critical Reviews in Plant Sciences* **23**: 401–413.

Frelich, L., C. Hale, S. Scheu, et al. 2006. Earthworm invasion into previously earthworm-free temperate and boreal forests. *Biological Invasions* **8**: 1235–1245.

Fuester, R. W., A. E. Hajek, J. S. Elkinton, and P. W. Schaefer. 2014. Gypsy moth (*Lymantria dispar* L.) (Lepidoptera: Erebidae: Lymantriinae), pp. 49–82. *In*: Van Driesche, R. G. and R. Reardon (eds.). *The Use of Classical Biological Control to Preserve Forests in North America*. FHTET- 2013-02. USDA Forest Service, Forest Health Technology Enterprise Team. Morgantown, West Virginia, USA. Available from: http://www.fs.fed.us/foresthealth/technology/pub_titles.shtml [Accessed January 2016].

Gassmann, A., I. Toševski, J. Jovic, et al. 2010. *Biological control of backthorns, Rhamnus catartica and Frangula alnus*. Report 2008–2009, CABI. Delémont, Switzerland.

Gleditsch, J. M. and T. A. Carlo. 2011. Fruit quantity of invasive shrubs predicts the abundance of common native avian frugivores in central Pennsylvania. *Diversity and Distributions* **17**: 244–253.

Gotelli, N. and A. Arnett. 2000. Biogeographic effects of red fire ant invasion. *Ecology Letters* **3**: 257–261.

Green, E. K. and S. M. Galatowitsch. 2002. Effects of *Phalaris arundinacea* and nitrate addition on the establishment of wetland plant communities. *Journal of Applied Ecology* **39**: 134–144.

Greet, J., R. D. Cousens, and J. A. Webb. 2013. More exotic and fewer native plant species: riverine vegetation patterns associated with altered seasonal flow patterns. *River Research and Applications* **29**: 686–706.

Grevstad, F., R. Shaw, R. Bourchier, et al. 2013. Efficacy and host specificity compared between two populations of the psyllid *Aphalara itadori*, candidates for biological control of invasive knotweeds in North America. *Biological Control* **65**: 53–62.

Gurnell, A. M., W. Bertoldi, and D. Corenblit. 2012. Changing river channels: The roles of hydrological processes, plants and pioneer fluvial landforms in humid temperate, mixed load, gravel bed rivers. *Earth-Science Reviews* **111**: 129–141.

Hale, C. M., L. E. Frelich, and P. B. Reich. 2006. Changes in hardwood forest understory plant communities in response to European earthworm invasions. *Ecology* **87**: 1637–1649.

Harris, P. T., G. H. Cannon, N. E. Smith, and N. Z. Muth. 2013. Assessment of plant community restoration following Tree-of-Heaven (*Ailanthus altissima*) control by *Verticillium albo-atrum*. *Biological Invasions* **15**: 1–7.

Herms, D. A. and D. G. McCullough. 2014. Emerald ash borer invasion of North America: history, biology, ecology, impacts, and management. *Annual Review of Entomology* **59**: 13–30.

Heybroek, H. M. 2000. Notes on elm breeding and genetics, pp. 249–258. *In*: Dunn, C. P. (ed.). *The Elms: Breeding, Conservation, and Disease Management*. Kluwer. Boston, Massachusetts, USA.

HilleRisLambers, J., S. G. Yelenik, B. P. Colman, and J. M. Levine. 2010. California annual grass invaders: the drivers or passengers of change? *Journal of Ecology* **98**: 1147–1156.

Hosner, J. F. and L. S. Minckler. 1963. Bottomland hardwood forests of southern Illinois – regeneration and succession. *Ecology* **44**: 29–41.

Hough-Goldstein, J., E. Lake, and R. Reardon. 2012. Status of an ongoing biological control program for the invasive vine, *Persicaria perfoliata* in eastern North America. *Biological Control* **57**: 181–189.

Hultine, K. R., J. Belnap, C. Van Riper III, et al. 2009. Tamarisk biological control in the western United States: ecological and societal implications. *Frontiers in Ecology and the Environment* **8**: 467–474.

Hupp, C. R. 2000. Hydrology, geomorphology and vegetation of Coastal Plain rivers in the southeastern USA. *Hydrological Processes* **14**: 2991–3010.

Jacobs, D. F. 2007. Toward development of silvical strategies for forest restoration of American chestnut (*Castanea dentata*) using blight-resistant hybrids. *Biological Conservation* **137**: 497–506.

Jesse, L. C., K. A. Moloney, and J. J. Obrycki. 2006. Abundance of arthropods on the branch tips of the invasive plant, *Rosa multiflora* (Rosaceae). *Weed Biology and Management* **6**: 204–211.

Jesse, L. C., J. D. Nason, J. J. Obrycki, and K. A. Moloney. 2010. Quantifying the levels of sexual reproduction and clonal spread in the invasive plant, *Rosa multiflora*. *Biological Invasions* **12**: 1847–1854.

Johnson, W. C., R. L. Burgess, and W. R. Keammerer. 1976. Forest overstory vegetation and environment on Missouri River floodplain in North Dakota. *Ecological Monographs* **46**: 59–84.

Junk, W. J., P. B. Bayley, and R. E. Sparks. 1989. The flood pulse concept in river-floodplain systems. *Canadian Special Publication of Fisheries and Aquatic Sciences* **106**: 110–127.

Kasson, M. T., D. P. G. Short, E. S. O'Neal, et al. 2014. Comparative pathogenicity, biological control efficacy, and multi locus sequence typing of *Verticillium nonalfalfae* from the invasive *Ailanthus altissima* and other hosts. *Phytopathology* **104**: 282–292.

Katz, G. L. and P. B. Shafroth. 2003. Biology, ecology and management of *Elaeagnus angustifolia* L. (Russian olive) in western North America. *Wetlands* **23**: 763–777.

Keeley, J. E. 2006. Fire management impacts on invasive plants in the western United States. *Conservation Biology* **20**: 375–384.

Kercher, S. M., A. Herr-Turoff, and J. B. Zedler. 2007. Understanding invasion as a process: the case of *Phalaris arundinacea* in wet prairies. *Biological Invasions* **9**: 657–665.

King, J. R. and W. R. Tschinkel. 2008. Experimental evidence that human impacts drive fire ant invasions and ecological change. *Proceedings of the National Academy of Sciences* **105**: 20339–20343.

Knight, K. S., J. S. Kurylo, A. G. Endress, et al. 2007. Ecology and ecosystem impacts of common buckthorn (*Rhamnus cathartica*): a review. *Biological Invasions* **9**: 925–937.

Knight, K. S., J. M. Slavicek, R. Kappler, et al. 2012. Using Dutch elm disease-tolerant elm to restore floodplains impacted by emerald ash borer, pp. 317–323. *In: Proceedings of the 4th International Workshop on Genetics of Host-Parasite Interactions in Forestry, Eugene, Oregon.* U.S.A General Technical Report 240. Pacific Southwest Research Station, USDA Forest Service. Washington, DC.

Knight, K. S., J. P. Brown, and R. P. Long. 2013. Factors affecting the survival of ash (*Fraxinus* spp.) trees infested by emerald ash borer (*Agrilus planipennis*). *Biological Invasions* **15**: 371–383.

Knight, T. M., J. L. Dunn, L. A. Smith, et al. 2009. Deer facilitate invasive plant success in a Pennsylvania forest understory. *Natural Areas Journal* **29**: 110–116.

Knutson, M. G. and E. E. Klaas. 1998. Floodplain forest loss and changes in forest community composition and structure in the Upper Mississippi River: a wildlife habitat at risk. *Natural Areas Journal* **18**: 138–150.

Koch, J. L., D. W. Carey, K. S. Knight, et al. 2011. Breeding strategies for the development of emerald ash borer-resistant North American ash. pp. 235–239. *In: Proceedings of the 4th International Workshop on Genetics of Host-Parasite Interactions in Forestry, Eugene, Oregon.* U.S.A General Technical Report 240. Pacific Southwest Research Station, USDA Forest Service. Washington, DC.

Kowarik, I., P. Pyšek, K. Prach, et al. 1995. Time lags in biological invasions with regard to the success and failure of alien species, pp. 15–38. *In: Plant Invasions: General Aspects and Special Problems. Workshop held at Kostelec nad Černými lesy, Czech Republic, 16–19 September 1993.* SPB Academic Publishing. The Netherlands.

Kuhman, T. R., S. M. Pearson, and M. G. Turner. 2010. Effects of land-use history and the contemporary landscape on non-native plant invasion at local and regional scales in the forest-dominated southern Appalachians. *Landscape Ecology* **25**: 1433–1445.

Kuhman, T. R., S. M. Pearson, and M. G. Turner. 2011. Agricultural land-use history increases non-native plant invasion in a southern Appalachian forest a century after abandonment. *Canadian Journal of Forest Research* **41**: 920–929.

Kumschick, S., M. Gaertner, M. Vilà, et al. 2015. Ecological impacts of alien species: quantification, scope, caveats, and recommendations. *Bioscience* **65**: 55–63.

Lake, E. C., J. Hough-Goldstein, and V. D'Amico. 2014. Integrating management techniques to restore sites invaded by mile-a-minute weed, *Persicaria perfoliata*. *Restoration Ecology* **22**: 127–133.

Lathrop, R. G., L. Windham, and P. Montesano. 2003. Does *Phragmites* expansion alter the structure and function of marsh landscapes? Patterns and processes revisited. *Estuaries* **26**: 423–435.

Lee, T. D. and J. H. Thompson. 2012. Effects of logging history on invasion of eastern white pine forests by exotic glossy buckthorn (*Frangula alnus* P. Mill.). *Forest Ecology and Management* **265**: 201–210.

Leyer, I. 2006. Dispersal, diversity and distribution patterns in pioneer vegetation: The role of river floodplain connectivity. *Journal of Vegetation Science* **17**: 407–416.

Lindsey, A. A., D. K. Sterling, R. O. Petty, and W. Vanasdal. 1961. Vegetation and environment along Wabash and Tippecanoe Rivers. *Ecological Monographs* **31**: 105–156.

MacDougall, A. S. and R. Turkington. 2005. Are invasive species the drivers or passengers of change in degraded ecosystems? *Ecology* **86**: 42–55.

MacDougall, A. S. and R. Turkington. 2007. Does the type of disturbance matter when restoring disturbance-dependent grasslands? *Restoration Ecology* **15**: 263–272.

Mahoney, J. M. and S. B. Rood. 1998. Streamflow requirements for cottonwood seedling recruitment – An integrative model. *Wetlands* **18**: 634–645.

Marks, C. O. and C. D. Canham. 2015. A quantitative framework for demographic trends in size-structured populations: analysis of threats to floodplain forests. *Ecosphere*. DOI: 10.1890/ES15-00068.1.

Marks, C. O., K. H. Nislow, and F. J. Magilligan. 2014. Quantifying flooding regime in floodplain forests to guide river restoration. *Elementa: Science of the Anthropocene* **2**: 000031.

Martin, L. J. and B. Blossey. 2013. The runaway weed: costs and failures of *Phragmites australis* management in the USA. *Estuaries and Coasts* **36**: 626–632.

Martin, P. H., C. D. Canham, and R. K. Kobe. 2010. Divergence from the growth–survival trade-off and extreme high growth rates drive patterns of exotic tree invasions in closed-canopy forests. *Journal of Ecology* **98**: 778–789.

Mascaro, J. and S. A. Schnitzer. 2007. *Rhamnus cathartica* L. (common buckthorn) as an ecosystem dominant in southern Wisconsin forests. *Northeastern Naturalist* **14**: 387–402.

McDonald, R. I., G. Motzkin, and D. R. Foster. 2008. Assessing the influence of historical factors, contemporary processes, and environmental conditions on the distribution of invasive species *The Journal of the Torrey Botanical Society* **135**: 260–271.

McNab, W. H. and M. Meeker. 1987. Oriental bittersweet: a growing threat to hardwood silviculture in the Appalachians. *Northern Journal of Applied Forestry* **4**: 174–177.

McNab, W. H. and D. L. Loftis. 2002. Probability of occurrence and habitat features for oriental bittersweet in an oak forest in the southern Appalachian mountains, USA. *Forest Ecology and Management* **155**: 45–54.

Meitzen, K. M. 2009. Lateral channel migration effects on riparian forest structure and composition, Congaree River, South Carolina, USA. *Wetlands* **29**: 465–475.

Merritt, D. M. and N. L. Poff. 2010. Shifting dominance of riparian *Populus* and *Tamarix* along gradients of flow alteration in western North American rivers. *Ecological Applications* **20**: 135–152.

Metz, M. R., J. M. Varner, K. M. Frangioso, et al. 2013. Unexpected redwood mortality from synergies between wildfire and an emerging infectious disease. *Ecology* **94**: 2152–2159.

Milgroom, M. G. and P. Cortesi. 2004. Biological control of chestnut blight with hypovirulence: a critical analysis. *Annual Review of Phytopathology* **42**: 311–338.

Minkin, P. and R. Ladd. 2003. *Success of Corps-required wetland mitigation in New England. US Army Corps of Engineers.* New England District, Concord, Massachusetts, USA.

Mittempergher, L. and A. Santini. 2004. The history of elm breeding. *Investigación Agraria: Sistemas y Recursos Forestales* **13**: 161–177.

Moreno-Mateos, D., M. E. Power, F. A. Comín, and R. Yockteng. 2012. Structural and functional loss in restored wetland ecosystems. *PLoS Biology* **10**: e1001247.

Mosher, E. S., J. A. Silander Jr., and A. M. Latimer. 2009. The role of land-use history in major invasions by woody plant species in the northeastern North American landscape. *Biological Invasions* **11**: 2317–2328.

Nanson, G. C. and H. F. Beach. 1977. Forest succession and sedimentation on a meandering-river floodplain, northeast British Columbia, Canada. *Journal of Biogeography* **4**: 229–251.

Nichols, G. E. 1916. The vegetation of Connecticut V: plant societies along rivers and streams. *Bulletin of the Torrey Botanical Club* **43**: 235–264.

Nislow, K. H., C. O. Marks, and K. A. Lutz. 2010. Aquatic conservation planning at a landscape scale, pp. 99–119. *In:* Trombulak, S. C. and R. F. Baldwin (eds.). *Landscape-scale Conservation Planning.* Springer Dordrecht. The Netherlands.

Nowak, D. J., J. E. Pasek, R. A. Sequeira, et al. 2001. Potential effect of *Anoplophora glabripennis* (Coleoptera: Cerambycidae) on urban trees in the United States. *Journal of Economic Entomology* **94**: 116–122.

Nuttle, T., T. E. Ristau, and A. A. Royo. 2014. Long-term biological legacies of herbivore density in a landscape-scale experiment: forest understoreys reflect past deer density treatments for at least 20 years. *Journal of Ecology* **102**: 221–228.

Nuzzo, V. A., J. C. Maerz, and B. Blossey. 2009. Earthworm invasion as the driving force behind plant invasion and community change in northeastern North American forests. *Conservation Biology* **23**: 966–974.

Pavlovic, N. B. and S. A. Leicht-Young. 2011. Are temperate mature forests buffered from invasive lianas? *The Journal of the Torrey Botanical Society* **138**: 85–92.

Perry, L. G., S. M. Galatowitsch, and C. J. Rosen. 2004. Competitive control of invasive vegetation: a native wetland sedge suppresses *Phalaris arundinacea* in carbon-enriched soil. *Journal of Applied Ecology* **41**: 151–162.

Porter, S. D. and D. A. Savignano. 1990. Invasion of polygyne fire ants decimates native ants and disrupts arthropod community. *Ecology* **71**: 2095–2106.

Porter, S. D., L. A. Nogueira de Sá, and L. W. Morrison. 2004. Establishment and dispersal of the fire ant decapitating fly *Pseudacteon tricuspis* in North Florida. *Biological Control* **29**: 179–188.

Rebek, E. J., D. A. Herms, and D. R. Smitley. 2008. Interspecific variation in resistance to emerald ash borer (Coleoptera: Buprestidae) among North American and Asian ash (*Fraxinus* spp.). *Environmental Entomology* **37**: 242–246.

Reynolds, L. V., D. J. Cooper, and N. T. Hobbs. 2014. Drivers of riparian tree invasion on a desert stream. *River Research and Applications* **30**: 60–70.

Richard, G. and P. Julien. 2003. Dam impacts on and restoration of an alluvial river, Rio Grande, New Mexico. *International Journal of Sediment Research* **18**: 89–96.

Richardson, D. M., P. M. Holmes, K. J. Esler, et al. 2007. Riparian vegetation: degradation, alien plant invasions, and restoration prospects. *Diversity and Distributions* **13**: 126–139.

Rooth, J. and J. Stevenson. 2000. Sediment deposition patterns in *Phragmites australis* communities: implications for coastal areas threatened by rising sea-level. *Wetlands Ecology and Management* **8**: 173–183.

Rooth, J. E., J. C. Stevenson, and J. C. Cornwell. 2003. Increased sediment accretion rates following invasion by *Phragmites australis:* the role of litter. *Estuaries* **26**: 475–483.

Royo, A. A., S. L. Stout, D. S. deCalesta, and T. G. Pierson. 2010. Restoring forest herb communities through landscape-level deer herd reductions: is recovery limited by legacy effects? *Biological Conservation* **143**: 2425–2434.

Sanders, N. J., N. J. Gotelli, N. E. Heller, and D. M. Gordon. 2003. Community disassembly by an invasive species. *Proceedings of the National Academy of Sciences* **100**: 2474–2477.

Sauer, L. 1998. *The Once and Future Forest: A Guide to Forest Restoration Strategies*. Island Press. Washington, DC.

Schall, M. J. and D. D. Davis. 2009a. *Verticillium* wilt of *Ailanthus altissima*: susceptibility of associated tree species. *Plant Disease* **93**: 1158–1162.

Schall, M. J. and D. D. Davis. 2009b. *Ailanthus altissima* wilt and mortality: etiology. *Plant Disease* **93**: 747–751.

Schmiedel, D. and O. Tackenberg. 2013. Hydrochory and water-induced germination enhance invasion of *Fraxinus pennsylvanica*. *Forest Ecology and Management* **304**: 437–443.

Schnitzler, A. 1995. Successional status of trees in gallery forest along the River Rhine. *Journal of Vegetation Science* **6**: 479–486.

Secor, E., D. Ross, and C. Balling. 2013. *Japanese knotweeds effect on erosion rates in riparian corridors.* Unpublished Report, University of Vermont, Burlington, Vermont, USA.

Shafroth, P. B., J. R. Cleverly, T. L. Dudley, et al. 2005. Control of *Tamarix* in the western United States: implications for water salvage, wildlife use, and riparian restoration. *Environmental Management* **35**: 231–246.

Shankman, D. 1993. Channel migration and vegetation patterns in the Southeastern Coastal Plain. *Conservation Biology* **7**: 176–183.

Shaw, R. H., S. Bryner, and R. Tanner. 2009. The life history and host range of the Japanese knotweed psyllid, *Aphalara itadori* Shinji: potentially the first classical biological weed control agent for the European Union. *Biological Control* **49**: 105–113.

Shelford, V. E. 1954. Some lower Mississippi Valley flood plain biotic communities: their age and elevation. *Ecology*: **126–142**.

Sher, A. and M. F. Quigley. 2013. *Tamarix: A Case Study of Ecological Change in the American West*. Oxford University Press. Oxford, UK.

Silliman, B. R. and M. D. Bertness. 2004. Shoreline development drives invasion of *Phragmites australis* and the loss of plant diversity on New England salt marshes. *Conservation Biology* **18**: 1424–1434.

Simberloff, D. 2011. How common are invasion-induced ecosystem impacts? *Biological Invasions* **13**: 1255–1268.

Slavicek, J. M., A. Boose, D. Balser, et al. 2005. Restoration of the American elm in forested landscapes, p. 74. *In*: Gottschalk, K. W. (ed.). *Proceedings, 16th U.S. Department of Agriculture Interagency Research Forum on Gypsy Moth and other Invasive Species, January 18–21, 2005, Annapolis, Maryland*. General Technical Report NE-337, USDA Forest Service. Northeast Research Station, Newtown Square, Pennsylvania, USA.

Slavicek, J. M. and C. O. Marks. 2011. Expansion of the American elm restoration effort to Vermont. p. 62. *In:* McManus, K. A. and K. W. Gottschalk (eds.). *Proceedings, 22nd U.S. Department of Agriculture Interagency Research Forum on Invasive Species 2011*. General Technical Report NRS-P-92, USDA Forest Service. Northeast Research Station, Newtown Square, Pennsylvania, USA.

Smith, D. M. and D. M. Finch. 2014. Use of native and nonnative nest plants by riparian-nesting birds along two streams in New Mexico. *River Research and Applications* **30**: 1134–1145.

Snyder, A. L., S. M. Salom, and L. T. Kok. 2014. Survey of *Verticillium nonalfalfae* (Phyllachorales) on tree-of-heaven in the southeastern USA. *Biological Control Science and Technology* **24**: 303–314.

Spiegel, K. S. and L. M. Leege. 2013. Impacts of laurel wilt disease on redbay (*Persea borbonia* (L.) Spreng.) population structure and forest communities in the coastal plain of Georgia, USA. *Biological Invasions* **15**: 2467–2487.

Stromberg, J. C., S. J. Lite, R. Marler, et al. 2007. Altered stream-flow regimes and invasive plant species: the *Tamarix* case. *Global Ecology and Biogeography* **16**: 381–393.

Suding, K. N., K. L. Gross, and G. R. Houseman. 2004. Alternative states and positive feedbacks in restoration ecology. *Trends in Ecology and Evolution* **19**: 46–53.

Tallamy, D. W. 2004. Do alien plants reduce insect biomass? *Conservation Biology* **18**: 1689–1692.

Tallamy, D. W., M. Ballard, and V. D'Amico. 2010. Can alien plants support generalist insect herbivores? *Biological Invasions* **12**: 2285–2292.

Tanentzap, A. J., L. E. Burrows, W. G. Lee, et al. 2009. Landscape-level vegetation recovery from herbivory: progress after four decades of invasive red deer control. *Journal of Applied Ecology* **46**: 1064–1072.

Tanentzap, A. J., D. R. Bazely, S. Koh, et al. 2011. Seeing the forest for the deer: Do reductions in deer-disturbance lead to forest recovery? *Biological Conservation* **144**: 376–382.

Tanentzap, A. J., J. Zou, and D. A. Coomes. 2013. Getting the biggest birch for the bang: restoring and expanding upland birchwoods in the Scottish Highlands by managing red deer. *Ecology and Evolution* **3**: 1890–1901.

Telford, C. J. 1926. *Third report on a forest survey of Illinois*. State of Illinois, Department of Registration and Education, Division of Natural History Survey.

Terwei, A., S. Zerbe, A. Zeileis, et al. 2013. Which are the factors controlling tree seedling establishment in North Italian floodplain forests invaded by non-native tree species? *Forest Ecology and Management* **304**: 192–203.

Tewksbury, L., R. Casagrande, B. Blossey, et al. 2002. Potential for biological control of *Phragmites australis* in North America. *Biological Control* **23**: 191–212.

Townsend, A., S. Bentz, and L. Douglass. 2005. Evaluation of 19 American elm clones for tolerance to Dutch elm disease. *Journal of Environmental Horticulture* **23**: 21–24.

Tracy, J. L. and C. J. DeLoach. 1999. Biological control of saltcedar in the United States: Progress and projected ecological effects, pp. 111–154. *In*: Bell, C.E. (ed.). *Proceedings of the Arundo and Saltcedar Workshop, Arundo and Saltcedar: The Deadly Duo, 17 June 1998, Ontario, California*. University of California Cooperative Extension Service. Holtville, California. Available from: http://www.cal-ipc.org/symposia/archive/pdf/Arundo_Saltcedar1998_1-71.pdf [Accessed January 2016].

Urgenson, L. S., S. H. Reichard, and C. B. Halpern. 2012. Multiple competitive mechanisms underlie the effects of a strong invader on early to late seral tree seedlings. *Journal of Ecology* **100**: 1204–1215.

Van Driesche, R., R. Carruthers, T. Center, et al. 2010. Classical biological control for the protection of natural ecosystems. *Biological Control* **54**: S2–S33.

van Hengstum, T., D. A. P. Hooftman, J. G. B. Oostermeijer, and P. H. van Tienderen. 2014. Impact of plant invasions on local arthropod communities: a meta-analysis. *Journal of Ecology* **102**: 4–11.

Wagner, D. L. and K. J. Todd. 2015. Conservation implications and ecological impacts of the emerald ash borer in North America, pp. 15–62. *In*: Van Driesche, R. G. and R. Reardon (eds.). *Biology and Control of Emerald Ash Borer*. Technical Bulletin FHTET 2014-09, USDA Forest Service. Morgantown, West Virginia, USA.

Warner, A. T., L. B. Bach, and J. T. Hickey. 2014. Restoring environmental flows through adaptive reservoir management: planning, science, and implementation through the Sustainable Rivers Project. *Hydrological Sciences Journal* **59**: 770–785.

Westgate, M. J., G. E. Likens, and D. B. Lindenmayer. 2013. Adaptive management of biological systems: A review. *Biological Conservation* **158**: 128–139.

Whisenant, S. 2005. Managing and directing natural succession, pp. 257–261. *In*: S. Mansourian and D. Vallauri (eds.). *Forest Restoration in Landscapes*. Springer. New York

Wilkinson, S. R., M. A. Naeth, and F. K. A. Schmiegelow. 2005. Tropical forest restoration within Galapagos National Park: application of a state-transition model. *Ecology and Society* **10**: 28–43.

Wilson, L. M., M. Schwarzländer, B. Blossey, and C. Bell Randall. 2009. *Biology and biological control of purple loosestrife*. FHTET-2004-12, USDA Forest Service. Morgantown, West Virginia, USA.

Wilson, S. D. and B. D. Pinno. 2013. Environmentally contingent behaviour of invasive plants as drivers or passengers. *Oikos* **122**: 129–135.

Wojcik, D. P., C. R. Allen, R. J. Brenner, E. A. Forys, et al. 2001. Red imported fire ants: impact on biodiversity. *American Entomologist* **47**: 16–23.

Zedler, J. B. and S. Kercher. 2004. Causes and consequences of invasive plants in wetlands: opportunities, opportunists, and outcomes. *Critical Reviews in Plant Sciences* **23**: 431–452.

Zedler, J. B. and S. Kercher. 2005. Wetland resources: status, trends, ecosystem services, and restorability. *Annual Review of Environment and Resources* **30**: 39–74.

CHAPTER 3

Matching tools to management goals

Charlotte Causton[1] and Roy G. Van Driesche[2]

[1] Charles Darwin Foundation, Galápagos Islands, Ecuador
[2] Department of Environmental Conservation, University of Massachusetts, USA

Introduction

Once it has been determined that an invasive species is a target for management (Chapter 2), the next step is to decide on the goal of the management program. Goals for invasive species suppression programs may include: (1) eradication, (2) limiting the invader's further spread (containment), (3) temporary suppression, either locally or area-wide, (4) permanent, area-wide suppression through modification of ecosystem processes, and (5) permanent, area-wide suppression through natural enemy introductions. Table 3.1 lists some of the factors that may affect achievement of these goals and that must be considered during the planning phase if a management program is to succeed.

Eradication or suppression efforts against an invasive species for ecosystem restoration or protection may involve a variety of tools. The choice of which tools are best for particular cases will be driven by the management goal, the invasive taxa targeted (e.g., vertebrates, insects, plants, etc.), and the conservation value and the resilience of the area to be managed; for example, tools used in areas with threatened or susceptible non-target species should minimize potential effects of interventions. The size of the area over which control is needed also influences tool selection; intensive control methods such as hand-pulling or pesticide application are most appropriate in small areas, while self-sustaining methods (e.g., biological control, changes in processes) are often most cost effective for large areas. The control method selected will also depend on the available funding and resources, whether proposed control measures are permitted according to local/national regulations, and the degree of support from

stakeholders and other community members (Table 3.1). Finally, as projects develop, regular evaluations using adaptive management methods are needed to assess whether objectives are being met and whether tools need to be changed.

Various tools exist for management of invasive species (Table 3.2; see also tools for invasive species in general [Wittenberg and Cock, 2001; Clout and Williams, 2009]; plants [Tu et al., 2001]; vertebrates [Orueta and Ramos, 2001; Courchamp et al., 2003], invertebrates [Green and O'Dowd, 2009; Brockerhoff et al., 2010]; and marine species [Hilliard, 2005; Thresher and Kuris, 2004]). To achieve restoration goals, it is common to use a management approach that integrates several tools (Hulme, 2006). In this chapter, many examples are discussed where an integrative approach was used to suppress an invasive species. Such multiple control efforts against invasive species may be particularly necessary, for example, if a control method depends on population numbers already being low (e.g., as does the sterile insect release method). In some cases, urgency may influence the choice and sequence of pest control actions. For example, fast-acting but expensive or difficult-to-implement control methods might be used initially against an invasive species to provide immediate relief to a species threatened with extinction and gain time to develop long-lasting, sustainable control methods. On the Galápagos Islands, for example, populations of endemic birds, including several critically endangered Darwin's finches, have been drastically reduced by the invasive parasitic fly *Philornis downsi* Dodge & Aitken (Diptera: Muscidae) (Causton et al., 2013). Measures that can be implemented immediately, such as treating nests with low toxicity insecticides,

Integrating Biological Control into Conservation Practice, First Edition. Edited by Roy G. Van Driesche, Daniel Simberloff, Bernd Blossey, Charlotte Causton, Mark S. Hoddle, David L. Wagner, Christian O. Marks, Kevin M. Heinz, and Keith D. Warner.

© 2016 John Wiley & Sons, Ltd. Published 2016 by John Wiley & Sons, Ltd.

Table 3.1 Criteria for determining safety and feasibility of invasive species management.

Criteria	Key questions
Biology and ecology of the target species	• Is it driving change? • Is it highly mobile, spread by animals or humans? • Are there interactions with other species (natives and invasives)? • Does it have mechanisms for persisting in the ecosystem (e.g., seed banks)? • Are generation times short? • Are there aspects of its reproductive biology that could make control harder (e.g., plants that spread vegetatively as well as sexually or have animal-dispersed seeds)?
Taxonomy and systematics	• What is the taxonomic certainty of the species (target species and any species proposed for use in a biological control program)? • Is the target species isolated taxonomically from native species or does it have phylogenetically close native relatives? • What vouchers will be needed for biological control efforts?
Ecological and geographical characteristics of the ecosystem	• Historically have changes in ecosystem processes facilitated invasions and can these be modified? • Level of degradation – is the area disturbed? Does it have a high number of introduced species? • Are there species that could be released from competition or predation that could become problematic once target species is suppressed?
Conservation value of the area	• Are there endemic species, or other species with restricted distributions, in the management area that could be at risk?
Size of the area invaded or that needs to be controlled	• Scale influences choice of management goal and tools • Can the area be subdivided to facilitate control?
Potential risks of management actions in short term to ecosystem health	• Is a risk assessment required before use of control tool? • Could the control methods negatively affect biodiversity and ecosystem health and/or health and livelihood of people? • Is it likely that other invasive species would thrive following removal of the target species? • Could there be indirect effects on the ecosystem?
Long-term effects of management program on environment	• Are any effects of species control irreversible?
Isolation	• Determines whether eradication or containment are options
Likelihood of re-invasion	• Is there a high risk that the species will be introduced again? • Are there measures in place to reduce the risk of re-invasion (early detection programs etc.)?
Are there good monitoring tools for evaluating success of program and response of ecosystem	• Can monitoring tools detect the target pest at very low densities? • Are there plans to conduct ecosystem monitoring both before and after management actions to evaluate the success of the program?
Urgency	• Is action needed quickly, i.e., new species arrival?
Feasibility	• Are techniques available and are they efficient in reducing invasive species numbers? • Is the rate of removal likely to exceed population growth of the invasive? • Will the tools be enough to reduce numbers to acceptable levels or are more tools needed?
Legislation	• Does the management program and choice of tools comply with local and national regulations?
Costs/benefits of control program	• How much will it take to achieve the goal and how much will it cost? • Do the damages caused by the invasive species outweigh the costs of the management program? • What is the optimal cost : benefit ratio?
Available funding and resources for the project	• Is there sufficient funding to achieve goals (this is especially important for eradication programs where resources should be available to ensure completion of the program)?
Social and political agreement	• Are stakeholders and community on board and engaged? • Are there any areas where conflicts could arise that could affect success of program?
Coordination between regulatory bodies	• Are all agencies in agreement?

Table 3.2 Summary of management options for invasive species.

GOALS: ERADICATION, CONTAINMENT OR LOCAL/AREA-WIDE TEMPORARY PEST SUPPRESSION

Management tools		Target group	Advantages	Disadvantages
Physical control	Hand collecting/ pulling or hand cutting	Invertebrates, plants, amphibian, pathogens	• Fairly specific to target pest • Can be used in ecologically sensitive areas • Efficient method for slow-growing species	• Can cause disturbance that can encourage invasion by other invasive species (weeds) • Labor intensive, best for small areas • Often requires repeat treatments
	Mechanical removal, cutting, or mowing	Plants, plant pathogens	• Can sometimes be species specific • Temporarily suppresses target • Can be used over large areas • Can be effective in reducing seed production	• Can cause disturbance • Often requires repeat treatments • May not be appropriate for ecologically sensitive areas or small areas • Can encourage invasion by other invasive species (weeds)
	Mulching or solarization	Plants	• Can kill seeds • Can be used in ecologically sensitive areas	• Labor intensive • Can only be used in small areas
	Girdling and stabbing	Plants	• Specific to target pest • Disturbance minimal • Efficacy increased when combined with herbicides • Can be used in ecologically sensitive areas	• Labor intensive
	Exclusion: fences, enclosures, and barriers	Vertebrates, invertebrates, terrestrial and aquatic weeds, fish	• Effective in containing an invasive species • Especially effective for large vertebrates • Effective for protecting threatened native species from invasive species	• Only cost effective for small areas • Require constant surveillance and maintenance
	Trapping with pheromones or other attractants	Insects, mammals, fish, amphibians	• Effective method • Can be used in ecologically sensitive areas when attractants are used • Important tool for monitoring populations • Can be especially effective for small, mobile species, i.e., flying insects	• Labor intensive • Requires good accessibility • Need to deploy on a regular basis • Not all methods are species specific • Traps for larger animals can cause animals to suffer
	Hunting/ culling/ fishing	Vertebrates, aquatic species, invertebrates	• Species selective • Can be very efficient for large species • Can be used over large areas	• Use of bounties may cause the species to become a valuable commodity and may encourage captive breeding • Issues with animal rights groups • May require additional control techniques

Table 3.2 (*Continued*)

GOALS: ERADICATION, CONTAINMENT OR LOCAL/AREA-WIDE TEMPORARY PEST SUPPRESSION

Management tools		Target group	Advantages	Disadvantages
Chemical control with toxins	Synthetic pesticides	Invertebrates, fish, mammals amphibians, reptiles, plants, pathogens	• Can be effective in reducing pest numbers • Some pesticides can be used in combination with biological control and other control methods • Can be used in ecologically sensitive areas if they are combined with an attractant or are targeted at a species	• Many are broad spectrum and safety would depend on technique used to deliver insecticide • Labor intensive • Often requires repeated applications and coverage over entire area to be effective • Resistance can be developed over time • Residues can enter ecosystem
	Botanical pesticides	Insects, plants, amphibians	• Can be effective in reducing pest numbers • Can be ecologically safe if used correctly	• Many are broad spectrum and safety would depend on what technique is used to deliver insecticide • Can be labor intensive • Often requires repeated applications and coverage over entire area to be effective • Little known about long-term resistance and effects on ecosystem
	Biopesticides	Insects, plants (only a few)	• Some are safer and more taxa specific • Some can be broadcast over large areas	• Many are broad spectrum and safety would depend on what technique is used to deliver insecticide • Can require repeated applications • Little known about long-term resistance and effects on ecosystem
Behavior-modifying chemicals	Pheromones (mass trapping and mating disruption)	Insects, fish, amphibians	• Species specific and ecologically safe • Most effective when controlling low to moderate pest densities • Works best if large areas are treated • Can treat inaccessible areas • Can be especially effective for small, mobile species, i.e., flying insects	• Can be labor intensive
	Repellents	Invertebrates, mammals, fish, amphibians, reptiles	• Disturbance minimal • Efficacy increased when combined with other methods • Can be used in ecologically sensitive areas • Some are species specific	• Effective only for small areas • Labor intensive • Need to apply on a regular basis • Not all methods are species specific
Sterile Insect Technique (SIT) or autocidal control using genetic manipulations		Insects, fish	• Species-specific control method • Can be applied over difficult topography • Inversely density dependent • Integrates well with other methods • Can result in eradication • Insects can be tracked with genetic markers	• SIT requires expensive infrastructure (irradiation) • Potential for performance reduction in insects (SIT-irradiation)

(*Continued*)

Table 3.2 (*Continued*)

GOALS: AREAWIDE, PERMANENT SUPPRESSION THROUGH ECOSYSTEM MODIFICATION

Management tools	Target group	Advantages	Disadvantages
Planting competitive native species	Plants	• Efficacy increased when combined with other methods • Can be used in ecologically sensitive areas • Increases competiveness of natives	• Often labor intensive
Changes in grazing regimes	Plants	• Can be cheap • Efficacy increased when combined with other methods • Increased grazing can favor native plants	• Causes disturbance • Cannot be used in ecologically sensitive areas • Many non-target risks
Controlled burning	Plants	• Works especially well for annual plants with short-lived seed banks	• Often requires repeat treatments • Cannot be used in ecologically sensitive areas • May promote further invasion of exotics by reducing competitive abilities of native plants • Many non-target risks
Changes in flooding regimes	Plants, fish	• Can inhibit invasive plant growth and survival • Can encourage growth of native species	• Potential conflicts may occur with landowners or in developed areas
Alteration of nutrient levels	Plants	• Can inhibit invasive plant growth • Encourages growth of native species	• Labor intensive

GOALS: AREA-WIDE, PERMANENT SUPPRESSION USING NATURAL ENEMIES

Management tools	Target group	Advantages	Disadvantages
Biological control	Insects and plants	• Can be genus or species specific • Ecologically safe • Used over large areas • Can be applied over difficult topography • Permanent and self-sustaining • Good benefit : cost ratios • Effective against invasives	• Takes longer to develop • Large, initial investment • Hard to know level of control until it is released

are being considered to protect declining species while research on biological control or trapping of flies is conducted to develop better tools for sustainable, area-wide control (Causton et al., 2013).

Scale of application, spatial overlap, or sequence of treatments may all affect the success of efforts to integrate two or more control methods. For example, biological control and chemical control may be incompatible at the local site level because of harm from pesticides to natural enemies, but at the landscape level they might be combined if each functions at somewhat different times or in different places. For example, in southern Florida, Australian paperbark trees (*Melaleuca quinquenervia* [Cav.] S.T. Blake) were controlled by combining (1) biological control to suppress seed production and kill seedlings and stump sprouts, (2) herbicides, and (3) cutting – the latter methods to remove existing mature trees (Center et al., 2012).

In the following sections we discuss the principal management goals and some of the tools that can be used to achieve these goals. Following this, we discuss key factors that can affect the efficacy of control efforts. Lastly, we end with a discussion about when biological control may be the right choice as a management option.

Eradication

The goal of eradication is to eliminate the whole invasive species population in a given area. The tools used for eradication programs are quite varied and overlap broadly with those used for temporary pest suppression at the local or area-wide level, but the two types of programs have different requirements because of the goal of eliminating every last individual in the population. Management based on modification of ecological processes or introduction of natural enemies (biological control), while they are suitable for pest suppression at the landscape level, will not lead to pest eradication.

The probability of success in eradication programs increases when invaders are detected early and the control response is mounted quickly, if infested areas are small, when re-invasion can be prevented (such as on islands), and if there is public acceptance (Panetta and Timmins, 2004; Atkinson et al., 2012; Pluess et al., 2012; Glen et al., 2013). To eradicate a species successfully, managers need good judgment as to when this goal is feasible because much time and money can be lost on projects with little chance of success. Table 3.1 lists criteria that should be evaluated before embarking on an eradication program. Only good survey work, plus knowledge of the pest's biology, the nature of the infested area, and efficacy of eradication methods can help managers determine if eradication is feasible. Furthermore, eradication should be attempted only when feasibility studies show that the benefits outweigh the costs of the program and when there are sufficient resources to see the program through to completion (Table 3.1). Lastly, good monitoring tools are essential for detecting remaining infestations and for assessing eradication success.

Technological advances have made eradication a feasible tool for many vertebrate species, especially on uninhabited islands, but eradication still has limited application for invertebrates, plants, and pathogens (Glen et al., 2013). Islands as large as 11,300 ha have been successfully cleared of rodents through the use of aerially deployed poisoned baits (McClelland, 2011).

Larger invasive species, such as cats (Campbell et al., 2011) and grazing mammals (e.g., goats, see Campbell and Donlan [2005] and Carrion et al. [2011]), have been eradicated from islands through combinations of poison baits, aerial shooting, trapping, and fencing. Aerial hunting of large vertebrates (goats, donkeys, etc.) on islands has been particularly successful at eradication (e.g., goats were eliminated from 240,000 ha of northern Isabela in the Galápagos Islands) (Carrion et al., 2011; Atkinson et al., 2012). Such programs require public acceptance and considerable planning. Additional techniques may be needed to find individuals once populations have been reduced, such as the Judas goat technique, where sterilized goats in estrous are used to attract other goats (Cruz et al., 2009).

For invasive plants, the probability of eradication declines with increasing size of infestations, with infestations >100 ha having <33% chance of success (Rejmánek and Pitcairn, 2003). Indeed, the majority of successful plant eradications have been in areas of <1 ha and have involved recent plant incursions, cultivated and non-naturalized plants, or naturalized plants with limited distributions (Panetta, 2009; Glen et al., 2013). The largest successful plant eradication program has been against kochia (*Kochia scoparia* [L.] Schrad.) (2480 ha), for which intervention was early and infested locations were known (Dodd and Randall, 2002; Panetta, 2009). Criteria that should be considered before deciding whether eradication is an option for invasive plants are outlined by Panetta and Timmins (2004). Factors that strongly influence the probability of successful eradication of plants include life-history traits such as time to maturity and the duration of plant seed banks (Panetta, 2004, 2009; Gardener et al., 2010).

Eradication of invertebrates has had variable success. Isolated populations of certain insects have been eradicated with pesticides or release of sterile males (SIT). Screwworm flies (*Cochliomyia hominivorax* [Coquerel]) (Diptera: Calliphoridae) were successfully eradicated from the southwestern United States and northern Mexico in the 1960s after the release of millions of sterile males (Klassen and Curtis, 2005). When wild females mate with males that have been sterilized by exposure to radiation or chemical mutagens, female reproductive capacity is reduced, and over several generations the population shrinks and dies out (Vreysen et al., 2007). SIT is highly species specific (effective against only the target pest) and environmentally benign.

Recent research developments suggest that genetic manipulations of the sex of organisms such as insects and fish also have potential as autocidal control techniques (Harris et al., 2011; Morrison and Alphey, 2012; Teem and Gutierrez, 2013).

Insecticides and biopesticides have been used to eradicate isolated populations of gypsy moth (*Lymantria dispar* [L.]) (Lepoptera: Erebidae), Asian longhorned beetle (*Anoplophora glabripennis* [Motchulsky]) (Coleoptera: Cerambycidae), and painted apple moth (*Teia anartoides* Walker) (Lepidoptera: Erebidae) (Dreistadt and Dalhsten, 1989; Suckling et al., 2007; Haack et al., 2010). Some of these eradication programs required the use of a combination of tools including trapping using attractants, SIT, and host plant removal.

Even marine organisms may sometimes be eradicated, as in the use of bleach against an isolated infestation of the alga *Caulerpa taxifolia* (M. Vahl) C. Agardh in coastal California (Anderson, 2005), but this is more likely in enclosed areas such as marinas.

Finally, it is important to recognize that sometimes eradication of an invader may not promote recovery of the affected ecosystem. Native species or ecosystems may be unable to recover without additional intervention, or in some cases other invaders, suppressed by the invasive species targeted for eradication, may be released from competition or predation (Zavaleta et al., 2001; Glen et al., 2013). For example, eradication of large, invasive herbivores on some islands has resulted in rapid population growth of invasive plants resulting in suppression of native species and facilitating dispersal of invasive species (e.g., invasive blackberry [*Rubus* sp.] on the Galápagos Islands, Atkinson et al., 2012). The removal of vertebrates at the top of the food chain, in particular, may have greater undesirable side effects; such as the increase of rodents following cat eradication (Phillips, 2010). Also, the removal of competitors may provide the chance for other invasive species with similar ecological niches to flourish, for example the elimination of rats promoting an increase in mice (Harris, 2009; Glen et al., 2013). To ensure eradication achieves or contributes to the goal of ecosystem restoration, it is important to try and predict as best as possible the consequences of removing an invasive species from an ecosystem and implement measures to prevent unexpected outcomes. This may mean carrying out removal of several species before, during, or soon after the removal of the targeted species,

or implementation of other preventative measures (Glen et al., 2013). Very importantly, biodiversity monitoring before and after the eradication program is crucial to evaluate the success of the program and to allow responses to any unexpected consequences that arise.

Limiting spread

If eradication is not feasible, it may still be possible to stop or slow the spread of the pest from the initial area of invasion. This strategy is known as containment (Hulme, 2006), and it is most effective against species whose natural spread is slow and whose populations are mainly moved by humans (e.g., scale insects or other herbivorous insects attached to nursery stock or wood-boring insects in infested firewood). If strong ecological or geographical barriers exist that prevent natural spread, it may be possible to control human-assisted spread with sub-national quarantines or product inspections. For example, both gypsy moth and Japanese beetle (*Popillia japonica* Newman) (Coleoptera: Scarabaeidae) have been present in the eastern United States for more than a century, but their spread to the western United States has been prevented or, where it occurred, outlying populations were eradicated (Gammon, 1962; Sharov et al., 2002). In North America, gypsy moth females are flightless, making long-distance spread reliant on humans, while spread of Japanese beetles is limited by geographical barriers such as mountains and deserts. Such efforts, however, are less likely to prevent spread when no ecological or geographical barriers occur as in, for example, the failure of a similar approach to stop the spread of emerald ash borer (*Agrilus planipennis* Fairmaire) (Coleoptera: Buprestidae) within the deciduous forest region of the eastern United States (McCullough, 2015).

Containment can also be useful to slow quickly dispersing species until mechanisms are found to control or eradicate them. In Australia, for example, the parasitic plant branched broomrape (*Orobanche ramosa* L.) was contained in a quarantine area where it was treated with herbicides and host plants were physically removed. Strict quarantine measures and regular surveys outside the containment area were essential to success (Panetta, 2012). In another case, the soil-borne plant pathogen *Phytophthora cinnamomi* Rands was contained in a 250 ha area in Australia to protect the Fitzgerald River National Park, a 3300 km² World

Biosphere Reserve. The pathogen was suppressed with host plant removal, fungicide applications, and root barriers, while fences were used to prevent large animals, capable of dispersing the pathogen, from entering the infested area (Dunstan et al., 2010). Fences can also be used to slow the spread of invasive vertebrate species. In arid parts of northern Australia, fences were effective in denying cane toads (*Rhinella marina* [L.]) access to artificial water sources (such as stock ponds) and are potential tools for slowing regional spread (Florance et al., 2011). In areas where an invader's eradication over the whole landscape is not feasible or is not socially acceptable (e.g., management of feral pigs on Hawaii [Cole et al., 2012]), fences can exclude the invader from large remnants of native communities, as for example the Maungatautari Reserve in New Zealand (see section "Options for large preserves or islands with reinvasion risk").

Containment can also be achieved using other tools. For example, large weekly releases of sterile males have kept the border of Panama free from invasions of the screwworm fly from Colombia. This requires considerable efforts and resources; however, the investment is low compared to the costs of the damages that would be incurred if the fly were able to get through Panama and travel through Central America to the United States (Robinson et al., 2009).

Local, or area-wide, temporary suppression of invaders

When the management goal is invader suppression in specific limited preserves or areas, a wide variety of tools may be used, including heavy machinery, cutting and rouging tools, poison baits, fences, traps, and pesticides. The choice of tools will depend on the characteristics of the ecosystem, the scale of the infestation, and the resources available (see section "Effects of treatment area"). Some of these tools may require particular types of expertise, but in general are quickly mobilized and within reach of most land stewards. Volunteers, student interns, or prison crews are often used to reduce the costs of such efforts where areas are small enough, but for larger areas professional contractors are typically hired. The success of these efforts varies widely and depends on the particular invader targeted and the local ecological context. If control is accompanied by a shift in ecological conditions toward ones that are unsuitable for the invading species, the effect of suppression may be long term. However, in many cases, suppression is temporary and cessation of control efforts often permits regrowth or re-invasion of the species being controlled, or invasion by another species (Jäger and Kowarik, 2010). In such cases, there is an ongoing need for continuous funding to sustain the suppression efforts.

Manual or mechanical removal

Hand removal is typically used to control small infestations or those in ecologically sensitive areas, particularly infestations of invasive plants in areas where use of herbicides is not feasible. For example, uprooting of European beach grass (*Ammophila arenaria* [L.]) was successful in wildlife refuges where herbicide application was not acceptable (Pickart, 2008). Manual removal can be a low-cost tool in places where paid labor is cheap (Simberloff, 2009). Mechanical control of weeds (such as mowing) (Figure 3.1) causes disturbance and these methods should be used with caution. In some instances, such as with beach grass removal, physical disturbance may facilitate restoration of native grass species (Pickart, 2008); however, in many cases disturbance may facilitate invasion or release seed banks of invasive species, even the one being removed (Mason and French, 2007; Jäger and Kowarik, 2010).

Stabbing or girdling can be effective means of control of some invasive plants, such as baby's breath (*Gypsophila paniculata* L.), which can be killed by stabbing the taproot with a spade or knife. This reduced weed ground cover from 50 to <5% in a sand dune habitat in Michigan, USA (Emery et al., 2013). Some trees and woody shrubs that have a single trunk can be controlled by girdling, particularly with the application of herbicides into the cut area or directly onto bark. Such methods are inexpensive and can be highly selective. Girdling of fire trees (*Morella faya* [Aiton] Wilbur) in Hawaii was effective because it caused minimal disturbance. The gradual disintegration of dead trees allowed shade-tolerant native species to re-establish, without opening gaps for sun-loving invasive plants (Loh and Daehler, 2008).

In some ecosystems, especially those that are highly degraded, heavy machinery can be used. Mulching machines and spot application of glyphosate suppressed Chinese privet, *Ligustrum sinense* Lour., in riparian forests of the southeastern United States and resulted in an

Figure 3.1 Mechanical removal of non-native gorse (*Ulex europaeus*) for wildlife habitat improvement at the Coquille Unit of Oregon Islands National Wildlife Refuge, Coos County, Oregon, USA. Photo courtesy of David Ledig, U.S. Fish and Wildlife Service, Pacific Region.

increase in plant and invertebrate richness, although additional intervention was sometimes required to return the forests to historical conditions (Hanula et al., 2009; Ulyshen et al., 2010). Mowing and cutting can be used to reduce seed production and restrict weed growth (Tu et al., 2001); semi-annual mowing of *Molinia arundinacea* Schrank in Carpathian fens suppressed the pest and increased plant diversity and evenness (Hájková et al., 2009).

Mechanical removal of invaders other than plants is more difficult, and this approach is generally not effective against insects. However, in some cases other groups of invertebrates may be controlled on a small scale. For example, giant African snail (*Achatina fulica* Bowditch) was suppressed in a 1 ha protected area in the Galápagos Islands by clearing of vegetation and collection and destruction of adults and eggs (Charles Darwin Foundation, 2014). In some cases, pathogens or herbivorous insects can be controlled indirectly by removing their host plants. This is particularly effective if the primary host plant is also an introduced species. The probability of success increases if the target pest has a limited host plant range or is restricted to a small area. Tree felling has been used widely to eradicate small infestations of Asian longhorned beetle, a polyphagous wood borer

(Haack et al., 2010). The Dutch elm disease pathogen (*Ophiostoma ulmi* [Buisman] Nannf.) was contained in Auckland, New Zealand by removing and destroying infected trees, controlling its vector (*Scolytus multistriatus* [Marsham]) (Coleop: Curculionidae: Scolitinae), and using quarantine and surveillance measures. Identification of areas with infected elm trees was facilitated by pheromone trapping of the beetle vector and examination of the beetle for spores of the fungal pathogen (Gadgil et al., 2000).

Mass trapping

Costs of deploying and monitoring traps limit the area that can be treated via mass trapping; also non-target species may be killed if attractants are not sufficiently selective. Traps with attractants that target the invasive species can be used in areas of high conservation value to suppress invasive species of special concern, especially those that exist at low density or as isolated populations. Non-selective traps, such as light traps or sticky yellow traps, may not be suitable for use in areas of ecological importance. Black rats were eliminated from Feno Islet (1.6 ha) in the Azores using cheese-baited rat traps set in grids, allowing recolonization by an endangered tern (Amaral et al., 2010). Mass trapping can also be combined

(a)   (b)

Figure 3.2 (a) Flightless female gypsy moth, *Lymantria dispar*; (b) lure traps using female pheromones effectively capture male adult gypsy moths. Photos courtesy of Daniel Herms, The Ohio State University, Bugwood.org.

with other approaches for the control of reptiles, amphibians, vertebrates, fish, and invertebrates (e.g., Courchamp et al., 2003; Ling, 2009). Trapping and baiting can also be used to estimate efficacy of control or eradication programs, in particular for rodents, whose sign at traps or track boards indicates missed animals or re-invasion.

For some insects, mass trapping can be cost effective even over large areas and has been used successfully to control or eradicate some species, alone or in combination with other techniques (e.g., fruit flies [Tephritidae] and boll weevil, *Anthonomus grandis grandis* [Boheman] [El-Sayed et al., 2006]). Trapping appears to work best for insects that are monophagous, mate only once, and have a single generation per year (El-Sayed et al., 2006). Attractants often include the preferred food of the pest or its pheromone. Pheromones have been used to trap gypsy moths (Figure 3.2) and bark beetles in natural areas (Schlyter et al., 2003; El-Sayed et al., 2006; Brockerhoff et al., 2010). The number and location of traps and lure strength must be sufficient to catch a significant number of the insects if damage is to be reduced to acceptable levels.

Hunting and bounties

Where eradication is not the goal, recreational hunting is sometimes used to control some invasive vertebrates (e.g., feral pigs and deer) but can be problematic if the pest has economic or cultural value (Gherardi and Angiolini, 2004). Bounties for killing, collecting, or reporting invasive species have been used in combination with other techniques in control programs, for example the eradication of the house crow (*Corvus splendens* Vieillot) in Yemen (Suliman et al., 2011), but they can be costly and may not be successful. In some cases, the bounty may even encourage local people to breed the invasive species, as occurred with the giant African snail in Samoa (Shine et al., 2000; Zabel and Roe, 2009).

Pesticides

Pesticides can be used for both eradication and local suppression of invasive vertebrates (including fish), invertebrates, and weeds from conservation areas, provided potential risks of pesticides or their residues to non-target species (including insect pollinators and

Figure 3.3 To target individual trees, cuts can be made into the trunk of a tree and a herbicide applied. Photo courtesy of Steve Manning, Invasive Plant Control, Bugwood.org.

natural enemies) are carefully considered beforehand. Additionally, the potential of pesticides to enter aquatic ecosystems by runoff or soil erosion (Pimentel, 2005) must be considered. Size of the treated area, material applied, and application methods will determine the feasibility of use. An *a priori* non-target risk assessment should, and typically is, conducted to determine whether any anticipated effects of pesticide use are unacceptable to managers. Safety may be increased by the use of attractants that lure the target species to the pesticide (e.g., attract and kill) or by targeting the application precisely, such as the application of herbicides to the cut stumps of invasive plants. In many instances when pesticides are used, repeated applications are needed, increasing cost and potential damage to ecosystems (Tu et al., 2001; Pimentel, 2005).

Some pesticides are selective, with low ecotoxicological risk, or they may be applied in ways that reduce exposure of non-target species, including natural enemies used in biological control (e.g., Gentz et al., 2010; Roubos et al., 2014). Nevertheless, unforeseen effects may occur if large areas are treated; for example, there is concern that imidacloprid, which has low mammalian toxicity, may be able to affect the reproduction and behavior of nectar-feeding insects because of residues in nectar of treated plants. This impact becomes important when the product is applied over large areas (Gentz et al., 2010; Godfray et al., 2014).

Herbicides suitable for use in natural areas exist (Tu et al., 2001) but vary in their potential to damage non-target vegetation. The timing of applications and the application techniques will influence efficacy (Figure 3.3). Pre-emergent herbicides that kill plants just as seeds are germinating can be a useful tool to eliminate large seed banks of invasive species. For example, a combination of pre-emergent herbicide, biological control, and re-vegetation suppressed mile-a-minute weed, *Persicaria perfoliata* (L.) H. Gross, prevented the invasion of Japanese stiltgrass (*Microstegium vimineum* [Trin.] A. Camus), and increased native plant cover (Lake et al., 2014). In some cases, herbicides can have long-lasting effects on native plants; Rinella et al. (2009) found that one application of picloram on native grasslands suppressed two native forbs, which remained rare for 16 years. Herbicides may also affect weed biological control agents by killing their host plants or affecting the agents themselves. In South Africa, herbicide/surfactant combinations used to control waterhyacinth (*Eichhornia crassipes* [Martius] Solms-Laubach), caused mortality in the laboratory of the biological control agents *Neochetina eichhorniae* Warner (Coleoptera: Curculionidae) and *Eccritotarsus catarinensis* (Carvalho) (Hemiptera: Miridae) and might affect their efficacy in the field, particularly that of the mirid (Hill et al., 2012).

Formulation of pesticides as baits can improve their safety, especially if they are developed to attract specific species. Baits may be used at bait stations, hand-broadcast, or applied aerially. If native species at treatment sites are not attracted to baits, safety is ensured. Otherwise, measures must be taken to ensure safety; for example, bait might be placed in dispensers not entered by non-target species or applied when key non-target species are not active, or wildlife might be captured and held off site during the risk period. Brodifacoum formulated in cereal pellets or wax blocks has been successfully used to clear rodents from

Figure 3.4 Island Conservation, Parks Canada, and Haida Nation worked together in 2013 to protect Ancient Murrelets by removing invasive rats. This image shows an aerial broadcast of cereal pellets on Murchison and Faraday Islands, Haida Gwaii, British Columbia. Photo courtesy of David Will/Island Conservation.

smaller islands or locations being restored as native bird refuges (Figure 3.4), though its broad-spectrum impacts, persistence, and ability to biomagnify (toxicant accumulates as it moves up the food chain) means that measures need to be in place to minimize non-target impacts (Towns and Broome, 2003; Campbell et al., 2015). Some invasive ants can be effectively controlled with baits (Rabitsch, 2011; Plentovich et al., 2011; Hoffman et al., 2011) because ants carry bait back to the nest, killing the entire colony. This minimizes non-target impacts. In some cases, a few applications result in eradication (Causton et al., 2005; Hoffman et al., 2011).

Botanical pesticides are compounds derived from plants. A non-selective botanical pesticide used for elimination of unwanted fish populations is rotenone. Although effective, this pesticide is highly toxic to all types of fish and many invertebrates and, because of this, in some situations (such as lakes) where recolonization of natives is not possible, susceptible non-target species have been temporarily removed from the water sources while the invasive pest is controlled (Nico and Walsh, 2011). Another widely used botanical pesticide

is azadirachtin from the neem tree (*Azadirachta indica* A. Juss.), which is effective against many insect pest species (Stark, 2007). While not suitable for broadcast application in natural areas, it may be used with selective delivery methods, such as tree injection. When injected into ash trees, azadirachtin appears to control emerald ash borer (*A. planipennis*) and pest cerambycids with no measurable effect on litter-dwelling arthropods or aquatic species that feed on fallen leaves (Mckenzie et al., 2010; Kreutzweiser et al., 2011). Botanical pesticides have also been used successfully to control some invasive amphibians. For example, citric acid application, by itself or in combination with vegetation removal, can help reduce populations of coqui frogs (*Eleutherodactylus coqui* Thomas) in Hawaii (Pitt et al., 2012).

Biopesticides are microorganisms formulated and applied to control pests. The best known biopesticide is the bacterium *Bacillus thuringiensis* Berliner (Bt), which both infects its hosts with spores and produces a highly specific gut toxin affecting some groups of insects, but with no risk to vertebrates (Lacey and Siegel, 2000). Different strains are available that attack mosquitoes and blackflies (Bti), caterpillars of some lepidopterans (Btk), and some beetle larvae (Btm). Formulations of this bacterium have been used to eradicate some invasive insects (e.g., gypsy moth [Hajek and Tobin, 2010]); however, this biopesticide can affect non-target species (Strazanac and Butler, 2005) and may not be appropriate for widespread application to high-value conservation areas. Another microbially derived material is spinosad, a mixture of toxins produced during fermentation of the soil actinobacterium *Saccharopolyspora spinosa* Mertz and Yao. Spinosad appears to be safer to beneficial insects than many other products (Gentz et al., 2010); however, non-target effects may be broader than originally thought (Biondi et al., 2012). When it is applied with species-specific lures (e.g., Specialized Pheromone and Lure Application Technology: SPLAT), non-target affects are minimized (Hardy, 2011). Another group of biopesticides consists of certain plant pathogens that have been developed for use as mycoherbicides (Barreto et al., 2012); however, their high specificity often leads to economic failure because demand is too small to support production. Nevertheless, interest in such products remains (Breeÿen and Charudattan, 2009; Barreto et al., 2012).

Behavior-modifying chemicals

Semiochemicals are compounds emitted by an organism that influence the physiology or behavior of an organism of the same or a different species. These include pheromones (attractants) and repellents. Pheromones of pest insects are often species specific and can be used in monitoring traps (Figure 3.2). Broadcast application of insect sex pheromones to reduce mating by interfering with normal mate attraction (called "mating disruption") is used in orchards and agricultural areas to reduce attack by pests such as the codling moth (*Cydia pomonella* L.) (Witzgall et al., 2008), and this approach can work in natural ecosystems to control or contain some invasive species, such as gypsy moth (Sharov et al., 2002; Brockerhoff et al., 2010). Semiochemicals are non-toxic to vertebrates and beneficial insects and target or interfere with the pest animal's life cycle. For example, pheromones emitted by male sea lampreys (*Petromyzon marinus* L.) are being tested to attract females to traps or streams of low ecological value (Luehring et al., 2011). Synthetic versions of ant trail pheromones are being evaluated against Argentine ant (*Linepithema humile* [Mayr])

(Suckling et al., 2008; Choe et al., 2012) with the goal of interfering with communication within colonies, thereby disrupting foraging and nest relocation.

Chemical repellents may also be used for invasive species management (see Orueta and Ramos, 2001). Repellents are derived from plant extracts, synthetic chemicals, predator odors, or even alarm cues of the target species; for example, alarm cues produced by injured tadpoles of the cane toad show potential for use in repelling conspecifics (Hagman and Shine, 2008). Some compounds in plants repel herbivorous mammals, and studies have been conducted to test whether some of these repellents could be applied to native plants to deflect herbivory onto invasive weeds (Kimball and Taylor, 2010). In choice tests, Kimball and Perry (2008) found that captive American beavers (*Castor canadensis* Kuhl) preferred feeding on saltcedar (*Tamarix ramosissima* Ledeb.) to feeding on native riparian plant cuttings painted with casein hydrolysate. Many repellents are not long lasting, however, and require repeated applications.

Area-wide, permanent suppression through modification of ecosystem processes

When an invasive species has spread over wide areas and is past the point of eradication, the question should be asked "Is this invasion being driven by human mismanagement in ways that could be changed?" If the answer is yes, then modification of fire, flood or grazing regimes, nutrient levels, or replanting of native species may be a method to suppress an invasive species permanently on an area-wide basis. Invasive species that prosper under disrupted environmental conditions are best thought of as "passengers" being pulled along by the underlying processes of ecological degradation rather than being the "driver" of those changes themselves (Didham et al., 2005; MacDougall and Turkington, 2005, see Chapter 2). In some cases, restoration of those underlying processes closer to historical norms may suppress this sort of invasive species, unless they have further altered (transformed) the system making such a return difficult. These invasive species are called transformer species and include such species as gamba grass (*Andropogon gayanus* Kunth var. bisquamulatus (Hochst.) Hack), which has invaded Australian savannas. There, the presence of dense, 4-m high swards of gamba grass has increased the fuel loads for fires sevenfold, which has altered the frequency and increased the intensity of fires. These fires have substantially decreased native grass and shrub recruitment and altered the function of the savanna ecosystem (Gallagher and Leishman, 2014).

Changes in fire regimes

These may promote plant invasions because native plants are usually adapted to the historical fire regime and are less competitive when this is altered. For example, an increase in the intensity of prescribed burns in one area may facilitate invasion by non-native plants (Masocha et al., 2011), while, on the other hand, in another area a reduction in fire events may promote the growth of an invasive plant species (Witt and Nongogo, 2011). Both human suppression of fire or invasion by fire-promoting plants may alter fire intensity or seasonality, promoting plant invasions (D'Antonio and Vitousek, 1992; Brooks et al., 2004).

Use of controlled burns to restore fire to fire-deprived communities, including grasslands invaded by woody plants, is a common practice (DiTomaso et al., 2006). Annual invasive weeds that seed after the fire season begins and have short-lived seeds that are exposed to fire may be easiest to control (DiTomaso et al., 2006). The effects of variation in timing and frequency of the burns needs to be assessed to ensure that the burns will achieve the goal of restoring native plant communities. Furthermore, additional management techniques such as replanting of native species may be needed. In some cases controlled burns, if not carried out correctly, may instead promote invasive plants by lowering native plant competition and opening up areas to invasion (see Dodson and Fiedler, 2006, and examples in Hulme, 2006). Even when successful, prescribed burns may cause shifts in vegetation type (e.g., Beest et al., 2012).

Changes in flood level or duration

Flooding affects the success of some invasive species (e.g., Merritt and Poff, 2010). In the western United States, in saltcedar-infested areas (*Tamarix* spp.) restoration of natural surface and groundwater flows enhances competitiveness of native plant growth (Stromberg et al., 2007). However, lack of floods or floods at arbitrary times can produce both high *Tamarix* mortality and massive *Tamarix* recruitment (Mortenson et al., 2012). In floodplain forests along the Connecticut River in New England, USA, shortening of flood durations promotes invasion of floodplains by flood-intolerant invasive plants (see Chapter 2 for details).

Changes in grazing regimes

Large grazing species influence vegetation by changing competiveness between native and invasive plants, and increases in grazing intensity may favor invasive species, particularly those unpalatable to the grazing animals (i.e., toxic or thorny) (DiTomaso, 2000). A reduction in the nomadic habits of Maasai pastoralists in Kenya resulted in intensified grazing and consequent reduction of native groundcover, facilitating invasion of the invasive cactus *Opuntia stricta* (var. *stricta*) (Strum et al., 2015). This invasive plant had been in the area for 50 years yet had not been a problem until grazing pressure increased. However, at this point the effects are not irreversible; exclosure experiments suggest that revegetation with native plants and trees could reduce density and spread of *O. stricta* and promote restoration (Strum et al., 2015).

Conversely, grazing can sometimes be used to suppress invasive species in communities that are well adapted to grazing (e.g., Blumenthal et al., 2012), especially in areas where herbicides cannot be applied (e.g., near water) or against large stands of invasive weeds such as in rangelands (DiTomaso, 2000). Grazing is unlikely to be an option for high conservation value areas owing to lack of adequate selectivity by grazing animals.

Changes in soil fertility levels

Aerial deposition of anthropogenic nitrogen or phosphorous is widespread in some regions, encouraging invasions by plants that need soils with enhanced nutrition (e.g., Brooks, 2003). Also invasions by nitrogen-fixing plants into areas with low-nutrient soil (e.g., wet lowland Hawaiian forests [Hughes and Denslow, 2005] and South African fynbos [Stock et al., 1995]) can enrich soils to the detriment of native plants, stimulating additional exotic plant invasions. In some cases, lowering of nutrient levels to historical norms may slow invasion by non-nitrogen-fixing invasive species. The addition of carbon to the soil (as woodchips, activated carbon, or sucrose) can lower nitrogen availability and can reduce biomass of invasive weeds such as *Centaurea diffusa* L. and prevent the invasion of others (e.g., baby's breath, *G. paniculata*) (Blumenthal, 2009). Carbon enrichment can also restore a competitive advantage to native plants over invasive species (Perry et al., 2004). Suppression of invasive nitrogen-fixing plants with biological control can also be a tool to restore or protect historical soil conditions adversely affected by such plants (see "Maintenance of soils" in Chapter 14).

Replanting with native plants

This approach is commonly used to reduce density, recruitment, or spread of invasive plants. For example, replanting the early successional, native ectomycorrhizal shrub *Leptolena bojeriana* (Baill.) Cavaco. in Madagascar promoted growth of the endemic tree *Uapaca bojeri* L. and reduced soil damage from the invasive plants *Eucalyptus camaldulensis* Dehnh. and *Pinus patula* Schiede ex Schltdl. & Cham. (Baohanta et al., 2012). In forests, revegetation with native species that create shade or with shade-tolerant species can be particularly effective in slowing invasions by sun-loving invasive species (e.g., Funk and McDaniel, 2010).

Area-wide, permanent control through natural enemy introductions

If control over a large area is needed, but adjustments of ecological processes as mentioned above do not sufficiently reduce pest density, then introducing specialized natural enemies from the invasive species' native range (classical biological control) should be considered. Classical biological control has been used effectively to suppress about 40 species of invasive plants and 200 invasive insects (Van Driesche et al., 2008) in various countries but has little application to invasive vertebrates, marine species, or pathogens. Good examples of biological control programs against invasive plants in wildlands include suppression of melaleuca (*M. quinque-nervia*) in wetlands in southern Florida, USA (Center et al., 2012) and several woody trees in the fynbos of South Africa (Moran and Hoffmann, 2012). Biological control has also been used to suppress invasive insects damaging endemic plants on islands, for example orthezia scale, *Orthezia insignis* Browne (Hemiptera: Ortheziidae), on St. Helena (Fowler, 2004) and cottony cushion scale, *Icerya purchasi* Maskell (Hemiptera: Monophlebidae), in the Galápagos Islands (Calderón-Alvarez et al., 2012; Hoddle et al., 2013). The steps in classical biological control projects (Harley and Forno, 1992; Van Driesche et al., 2008) are generally similar to those listed in Table 3.3.

Table 3.3 Steps in classical biological control.

1 **Target selection**: ecological rationale for engaging in project; obtaining financing
2 **Species confirmations**: verify identity of pest and, later, its potential natural enemies
3 **Surveys for natural enemies**: carried out initially in the invaded area
4 **Identifying pest's native range**: to locate areas to search for natural enemies
5 **Collecting natural enemies in native range**: to obtain candidate natural enemies
6 **Judging potential efficacy and host range**: field estimates to guide initial selection of natural enemies
7 **Establishing quarantine colonies**: to preserve stocks of natural enemies of likely value
8 **Estimating host ranges**: done for natural enemies of interest, usually in quarantine laboratories in receiving countries
9 **Petitioning for release**: done for natural enemies believed safe and potentially effective
10 **Release and establishment:** carried out in various parts of the invaded range
11 **Post-release monitoring**: to determine impacts on pest and non-target species
12 **Assessing program outcomes**: assessment of the completeness and value of results via monitoring of pest and recovery of the invaded community

Benefits of the use of classical biological control in various natural ecosystems are reviewed by Van Driesche et al. (2010) and De Clercq et al. (2011); potential risks are discussed by De Clercq et al. (2011). This approach requires social and political agreement on suppressing the pest throughout its invasive range as natural enemies cannot be restricted to particular areas. Also, start-up costs are large, and success, while likely to be substantial over a series of projects, cannot be guaranteed against any individual target species. Natural enemies of invasive pests can be safely introduced into new regions, provided they are highly specialized (so as to pose little or no threat to native species), and they can be highly effective in reducing pests permanently to non-damaging levels over large areas (Van Driesche et al., 2010). Most natural enemies used in modern projects are genus or species specific and are ecologically safe when proper risk analysis and procedures are followed. Safety depends on finding species unlikely to have population-level effects (depression of growth rates, see Chapters 8 and 9) on non-target species taxonomically or ecologically related to the target pest. Finding safe natural enemies is most easily done when the invasive species is distantly related to native species in the invaded area, as this reduces risk to native species, facilitates agent screening, often makes the project cheaper, and also makes it more likely to succeed. Such taxonomic target separation is more commonly found for invasive plants than insects. The likelihood of successful suppression of a pest is also greatly increased when the invasive species of interest has already been the successful target of biological control elsewhere (e.g., Calderón-Alvarez et al., 2012; Hoddle et al., 2013), as likely effective natural enemies may already be known and be available for introduction.

Over time, natural enemies can colonize large areas, including zones that are hard to reach with other control methods. Compared with many control techniques, classical biological control is especially suitable for large infestations because project costs are independent of the total area treated owing to natural dispersal of control agents. Natural enemy reproduction makes control permanent and self-sustaining. Efficacy may vary among habitats, however, if the agent's habitat requirements are narrower than those of the target pest. Also, for control of polyphagous herbivorous insects, such as cottony cushion scale, control may vary among the pest's plant host species (Hoddle et al., 2013). In such cases, multiple agents may be needed to match different parts of the pest's range or other tools must also be used (Center et al., 2012). For control of exotic insects invasive in natural areas, biological control by itself usually suffices, but for recovery of ecosystems damaged by invasive plants, replanting of native species (among other restoration techniques) may also be required.

Biological control in wildlands should take a cooperative and holistic approach. Effective projects must be partnerships between biological control scientists and conservation biologists, and biological control activities should be done within a comprehensive restoration plan to restore the degraded ecosystem.

Factors affecting control efficacy

Tools used for eradication or suppression of invasive species must fit the task. Both the particular biology of the invader and the ecological or geographic features of the invaded ecosystem may limit the application of tools, making some tools especially favorable in some circumstances or making some tools inapplicable to other problems. While the details of particular cases will trump general relationships, below we present principles affecting efficacy of tools for various contexts.

Invader biology

The characteristics that make a species a successful invader can also make it hard to control. Traits that can affect the utility of particular control options include the dispersal rate of an organism (affecting rate of colonization and re-infestation), its body size (large species are easier to spot and attack, while different strategies are needed for smaller species), and its taxonomic group (for example, control options for vertebrates and fungi differ dramatically). Also, some species promote the invasion of other species, so community-level factors may also affect control.

Dispersal capability

The dispersal capacity of an invasive species strongly influences the chances of successful control and choice of methods. Invasive species with effective long-distance dispersal mechanisms (wind and water currents, flight, bird-dispersed seeds) are harder to control. Spread can also be facilitated by people (intentional introductions, movement in vehicles or on products). Species with low intrinsic mobility are easier to contain if they are not easily spread by human activities. The little fire ant, *Wasmannia auropunctata* (Roger), for example, spreads by budding, in which queens break off from the colony and walk only a few meters before forming a new colony. Hydramethylnon baits deployed in grids eradicated this species from an isolated area on Marchena Island (Galápagos Islands) (Causton et al., 2005). In contrast, queens of the red imported fire ant (*Solenopsis invicta* Buren) disperse during nuptial flights over large distances (Tschinkel, 2006), and more distant, isolated infestations are easily missed. Because toxic baits are inadequate at this larger scale, self-dispersing natural enemies (phorid flies or pathogens) are being investigated because they can better discover isolated infestations (Morrison and Porter, 2005; Tschinkel, 2006).

Dispersal of many invasive species, especially plants, is increased by other species (McConkey et al., 2011). Birds and other vertebrates spread seeds of invasive plants, and for effective control this pathway must be minimized (Buckley et al., 2006; Davies and Sheley, 2007). If seed dispersers are few, non-native, and subject to control, then their suppression may help slow the spread of the invasive plants. If dispersers are native species (i.e., in the eastern United States, native birds spread the seeds of many invasive shrubs, such as autumn olive, *Elaeagnus umbellata* Thunb. [Lafleur et al., 2007]), then little or nothing can be done to slow seed movement.

Organism size

Larger-bodied animals (goats, pigs, etc.) produce fewer offspring that take longer to develop than smaller species such as rodents or insects, and consequently the control measures useful against these groups differ. The individual size of the invader will also influence the scale at which a management program can be conducted. For example, because they are easy to detect, goats can be tracked and shot over extremely large areas (>200,000 km²) (e.g., Carrion et al., 2011). Larger-bodied species are typically easier to control using targeted control methods such as hunting and trapping. Furthermore, the containment or exclusion of larger animals with fences is likely to cost less than fencing of smaller mammals (cats, brushtail possums, rats, mice, stoats, weasels), which require smaller mesh and intensive surveillance to ensure that species do not break through the barriers (Speedy et al., 2007).

Smaller animal species such as insects are harder to find, are typically more numerous, have shorter generation times, and often recover more rapidly after control efforts end. The placement of traps with attractant baits or sentinel plants can be used to locate and control small animal species. The grid size for trap placement will depend on the objective of the program, with eradication programs generally requiring small distances between traps to ensure that the last remaining individuals are detected. For example, confirmation of eradication of the ant *W. auropunctata* from 21 ha on Marchena Island required a grid size of 3 m (Causton et al., 2005) (Figure 3.5). Self-sustaining control methods such as classical biological control are especially able to target small species, even those that are efficient dispersers or are hard to find. The parasitic wasp *Eurytoma erythrinae* Gates & Delvare appears to be effectively suppressing the erythina gall wasp, *Quadrastichus erythrinae* Kim, a species that forms galls that seriously damage endemic wili-wili trees (*Erythrina sandwicensis* O. deg.) in Hawaii (see Van Driesche et al., 2010) (see Chapter 10).

Larger plant species (trees, shrubs, large vines) are easily targeted with herbicides and manual or mechanical removal because they are easy to find, generally take considerable time to mature, and may not regrow from fragments remaining after control. In contrast, invasive plants that are small but numerous (e.g., Japanese stiltgrass) or those that reproduce vigorously from small fragments (e.g. Japanese knotweed, *Fallopia japonica* [Hout.] Ronse Decr.) may be more difficult to control through manual methods but may still be amenable to control with herbicides. Because of the abundance of specialized insect herbivores, biological control agents are ideal for the control of many invasive plants.

Invader taxon

It is fairly obvious that some tools are better for some organisms than others, as suggested by reviews for the management of invasive plants (Gardener et al., 2010; Kettenring and Adams, 2011), vertebrates (Courchamp et al., 2003; Towns et al., 2013), amphibians and reptiles (Pitt et al., 2005), and insects (Green and O'Dowd, 2009). There is not a specific toolset for each group of organisms, but rather the choice will depend on the life history of individual species and the ecosystem that is invaded. For example, a plant species with a long-lived seed bank would require different control strategies than one with short-lived seeds. Reduction of seed banks is crucial for controlling plants with persistent seeds, and this requires a combination of techniques to attack both the plant and the seeds in the soil. Management of Australian *Acacia* in South Africa included the use of fire, tillage, solarization, and litter removal to destroy seeds in the soil; however, even these techniques are unable to reach deeply buried seeds. Natural enemies were therefore also used to prevent seed production through attacks on buds, flowers, and seed pods (Wilson et al., 2011).

Methods for controlling invasive vertebrates (Courchamp et al., 2003) focus on tracking and shooting for ungulates and other large vertebrates. In contrast, smaller vertebrates often require a combination of poisoning, hunting, and trapping, except for rats and mice, which can be controlled by poisoning alone. Sterilization is used infrequently as a management tool for invasive vertebrates, though novel techniques for inducing sterility have renewed interest in this option (Courchamp et al., 2003; Saunders et al., 2010; Campbell et al., 2015). Biological control through the release of generalist vertebrate predators, as was done in the past, often has devastating effects on native fauna and is strongly discouraged (Hoddle, 2002). Use of pathogens for vertebrate control is also uncommon because of ethical

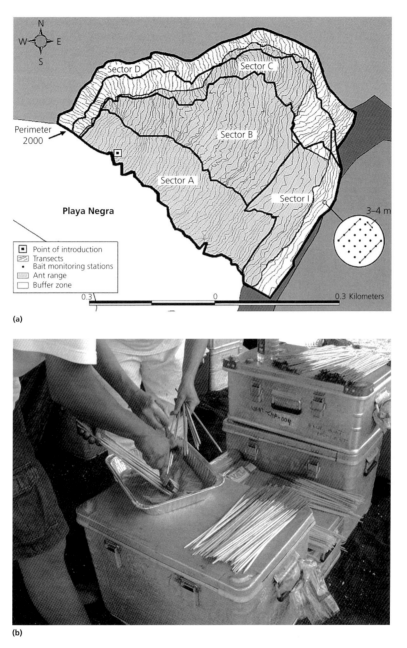

Figure 3.5 (a) Treatment area for eradicating *Wasmannia auropunctata* from a 21 ha area on Marchena Island in the Galápagos Islands archipelago. The inset is a closer view of the 3–4 m grid of bait monitoring stations (Causton et al., 2005); (b) preparation of peanut butter bait sticks for monitoring (up to 44,000 baits were put out in grids). Photo credit: Charlotte Causton.

issues (animals suffer before dying) and safety concerns (risks to non-target species or moving infected animals to other countries), but innovative techniques are being investigated (Saunders et al., 2010; Campbell et al., 2015).

For terrestrial invertebrates (insects, mites), direct kill methods may be used on a small scale in natural ecosystems to protect specific plants temporarily, but on a large scale they are impossible logistically or too polluting (e.g., broadcasting synthetic insecticides). One exception is the control of invasive ants by hand application or aerial broadcast of toxic baits. Supercolonies of yellow crazy ant, *Anoplolepis gracilipes* (Smith), were successfully controlled on Christmas Island, including areas that were inaccessible to conservation workers, by helicopter application of baits with low concentrations of fipronil (Boland et al., 2011) with apparently minimal impacts on vertebrate and invertebrate fauna (Stork et al., 2014). Indirect methods, such as classical biological control using predators and parasitoids, the aerial application of pheromones for mating disruption, mass trapping using lure and kill traps, and sterile male release programs, are potentially effective over large areas (El-Sayed et al., 2006; Van Driesche et al., 2008; Brockerhoff et al., 2010, 2012).

Parasites and pathogens are often small or internal to the host, which makes them poor targets for control by most methods, including biological control. In some instances larger parasites such as the sea lamprey (*P. marinus*) may be controlled using traps with or without lures (McLaughlin et al., 2007). Plant pathogens, on the other hand, may sometimes be controlled by removing their host or controlling their vector (see examples presented earlier in the chapter). Some internal parasites or pathogens of large animals may be suppressed by treating or eliminating the host. The sabellid worm (*Terebrasabella heterouncinata* [Fitzhugh and Rouse]), a parasite of abalone (*Haliotis rufescens* Swainson) and other native gastropods, was eliminated from an infested bay in California by the complete manual removal of its preferred host, the black turban snail (*Tegula funebralis* [Adams]) (Culver and Kuris, 2000).

Management of invasive fish and aquatic invertebrates in fresh or marine waters is a developing field (see Keller et al., 2015 for new examples). Techniques that have been used include chemical, physical, and mechanical removal, and electric barriers. Other techniques under investigation for control of pest fish include de-stratification of lakes, pathogens, pheromones, and biological control (Ling, 2009). Biological control has been proposed for one marine species, the green crab (*Carcinus maenas* L.), and while preliminary work has been done (Kuris et al., 2005), no releases have been made. Indeed, in general, classical biological control may be intrinsically less suitable for marine species than for insects and plants. These latter groups are both affected by insects, a species-rich taxon, which makes finding specialized agents much easier. In contrast, for groups like marine animals or algae, relatively few natural enemies occur on any given host, making it difficult to find highly effective and specialized natural enemies.

Interactions among invaders

In some instances, interactions between introduced species may increase their efficiency as invaders. When multiple invasive species interact positively, collectively they may cause substantial change to the native communities that they invade, even their transformation. In these situations, choice of management tools and timing of control actions need to consider these interactions, and targeting several species simultaneously may be necessary. Simberloff and Von Holle (1999) termed these interactions "invasional meltdown," though in a later review Simberloff (2006) cautioned that few such scenarios have been proven. Given that, complex interactions among a set of invaders may occur but may not be easily detected (e.g., Montgomery et al., 2012).

Invasive plants often rely on vectors for dispersal and, in some cases, these can be other invasive species (McConkey et al., 2011). For example, wild black rats (*Rattus rattus* L.) feed on seeds of the most damaging invasive plants in Hawaii (including *Miconia calvescens* DC and *Clidemia hirta* [L.] D. Don), suggesting that rats are major contributors to their dispersal (Shiels, 2011). Many invasive pathogens are spread by invasive insects, as for example exotic elm pathogens (*Ophiostoma* spp.), which are transmitted to American elm (*Ulmus americana* L.) by bark beetles, such as *S. multistriatus* (Gadgil et al., 2000).

A well-known example of invader synergism is the relationship between invasive ants and honeydew-producing hemipterans, in which ants feed on the sugary waste (honeydew) of the hemipterans and defend them against their natural enemies. Sustained high densities of foraging ants result in high population densities of hemipterans and the consequent growth of sooty molds, leading to stunted growth and even death of trees. Such interactions also affect native ants and native butterflies that are either denied access to their host plants by the hemipteran-tending exotic ant or lose their native ant associates owing to competition from the invasive ant (Waterhouse and Sands, 2001; Van Driesche et al., 2010). On Christmas Island, the mutualistic interactions between invasive scale insects and the yellow crazy ant (*A. gracilipes*) have increased ant numbers, which in turn has facilitated the invasion of yet another invasive species, the giant African land snail. Green et al. (2011) demonstrated that ant supercolonies selectively kill land crabs but not land snails, thus eliminating one of the principal predators (land crabs) formerly keeping the snail population in check.

Ecological or geographic features of the invaded ecosystem

Small preserves or small islands

Mechanical and manual methods are particularly applicable for some species in smaller areas, including time-consuming methods such as hand pulling or stump application of herbicides. Such approaches can even be used quite close to ecologically sensitive areas (e.g., water ways) or threatened species. Control of invasive plants can be as simple as hand-weeding, which in very small, rare habitats (e.g., remaining native forest in Mauritius, patches of <20 ha) can help native plants escape extinction (e.g., Baider and Vincent Florens, 2011). However, even in small preserves, labor may become a serious limitation. For example, removal of red mangrove (*Rhizophora mangle* L.) and pickleweed (*Batis maritima* L.) from an 8 ha Hawaiian mudflat to protect habitat of endangered black-necked stilts [*Himantopus mexicanus* Müller, ssp. *knudseni*]) required thousands of hours of volunteer labor, over US$2.5 million of contracted labor, and 20 years of systematic effort (Rauzon and Drigot, 2003), although costs may be lower for projects against terrestrial plants.

Whether classical biological control can work on small islands (i.e., atolls) is debated, as they typically lack environmental heterogeneity and have only a finite population of the target species. Consequently there is a risk that the agent may drive the pest to extinction and then die out, leaving the ecosystem vulnerable to re-invasion. However, this does not seem to be the case for *Gonatocerus ashmeadi* (Hymenoptera: Mymaridae), a parasitoid of the glassy-wing sharpshooter, *Homalodisca vitripennis* (Germar), both of which have spread quickly among small Pacific islands (Petit et al., 2009), with biological control of this invasive insect benefiting many native plants.

Islands with low re-invasion risk

Islands differ from continental settings in ways that make them more vulnerable to invasions, but their isolation may make them better candidates for invasive species management, in particular eradication (Reaser et al., 2007; Walsh et al., 2012; Glen et al., 2013). Re-invasion risk is a critical factor that must be assessed, as high re-invasion rates would make eradication efforts, even if initially successful, ultimately pointless. Re-invasion may be due to natural events (e.g., rats swimming, wind dispersal of insects and seeds) or human actions, either deliberate or accidental. Strict enforcement of quarantine measures and early warning monitoring systems can be used to limit human-assisted re-invasions of islands that are otherwise suitable for pest eradication programs. Lakes on continents, by extension, are like oceanic islands with respect to aquatic invaders, which may be isolated in particular lakes. Spread, however, can occur through rivers or, artificially, through boat traffic. Nevertheless, if eradication efforts are mounted soon after initial detection and the infested area is a manageable size, elimination of some categories of aquatic invaders (fish, plants) is possible (e.g., Britton and Brazier, 2006).

Islands or preserves with restricted-range endemics

Endangered endemic species on islands or in small preserves are particularly susceptible to harm from both invasive species and management techniques used to control them (Reaser et al., 2007; Walsh et al., 2012). Although invasive species removal may be beneficial in the long run, there may be short-term risks from control programs, especially if they are not carefully planned. Such risks may be critical for endemic species with highly restricted ranges and no other populations or for unique, non-resilient communities.

When non-selective techniques are the only options available and control is essential, measures should be put in place to protect threatened species from the baits or other controls. On Buck Island Reef in the Caribbean, rat baits were placed inside plastic containers mounted on platforms or stakes to reduce bait consumption by native birds and crabs (Witmer et al., 2007). When no techniques exist to screen sensitive species from control measures, temporarily removing at-risk species may be an option, as was done with raptors and the Anacapa deer mouse (*Peromyscus maniculatus anacapae* von Bloeker) on Anacapa island before control of black rats (Howald et al., 2010). Other examples include removal of aquatic invertebrates from Lake Junco in the Galápagos Islands before *Tilapia* eradication (Nico and

Walsh, 2011) and capture of some individuals of the endemic lizard *Sphenodon punctatus* Evans before use of baits against rabbits and Polynesian rats (*Rattus exulans* [Peale]) on islands off the coast of New Zealand (Towns and Broome, 2003).

When biological control is used in areas of high conservation value, only natural enemies with high specificity should be used to avoid harm to native species. Risk assessments should include food web analyses to identify potential indirect effects from release of a biological control agent (see Chapter 7). On the Galápagos Islands, a risk assessment of *Rodolia cardinalis* (Mulsant) (Coleoptera: Coccinellidae) for control of cottony cushion scale concluded the ladybird would not attack native invertebrates, nor would its defensive alkaloids (Dixon, 2000) harm local finches (Lincango et al., 2011; see Chapter 10).

Large preserves or islands, with high re-invasion risk

When eradication is not an option because of a high probability of recolonization, or the invaded areas are too large, the most cost-effective methods are those that are self-sustaining, such as classical biological control, or that prevent further spread of a species, such as land management, or that limit management to key areas. It may also be useful to divide the infested area into manageable subunits using fences or natural barriers, which can then be prioritized for treatment to maximize success (Adams et al., 2014).

Predator-proof fences have been used in some cases to isolate areas within which invasive species can be eliminated and native species restored. The Maungatautari Reserve in New Zealand is a 3400 ha mountainous area that was enclosed by a 47 km-long fence, inside of which 15 species of invasive vertebrates were targeted for eradication using aerially distributed baits, trapping, and bait stations (Speedy et al., 2007). Forty native species – mostly birds, but also reptiles, invertebrates, fish, and one amphibian – were re-introduced to the reserve (Burns et al., 2012).

Natural features (glaciers, rivers, and mountains) are also sometimes effective barriers that slow the spread of an invasive species. Such natural barriers can also assist in dividing an area into smaller units for control purposes. For example, on South Georgia Island (4000 km²), genetic studies of the brown rat, *Rattus norvegicus* (Berkenhout), showed that rats were isolated by glaciers,

which delimited areas that formed natural eradication units of a manageable size (Robertson and Gemmell, 2004).

Ecologically degraded areas

In contrast to ecologically intact areas of high value, where control tools should have minimal impact, in highly degraded areas more aggressive control methods (tillage of the soil, fire, or herbicides) may be acceptable.

Disturbed habitats often have many established invaders and disturbance itself may favor invasions (see Catford et al., 2012; Walsh et al., 2012), especially of disturbed-habitat specialists, such as the red imported fire ant, *S. invicta* (King and Tschinkel, 2009). Many invasive species are opportunists and identification of the disturbance factors that trigger invasion can be hard, but is necessary to identify what management actions are most appropriate (Hulme, 2006). Project planning should not focus just on the target species but also consider all major invasive species present, whether there are interactions between them, and how they might benefit from the eradication of the target species. In some cases, single-species management can exacerbate damage from other introduced species (Zavaleta et al., 2001).

Effects of treatment scale on control operations

Cost per unit area

Economy of scale applies to some control measures and not others. For a given type of control, the benefit : cost ratio may depend on the biology of the invader, its distribution, ability to spread, and other factors. Several different control strategies may be possible. One might, for example, choose to control only the highest density patches of the invader or to suppress invaders only in areas of special concern. Or programs might focus on the boundary of the infestation to slow spread. These choices may yield different benefit : cost ratios (see Epanchin-Niell and Hastings, 2010). Zull et al. (2006) and Cacho et al. (2008) discuss use of models to optimize invasive species control to protect economic values. Such models do not deal well with the non-economic values at risk from invasive species, such as biodiversity, because such effects lack well-defined economic costs (see also Chapter 14).

For chemical control, the amount of pesticide applied is directly proportional to the area treated. The total cost, however, may not scale directly because different types of application machinery may be used at different scales. Pesticide programs also entail costs in applicator education, regulation compliance, and bookkeeping.

In contrast to the above approach with positive scale-dependent cost structure, biological control has a negative scale-dependent cost structure, with the cost per unit area falling as the area treated increases because biological control agents reproduce and spread themselves. Costs, however, are front-end loaded (i.e., foreign exploration, quarantine studies, host-range testing, mass rearing, release, and establishment costs all happening at the beginning of the project) and must be borne regardless of the size of the infested area. As treated areas enlarge, new costs (mass rearing, transport, release, and monitoring of releases) occur, but these are relatively small compared to initial costs. Therefore, biological control is well suited for use against invasive species infesting large areas (thousands to millions ha).

Conflicts with abutters and other groups

For control activities that are easily visible to observers, the number and diversity of third party conflicts increase as treated areas become larger because there are usually more abutters and more opportunities for the general public to observe events. Also, when larger areas are treated, the number of surrounding areas with special ecological value increases, creating the potential for more conflicts should the control operations harm species on adjacent land (see Chapter 12).

Scale-related issues for biological control

Living agents released against invasive insects or plants may spread to areas far from the release point. In general this is a desirable feature of all biological control projects. However, in some cases agents may cross barriers (oceans, etc.) and invade wholly new regions quite distant from their area of release, acting in such instances like any other invasive insect (e.g., Petit et al., 2009; Pratt and Center, 2012). While such spread is often not damaging, one could imagine cases in which the agent might spread to areas with native species similar to the target pest that were not considered in the initial safety evaluation. To protect against

such an event in North America, biological control projects take a continental view of the biota during safety estimation, and release must be approved by all NAPPO (North American Plant Protection Organization) members (Canada, the USA, and Mexico). The need for such coordination is shown by some past releases in which a release that was safe for one region of North America proved less so in another. The weevil *Mogulones cruciger* (Herbst) was released in Canada against houndstongue (*Cynoglossum officinale* L.), but was not approved in the United States (Andreas et al., 2008). Spread of this weevil eventually raised concerns for the safety of plants native only to the southern United States. Such continental coordination would, of course, be considerably more difficult in regions with more countries per continent or areas where movement from a continent to nearby island nations was likely (e.g., USA to the Caribbean). Even within single countries, conflicts may arise as biological control agents enter new areas, as occurred when the saltcedar beetle (*Diorhabda carinulata* Desbrochers) was moved legally within-state to part of the Virgin River drainage near St. George in southern Utah, an area supporting the endangered southwestern willow flycatcher (*Empidonax traillii extimus* Phillips), which nests in saltcedar in areas no longer supporting sufficient native cottonwoods and willows (Dudley and Bean, 2012).

Projects that span international borders pose a higher level of difficulty because of the need to coordinate multiple sets of laws and agencies, and potentially different perspectives and interests. Also, non-target plants may be valued differently in different countries, which may create conflicts of interest between countries (e.g., Milbrath and DeLoach, 2006). Coordination at this scale requires more time, resources, and political acumen.

When is biological control the right choice?

The preceding portion of this chapter has reviewed available methods for control of invasive species. However, one goal of this book is bring about greater consideration of biological control as an appropriate and effective tool for some invasive pests of wildlands. Here we summarize our ideas of when this might be the case and, conversely, what organisms or situations would not be favorable for use of biological control.

When biological control is the right choice

In general, biological control is the right choice when a non-native invasive insect or plant is a primary driver of ecosystem degradation or strongly threatens one or more native species, when eradication is not possible, and when permanent pest reduction is desired over the whole landscape, not just in local preserves.

Project cost is always a consideration when a new biological control project is being developed, but cost may be reduced if the species of concern has been targeted successfully elsewhere, as previous work will suggest potentially effective agents and reduce the amount of necessary research. However, regardless of past work, additional host-range testing may be needed to evaluate the risk to unique species that occur in the new area. For example, when cottony cushion scale (*I. purchasi*) threatened native plants on the Galápagos Islands, it was clear from the beginning that the lady beetle *R. cardinalis* would likely be effective based on the many past success uses of the beetle against that pest elsewhere (Caltagirone and Doutt, 1989), as was later proven to be correct (Hoddle et al., 2013); however, the safety of this natural enemy to native fauna of the Galápagos Islands still had to be determined by local work before the beetle's introduction.

When biological control may not be a good fit

Several features argue against selecting biological control as the strategy for the control of some pests, including (1) the pest taxon, (2) limited pest importance, (3) presence of many native congeners, (4) pests enhanced by altered ecosystem processes, or (5) pests in groups with few effective natural enemies.

Invaders other than insects and plants

The historical record of classical biological control is overwhelmingly focused on control of invasive insects and plants (Van Driesche et al., 2008). A key reason for this is that in both cases, the natural enemies employed are insects (or sometimes pathogens), and insects are a group with a uniquely high number of species, literally millions and counting. Given that, it is common to find dozens, sometimes hundreds of species of insects attacking a target species in its native range. This creates a huge pool of potential natural enemies, from which the odds of finding one that is specific enough for introduction, and plausibly effective, is relatively good. This is not the case with vertebrate, mollusk, or marine pests, for which only a few specialized agents are likely to exist.

A second reason why arthropods and plants are suitable targets while marine species, for example, seem not to be, is that our understanding of population dynamics is rooted in land vertebrates, insects, and plants. For other groups, such as marine crustaceans or algae, population dynamical forces are much less well understood because many fewer studies on those groups have been attempted. Therefore, for example, we are not even certain that natural enemies are key factors setting average density in these groups, potentially invalidating the central premise of biological control.

Pests of limited importance

Species of only local importance often may be suppressed using other tools. Invasive plants found only in a few specific habitats – if not highly destructive or rapidly spreading – might best be treated with herbicides or manual removal. However, some species with small distributions may still merit biological control, such as lobate lac scale (*Paratachardina pseudolobata* Kondo & Gullan) (Hemiptera: Kerriidae) in tropical hardwood hammocks of the Everglades, a World Heritage site (Schroer et al., 2008), if the damage they cause has great ecological importance and the pest or its environment is poorly suited for pesticide use.

Pests with many native congeners

Invasive species with many native congeners in the invaded area are more difficult targets because species-level specificity (as opposed to genus- or family-level specificity) is required, which would likely mean screening more candidates, with lower chances of finding agents with adequate specificity. Invasive plants only distantly related to native species are less costly targets and agents are more likely to succeed. Melaleuca, for example, has few or no close relatives in the native US flora (Serbesoff-King, 2003), making discovery of safe agents easier. In contrast, invasive thistles in North America (*Carduus, Cirsium,* and *Silybum*) are closely related to many native thistles (Schroder, 1980), some of which are endangered. In such cases, the risk increases of releasing insufficiently specialized agents (e.g., see section on *Rhinocyllus coni-cus* Förster for thistle control in Chapter 4). For some invasive phytophagous insects, the presence of many native congeners may be less of a barrier, provided they feed on different host plants than the pest. For example, emerald ash borer (*A. planipennis*) has over 100 native congeners in North America, but few of these attack ash (the pest's host). Since parasitoids, the type of agent used, frequently use volatiles from a pest's host plant to find the host, host plant fidelity may be a stronger filter affecting host range than taxonomic relatedness at the genus level (e.g., Yang et al., 2008).

Invaders promoted by altered processes

Invasive species whose densities have been increased by human effects on ecosystems, for example species that are "passengers" of ecological degradation such as altered fire, flood, or grazing regimes, or nutrient loading (MacDougall and Turkington, 2005), may be inappropriate targets for biological control. If the fundamental reason for the high density of the invader has to do with such changes, then those conditions should be returned to normal levels first, if possible, before considering natural enemy introductions. For example, in the 1950s, several species of native and invasive *Opuntia* cacti (Pemberton and Liu, 2007) became pests on Nevis and St. Kitts, largely because of overgrazing by goats (Simmonds and Bennett, 1966). It was therefore a mistake to introduce the moth *C. cactorum*, even though it is a species known from previous work in Australia to be highly effective at controlling such cacti (Dodd, 1940). A proposal to release *C. cactorum* on Santa Cruz Island off the coast of California in response to a similar situation was rejected by the biological control scientists contacted out of concern for native cacti on the mainland (Goeden et al., 1967).

Groups with few natural enemies

Pests in groups with few past biological control successes are likely to be more difficult targets than pests in groups from which many species have been suppressed through natural enemy introductions. Thus, for insects, many scales, mealybugs, psyllids, and whiteflies have been suppressed using biological control (Clausen, 1978), while species in concealed habitats (e.g., wood borers or insects that live in soil) are generally more difficult to control with this method (Gross, 1991).

For plants, invasive grasses have rarely been targeted with natural enemy releases, in part because most grasses were felt to have economic value (many having been introduced as forage) and because grass-feeding insects were thought to be insufficiently specialized. However, this perception is changing, and some grasses are currently targets of biological control, such as *Arundo donax* L. (Goolsby and Moran, 2009) and *Phragmites australis* L. (Tewksbury et al., 2013). Trees are another group of species that has only in the last few decades been effectively targeted for biological control. While invasive insects often kill mature trees, there are few examples of species introduced for biological control doing so. Also, social acceptance of programs that kill widely planted trees that also invade natural areas may be low as societies may want to continue enjoying the benefits of the trees. However, the ecological damage from such invasive trees in natural areas may be reduced by lowering their seed production and seedling (or stump sprout) survival, and this approach has led to success in several cases, for example, *M. quinquenervia* in Florida (Center et al., 2012) and species of *Acacia*, *Sesbania*, and *Hakea* in the South African fynbos (Moran and Hoffmann, 2012).

Key messages

Tools for control of invasive pests in natural ecosystems must fit the project's goals, the pest biology, the site conditions, and the ecological value of the site. Some guiding concepts include the following points:

- Control efforts should not be adopted *ad hoc* but rather should be part of a restoration plan for the site that has been developed jointly by conservation biologists, biological control scientists, and land stewards, taking into account the complexities of the ecosystem to be managed.
- Before attempting direct reduction of the invasive species, project planners should consider the possibility that the pest has become abundant because of altered processes (fire, flood, grazing, nutrient levels, species removal), and, if this is the case, efforts to change those conditions should be investigated and changes made if feasible before other control methods are attempted.
- Since knowledge of how the system will respond to control measures will always be incomplete, projects

should be monitored before, during, and after management to assess whether management objectives are being met, and methods adapted as needed as the program progresses.

- When control is needed only at the local level, mechanical and chemical control should be considered first, recognizing that re-treatment will likely be required unless the pest is eradicated (as on small islands) and the area is not subject to re-invasion.

- For permanent landscape-level suppression of pest species whose density is not driven by altered ecological processes, biological control is likely to be the only feasible approach.

- In some cases biological control may be necessary but not sufficient, and integration with mechanical or chemical control or other techniques, such as native species replanting, may be required to control the target pest and restore the ecosystem.

Acknowledgments

We thank Dan Simberloff, Bernd Blossey, Mark Gardener, Rachel Atkinson, Mark Hoddle, David Wagner, and Karl Campbell for reviews of Chapter 3.

References

Adams, A. L., Y. van Heezik, K. J. M. Dickinson, and B. C. Robertson. 2014. Identifying eradication units in an invasive mammalian pest species. *Biological Invasions* **16**: 1481–1496.

Amaral J., S. Almeida, M. Sequeira, and V. Neves. 2010. Black rat *Rattus rattus* eradication by trapping allows recovery of breeding roseate tern *Sterna dougallii* and common tern *S.hirundo* populations on Feno Islet, the Azores, Portugal. *Conservation Evidence* **7**: 16–20.

Anderson, L. W. J. 2005. California's reaction to *Caulerpa taxifolia*: a model for invasive species rapid response. *Biological Invasions* **7**: 1003–1016.

Andreas, J. E., M. Schwarzländer, and R. de Clerck-Floate. 2008. The occurrence and potential relevance of post-release, nontarget attack by *Mogulones cruciger*, a biological control agent for *Cynoglossum officinale* in Canada. *Biological Control* **46**: 304–311.

Atkinson, R., M. R. Gardener, G. Harper, and V. Carrion V. 2012. Fifty years of eradication as a conservation tool in Galápagos: what are the limits? pp. 183–198. *In*: Wolff, M. and M. R. Gardener (eds.). *The Role of Science for Conservation*. Routledge. Oxon, UK.

Baider, C. and F. B. Vincent Florens. 2011. Control of invasive alien weeds averts imminent plant extinction. *Biological Invasions* **13**: 2641–2646.

Baohanta, R., J. Thioulouse, H. Ramanankierana, et al. 2012. Restoring native forest ecosystems after exotic tree plantation in Madagascar: combination of the local ectotrophic species *Leptolena bojeriana* and *Uapaca bojeri* mitigates the negative influence of the exotic species *Eucalyptus camaldulensis* and *Pinus patula*. *Biological Invasions* **14**: 2407–2421.

Barreto, R. W., C. A. Ellison, M. K. Seier, and H. C. Evans. 2012. Biological control of weeds with plant pathogens: four decades on, pp. 299–350. *In*: Abrol, D. P. and U. Shankar (eds.). *Integrated Pest Management*. CABI Publishing. Wallingford, UK.

Beest, M. J. P., G. M. Cromsigt, J. Ngobese, and H. Olff. 2012. Managing invasions at the cost of native habitat? An experimental test of the impact of fire on the invasion of *Chromolaena odorata* in a South African savanna. *Biological Invasions* **14**: 607–618.

Biondi, A., V. Mommaerts, G. Smagghe, et al. 2012. The non-target impact of spinosyns on beneficial arthropods. *Pest Management Science* **68**: 1523–1536.

Blumenthal, D. M. 2009. Carbon addition interacts with water availability to reduce invasive forb establishment in a semi-arid grassland. *Biological Invasions* **11**: 1281–1290.

Blumenthal, D. M., A. P. Norton, S. E. Cox, et al. 2012. *Linaria dalmatica* invades south-facing slopes and less grazed areas in grazing-tolerant mixed-grass prairie. *Biological Invasions* **14**: 395–404.

Boland, C. R. J., M. J. Smith, D. Maple, et al. 2011. Heli-baiting using low concentration fipronil to control invasive yellow crazy ant supercolonies on Christmas Island, Indian Ocean, pp. 152–156. *In*: Veitch, C. R., M. N. Clout, and D. R. Towns (eds.). *Island Invasives: Eradication and Management*. International Union for Conservation of Nature (IUCN). Gland, Switzerland and Centre for Biodiversity and Biosecurity (CBB), Auckland, New Zealand.

Breeÿen, A. D. and R. Charudattan. 2009. Biological control of invasive weeds in forests and natural areas by using microbial agents, pp. 189–209. *In*: Inderjit (ed.). *Management of Invasive Weeds*. Springer. Dordrecht, The Netherlands.

Britton, J. R., and M. Brazier. 2006. Eradicating the invasive topmouth gudgeon, *Pseudorasbora parva*, from a recreational fishery in northern England. *Fisheries Management and Ecology* **13**: 329–335.

Brockerhoff, E. G., A. Liebhold, B. Richardson, and D. M. Suckling. 2010. Eradication of invasive forest insects: concepts, methods, costs and benefits. *New Zealand Journal of Forest Science* **40**: S117–S135.

Brockerhoff, E. B., D. M. Suckling, M. Kimberley, et al. 2012. Aerial application of pheromones for mating disruption of an invasive moth as a potential eradication tool. *PLoS One* **7**(8): e43767. doi:10.1371/journal.pone.0043767.

Brooks, M. L. 2003. Effects of increased soil nitrogen on the dominance of alien annual plants in the Mojave Desert. *Journal of Applied Ecology* **40**: 344–353.

Brooks, M. L., C. M. D'Antonio, D. M. Richardson, et al. 2004. Effects of invasive alien plants on fire regimes. *BioScience* **54**: 677–688.

Buckley, Y. M., S. Anderson, C. P. Catterall, et al. 2006. Management of plant invasions mediated by frugivore interactions. *Journal of Applied Ecology* **43**: 848–857.

Burns, B., J. Innes, and T. Day. 2012. The use and potential of pest-proof fencing for ecosystem restoration and fauna conservation in New Zealand, pp. 65–90. *In*: Somers, M. J. and M. Hayward (eds.). *Fencing for Conservation: Restriction of Evolutionary Potential or a Riposte to Threatening Processes?* Springer. New York.

Cacho, O. J., R. M. Wise, S. M. Hester, and J. A. Sinden. 2008. Bioeconomic modeling for control of weeds in natural environments. *Ecological Economics* **65**: 559–568.

Calderón-Alvarez, C., C. E. Causton, M. S. Hoddle, et al. 2012. Monitoring the effects of *Rodolia cardinalis* on *Icerya purchasi* populations on the Galápagos Islands. *Biological Control* **57**: 167–179.

Caltagirone, L. E. and R. L. Doutt. 1989. The history of the vedalia beetle importation to California and its impact on the development of biological control. *Annual Review of Entomology* **34**: 1–16.

Campbell, K. and C. J. Donlan. 2005. Feral goat eradications on islands. *Conservation Biology* **19**: 1362–1374.

Campbell, K. J., G. Harper, D. Algar, et al. 2011. Updated review of feral cat eradications, pp. 37–46. *In*: Veitch, C. R., M. N. Clout, and D. R. Towns (eds.). *Island Invasives: Eradication and Management*. IUCN (International Union for Conservation of Nature). Gland, Switzerland, and Centre for Biodiversity and Biosecurity (CBB), Auckland, New Zealand.

Campbell, K. J., J. Beek, C. T. Eason, et al. 2015. The next generation of rodent eradications: innovative technologies and tools to improve species specificity and increase their feasibility on islands. *Biological Conservation* **185**: 47–58.

Carrion, V., C. J. Donlan, K. J. Campbell, et al. 2011. Archipelago-wide island restoration in the Galápagos Islands: reducing costs of invasive mammal eradication programs and reinvasion risk. *PLoS One* **6**(5): e18835 doi:10.1371/journal.pone.0018835.

Catford, J. A., C. C. Daehler, H. T. Murphy, et al. 2012. The intermediate disturbance hypothesis and plant invasions: implications for species richness and management. *Perspectives in Plant Ecology, Evolution and Systematics* **14**: 231–241.

Causton, C. E., C. R. Sevilla, and S. D. Porter. 2005. Eradication of the little fire ant, *Wasmannia auropunctata* (Hymenoptera: Formicidae), from Marchena Island, Galápagos: on the edge of success? *Florida Entomologist* **88**: 159–168.

Causton, C. E., F. Cunninghame, and W. Tapia. 2013. Management of the avian parasite *Philornis downsi* in the Galápagos Islands: a collaborative and strategic action plan, pp. 167–173. *In*: Cayot L. J. (ed.). *Galápagos Report 2011–2012*. Charles Darwin Foundation, Galápagos National Park, and INGALA, Puerto Ayora, Galápagos, Ecuador.

Center, T. D., M. F. Purcell, P. D. Pratt, et al. 2012. Biological control of *Melaleuca quinquenervia*: an Everglades invader. *Biological Control* **57**: 151–165.

Charles Darwin Foundation. 2014. Galapagos Species Checklist. *Lissachatina fulica*. Available from: http://www.darwinfoundation.org/datazone/checklists/18132/ [Accessed January 2016].

Choe, D-H., D. B. Villafuerte, and N. D. Tsutsui. 2012. Trail pheromone of the Argentine ant, *Linepithema humile* (Mayr) (Hymenoptera: Formicidae). *PLoS One* **7**: e45016. doi:10.1371/journal.pone.0045016.

Clausen, C. P. (1978). *Introduced Parasites and Predators of Arthropod Pests and Weeds: A World Review*. Agricultural Handbook No. 480, U.S. Department of Agriculture. Washington, DC. 545 pp.

Clout, M. N. and P. A. Williams (eds.). 2009. *Invasive Species Management: A Handbook of Principles and Techniques*. Oxford University Press. Oxford, UK. 308 pp.

Cole, R. J., C. M. Litton, M. J. Koontz, and R. K. Loh. 2012. Vegetation recovery 16 years after feral pig removal from a wet Hawaiian forest. *Biotropica* **44**: 463-471.

Courchamp, F., J. L. Chapuis, and M. Pascal. 2003. Mammal invaders on islands: impact, control and control impact. *Biological Reviews* **78**: 347–383.

Cruz, F., V. Carrion G., K. J. Campbell, et al. 2009. Bio-economics of large-scale eradication of feral goats from Santiago Island, Galápagos. *Journal of Wildlife Management* **73**: 191–200.

Culver, C. S. and A. M. Kuris. 2000. The apparent eradication of a locally established introduced marine pest. *Biological Invasions* **2**: 245–253.

D'Antonio, C. M. and P. M. Vitousek. 1992. Biological invasions by exotic grasses, the grass/fire cycle, and global change. *Annual Review of Ecology and Systematics* **23**: 63–87.

Davies, K. W. and R. L. Sheley. 2007. A conceptual framework for preventing the spatial dispersal of invasive plants. *Weed Science* **55**: 178–184.

De Clercq, P., P. Mason, and D. Babendreier. 2011. Benefits and risks of exotic biological control agents. *Biological Control* **56**: 681–698.

Didham, R. K., J. M. Tylianakis, M. A. Hutchison, et al. 2005. Are invasive species the drivers of ecological change? *Trends in Ecology & Evolution* **20**: 470–474.

DiTomaso, J. M. 2000. Invasive weeds in rangelands: species, impacts, and management. *Weed Science* **48**: 255–265.

DiTomaso, J. M., M. L. Brooks, E. B. Allen, et al. 2006. Control of invasive weeds with prescribed burning. *Weed Technology* **20**: 535–548.

Dixon, A. F. G. 2000. *Insect Predator–Prey Dynamics: Ladybird Beetles and Biological Control*. Cambridge University Press. Cambridge, UK.

Dodd, A. P. 1940. *The Biological Campaign against Prickly Pear. Commonwealth Prickly Pear Board Bulletin*. Brisbane, Australia. 177 pp.

Dodd, J. and R. P. Randall. 2002. Eradication of kochia [*Bassia scoparia* (L.) A. J. Scott, Chenopodiaceae] in Western Australia, pp. 300–303. *In:* Jacob, H. S., J. Dodd, and J. H. Moore (eds.). *Proceedings of the 13th Australian Weeds Conference, Perth, Western Australia, 8–13 September 2002*. Plant Protection Society of Western Australia. Perth, Australia.

Dodson, E. K. and C. E. Fiedler. 2006. Impacts of restoration treatments on alien plant invasion in *Pinus ponderosa* forests, Montana, USA. *Journal of Applied Ecology* **43**: 887–897.

Dreistadt, S. H. and D. L. Dalhsten. 1989. Gypsy moth eradication in Pacific Coast states: history and eradication. *Bulletin of the Entomological Society of America* **35**: 13–19.

Dudley, T. L. and D. W. Bean. 2012. Tamarisk biological control, endangered species risk, and resolution of conflict through riparian restoration. *Biological Control* **57**: 331–347.

Dunstan, W. A., T. Rudman, B. L. Shearer, et al. 2010. Containment and spot eradication of a highly destructive, invasive plant pathogen (*Phytophthora cinnamomi*) in natural ecosystems. *Biological invasions* **12**: 913–925.

Emery, S. M., P. J. Doran, J. T. Legge, et al. 2013. Aboveground and belowground impacts following removal of the invasive species baby's breath (*Gypsophila paniculata*) on Lake Michigan sand dunes. *Restoration Ecology* **21**: 506–514.

El-Sayed, A. M., D. M. Suckling, C. H. Wearing, and J. A. Byers. 2006. Potential of mass trapping for long-term pest management and eradication of invasive species *Journal of Economic Entomology* **99**: 1550–1564.

Epanchin-Niell, R. S. and A. Hastings. 2010. Controlling established invaders: integrating economics and spread dynamics to determine optimal management. *Ecology Letters* **13**: 528–541.

Florance, D., J. K., Webb, T. Dempster, et al. 2011. Excluding access to invasion hubs can contain the spread of an invasive vertebrate. *Proceedings of the Royal Society of London B* **278**: 2900–2908.

Fowler, S. V. 2004. Biological control of an exotic scale, *Orthezia insignis* Browne (Homoptera: Ortheziidae), saves the endemic gumwood tree, *Commidendrum robustum* (Roxb.) DC. (Asteraceae) on the island of St. Helena. *Biological Control* **29**: 367–374.

Funk, J. L. and S. McDaniel. 2010. Altering light availability to restore invaded forest: the predictive role of plant traits. *Restoration Ecology* **18**: 865–872.

Gammon, E. T. 1962. The Japanese beetle in Sacramento. *Bulletin of the Department of Agriculture of California* **50** (4): 221–235.

Gardener, M. R., R. Atkinson, and J. L. Rentería. 2010. Eradications and people: lessons from the plant eradication program in Galápagos. *Restoration Ecology* **18**: 20–29.

Gadgil, P. D., L. S. Bulman, M. A. Dick, and J. Bain. 2000. Dutch elm disease in New Zealand, pp. 189–199. *In*: Dunn, C. P. (ed.).

The Elms: Breeding, Conservation and Disease Management. Kluwer Academic Publishers. Boston, Massachusetts, USA.

Gallagher, R.V. and M. R. Leishman. 2014. Invasive plants and invaded ecosystems in Australia: implications for biodiversity, pp. 105–133. *In*: Stow, A., N. Maclean, and G. Holwell (eds.). *Austral Ark: The State of Wildlife in Australia and New Zealand*. Cambridge University Press. Cambridge, UK.

Gentz, M. C., G. Murdoch, and G. F. King. 2010. Tandem use of selective insecticides and natural enemies for effective, reduced-risk pest management. *Biological Control* **52**: 208–215.

Gherardi, F. and C. Angiolini. 2004. Eradication and control of invasive species. *In:* Gherardi, F., M. Gualtieri, and C. Corti (eds.). *Biodiversity Conservation and Habitat Management*. Encyclopedia of Life Support Systems (EOLSS), UNESCO, Eolss Publishers, Oxford, U.K. [http://www.eolss.net].

Glen, A. S., R. Atkinson, K. J. Campbell, et al. 2013. Eradicating multiple invasive species on inhabited islands: the next big step in island restoration? *Biological Invasions* **15**: 2589–2603.

Godfray, H. C., J. Tjeerd Blacquiere, L. M. Field, et al. 2014. A restatement of the natural science evidence base concerning neonicotinoid insecticides and insect pollinators. *Proceedings of the Royal Society B: Biological Sciences* **281**: 20140558. doi: 10.1098/rspb.2014.0558.

Goeden, R. D., C. A. Fleschner, and C. W. Ricker. 1967. Biological control of prickly pear cacti on Santa Cruz Island, California. *Hilgardia* **38**: 579–606.

Goolsby J. A. and P. J. Moran. 2009. Host range of *Tetramesa romana* Walker (Hymenoptera: Eurytomidae), a potential biological control of giant reed, *Arundo donax* L. in North America. *Biological Control* **49**: 160–168.

Green, P. T. and D. J. O'Dowd. 2009. Management of invasive invertebrates: lessons from the management of an invasive alien ant, pp. 153–172. *In:* Clout, M. N. and P. A. Williams (eds.). *Invasive Species Management: A Handbook of Principles and Techniques*. Oxford University Press. Oxford, UK.

Green, P. T., D. J. O'Dowd, K. L. Abbott, et al. 2011. Invasional meltdown: invader-invader mutualism facilitates a secondary invasion. *Ecology* **92**: 1758–1768.

Gross, P. 1991. Influence of target pest feeding niche on success rates in classical biological control. *Environmental Entomology* **20**: 1217–1227.

Haack, R. A., F. Herard, J. Sun, and J. J. Turgeon. 2010. Managing invasive populations of Asian longhorned beetle and citrus longhorned beetle: a worldwide perspective. *Annual Review of Entomology* **55**: 521–546.

Hagman, M. and R. Shine. 2008. Understanding the toad code: behavioural responses of cane toad (*Chaunus marinus*) larvae and metamorphs to chemical cues. *Austral Ecology* **33**: 37–44.

Hajek, A. E. and P. C. Tobin. 2010. Micro-managing arthropod invasions: eradication and control of invasive arthropods with microbes. *Biological Invasions* **12**: 2895–2912.

Hájková, P., M. Hájek, and K. Kintrova. 2009: How can we effectively restore species richness and natural composition of a *Molinia*-invaded fen? *Journal of Applied Ecology* **46**: 417–425.

Hanula, J. L., S. Horn, and J. W. Taylor. 2009. Chinese privet (*Ligustrum sinense*) removal and its effect on native plant communities of riparian forests. *Invasive Plant Science and Management* **2**: 292–300.

Hardy, M. C. 2011. Using selective insecticides in sustainable IPM. *CAB Reviews: Perspectives in Agriculture, Veterinary Science, Nutrition and Natural Resources* **6**(039): 1–7.

Harley, K. L. S. and I. W. Forno. 1992. *Biological Control of Weeds. A Handbook for Practitioners and Students*. Inkata Press. Melbourne, Australia. 74 pp.

Harris, D. 2009. Review of negative effects of introduced rodents on small mammals on islands. *Biological Invasions* **11**: 1611–1630.

Harris, A. F., A. R. McKemey, D. Nimmo, et al. 2011. Successful suppression of a field mosquito population by sustained release of engineered male mosquitoes. *Nature Biotechnology* **30**: 828–830.

Hill, M. P., J. A. Coetzee, and C. Ueckermann. 2012. Toxic effect of herbicides used for water hyacinth control on two insects released for its biological control in South Africa. *Biological Control Science and Technology* **22**: 1321–1333.

Hilliard, R. 2005. *Best Practice for the Management of Introduced Marine Pests – A Review*. Publisher Global Invasive Species Programme. Cape Town, South Africa.

Hoddle, M. S. 2002. Restoring balance: using exotic species to control invasive exotic species. *Conservation Biology* **18**: 38–49.

Hoddle, M. S., C. Crespo Ramírez, C. D. Hoddle, et al. 2013. Post release evaluation of *Rodolia cardinalis* (Coleoptera: Coccinellidae) for control of *Icerya purchasi* (Hemiptera: Monophlebidae) in the Galápagos Islands. *Biological Control* **67**: 262–274.

Hoffmann, B., P. Davis, K. Gott, et al. 2011. Improving ant eradications: details of more successes, a global synthesis, and recommendations. *Aliens* **31**:16–23.

Howald, G., C. J. Donlan, K. R. Faulkner, et al. 2010. Eradication of black rats *Rattus rattus* from Anacapa Island. *Oryx* **44**: 30–40.

Hughes, R. F. and J. S. Denslow. 2005. Invasion by a N_2-fixing tree alters function and structure in wet lowland forests of Hawaii. *Ecological Applications* **15**: 1615–1628.

Hulme, P. E. 2006. Beyond control: wider implications for the management of biological invasions. *Journal of Applied Ecology* **43**: 835–847.

Jäger, H. and I. Kowarik. 2010. Resilience of native plant community following manual control of invasive *Cinchona pubescens* in Galápagos. *Restoration Ecology* **18**: 103–112.

Keller, R. P., M. W. Cadotte, and G. Sandifor. 2015. *Invasive Species in a Globalized World: Ecological, Social, and Legal Perspectives on Policy*. University of Chicago Press. Chicago, Illinois, USA. 410 pp.

Kettenring, K. M. and C. R. Adams. 2011. Lessons learned from invasive plant control experiments: a systematic review and meta-analysis. *Journal of Applied Ecology* **48**: 970–979.

Kimball, B. A. and K. R. Perry. 2008. Manipulating beaver (*Castor canadensis*) feeding responses to invasive tamarisk (*Tamarix* spp.). *Journal of Chemical Ecology* **34**: 1050–1056.

Kimball, B. A. and J. D. Taylor. 2010. Mammalian herbivore repellents: tools for altering plant palatability. *Pest Management Science* **21**: 181–187.

King, J. R. and W. R. Tschinkel. 2008. Experimental evidence that human impacts drive fire ant invasions and ecological change. *Proceedings of the National Academy of Sciences* **105**: 20339–20343.

Klassen, W. and C. F. Curtis. 2005. History of the sterile insect technique, pp. 3–36. *In*: V. A. Dyck, J. Hendrichs, and A. S. Robinson (eds.). *Sterile Insect Technique. Principles and Practice in Area-Wide Integrated Pest Management*. Springer. Dordrecht, The Netherlands.

Kreutzweiser, D., D. Thompson, S. Grimalt, et al. 2011. Environmental safety to decomposer invertebrates of azadirachtin (neem) as a systemic insecticide in trees to control emerald ash borer. *Ecotoxicology and Environmental Safety* **74**: 1734–1741.

Kuris, A. M., K. D. Lafferty, and M. E. Torchin. 2005. Biological control of the European green crab, *Carcinus maenas*: natural enemy evaluation and analysis of host specificity, pp. 102–115. *In*: Hoddle, M. S. (ed.). *Second International Symposium on the Biological Control of Arthropods, Vol. I, Davos, Switzerland, September 12–16, 2005*. FHTET-2005-08, USDA Forest Service. Morgantown, West Virginia, USA.

Lacey, L. A. and J. P. Siegel. 2000. Safety and ecotoxicology of entomopathogenic bacteria, pp. 253–273. In: Charles, J. F., A. Délécluse, and C. Nielsen-le Roux (eds.). *Entomopathogenic Bacteria: From Laboratory to Field Application.* Kluwer. Dordrecht, The Netherlands. 524 pp.

Lafleur, N. E., M. A. Rubega, and C. S. Elphick. 2007. Invasive fruits, novel foods, and choice: an investigation of European Starling and American Robin frugivory. *Wilson Journal of Ornithology* **119**: 429–438.

Lake, E. C., J. Hough-Goldstein, and V. D'Amico. 2014. Integrating management techniques to restore sites invaded by mile-a-minute weed, *Persicaria perfoliata*. *Restoration Ecology* **22**: 127–133.

Lincango, M. P., C. E. Causton, C. Calderón Alvarez, and G. Jiménez-Uzcátegui. 2011. Evaluating the safety of *Rodolia cardinalis* to two species of Galápagos finch; *Camarhynchus parvulus* and *Geospiza fuliginosa*. *Biological Control* **56**: 145–149.

Ling, N. 2009. Management of invasive fish, pp. 185–203. *In*: Clout, M. N. and P. A. Williams (eds.). *Invasive Species Management: A Handbook of Principles and Techniques.* Oxford University Press. Oxford, UK.

Loh, R. K. and C. C. Daehler. 2008. Influence of woody invader control methods and seed availability on native and invasive species establishment in a Hawaiian forest. *Biological Invasions* **10**: 805–819.

Luehring, M. A., C. M. Wagner, and W. Li. 2011. The efficacy of two synthesized sea lamprey sex pheromone components as a trap lure when placed in direct competition with natural male odors. *Biological Invasions* **13**: 1589–1597.

McClelland, P. J. 2011. Campbell Island – pushing the boundaries of rat eradications, pp. 204–207. *In*: Veitch, C. R., M. N. Clout, and D. R. Towns (eds.). *Island Invasives: Eradication and Management*. International Union for Conservation of Nature (IUCN). Gland, Switzerland and Centre for Biodiversity and Biosecurity (CBB), Auckland, New Zealand.

McConkey, K. R., S. Prasad, R. T. Corlett, et al. 2011. Seed dispersal in changing landscapes. *Biological Conservation* **146**: 1–13.

McLaughlin, R. L., A. Hallett, T. C. Pratt, et al. 2007. Research to guide use of barriers, traps, and fishways to control sea lamprey. *Journal of Great Lakes Research* **33**: 7–19.

Mckenzie, N., B. Helson, D. Thompson, et al. 2010. Azadirachtin: an effective systemic insecticide for control of *Agrilus planipennis* (Coleoptera: Buprestidae). *Journal of Economic Entomology* **103**: 708–717.

MacDougall, A. S. and R. Turkington. 2005. Are invasive species the drivers or passengers of change in degraded ecosystems? *Ecology* **86**: 42–55.

Masocha, M., A. K. Skidmore, X. Poshiwa, and H. H. T. Prins. 2011. Frequent burning promotes invasions of alien plants into a mesic African savanna. *Biological Invasions* **13**: 1641–1648.

Mason, T. J. and K. French. 2007. Management regimes for a plant invader differentially impact resident communities. *Biological Conservation* **136**: 246–259.

McCullough, D. G. 2015. Other options for emerald ash borer management: eradication and chemical control, pp. 75–82. *In*: Van Driesche, R. G. and R. Reardon (eds.). *The Biology and Control of Emerald Ash Borer*. FHTET 2014-09. USDA Forest Service. Morgantown, West Virginia, USA.

Merritt, D. M. and N. Le Roy Poff. 2010. Shifting dominance of riparian *Populus* and *Tamarix* along gradients of flow alteration in western North American rivers. *Ecological Applications* **20**: 135–152.

Milbrath, L. R. and C. J. Deloach. 2006. Acceptability and suitability of athel, *Tamarix aphylla*, to the leaf beetle *Diorhabda elongata* (Coleoptera: Chrysomelidae), a biological control agent of saltcedar (*Tamarix* spp.). *Environmental Entomology* **35**: 1379–1389.

Montgomery, W. I., M. G. Lundy, and N. Reid. 2012. Invasional meltdown: evidence for unexpected consequences and cumulative impacts of multispecies invasions. *Biological Invasions* **14**: 1111–1125.

Moran, V. C. and J. H. Hoffmann. 2012. Conservation of the fynbos biome in the Cape Floral Region: the role of biological control in the management of invasive alien trees. *Biological Control* **57**: 139–149.

Morrison, N. I. and L. Alphey. 2012. Genetically modified insects for pest control: an update. *Outlooks on Pest Management* **23**: 65–68.

Morrison, L. W. and S. D. Porter. 2005. Testing for population-level impacts of introduced *Pseudacteon tricuspis* flies, phorid parasitoids of *Solenopsis invicta* fire ants. *Biological Control* **33**: 9–19.

Mortenson, S. G., P. J. Weisberg, and L. E. Stevens. 2012. The influence of floods and precipitation on *Tamarix* establishment in Grand Canyon, Arizona: consequences for flow regime restoration. *Biological Invasions* **14**: 1061–1076.

Nico, L. G. and S. J. Walsh. 2011. Nonindigenous freshwater fishes on tropical Pacific islands: a review of eradication efforts, pp. 97–107. *In*: C. R. Veitch, M. N. Clout, and D. R. Towns (eds.). *Island Invasives: Eradication and Management*. International Union for Conservation of Nature (IUCN). Gland, Switzerland and Centre for Biodiversity and Biosecurity (CBB), Auckland, New Zealand.

Orueta, J. F. and Y. A. Ramos. 2001. Methods to Control and Eradicate Non-native Terrestrial Vertebrate Species. Nature & Environment No. 118. Council of Europe Pub. 64 pp.

Panetta, F. D. 2004. Seed banks: the bane of the weed eradicator, pp. 523–526. *In*: Sindel, B. M. and S. B. Johnson (eds.). *Proceedings of the Fourteenth Australian Weeds Conference*. Wagga Wagga, New South Wales, 6–9 September 2004. Australian Weed Society of New South Wales. Wagga Wagga, New South Wales, Australia.

Panetta, F. D. 2009. Weed eradication – an economic perspective. *Invasive Plant Science Management* **2**: 360–368.

Panetta, F. D. 2012. Evaluating the performance of weed containment programmes. *Diversity and Distributions* **18**: 1024–1032.

Panetta, F. D. and S. M. Timmins. 2004. Evaluating the feasibility of eradication for terrestrial weed incursions. *Plant Protection Quarterly* **19**: 5–11.

Pemberton, R. W. and H. Liu. 2007. Control and persistence of native *Opuntia* on Nevis and St. Kitts 50 years after the introduction of *Cactoblastis cactorum*. *Biological Control* **41**: 272–282.

Perry, L. G., S. M. Galatowitsch, and C. J. Rosen. 2004. Competitive control of invasive vegetation: a native wetland sedge suppresses *Phalaris arundinacea* in carbon-enriched soil. *Journal of Applied Ecology* **41**: 151–162.

Petit, J. N., M. S. Hoddle, J. Grandgirard, et al. 2009. Successful spread of a biological control agent reveals a biosecurity failure: elucidating long distance invasion pathways for *Gonatocerus ashmeadi* in French Polynesia. *Biological Control* **54**: 485–495.

Phillips, R. A. 2010. Eradications of invasive mammals from islands: why, where, how and what next? *Emu* **110**: 1–7.

Pickart, A. J. 2008. Restoring the grasslands of northern California's coastal dune. *Grasslands* **17**: 3–9.

Pimentel, D. 2005. Environmental and economic costs of the application of pesticides primarily in the United States. *Environment, Development and Sustainability* **7**: 229–252.

Pitt, W., D. Vice, and M. Pitzler. 2005. Challenges of invasive reptiles and amphibians, pp. 112–119. *In*: Nolte, D. L. and K. A. Fagerstone (eds.). *Proceedings of the 11th Wildlife Damage Management Conference, May 16–19, 2005 Traverse City, Michigan, USA*. University of Nebraska. Lincoln, Nebraska, USA.

Pitt, W. C., K. H. Beard, and R. E. Doratt. 2012. Management of invasive coqui frog populations in Hawaii. *Outlooks on Pest Management* **23**: 166–169.

Plentovich, S., J. Eijzenga, H. Eijzenga, and D. Smith. 2011. Indirect effects of ant eradication efforts on offshore islets in the Hawaiian Archipelago. *Biological Invasions* **13**: 345–557.

Pluess, T., R. Cannon, V. Jarošík, et al. 2012. When are eradication campaigns successful? A test of common assumptions. *Biological Invasions* **14**: 1365–1378.

Pratt, P. D. and T. D. Center. 2012. Biological control without borders: the unintended spread of introduced weed biological control agents. *Biological Control* **57**: 319–329.

Rabitsch, W. 2011. The hitchhiker's guide to alien ant invasions. *Biological Control* **56**: 551–572.

Rauzon, M. J. and D. C. Drigot. 2003. Red mangrove eradication and pickleweed control in a Hawaiian wetland, water-bird responses, and lessons learned, pp. 240–248. *In*: Veitch, C.R. and M. N. Clout (eds.). *Turning the Tide: Proceedings of the International Conference on Eradication of Island Invasives*. Occasional paper of the IUCN Species Survival Commission No. 27. International Union for Conservation of Nature (IUCN), Gland, Switzerland.

Reaser, J. K., L. A. Meyerson, Q. Cronk, et al. 2007. Ecological and socioeconomic impacts of invasive alien species in island ecosystems. *Environmental Conservation* **34**: 98–111.

Rejmánek M. and M. J. Pitcairn. 2003. When is eradication of exotic pest plants a realistic goal? pp. 249–253. *In*: Veitch, C.R. and M. N. Clout (eds.). *Turning the Tide: Proceedings of the International Conference on Eradication of Island Invasives*. Occasional paper of the IUCN Species Survival Commission No. 27. International Union for Conservation of Nature (IUCN), Gland, Switzerland.

Rinella, M. J., B. D. Maxwell, P. K. Fay, et al. 2009. Control effort exacerbates invasive-species problem. *Ecological Applications* **19**: 155–162.

Robertson, B. C. and N. J. Gemmell. 2004. Defining eradication units to control invasive pests. *Journal of Applied Ecology* **41**: 1042–1048.

Robinson, A. S., M. J. B. Vreysen, J. Hendrichs, and U. Feldmann. 2009. Enabling technologies to improve area-wide integrated pest management programmes for the control of screwworms. *Medical and Veterinary Entomology* **23**: 1–7.

Roubos, C. R., C. Rodriguez-Saona, and R. Isaacs. 2014. Mitigating the effects of insecticides on arthropod biological control at field and landscape scales. *Biological Control* **75**: 28–38.

Saunders, G., B. Cooke, K. McColl, et al. 2010. Modern approaches for the biological control of vertebrate pests: an Australian perspective. *Biological Control* **52**: 288–295.

Schroder, D. 1980. The biological control of thistles. *Biological Control News and Information* **1**(1): 9–26.

Schroer, S., R. W. Pemberton, L. G. Cook, et al. 2008. The genetic diversity, relationships, and potential for biological control of the lobate lac scale, *Paratachardina pseudolobata* Kondo & Gullan (Hemiptera: Coccoidea: Kerriidae). *Biological Control* **46**: 256–266.

Serbesoff-King, K. 2003. Melaleuca in Florida: a literature review on the taxonomy, distribution, biology, ecology, economic importance and control measures. *Journal of Aquatic Plant Management* **41**: 98–112

Sharov, A. A., D. Leonard, A. M. Liebhold, et al. 2002. "Slow the spread," a national program to contain the gypsy moth. *Journal of Forestry* **100**: 30–35.

Schlyter, F., Q. H. Zhang, G. T. Liu, and L. Z. Ji. 2003. A successful case of pheromone mass trapping of the bark beetle *Ips duplicatus* in a forest island, analysed by 20-year time-series data. *Integrated Pest Management Reviews* **6**: 185–196.

Shiels, A. B. 2011. Frugivory by introduced black rats (*Rattus rattus*) promotes dispersal of invasive plant seeds. *Biological Invasions* **13**: 781–792.

Shine, C., N. Williams, and L. Gundling. 2000. *A Guide to Designing Legal and Institutional Frameworks on Alien Invasive Species*. International Union for Conservation of Nature (IUCN). Gland, Switzerland. 138 pp.

Simberloff, D. 2006. Invasional meltdown 6 years later: important phenomenon, unfortunate metaphor, or both? *Ecology Letters* **9**: 912–919.

Simberloff, D. 2009. We can eliminate invasions or live with them. Successful management projects. *Biological Invasions* **11**: 149–157.

Simberloff, D. and B. Von Holle. 1999. Positive interactions of nonindigenous species: Invasional meltdown? *Biological Invasions* **1**: 21–32.

Simmonds, F. J. and F. D. Bennett. 1966. Biological control of *Opuntia* spp. by *Cactoblastis cactorum* in the Leeward Islands (West Indies). *Entomophaga* **11**: 183–189.

Speedy, C., T. Day, and J. Innes. 2007. Pest eradication technology – the critical partner to pest exclusion technology: the Maungatautari experience, pp. 115–126. *In*: Witmer, G. W., W. C. Pitt, and K. A. Fagerstone (eds.). *Managing Vertebrate Invasive Species: Proceedings of an International Symposium*. USDA/APHIS/WS, National Wildlife Research Centre. Fort Collins, Colorado, USA.

Stark, J. D. 2007. Ecotoxicology of neem, pp. 275–286. *In*: A. S. Felsot and K. D. Racke (eds.). *Crop Protection Products for Organic Agriculture*. Symposium Series 947. American Chemical Society.

Stock, W. D., K. T. Wienand, and A. C. Baker. 1995. Impacts of invading N_2 – fixing *Acacia* species on patterns of nutrient cycling in two Cape ecosystems: evidence from soil incubation studies and ^{15}N natural abundance values. *Oecologia* **101**: 375–382.

Stork, N. E., R. L. Kitching, N. E. Davis, and K. L. Abbott. 2014. The impact of aerial baiting for control of the yellow crazy ant, *Anoplolepis gracilipes*, on canopy-dwelling arthropods and selected vertebrates on Christmas Island (Indian Ocean). *Raffles Bulletin of Zoology* **30**: 81–92.

Strazanac, J. S. and L. B. Butler (eds.). 2005. Long-term evaluation of the effects of *Bacillus thuringiensis kurstaki*, gypsy moth nucleopolyhedrosis virus product *Gypchek*, and *Entomophaga maimaiga* on nontarget organisms in mixed broadleaf-pine forests in the central Applachians. FHTET-2004-14. USDA Forest Service. Morgantown, West Virginia, USA.

Stromberg, J. C., S. J. Lite, R. Marler, et al. 2007. Altered streamflow regimes and invasive plant species: the *Tamarix* case. *Global Ecology and Biogeography* **16**: 381–393.

Strum, S. C., G. Stirling, and S. Kalusi Mutunga. 2015. The perfect storm: land use change promotes *Opuntia stricta*'s invasion of pastoral rangelands in Kenya. *Journal of Arid Environments* **118**: 37–47.

Suckling, D. M., A. M. Barrington, A. Chhagan, et al. 2007. Eradication of the Australian painted apple moth, *Teia anartoides*, in New Zealand: trapping, inherited sterility, and male competitiveness, pp. 603–615. *In*: Vreysen, M. J. B., A. S. Robinson, and J. Hendrichs (eds.). *Area-Wide Control of Insect Pests: From Research to Field Implementation*. Springer. Dordrecht, The Netherlands.

Suckling, D. M., R. W. Peck, L. M. Manning, et al. 2008. Pheromone disruption of Argentine ant trail integrity. *Journal Chemical Ecology* **34**: 1602–1609.

Suliman, A. S., G. G. Meier, and P. J. Haverson. 2011. Eradication of the house crow from Socotra Island, Yemen, pp. 361–363. *In*: Veitch, C. R., M. N. Clout, and D. R. Towns (eds.). *Island Invasives: Eradication and Management*. IUCN (International Union for Conservation of Nature). Gland, Switzerland and Centre for Biodiversity and Biosecurity (CBB), Auckland, New Zealand.

Teem, J. L. and J. B. Gutierrez. 2013. Combining the Trojan Y chromosome and daughterless carp eradication strategies. *Biological Invasions* **16**: 1231–1240.

Tewksbury, L., R. Casagrande, P. Häfliger, et al. 2013. Host specificity testing of *Archanara geminipuncta* and *A. neurica* (Lepidoptera: Noctuidae), candidates for biological control of *Phragmites australis* (Poaceae), p. 69. *In*: Wu, Y., T. Johnson, S. Sing, et al (eds.). *Proceedings of the XIII International Symposium on Biological Control of Weeds, Waikoloa, Hawaii, USA, 11–16 September, 2011.* FHTET- 2012-07 USDA Forest Service. Morgantown, West Virginia, USA.

Thresher, R. E., and A. M. Kuris. 2004. Options for managing invasive marine species. *Biological Invasions* **6**: 295–300.

Towns, D. R. and K. G. Broome. 2003. From small Maria to massive Campbell: forty years of rat eradications from New Zealand islands. *New Zealand Journal of Zoology* **30**: 377–398.

Towns, D. R., C. J. West, and K. G. Broome. 2013. Purposes, outcomes and challenges of eradicating invasive mammals from New Zealand islands: an historical perspective. *Wildlife Research* **40**: 94–107.

Tschinkel, W. R. 2006. *The Fire Ants*. Belknap/Harvard University Press. Cambridge, Massachusetts, USA.

Tu, M., C. Hurd, and J. M. Randall. 2001. Weed Control Methods Handbook: Tools and Techniques for Use in Natural Areas. The Nature Conservancy. Available from: http://digitalcommons.usu.edu/govdocs/533/ [Accessed January 201].

Ulyshen, M. D., S. Horn, and J. L. Hanula. 2010. Response of beetles (Coleoptera) at three heights to the experimental removal of an invasive shrub, Chinese privet (*Ligustrum sinense*), from floodplain forests. *Biological Invasions* **12**: 1573–1579.

Van Driesche, R.G., M. Hoddle, and T. Center. 2008. *Control of Pests and Weeds by Natural Enemies: An Introduction to Biological Control*. Blackwell. Oxford, UK.

Van Driesche, R. G., R. I. Carruthers, T. Center, et al. 2010. Classical biological control for the protection of natural ecosystems. *Biological Control* **54**: S2–S33.

Vreysen, M. J. B., A. S. Robinson, and J. Hendrichs (eds.). 2007. *Area-Wide Control of Insect Pests*. Springer. Dordrecht, The Netherlands.

Walsh, J. C., O. Venter, J. E. M. Watson, et al. 2012. Exotic species richness and native species endemism increase the impact of exotic species on islands. *Global Ecology and Biogeography* **21**: 841–850.

Waterhouse, D. F. and D. P. A. Sands. 2001. *Classical Biological Control of Arthropods in Australia*. Australian Centre for International Agricultural Research. Canberra, Australia.

Wilson, J. R. U., C. Gairifo, M. R. Gibson, et al. 2011. Risk assessment, eradication, and biological control: global efforts to limit Australian acacia invasions. *Diversity and Distributions* **17**: 1030–1046.

Witmer, G. W., F. Boyd, and Z. Hillis-Starr. 2007. The successful eradication of introduced roof rats (*Rattus rattus*) from Buck Island using diphacinone, followed by an eruption of house mice (*Mus musculus*). *Wildlife Research* **34**: 108–115.

Witt, A. B. R. and A. X. Nongogo. 2011. The impact of fire, and its potential role in limiting the distribution of *Bryophyllum delagoense* (Crassulaceae) in southern Africa. *Biological Invasions* **13**: 125–133.

Wittenberg, R. and M. J. W. Cock (eds.). 2001. *Invasive Alien Species: A Toolkit of Best Prevention and Management Practices*. CAB International. Wallingford, Oxon, UK. 228 pp.

Witzgall P., L. Stelinski, L. Gut, and D. Thomson. 2008. Codling moth management and chemical ecology. *Annual Review Entomology* **53**: 503–522.

Yang, Z-Q., X-Y. Wang, J. R. Gould, and H. Wu. 2008. Host specificity of *Spathius agrili* Yang (Hymenoptera: Braconidae), an important parasitoid of the emerald ash borer. *Biological Control* **47**: 216–221.

Zabel, A. and B. Roe. 2009. Performance payments for environmental services: lessons from economic theory on the strength of incentives in the presence of performance risk and performance measurement distortion. IED Working Paper 09-07, Institute for Environmental Decisions, Eidgenössische Technische Hochschule, Zurich, Switzerland.

Zavaleta, E. S., R. J. Hobbs, and H. A. Mooney. 2001. Viewing invasive species removal in a whole-ecosystem context. *Trends in Ecology & Evolution* **16**: 454–459.

Zull, A., O. Cacho, and R. Lawes. 2006. A matrix model for the management of perennial weeds in the North Queensland rangelands system: application to *Ziziphus mauritiana* (Lam.), pp. 671–674. *In*: Anon. *Managing Weeds in a Changing Climate, 15th Australian Weeds Conference, Papers and Proceedings, Adelaide, South Australia, 24–28 September 2006*. Weed Management Society of South Australia. Victoria, Australia.

CHAPTER 4

Tools in action: understanding tradeoffs through case histories

Roy G. Van Driesche[1], Bernd Blossey[2], and Daniel Simberloff[3]

[1] *Department of Environmental Conservation, University of Massachusetts, USA*
[2] *Department of Natural Resources, Cornell University, USA*
[3] *Department of Ecology & Evolutionary Biology, University of Tennessee, USA*

Risk tradeoffs in invasive species management

All tools used for invasive species management in ecological restoration projects have potential risks (see Chapter 3), but all risks are context dependent and the relative risks of different actions must be assessed against each other for each particular case. Additionally, the potential damage from the safest approach should be compared to the damage likely to occur if the invader were not controlled or less well controlled. Effects of invasive species vary dramatically, and whether any particular species' effects rise to the level requiring its control must be determined (see Chapter 2). Most invasive species are ecologically insignificant, and even harmful species may be beneficial in some contexts. For example gorse, *Ulex europaeus* L., which in general is highly damaging, in some locations in New Zealand protects a rare insect, the Mahoenui giant weta (*Deinacrida* sp., Orthoptera), from predation by invasive rats (Stronge et al., 1997). However, for invasive species that are highly damaging to the biodiversity, function, or structure of ecosystems, control measures will be necessary. While several measures are available, each will have its merits and shortcomings, and risk tradeoffs will exist among different options.

The relative risks and values of various control actions will depend on the idiosyncratic features of the species involved and the context in which the invasion is occurring. The do-nothing option may be appropriate when damage from the invader is light or the invasion's

effects develop slowly, allowing time to determine how severe they are likely to become, or if the species becomes naturally limited to low densities by existing factors. Direct controls, such as burning (Figure 4.1), physical removal, or herbicidal treatment (Figure 4.2) of invasive plants, can often be devised that can kill individuals or stands of the target species, but this may or may not allow ecosystem or biodiversity recovery. When landscape-level invader suppression is needed for ecosystem recovery, introductions of biological control agents may be appropriate. Such biological control programs must ask critical questions about recovery goals, suitability of agents proposed for release, and risks from agents to non-target species in both the region of intended release and more distant areas where the agents might spread. Examples of such risks from invaders and potential control methods are discussed here via a series of case studies.

When is no action the right choice?

Strawberry guava

Psidium cattleianum Sabine is a woody tree or shrub native to coastal southeastern Brazil that threatens tropical forests on islands around the world, including Hawaii (Zimmerman et al., 2008), the Kartala forest on Ngazidja island in the Comoros Islands (Yahaya, 2003), and Mauritius (Florens et al., 2010). In Mauritius, invasive plants (especially strawberry guava) reduce native butterfly diversity in weedy vs. weeded forests (Florens

Integrating Biological Control into Conservation Practice, First Edition. Edited by Roy G. Van Driesche, Daniel Simberloff, Bernd Blossey, Charlotte Causton, Mark S. Hoddle, David L. Wagner, Christian O. Marks, Kevin M. Heinz, and Keith D. Warner.
© 2016 John Wiley & Sons, Ltd. Published 2016 by John Wiley & Sons, Ltd.

Figure 4.1 Prescribed fire to control Scotch broom on the Elizabeth Islands, Massachusetts. Scotch broom has invaded maritime grasslands. Photo courtesy of Julie Richburg.

Figure 4.2 Herbicide application is a common, usually non-selective form of control used against invasive plants. Here, an aerial herbicide application to kill *Phragmites australis*. Photo credit: Bernd Blossey.

et al., 2010). We consider strawberry guava a driver of ecological degradation in Hawaii; however, the ecological damage it causes there is enhanced by interactions with other invasive species, especially nitrogen-fixing plants and pigs.

In Hawaii, lowland wet forests develop on lava flows through succession. Invasion by strawberry guava is facilitated by the invasive nitrogen-fixing tree *Falcataria moluccana* (Miquel) Barneby & Grimes (Hughes and Denslow, 2005; Zimmerman et al., 2008), which enriches the soil (Hughes and Denslow, 2005). When *F. moluccana* was absent, native species nearly completely dominated the site, but when this invasive plant was present, alien species accounted for 68–99% of total basal area (Hughes and Denslow, 2005). Lava flows invaded by *F. moluccana* showed large increases of strawberry guava and decreases in the native tree *Metrosideros polymorpha* Gaudich (Hughes and Denslow, 2005). Given the limited remaining native-dominated wet lowland forests in Hawaii, it is expected that the extent of the *F. moluccana* invasion and its follow-on positive effects on strawberry guava and negative effects on *M. polymorpha* will determine the degree to which this unique forest ecosystem will continue to exist.

Densities of strawberry guava both affect and are affected by pigs (*Sus scrofa*). Pigs eat strawberry guava fruits, and thus pig numbers are enhanced by the tree; conversely, because pigs spread strawberry guava seeds in their feces, the plant also benefits. Invasive pigs degrade native Hawaiian forests, killing and consuming a variety of trees and tree ferns (Cole et al., 2012). A 16-year pig-exclusion experiment showed that both native and non-native understory vegetation responded strongly to feral pig removal, with the density of native woody plants rooted in mineral soil increasing sixfold and density of young tree ferns increasing by over 50% in pig-free sites (Cole et al., 2012).

Chemical control of strawberry guava is possible. Glyphosate (via hack and squirt application) and 2, 4,-D (via foliar applications) both effectively kill strawberry guava under field conditions in Hawaii (Motooka et al., 1983), and small infestations of strawberry guava can be eradicated, as was done on the New Zealand island of Raoul (2943 ha) (West, 2003). However, for larger infestations, especially in remote areas of difficult terrain, eradication via chemical control is too costly and difficult.

In Hawaii, once it was understood that eradication was not likely to work and that the costs of control with herbicides would be high, biological control was considered. Surveys in the plant's native range found a number of prospective agents, among which the leaf-galling scale insect *Tectococcus ovatus* Hempel (Hemiptera: Eriococcidae) was the most promising because of its damage and ease of rearing (Wikler et al., 2000). Wessels et al. (2007) showed that *T. ovatus* was specific enough to release in Florida for biological control of strawberry guava. In Hawaii, however, public understanding of the proposed biological control project against strawberry guava was manipulated by a few members of the public who represented pig hunters and had a general anti-government sentiment, who dismissed the harm from the plant and inflamed public feeling against the project (see Chapter 12) (Warner and Kinslow, 2013). This caused a three-year delay during which a greatly expanded, 700-page, state-level environmental impact statement was developed (OEQC, 2011). This led to approval of releases of the scale insect on state land, which began in 2011. Monitoring for scale establishment and impact are in progress. Failure to move forward with area-wide control of strawberry guava in Hawaii using biological control could have caused serious, long-lasting, and potentially irreversible harm to native forests and clearly would have been the wrong choice.

Garlic mustard

Garlic mustard (*Alliaria petiolata* [M. Bieb.] Cavara & Grande) has invaded the understories of many North American forests and continues to spread (Nuzzo, 1999; Blossey et al., 2001). This species is commonly listed among the top three invasive species of concern by land-management agencies in the northeastern United States (Acharya, 2010). Plant biodiversity declines in areas infested with garlic mustard, especially tree seedlings (Stinson et al., 2007), but ground beetles and their invertebrate prey are not affected (Dávalos and Blossey, 2004). The negative effects of garlic mustard on plant biodiversity are due to interspecific competition and to allelopathic chemicals exuded into the soil (Lankau, 2010; Wixted and McGraw, 2010). Benzyl isothiocyanate, the allelochemical exuded by garlic mustard, inhibits the ectomycorrhizal fungi of conifer seedlings (Wolfe et al., 2008), northern red oak (*Quercus rubra* L.) (Castellano and Gorchov, 2012), and potentially perennials such as false Solomon's seal (*Maianthemum racemosum* [L.] Link) (Hale et al., 2011). Levels of allelopathic exudates at sites infested with garlic mustard, however, decrease over time (1–50 years), apparently owing to intensive intraspecific competition (Lankau et al., 2009).

Effects of garlic mustard interact with those of introduced earthworms and white-tailed deer (*Odocoileus virginianus* [Zimmermann]). Garlic mustard invasions appear to be facilitated by the presence of invasive earthworms, which greatly modify the leaf litter food web (Nuzzo et al., 2009). Furthermore, seedling survival of many native plants is greatly reduced in the presence of earthworms (Dobson and Blossey, 2015), and the effect of deer on native herbaceous and woody species is well documented (Côté et al., 2004). Native understory species can thrive and recover once deer abundance is reduced, even when garlic mustard populations are high (Nuzzo, 2006). Furthermore, earthworm populations collapse as deer are excluded experimentally (Dávalos et al., 2015), and garlic mustard populations decline and population growth rates of native species increase when deer access is curtailed (Kalisz et al., 2014).

Land managers were unable to control garlic mustard using physical, mechanical, or chemical means (Nuzzo, 1991; Bowles et al., 2007; Slaughter et al., 2007) and consequently research to develop biological control was initiated in Europe (Blossey et al., 2001). A number of host-specific and promising control agents were identified, most importantly four weevils in the genus *Ceutorhynchus* (Coleoptera: Curculionidae) (Blossey et al., 2001). These species attack rosettes (*C. scrobicollis* Nerensheimer and Wagner), stems (*C. roberti* Gyllenhal, *C. alliariae* H. Brisout), and seeds (*C. constrictus* [Marsham]) of *A. petiolata* (Gerber et al., 2007). Demographic models suggest that targeting rosettes and seed production was of greatest importance, giving highest priority to *C. scrobicollis* followed by the stem miners (Davis et al., 2006; Gerber et al., 2007, 2008).

The available evidence suggests that garlic mustard is not a driver of ecological degradation but rather is taking advantage of conditions created by high deer populations and invasive earthworms. Native plant communities are able to thrive in its presence if deer abundance is reduced. However, while garlic mustard declines with time at a given site, this does not prevent colonization of new areas provided earthworm invasion

has occurred. In summary, while the strawberry guava example suggested that biological control was a prudent and important management tool, it does not appear to be needed for garlic mustard. At already degraded sites, habitat restoration should focus on deer reductions and replanting native species.

Is herbicide use a good option for invasive plant control?

Herbicides are a powerful tool for management of invasive plants. They can be applied where and when needed, and a variety of products and application methods provide great flexibility that can be adapted for control of particular pests. However, the ability to kill targeted individual plants, while often high, is not the same as affecting plant demography or meeting goals for ecosystem restoration. For some plant species, regrowth from roots or seeds may lead to quick recovery of biomass of the targeted plant. Here we contrast the use of herbicides against two invaders of North American wetlands: melaleuca (*Melaleuca quinquenervia* [Cav.] S.T. Blake), introduced from Australia, and invasive genotypes of the common reed (*Phragmites australis* (Cav.) Trin. ex Steud.) that were introduced from Europe.

Phragmites australis (common reed)

Invasive genotypes of common reed, introduced in the early 1800s, have become a major invader over much of temperate North America, where they displace native vegetation and the North American endemic subspecies, *Phragmites australis americanus* Saltonstall, P.M. Peterson, and Soreng (Saltonstall, 2002; Saltonstall et al., 2004). Invaded wetlands include Atlantic and Pacific tidal marshes, Great Lakes wetlands, and isolated US wetlands and inland rivers. The combined effect of the rapid spread of these introduced genotypes and the resulting near monocultures was of great concern to wetland managers, and for decades various means were used to suppress *P. australis*, to restore native vegetation diversity and protect marsh-inhabiting birds. Control methods included burning, plastic mulch, soil disking, chemical control, cutting, grazing, dredging, draining, and the manipulations of water tables and salinity (Marks et al., 1994; Sun et al., 2007). Herbicides such as aminotriazole, dalapon, glyphosate, and imazapyr readily killed *P. australis* under field conditions (Beck,

1971; Riemer, 1976; Mozdzer et al., 2008), and semi-selective application techniques were developed (Anon., 1984). While herbicide suppression has temporary benefits for annual plants and some invertebrate taxa (Gratton and Denno, 2005; Kulesza et al., 2008), treatments usually need to be repeated every few years at enormous expense because of rapid re-invasion (Moreira et al., 1999; Blossey and McCauley, 2000). At some freshwater marsh sites, glyphosate application to *P. australis* followed by burning increased plant, but not animal, diversity (Ailstock et al., 2001). Marshes along the Connecticut River treated with herbicides to reduce *P. australis* showed increased plant diversity, but treatment did not improve use of treated sites by fish or crustaceans (Fell et al., 2003).

A review of expenditures versus outcomes for control of common reed by 285 US land managers' activities from 2005–09 found that these organizations combined spent US$4.6 million per year on *P. australis* management (Martin and Blossey, 2013). Herbicide treatment – the most commonly used method of control – was applied to ~80,000 ha over the study period. Few organizations, however, reported that they met their management objectives. Furthermore, there was no relationship between resources expended and management success (Martin and Blossey, 2013). In summary, benefits from *P. australis* control appear to be temporary and to accrue only to certain groups of organisms (plants, insects, spiders). In addition, few data exist for evaluation of potential long-term effects of repeated herbicide use on species of conservation interest in these wetlands.

Melaleuca

In contrast to the uncertain long-term outcomes following herbicide use against *P. australis*, chemical control of *M. quinquenervia* in Florida was decisive and provided long-term benefits, but only when combined with biological control in an integrated program. Melaleuca, an Australian swamp tree, was brought to Florida as an ornamental. Through both seed dispersal and deliberate planting, melaleuca came to dominate many marshes and other wetland and upland habitats in the region. At the peak of population abundance (before implementation of widespread control efforts), it occupied some 200,000 ha in southern Florida, displacing native plants in prairie marshes, and invading cypress swamps, pine flatwoods, hardwood bottomlands, and mangrove swamps (Buckingham, 2004). Of special concern was

impact on marshes, in particular the Everglades, a World Heritage site. In addition to forming extensive mono-specific stands with few native plants, melaleuca promoted soil accretion in invaded marshes, permanently reducing water depth and physically changing the habitat.

By the 1980s, efforts to kill stands of melaleuca were underway, using various combinations of herbicide application, cutting, and fire. Some of the earliest removal efforts were on levees on Lake Okeechobee (Stocker and Sanders, 1981), where trees had been planted earlier. At that location, foliar or pelletized applications of Velpar® (triazines) or Spike® (tebuthiuron) provided about 80% control of mature trees on terrestrial or seasonally wet sites, while on wet sites, tree injection was tried. At the same location, bromacil (Hyvar X®), tebuthiuron, or Velpar provided good control of seedlings (Stocker and Sanders, 1981). Woodall (1982) found that container-grown seedlings of melaleuca were killed by Hyvar (bromacil), Karmex® (diuron), Velpar (hexazinone), and Tordon® (picloram with 2,4-D amine), but that Ammate® (ammonium sulphamate) and Roundup® (glyphosate) had limited effects. By 1998, melaleuca had been removed from several management zones south of Lake Okeechobee through use of herbicides and cutting, but continuous monitoring and control efforts were required to suppress regrowth (Laroche, 1998). Use of herbicides, while helpful, entailed risks of non-target plants dying from herbicide leaching and, potentially, of applicator injury or contamination owing to accidents with spray machinery or cutting tools.

Because public land sites where melaleuca had been suppressed with cutting and herbicides needed ongoing re-treatment, with no foreseeable end, and because most private land sites left melaleuca stands untreated (and hence a source of seeds), a new approach for melaleuca suppression was developed based on the integration of biological control of melaleuca reproduction with tree clearance by cutting and herbicides (Silvers et al., 2007; Center et al., 2008, 2012). Biological control's role was to curtail seed production, especially on unmanaged private land, and lower seedling survival, while chemical/mechanical control's role was to clear mature trees from publically owned lands. Chemical/mechanical control was especially important in areas with permanent standing water, mainly artificial impoundments created to store water for urban use or flood prevention, as the seed-destroying weevil *Oxyops vitiosa* (Pascoe) needs a dry season to complete its life cycle (Center et al., 2012) and was less effective at such sites.

Field experiments showed the effectiveness of the biological control agents, which included the melaleuca weevil, *O. vitiosa* (affecting seed production), and the meleuca psyllid, *Boreioglycaspis melaleucae* Moore (destroying seedlings and stump sprouts). Tipping et al. (2009) used an insecticidal check method (spraying insecticides on melaleuca regrowth to exclude the biological control agents), creating plots with and without the biological control agents (see Chapter 9 for discussion of evaluation techniques). They found significant effects of agents on tree growth, tree survival, and seed production. Over the course of a five-year experiment, melaleuca stem density in sprayed plots (no natural enemies) held steady, but in plots with natural enemies, stem number was reduced nearly by half (47.9%). In sprayed plots, mortality of melaleuca trees never exceeded 6% per year, but was much higher (11–25% per year) in unsprayed plots with natural enemies present. Seed production occurred on sprayed trees but never on unsprayed trees, subject to attack by natural enemies. Mean tree height increased by 20% in sprayed plots while it decreased by 31% in unsprayed plots with natural enemies. In a separate experiment, Rayamajhi et al. (2010) showed that the biological control agents strongly suppressed survival of coppice suckers on melaleuca stumps. In a third experiment, in a seasonally inundated wetland in the western Everglades, Tipping et al. (2012) found that the population density of melaleuca seedlings or saplings per m^2 declined by 99% over a three-year study, from 64.8 to 0.5 stems per m^2. Elimination of introduced insect herbivores using insecticides did not reduce mortality of new melaleuca seedlings/sapling at the study site, indicating that direct herbivory was not responsible for the decline in seedling density. The authors concluded that reduced seed input (caused by the melaleuca weevil) was the most likely cause of the decreasing number of seedlings in the plots.

Reduction in melaleuca density in study plots conferred several benefits on native biodiversity. Rayamajhi et al. (2009), in plots monitored for 10 years (1995 to 2005), observed that reduction of melaleuca density was associated with a two- to fourfold increase in native plant species diversity, with more change in non-flooded habitats than seasonally flooded areas. Additionally, Martin et al. (2010) compared the effects of chemical vs. biological control of melaleuca on nutrient storage and availability, and found that herbicide use reduced both above- and belowground storage of nutrients both before and after a seasonal fire compared to a non-invaded area, while biological control increased storage. However, phosphorus availability was greatest in the herbicide-treated site post fire. These results suggest that biological control of melaleuca had less effect on nutrient storage and cycling than use of herbicides.

Changes in vegetation caused by invader suppression often affect animals using the modified habitats. In parts of the Big Cypress National Preserve, where virtually all melaleuca (99.9%) was removed manually, Julian et al. (2012) recorded large increases in total cover of native upland forest (227%), wetland forest (211%), and prairie (54%) communities. This vegetative change was associated with a 16% contraction in home-range sizes of Florida panthers using the area (a beneficial effect, meaning less habitat was needed per animal). Similar reductions in mean home-range size were not observed during the same time period for the regional population of radio-collared panthers occupying contiguous conservation lands in south Florida that were still invaded by melaleuca.

In conclusion, unlike the use of herbicides for suppression of *P. australis*, which in general failed to provide long-lasting benefits, the use of herbicides for melaleuca removal, when combined with plant neutering (permanent reductions in seed production and seedling survival via biological control), increased native plant biodiversity and improved habitat quality for an endangered mammal. These contrasting cases show that merely achieving short-term suppression of invasive plants may not lead to ecosystem recovery and that a longer-term view is required.

Are biological control projects high or low risk?

Biological control projects over the span of the discipline (1888 to present) have presented risks that have varied greatly, sometimes because agents of different risk levels were used and sometimes because the target or geographical location presented risks that were either not recognized or not treated seriously at the time.

Here we explore this risk spectrum by examining five case histories that we have ranked from low (rank 1) (prickly pear cactus in Australia in the 1920s; melaleuca in Florida, 1986 to present), to low-to-medium (rank 2) (saltcedar in the United States, 1990–2010; emerald ash borer, North America, 2002 to present), to moderate-to-high (rank 3) (invasive thistles, North America, 1960s), to high (rank 4) (*Opuntia* cacti, Caribbean Islands in the 1950s), to unacceptably high risk (rank 5) (spotted-wing *Drosophila*, in Hawaii, a hypothetical case).

Biological control, *in its entirety*, is widely perceived by conservation biologists as a risky activity because release of new biological control agents is irreversible and because, in the past, releases were sometimes made without adequate forethought and testing to predict consequences. This situation has changed greatly. As with any restoration tool, biological control *releases* have risks that fall along a spectrum. The highest risk releases are often not carried out (agents are rejected owing to perceived risk) and this fact may go unnoticed precisely because identification of potential high risk leads to agent rejection during testing (e.g., Heard et al., 1998; Dhileepan et al., 2005; Balciunas, 2007; Bennett and Pemberton, 2008; Yobo et al., 2009). These rejected agents themselves span a spectrum of estimated risk, with some species being rejected because they clearly included native, even endangered, species in their core host ranges (Heard et al., 1998; Bennett and Pemberton, 2008; Yobo et al., 2009). Less clear cases, but nevertheless ending in rejection, were agents whose use of a native species was more marginal (Dhileepan et al., 2005) or geographically contextual (Balciunas, 2007). The agent (*Lixus cardui* Olivier [Coleoptera: Curculionidae]) studied by Balciunas (2007), for example, was safe for introduction to Australia (a region with few close relatives to the target weed, Scotch thistle [*Onopordum acanthium* L.]), but was unsuitable for introduction to the United States, where many close relatives in the genus *Cirsium* exist.

Biological control projects (as opposed to individual agents) also present a spectrum of risk, due to the agents used, but also the geographical location and ecological context of the project. To illustrate this point, seven projects are discussed below that fall along such a risk spectrum (Table 4.1), some of which we rank as intrinsically low risk, while others are ranked higher, some as high risk or even as unacceptable.

Prickly pear cactus in Australia in the 1920s (Rank 1: little to no risk)

This project's details are summarized by Wilson (1960). Cacti, which as a family are native only to the Western Hemisphere, have been spread widely as ornamentals, including to Australia (Queensland and New South Wales, in or before the nineteenth century). Seven species of prickly pears (*Opuntia* spp.) became invasive in Australia, of which *Opuntia inermis* (DC.) DC. and *Opuntia stricta* (Haw.) Haw. became the most widespread, covering 24 million ha of land by 1925, half the area covered so densely that the land had no economic value. While more insects were introduced over a longer period of time than is generally recognized, the key introduction in hindsight proved to the pyralid moth *Cactoblastis cactorum* (Bergroth) from Argentina. This species was introduced and then established in Australia starting in 1925, and by 1930 this single species destroyed the two most invasive cacti over much of their invaded ranges. This impact in turn reduced populations of other agents introduced earlier and made them less important. *Cactoblastis cactorum* was subjected to host–plant feeding tests to ensure that crops would not be attacked. It was found to develop only on cacti. Because there were no native plants in this family in Australia, this was an adequate level of specificity to protect native plants fully.

Melaleuca in Florida, 1986 to present (Rank 1: little to no risk)

Melaleuca quinquenervia is an Australian swamp tree brought to Florida in the nineteenth century as an ornamental. It invaded and greatly altered natural habitats, including sawgrass marshes, cypress forests, and other wetland and upland plant communities (Serbesoff-King, 2003). Because of the ecological damage it caused to native plant communities, by the 1980s control measures (cutting, use of herbicides) were begun but proved insufficient. In 1986, a biological control program began, starting with surveys for natural enemies of the plant in its native range (e.g., Balciunas et al., 1994). An important advantage for the project, greatly increasing the ease of work and safety to native plants, was that melaleuca is in the subfamily Leptospermoideae, while the only native US plants in the family Myrtaceae are in the other subfamily (the Myrtoideae), in the four genera *Calyptranthes*, *Eugenia*, *Mosiera*, and *Myricanthes*. In the continental United States, these genera occur only in

Table 4.1 Biological control projects selected to illustrate a spectrum of risk due to characteristics, not of potential biological control agents, but of the project itself.

Project	Where carried out	Degree of risk	Reasons for risk level	
			Closeness of target to native species	**Rarity or abundance of closely related species**
Prickly pear cactus, 1920s	Australia	◯	No native species at family level	No rare or endangered closely related native species
Melaleuca, 1990–2010	Florida	◯	No native species at tribe level	No rare or endangered closely related native species
Saltcedar (*Tamarix*), 1990–present	Southwest United States	◯◯	No native species in subfamily (Baum, 1967; Crins, 1989)	Endangered bird subspecies using invasive plant for nesting sites
Emerald ash borer, 2004–present	United States and Canada	◯◯	Many native *Agrilus*	No native *Agrilus* of same size that attack ash trees
Invasive thistles, 1960s	Canada and United States	◯◯◯	Many native thistles in same genus/ tribe	Many closely related thistles, some that are rare, endangered, or occupy small specialized habitats
Opuntia cacti, 1950s	Caribbean islands	◯◯◯◯	Release location near center of evolution of target genus; Hundreds of native *Opuntia* in United States and Mexico	Native *Opuntia* species include rare or endangered cacti and economically or culturally important species; keystone taxon in Chihuahuan and Sonora Deserts.
Spotted-wing drosophila (hypothetical)	Hawaii	◯◯◯◯◯	Nearly 1000 endemic *Drosophila* in Hawaii	Large % of species in genus rare, endangered, or poorly studied. Hawaii species important example of evolutionary radiation

Risk rating categories: 1 oval, little or no risk; 2 safe agents available; some complications with native species; 3 difficult to find sufficiently safe agents, but potentially realistic; 4 extremely difficult to find sufficiently safe agents because of large number of non-target species; 5 unlikely to be able to screen non-target species adequately because of number and rarity and also non-target species have high value.

subtropical Florida. Furthermore, the essential oils of these US native Myrtoideae species show strong differences from the oils of the subfamily Leptospermoideae, in which melaleuca occurs (Serbesoff-King, 2003). Since the specialized herbivorous insects to be used as biological control agents against melaleuca would likely use the plant's essential oils as both host recognition cues and feeding stimulants, the probability was thus high that insects attacking melaleuca would not feed on native US myrtaceous species. Species introduced as ornamentals or crops, however, occur in both subfamilies and might incur some damage.

While the taxonomic distinctness of melaleuca from the native flora of Florida provided high potential for safety for the project in that location, two features of Florida's position relative to the Caribbean islands and

South America created new risks that had to be considered. The first of these was southern Florida's position relative to northward migrating birds in spring. To shorten their water crossing, many birds returning from their winter stay in the Caribbean or South America enter the United States through southern Florida. This path results in large numbers of hungry birds arriving in the area, searching for insect prey, and quickly eating as many insects as possible. One group of insects considered for biological control of melaleuca was the pergid sawflies (Purcell and Goolsby, 2005), particularly *Lophyrotoma zonalis* Rower (Center et al., 2012), whose larvae eat melaleuca foliage and in doing so sequester several plant-derived toxins (lophyrotomin, pergidin, and Val4-pergidin) (Oelrichs et al., 2001). Because cattle poisoning from consumption of larvae of related pergid sawflies was known from Australia (e.g., Tessele et al., 2012), there seemed to be some potential for bird poisoning, and release of this agent was rejected by scientists leading the effort (Center et al., 2012).

A second concern was that Florida's position stretching south towards the Caribbean would provide a launching pad from which melaleuca biological control agents would reach the Caribbean islands and South America, following the flow of tourists and goods between Miami and these locations. Melaleuca has been planted as an ornamental in many locations in the Caribbean, southern Mexico, and Central and South America (e.g., French Guiana) (Watts et al., 2009; Delnatte and Meyer, 2012), and climatically the tree is well adapted to spread across a large part of the region. Within this range, melaleuca is already invading ecologically important wetlands, such as Tortuguero Lagoon Natural Reserve in Puerto Rico (Pratt et al., 2005). Few opportunities existed to reduce this risk of spread of the introduced agents and, indeed, one of the most effective agents, the melaleuca psyllid, reached Puerto Rico by 2006 without any known releases (Pratt et al., 2006). While this insect appears to have genus-level specificity to *Melaleuca*, a group with no species native to the Americas, the myrtaceous flora of Puerto Rico is much larger than that of Florida and several genera occur there that were not included in host-range tests in Florida (Pratt et al., 2006). Field surveys in Puerto Rico are needed to see if the melaleuca psyllid damages any native species in these genera in Puerto Rico. Post-release field studies in Florida with plants receiving some melaleuca psyllid feeding or oviposition in laboratory tests have shown these species not to be hosts or to rank very low as hosts (Center et al., 2007), re-enforcing the view that this psyllid is limited to species of *Melaleuca*. Similarly, pre-release (Balciunas et al., 1994), laboratory (Wheeler, 2005) and post-release field studies (Pratt et al., 2009) show a high level of specificity for the other important melaleuca biological control agent, the weevil *O. vitiosa*.

The agents introduced for control of melaleuca in Florida appear to have a sufficiently high level of specificity to preclude damage to native plants even as they spread far beyond Florida. Indeed, the agents may contribute to lowering melaleuca's invasion potential in natural habitats in Puerto Rico and other Caribbean or South American locations. However, some damage to introduced myrtaceous plants is possible as the biological control agents come into contact with an ever increasing number of such species as their geographic ranges expand.

Saltcedar (tamarix) in the United States, 1990–2010 (Rank 2: low to moderate risk)

Tamarix ramossisima Ledeb., several other introduced Eurasian species of *Tamarix* (saltcedars), and their hybrids collectively have become invasive over large parts of the western United States (Tracy and DeLoach, 1999). The success of these plants has been greatly increased by alterations to flood regimes (extent and timing) caused by dams and water withdrawals, which decrease the recruitment and population-level competitiveness of native cottonwoods and willows. However, saltcedar has invaded even relatively pristine watersheds with natural hydrological regimes (Dudley and Bean, 2012, and references therein). Some 500,000 ha of riparian forests and shrublands in the western United States were largely replaced by saltcedars (Christensen, 1962; Robinson, 1965; Everitt, 1980), causing extensive ecological harm to native plants and animals (Tracy and DeLoach, 1999). Some species, however, have been able to use saltcedar as a food source or habitat (Sher and Quigley, 2013). Given this wide pattern of invasion and harm, a reduction in saltcedar and a corresponding increase in native vegetation would be desirable, but achieving this transition is complicated by the presence of an endangered subspecies of bird, the southwestern willow flycatcher (*Empidonax traillii extimus* Phillips), which in some areas nests in saltcedar. Because of this subspecies' protection under the US Endangered Species Act, reduction in its "habitat" (including saltcedar) is potentially a "taking," which would be prohibited unless authorized. Conflict thus arose between the legal mandate to protect all individuals of the protected species in the short term vs. the need to favor native vegetation by suppressing saltcedar. Unlike mechanical removal of saltcedar, which can be limited to areas where no endangered birds nest, the biological control agent could not be contained in this manner, but rather would spread naturally over large areas. Below we provide some background on this controversy by considering the ecological effects of saltcedar on plant communities, birds in general, and the southwestern flycatcher in particular.

In cottonwood/willow woodlands, saltcedar has become dominant along many rivers where spring floods have been reduced by dams (Tracy and DeLoach,

1999). On the lower Colorado River, when saltcedar's competition with native plants was reduced by mechanical removal of saltcedar, Gooding willow (*Salix goodingii* Ball) exhibited 62% greater stem growth and 88% greater leaf area (Busch and Smith, 1995). In herbaceous broadleaf meadows, saltcedar invasion leads to declines in many common species such as alkali heliotrope (*Heliotropium currasavicum* L.) and such rare species as Pecos sunflower (*Helianthus paradoxus* Heiser) and Red Rock tarplant (*Hemizonia arida* Keck) (Tracy and DeLoach, 1999, and references therein).

Effects of saltcedar on birds are generally negative because the tiny fruits and seeds of saltcedar are not eaten by North American birds. Also insect density is lower on saltcedar than on the native plants it replaces, creating "food deserts" for most insectivorous birds; along the lower Colorado River, in a five-year study, counts of insectivorous birds showed that two-thirds of all species censused were less abundant in areas dominated by saltcedar than in areas with native vegetation (Cohan et al., 1979). The biological control program against saltcedar is now far enough along in some areas that direct field evidence is becoming available on its effects on the saltcedar, recovery of native vegetation, and changes in quality of habitat for native birds. In areas where defoliation by the tamarix beetle has led to regrowth of native vegetation, food for insectivorous birds has increased (Longland and Dudley, 2008), increasing this aspect of habitat quality for the southwestern flycatcher.

Southwestern willow flycatchers, which normally nest in willows and cottonwoods, can nest in saltcedar stands or mixed stands; however, numbers of birds fledged per nest is twice as great in stands of native vegetation (1.89 fledglings per female per year) than in stands dominated by saltcedar (0.89) (Dudley et al., 2000). Indeed, this flycatcher has disappeared from many areas it historically occupied that are now dominated by saltcedar (Finch et al., 2002), and of 1000 known nesting sites surveyed by Finch et al. (2002), 91% were in mixed vegetation stands, while only 9% of nests occurred in areas where saltcedar comprised >90% of the vegetation. Because saltcedar progressively replaces willows and cottonwoods, reliance on mixed stands as breeding habitat for this flycatcher is an unsustainable strategy for the bird's long-term protection.

A biological control program was developed against saltcedar based on release of specialized herbivorous insects from several parts of the native range of tamarix (DeLoach et al., 1996). While many species were evaluated, only the saltcedar leaf beetle, *Diorhabda elongata* (Brullé) (sensu latu), was released. This beetle was later found to be a complex of slightly differentiated populations, at first considered subspecies, but later elevated into four separate species based on molecular evidence (Bean et al., 2013), although some difference of opinion about this remains. The form that established widely in Nevada, Utah, and surrounding areas is now known as *D. carinulata* Desbrochers. All four *Diorhabda* forms were later shown to have very similar fundamental host ranges (Milbrath and DeLoach, 2006a). This program has proceeded but has been slowed by conflict with conservation biologists concerned about the effects of loss of tamarix habitat on the southwestern willow flycatcher.

The fears articulated by wildlife biologists (e.g., Sogge et al., 2008) have been (1) that the biological control agents would kill saltcedar faster than native shrubs or trees would regrow, (2) that in some areas (e.g., the lower Colorado River) current artificial water regimes (lack of spring floods owing to dam control) would preclude native vegetation from ever regrowing, (3) that even if saltcedar plants were not killed, summer defoliation would expose nestlings to lethal temperatures caused by exposure to direct sunlight, or (4) that biological control insects feeding on saltcedar would sequester plant toxins at high levels and be toxic to birds that might eat large numbers of them. While these fears are plausible, evidence from field events following the implementation of the biological control program shows they are unfounded.

1 *Rate of saltcedar death.* Comparing areas where saltcedar had experienced various numbers of years of defoliation, Hudgeons et al. (2007) found that defoliation by the tamarisk beetle (*D. carinulata*) reduced levels of sugars and starch stored in root crowns from about 9% of total nonstructural carbohydrates to 2–3% in one set of years and from 14% to 2–8% in another depending on number of years of defoliation. In cage trials, fall defoliation reduced spring regrowth by about 35% (Hudgeons et al., 2007). At one of the sites (Humboldt River area in northern Nevada) where the biological control beetle established and caused defoliation earliest, Dudley and DeLoach (2004) noted that despite defoliation in 2002 and 2003, in 2004 there were still no plant deaths. However, Moran et al. (2009) note that after four

years of defoliation extensive branch dieback occurred, accompanied by some saltcedar mortality. Dudley and Bean (2012) report that mortality of *T. ramosissima* trees along the Humboldt River (Nevada) increased from ca. 5% two years after the first beetle-induced saltcedar defoliation to 40% after three years, 80% after five years, and 85% after seven years. These data suggest that impacts of the biological control program on saltcedar will be large but that tree death will occur over a protracted period, allowing three to seven years from the first defoliation for recovery of native vegetation or growth of planted vegetation to a size useful as habitat (which takes two to three years).

2 *Lack of regrowth of native willows/cottonwoods.* In the most degraded habitats, where dams have created hydroregimes unfavorable to natural regrowth of cottonwoods and willow, saltcedar heavily dominates and the southwestern willow flycatcher is no longer present as it does not persist well in pure saltcedar stands (DeLoach et al., 2000). Consequently, slowed or diminished regrowth of native plants following suppression of saltcedar by the biological control agent in such areas will have the least effect on the southwestern willow flycatcher because it is not there. However, even in areas such as the lower Colorado River, if some native willows were still present, willows responded to mechanical removal of saltcedar with increased growth, despite the unfavorable hydroregime (Busch and Smith, 1995). Also, Moran et al. (2009) note that, in some of the areas in Utah and Colorado where saltcedar beetles arrived in 2004 and caused defoliation in subsequent years, within two years of such defoliation, willows and other native plants increased greatly in abundance. At Elephant Butte Reservoir on the Rio Grande, a flood in 1995 killed a stand of saltcedar that then occupied the site and that had supported two breeding pairs of southwestern willow flycatchers. Native willows subsequently re-vegetated the area, and seven years later (2002) the southwestern willow flycatcher population had increased 27-fold to 54 pairs, dispersed over some 700 ha of suitable habitat (Ahlers and Moore, 2009). These data suggest that these birds can adjust and even prosper following abrupt loss of saltcedar if suitable native plants reclaim the site.

3 *Nestling death from sun exposure following defoliation.* In principle, defoliation might increase mortality of

flycatcher nestlings. However, if birds are given options, they are likely to relocate to other vegetation as beetles begin to defoliate a site. In the Utah portion of the Virgin River, active planting of willow and other native plants around St. George, Utah, was done concurrently with introduction of the tamarix leaf beetle. In 2009, nine out of ten females nested in tamarix, but in 2010 this changed to just 22% (two out of nine birds), with the rest nesting in willow. Also, nests in willow showed higher offspring survival, for a net tripling of the number of birds fledged (from two in 2009 to six in 2010) (Dudley and Bean, 2012). This fact suggests that birds actively select better habitat if it is available.

In 2013, it was observed (Dudley, pers. comm.) at the lower Virgin River in Utah that defoliation by *Diorhabda* occurred before return of southwest willow flycatchers from their wintering grounds in Central America, and that the birds avoided defoliated plants, nesting instead in nearby willow. This suggests that once beetles are well established, sudden loss of canopy cover after nest initiation would be unlikely to occur.

4 *Intoxication of birds eating saltcedar beetles.* Southwest willow flycatchers have recently been observed to feed heavily on saltcedar beetles (feces consisting almost entirely of *Diorhabda* parts) with no observable ill effects under field conditions (Dudley and Bean, 2012). Biological control has converted the "food desert" of saltcedar to a food source by adding an upper trophic level layer suitable for bird consumption.

Planning for native plant restoration needs to be part of any attempt to reduce saltcedar density, including efforts based on herbicides, fire, mechanical clearance, or biological control. In general terms, such planning requires site-specific goals that account for the degree to which desired native plants are still present at the site, the current hydrologic regime, and the presence of species of special concern (Shafroth et al., 2008). With respect to restoring willow and cottonwood to benefit birds, particularly the southwestern willow flycatcher, two situations must be considered. In one case, tamarix is reduced at sites where flycatchers are present. In this case, expansion of existing native plants or replanting them if none exist must be planned, allowing two to three years for replanted natives to reach sizes suitable for flycatcher nesting. In the other case, habitats that historically were used by this flycatcher but are not currently used might be rehabilitated by creating "propagule islands" (fenced areas replanted to cottonwood, where cows are prevented from grazing on cottonwood seedlings). Such areas can provide cottonwood and willow seeds to restock downstream river reaches (Dudley and Bean, 2012).

One further complication potentially of concern in tamarix biological control was the potential to harm two groups of non-target plants. One was an introduced tamarix species, *Tamarix aphylla* (L.) Karsten (athel), used as an ornamental tree but not considered invasive. This species, while not a preferred host of *D. elongata* (sensu lato), is part of its fundamental host range (Milbrath and DeLoach, 2006b). Damage to this commercially useful but exotic plant was expected. Efforts to deliberately establish saltcedar leafbeetle on athel by enclosing beetles in large field cages placed over athel plants resulted in about 50% defoliation, with green foliage recovering in ten weeks (Moran, 2010). When bags were placed over athel limbs and then stocked with the beetles, limbs were defoliated but the beetle did not establish, suggesting only limited potential for damage to this ornamental species.

The other non-target plant group recognized in host range testing as potentially at some risk was native species of *Frankenia* (because larvae, if placed on the plant, could complete their development) (Herr et al., 2009). However, subsequent open-field tests, in which key species of *Frankenia* were planted near tamarix stands heavily infested with saltcedar leaf beetles, showed <4% leaf damage to *Frankenia salina* (Molina) I. M. Johnst. (Dudley and Kazmer, 2005), likely because adult beetles have low-to-no preference to oviposit on *Frankenia* species (Lewis et al., 2003).

In summary, the saltcedar project can be rated as somewhat riskier than the melaleuca project, despite saltcedar species having a family-level degree of separation from North American plants (Tracy and DeLoach, 1999). This was because (1) an endangered subspecies (southwestern willow flycatcher) used saltcedar as a resource (albeit, a poor one), (2) because human-made changes to hydrologic conditions along rivers reduced native plant establishment and thus favored saltcedar, (3) because a valued exotic ornamental (athel) was fed on by the biological control agent, and (4) because a native *Frankenia* species supported beetle larval development in host-range tests. However, the risks

turned out to be lower than this list might suggest, as discussed above. And, as a final postscript, water savings (Pattison et al., 2010) and reduced fire risk (Drus et al., 2012) have been documented owing to biological control of saltcedar.

Emerald ash borer, North America, 2002 to present (Rank 2: low to moderate risk)

Emerald ash borer (*Agrilus planipennis* Fairmaire) (Coleoptera: Buprestidae) is an East Asian phloem-feeding borer affecting ash (*Fraxinus*) trees, which was first recognized as invasive in North America (Michigan and Ontario) in 2002 (Haack et al., 2002). It likely arrived in wooden packing material from China in the 1990s but went unrecognized until ash mortality reached very high levels. Most native ash in eastern North American are suitable hosts, especially green (*Fraxinus pennsylvanica* Marshall), white (*F. americana* L.), and black (*F. nigra* Marshall) ash (Liu et al., 2003). North America has some 7 billion ash trees, most of which are at risk of being killed by an expanding, unrestrained emerald ash borer population. Widespread mortality of ash, particularly of older ash in riparian areas, occurred in southern Michigan by 2010 (Emerald Ash Borer Information, 2013; USDA PLANTS Database, 2013).

The ecological effects of this loss of ash include (1) severe reduction in ash itself, especially of older trees, (2) loss of riparian communities in wet soils, dominated by pure stands of black ash, stands that provide browse and cover for deer and moose (Rockermann, 2011) and which may not remain forested following loss of black ash, (3) loss or density reductions of native arthropods, including 100 invertebrate species dependent on ash trees for either feeding or breeding purposes (Wagner, 2007; Wagner and Todd, 2015), and (4) changes in ecosystem structure and function because of cascading effects either through the disruption of ash-based food webs or changes in microclimatic or ecological conditions in the forests (reviewed in Wagner and Todd, 2015).

In response to the ecological and economic losses posed by emerald ash borer, eradication was attempted but failed (Emerald Ash Borer Information, 2013). Surveys for natural enemies in China (part of the native range of the pest and the likely source of the pest introduction) began in 2003 (Liu et al., 2003), followed later by surveys in the Russian Far East (Duan et al., 2012a).

Six species of parasitoids were found: (1) *Sclerodermus pupariae* Yang et Yao (Hymenoptera: Bethylidae), an ectoparasitoid of larvae, prepupae, and pupae from China (Wu et al., 2008); (2) *Spathius agrili* Yang (Hymenoptera: Braconidae), an ectoparasitoid of larvae from China (Liu et al., 2003; Yang et al., 2005); (3) *Tetrastichus planipennisi* Yang (Hymenoptera: Eulophidae), an endoparasitoid of larvae from China and Russia (Liu et al., 2003; Yang et al., 2006; Duan et al., 2012a; Belokobylskij et al., 2012); (4) *Oobius agrili* Zhang and Huang (Hymenoptera: Encyrtidae), an egg parasitoid from China (Zhang et al., 2005; Liu et al., 2007); (5) *Spathius galinae* Belokobylskij and Strazanac (Hymenoptera: Braconidae), an ectoparasitoid of larvae from Russia (Duan et al., 2012a; Belokobylskij et al., 2012); and (6) *Atanycolus nigriventris* Vojnovskaja-Krieger (Hymenoptera: Braconidae), an ectoparasitoid of larvae from Russia (Duan et al., 2012a; Belokobylskij et al., 2012).

Of these, the host range of *S. pupariae*, based on studies in China, was deemed too broad (Yang et al., 2012), and it was not considered further as a potential biological control agent. The other two Chinese larval parasitoids (*T. planipennisi* and *S. agrili*) and the egg parasitoid *O. agrili* were imported, and after host-range estimation studies, a petition for their release was written and approved (Gould, 2007). Those three species were the only ones released through 2014. Separately, similar work was done with the Russian parasitoid *S. galinae*, and its release has been approved for 2015. Importation of *A. nigriventris* is being considered, but host-range studies have not yet started.

In assessing both the need for and the risk of this project, the following questions must be answered: (1) was no-action unacceptable (i.e., was significant ecological damage being done), (2) was the invasive species the clear and sole cause of the damage, (3) was eradication or suppression through some combination of management tools possible, (4) were the biological control agents sufficiently specific that the ecological damage, if any, of their release was low compared to the likely damage of permitting the pest to go unsuppressed, (5) did the release of natural enemies contribute to lowering the pest population, (6) will lower pest populations reduce ecological damage from the invader, and (7) are any other ecological restoration actions other than pest suppression needed for ecosystem restoration?

The first three of these questions have been answered. Loss of large portions of riparian forests dominated by several species of ash clearly demonstrated significant ecological damage from the pest. This damage, coupled with rapid increase in the size of the invaded area, provided clear evidence of the need to lower the density of this invader. That the invader alone was the sole cause of the damage is self-evident, given that intense damage occurred in stable, protected forest communities with healthy stands of ash and that tree mortality increased rapidly after invasion. Eradication of the pest through ash removal was tried on a large scale in Michigan (McCullough, 2015), with prolonged effort and extensive funding, but failed for several reasons: (1) the infestation was widespread in forested areas when recognized, (2) female beetles fly extensively, and (3) traps for detection are not based on pheromones, but rather on color and host odors, greatly reducing their power for early detection and monitoring. Multi-factor management, while possible in urban areas (where resistant ash might be planted eventually and pesticides are able to protect specimen trees), may not be applicable in forests because of the size of the infested area, the cost of pesticide applications, and the manner of pesticide application required (chemical injection of individual trees) (McCullough, 2015). Questions (4, 5, and 6) about the biological control agents released (specificity, efficacy, and ability to promote ecological recovery) are addressed below. Whether further ecological restoration activities (question 7) will be needed is not clear at this time. First the degree of success with biological control must be learned and more information will be needed on the actual ecological damage emerald ash borer eventually causes.

Were the biological control agents sufficiently specific?

In weed biological control projects, heavy emphasis is placed on direct testing of native species in the same genus as the target pest under no-choice conditions of close confinement. With a few exceptions, seeds of the required plants can usually be purchased or field-collected. This approach is based on the commonly found trend that specialized herbivorous insects use the signature secondary (defense) compounds of plants for host recognition and as oviposition and feeding stimulants. Such secondary compounds are commonly conserved, with variation, within plant clades during speciation events. This produces a correlation between herbivore plant acceptance and plant phylogeny. Host ranges of parasitoids are less closely tied to host taxonomy, as insects generally lack defensive compounds analogous to those of plants, although some insect groups do sequester the defensive compounds of their host plants (e.g., Sime, 2002). Typically, however, parasitoid host ranges are driven by two filters: the ability of the adult parasitoid to find and recognize a generally suitable host and the ability of the immature parasitoid to withstand attempts of the host to defend itself. For external parasitoids, position alone (not being exposed to attack by the host's immune system) enhances survival of immature parasitoids. For internal parasitoids of host stages with immune systems (larvae, nymphs, and adults), encapsulation of parasitoid eggs or larvae by the host's hemocytes is an important risk, but one that is opposed by a variety of mechanisms mounted by the parasitoid, including polydnaviruses, teratocytes, venom, selective positioning of eggs, and deposition of supernumerary eggs. The first filter, host finding, is typically modulated by whatever volatiles or signals the adult parasitoid is responsive to. These typically are volatile compounds from the plant, with or without effects of feeding by the herbivore, or such signals may originate directly from the insect (pheromones or vibrations from feeding larvae). Under conditions of close confinement in laboratory tests, such as are typical for studying the host ranges of herbivorous insects, this important filter shaping parasitoid host ranges is often bypassed.

Evidence about host-range widths of parasitoids thus commonly comes from field collections of taxonomically related hosts from the pest's native range, laboratory tests of some taxonomically related hosts, and olfactometer tests of responses of adult parasitoids to volatiles produced by the insect's host plant directly, or during interactions between the insect and its plant host.

Spathius agrili

In the case of *S. agrili*, field collections were made of 17 species of wood-boring insects from Tianjin, China, and other locations where emerald ash borer *S. agrili* densities were high, including six species of *Agrilus* (*n* = 2074 individuals), and no cases of parasitism by *S. agrili*, apart from the target pest, were found (Yang et al., 2008). When field-collected larvae of various *Agrilus* species from China or the United States were presented to *S. agrili* females, under close confinement and no-choice conditions with larvae presented in

their natural host plants, three of nine *Agrilus* species (in addition to emerald ash borer) were attacked and successfully yielded adults of *S. agrili* (Yang et al., 2008). The final release of oviposition behavior is stimulated by the detection of mechanical vibrations caused by larval feeding within the infested log (Wang et al., 2010). While these species are within the physiological host range of the parasitoid, they are likely outside its ecological host range (i.e., not attacked in nature) because close-confinement testing skips the important host–plant odor filter. Indeed, when foliage of host trees was tested in an olfactometer, only the two species of ash were attractive to adult female *S. agrili*, while 12 other plants tested were neutral or repellant (Yang et al., 2008). Collectively, these data were taken as evidence of an adequately narrow host range for this species, with relatively low risk to native congeners despite the fact that there are at least 175 native *Agrilus* in the United States and 624 in Mexico. Release of this species was approved and releases made, although to date there have been few recoveries to evaluate efficacy and host range in the field.

Tetrastichus planipennisi

Three of six species of non-target *Agrilus* collected from various plant species in the field in China were found to be parasitized by other (not *T. planipennisi*), but unidentified, species of *Tetrastichus* (see Table 6 in Gould [2007]). When 11 non-target species of buprestid borer larvae (including five *Agrilus* and three *Chrysobothris* species) were presented to *T. planipennisi* females under conditions of close confinement in the natural hosts (ash, birch, oak, raspberry, maple, apple, or pine), none were attacked (Gould, 2007).

Oobius agrili

Filters shaping the host ranges of egg parasitoids are harder to define. No volatiles have been discovered that attract *O. agrili* to host eggs, although rates of parasitism of emerald ash borer in China by *O. agrili* are quite high (see Figure 4 in Gould [2007]), suggesting some efficient method of host detection. In laboratory no-choice tests under conditions of close confinement, *O. agrili* oviposited in eggs of three species of non-target *Agrilus* (of six offered) (see Table 7, Gould, 2007), but not in eggs of six other insects in other buprestid genera or families of borers.

Did the biological control agents lower the pest population?

To measure the ability of the above three Chinese parasitoids, in conjunction with native mortality factors (tree resistance, woodpecker predation, and native parasitoids – especially *Atanycolus* spp. and *Phasgonophora sulcata* Westwood), to lower densities of emerald ash borer, six pairs of plots (release and control) were established in south and central Michigan in 2008 and releases made of all three parasitoid species (Duan et al., 2012b). Establishment of *T. planipennisi* and *O. agrili* was demonstrated by 2009 (Duan et al., 2011, 2012b, c), but *S. agrili* was rarely recovered. Plots were monitored yearly. Factors (including death from host plant resistance, parasitism by native larval parasitoids, parasitism by the introduced *T. planipennisi*, predation by woodpeckers) affecting emerald ash borer larvae and pupae were measured by debarking selected trees in the 10–20 cm dbh range, examining the fates of either experimentally created cohorts of emerald ash borer or of naturally occurring larvae or pupae (Duan et al., 2010; 2012c; 2013, 2014). The effect of these mortality factors was summed by calculation of population rate of increase (R_0) for several generations, during which R_0 fell from 16–19 in 2009 (an exploding population) to 5 in 2010 (an expanding population) (Duan et al., 2014). By 2014, emerald ash borer larval density (per m^2 of phloem in sampled trees) declined ca. 80% relative to 2008 (Duan et al., 2015). Rates of mortality across years by factors showed that woodpecker predation was consistently high (40–60%) but with no trend. Rates of parasitism by native parasitoids increased in some years and then declined. In contrast, parasitism by the introduced parasitoids increased year by year: *T. planipennisi* increased from <5 to 30% and egg parasitism from <5 to ca. 20% (Duan et al., 2015), suggesting them as the cause of the decline in borer density.

Will lower pest populations allow damaged ecosystems to recover?

At the long-term study plots mentioned above, by fall 2014 only modest numbers of live large emerald ash borer larvae were detected in sampling (Duan et al., 2015). Large ash trees were mostly dead, but there remained a significant ash component to the stands in both the medium (15–20 cm dbh) and seedling/sapling (2–10 cm dbh) sizes. Low rates of infestation in 15–20 cm dbh-sized ash permitted many trees in this size range to survive, and these trees plus maturing saplings will form the basis for future ash stands. Forest recovery will require additional time and monitoring will be required.

Invasive thistles, North America, 1960s (Rank 3: moderate to high risk)

Eurasian species of *Carduus* and *Cirsium* thistles were accidentally introduced to North America and other temperate areas with European colonization. Some of these thistles become abundant pests of pastures, especially Canada thistle (*Cirsium arvense* [L.] Scop.), bull thistle (*Cirsium vulgare* [Savi] Ten.), musk or nodding thistle (*Carduus nutans* L.), plumless thistle (*Carduus acanthoides* L.), and slenderflower thistle (*Carduus tenuiflorus* Curtis). *Carduus nutans* in North America is a complex of cryptic species or subspecies (Desrochers et al., 1988). *Carduus* species in North America are introduced (McCarty, 1978), but there are many native species of *Cirsium* (Moore and Frankton, 1969, 1974). Surveys of natural enemies affecting these invasive thistles in Europe began in 1959 (Schroder, 1980), and a large number of insects with tribe-level specificity (Cynareae) were found. By 1980, six species had been introduced for biological control. While these had little or no effect on invasive *Cirsium* species (Cripps et al., 2011), several imported weevils did control some of the invasive *Carduus* species, especially *C. nutans* (Kok, 1976). While several released species were of importance in this process, the oligophagous weevil *Rhinocyllus conicus* Frölich (for bibliography, see Boldt and Kok, 1982) was especially effective (Rees, 1978; Surles and Kok, 1978; Kok et al., 2004; Roduner et al., 2003) but also proved able to attack a large number of native *Cirsium* species. This weevil was collected in several parts of Europe from various genera of thistles. Individuals from different source populations were established in North America and now are considered races whose host ranges vary somewhat (Unruh and Goeden, 1987). Host-specificity tests in Europe indicated that this weevil's host range includes various species of *Cirsium*, *Carduus*, and *Silybum* thistles (Zwölfer and Preiss, 1983). A subsequent re-examination of the host range of the *R. conicus* biotype attacking native *Cirsium* species in North America confirmed that there had been no change in the fundamental host range of the weevil (Arnett and Louda, 2002). While *R. conicus* prefers the invasive thistle *C. nudans* under laboratory choice conditions (Arnett and Louda, 2002), when it spreads into regions lacking this species it attacks additional species and reproduces successfully (Louda et al., 1997) and does so even when *C. nudans* overlaps with such lower-ranking native *Cirsium* species (Rand and Louda, 2004). Native North American thistles attacked include *Cirsium canescens* Nuttall (Nebraska), *C. pitcheri* (Torrey) Torrey and Gray (Michigan), *C. undulatum* (Nuttall) Spreng. (Nebraska) (Rand and Louda, 2004), *C. pumilum* var. *hillii* (Canby) B. Boivin (= *C. hillii*) (Wisconsin) (Sauer and Bradley, 2008), *C. ownbeyi* S. L. Welsh (Colorado/Utah/Wyoming) (DePringer-Levin et al., 2010), *C. carolinianum* (Walter) Fernald and Schubert (Tennessee) (Wiggins et al., 2010), *C. horridulum* Michaux (Tennessee) (Wiggins et al., 2010), *C. altissimum* (L.) Hill (Kansas) (Russell et al., 2010), and five (Herr, 2000) to 12 (Turner et al., 1987) native *Cirsium* species in California, albeit for these California species at levels not having apparent population effects. Most of these records are of plant-use, not population-level impact. (For discussion of this distinction, see Chapter 8.) Two rare species, however, have been studied well enough to show clear population impact of significant intensity via seed reduction: Platte thistle in Nebraska (Louda, 1998; Russell and Louda, 2004; Rose et al., 2005; Rand and Louda, 2006; Louda et al., 2011; Rand and Louda, 2012) and Pitcher's thistle in sand dunes of the western Great Lakes (Bevill et al., 1999; Louda et al., 2005a).

While most releases of *R. conicus* were made in the 1970s before these effects on non-target species were widely known, releases continued in some areas much later: for example, Texas 1984–87 (Boldt and Jackman, 1993); Tennessee 1989–90 (Grant and Lambdin, 1993); Georgia 1990–92 (Buntin et al., 1993).

This case has been widely discussed to determine the reasons that allowed such significant impacts to non-target plants to occur (Louda, 2000; Gassmann and Louda, 2001; Louda et al., 2003; Rand et al., 2004; Louda et al., 2005b). While the oligophagous nature of *R. conicus* (lacking specificity to *Carduus* spp.) was known, inferential arguments based on laboratory preference for the target species were used to suggest that less preferred hosts would not suffer population-level effects. This proved untrue for some species. Also lacking in the original host-range testing was inclusion of all but a few of the North American *Cirsium* species. These errors, plus a general lack of importance assigned to the target group ("thistles" being primarily known through their pest species) allowed insects to be released

that lacked adequate levels of specificity. The features of this case indicate higher risk both because the target plant had many closely related native species and because faulty ecological assumptions allowed a clearly oliphagous agent to be deemed, incorrectly, as relatively specific.

Opuntia cacti, Caribbean Islands in the 1950s (Rank 4: high risk)

Cacti, native only to the Americas, were widely introduced to the drier areas of Australia, Asia, South Africa, and various islands (including Madagascar) by European colonists. In Australia, as noted above in an earlier case history, several species of *Opuntia* became widely invasive. A biological control program was initiated based on collecting cacti-specific insects from Argentina. The program was highly successful and, because there are no native cacti in Australia, highly safe. At the time, only family-level (Cactaceae) specificity was sought, but laboratory tests done at the time showed that all but one (*O. sulphurea* Gillies ex Salm-Dyck) of the *Opuntia* species tested received eggs and supported larval development. Also it was noted that species in the *Cylindropuntia* section of *Opuntia* were rarely attacked (Dodd, 1936). Because of the obvious success of this species, it was quickly introduced into other regions, outside the Americas, where *Opuntia* species were invasive, including South Africa (Pettey, 1939), Madagascar (Decary, 1965), and Hawaii (Fullaway, 1958). In South Africa, the insect became a pest in crops of spineless *Opuntia* cacti grown as stock fodder (Annecke et al., 1976).

In 1952, the Commonwealth Institute of Biological Control was approached by the Colonial Development and Welfare Organization in Barbados for help suppressing stands of *Opuntia* cactus that had come to dominate pastures, particularly on Nevis. Use of *C. cactorum* and other cacti-feeding insects was discussed, and in 1956 funds were made available to begin introductions (Simmonds and Bennett, 1966). The decision seems to have been made solely in the context of pasture improvement with no thought of impact on native cacti or the production of spineless *Opuntia* for fodder or human consumption. Nor was the role of pasture overgrazing as the fundamental driver of cactus infestations considered.

Following its successful establishment on Nevis, *C. cactorum* was moved to Montserrat and Antigua in 1960

(Simmonds and Bennett, 1966) and Grand Cayman in 1970 (Bennett, 1971). Either by its own movement or by ranchers moving infested plants (to spread the insect) unofficially it was soon established in St. Kitts (1964) and the US Virgin Islands (1963) (Simmonds and Bennett, 1966). The moth reached Puerto Rico in the 1960s (Garcia-Tuduri et al., 1971), and by the 1990s it was reported in Florida (Dickel et al., 1991) and Cuba (Hernández and Emmel, 1993). The introduction to Florida may have been via plant importations from the Dominican Republic between 1981 and 1986 (Pemberton, 1995).

Following its detection in Florida, impacts of *C. cactorum* on native, especially rare, species of *Opuntia* became of conservation interest. In Florida, of particular relevance was the moth's effect on the rare cacti *Opuntia spinosissima* (Martyn) Mill. (Johnson and Stiling, 1996) and *Opuntia corallicola* (Small) Werderm. (the semaphore cactus) (Hight et al., 2002). With the invasion of the North American mainland, risks to cacti in Mexico and the southwestern United States increased greatly and became of political concern. In parts of Mexico *Opuntia* species are a major human food that is harvested commercially. Moreover members of the genus are of considerable ecological importance, supporting many specialist herbivores and serving as a base for many desert food webs.

Approximately 200 species of *Opuntia* are known worldwide, of which 114 occur in Mexico. Some of the species with greatest economic use that are likely to be attacked by the moth are *Opuntia compressa* (= *O. humifusa* [Raf.] Raf.), *O. ficus-indica* [L.] Mill., *O. megacantha* Salm-Dyck, *O. stricta* (Haw.) Haw., and *O. tomentosa* Salm-Dyck (Vigueras and Portillo, 2001). An estimated 79 platyopuntia (prickly pear) species are at risk, 51 of which are endemic to Mexico, 9 to the United States, and 19 common to both countries (Zimmermann et al., 2000). *Cactoblastis cactorum* was predicted to arrive first in Mexico overland via Texas (Soberon et al., 2001). However, the first detection in Mexico was from another direction, the southeast, where the moth was found near Cancun, on the Isla de las Mujeres, from which it was eradicated. To date (2015), no further invasions have been detected in Mexico.

Experience outside the native range provides some insight into possible suitability of some North American species as hosts. Under field conditions in South Africa, moth survival (egg to pupation) varied significantly among the six *Opuntia* naturalized there: *O. ficus-indica* (79.2%), *Opuntia engelmannii* Salm-Dyke (57.5%), *O. stricta* (55.0%), *Opuntia leucotrichta* DC. (29.2%), *O. fulgida* Englm. (24.2%), and *O. imbricata* (Haworth) de Candolle (16.7%) (Mafokoane et al., 2007)

Possible methods to reduce effects of *C. cactorum* on native cacti include classical biological control through the introduction of specialized parasitoids from its native range (Pemberton and Cordo, 2001) and the use of sterile insect techniques to eradicate populations from the infested parts of the continental United States (Carpenter et al., 2001; Hight et al., 2005), followed by island-by-island eradication efforts.

On Nevis and St. Kitts, where *C. cactorum* was introduced more than 50 years ago, *C. cactorum* was detected at 10 of 16 sites surveyed on the two islands, where it attacked 23% of total plants. Of the four *Opuntia* sensu lato species found, three were attacked, including *O. triacantha* (Willd.) Sweet (with 16% of plants and 9% of pads attacked), *O. stricta* (with 44% plants and 8% pads attacked), and *Opuntia cochenillifera* (L.) P. Mill. (with 19% plants and 1% pads attacked). The native non-target tree pear, *Consolea* (= *Opuntia*) *rubens* (Salm-Dyck ex DC.) Lemaire, was not attacked. Larger *Opuntia* plants in wildlands had significantly higher percentages of attack than landscape plants (Pemberton and Liu, 2007). However, in laboratory no-choice starvation tests with North American species of *Opuntia*, of 14 species tested, larval survival was highest on *C. rubescens* and *Opuntia streptacantha* Lemaire (Jezorek et al., 2010).

The likely future impacts on particular *Opuntia* species are still unclear. Marked *O. stricta* ($n = 253$) and *O. humifusa* ($n = 327$) plants along the west coast of Florida were checked for six years to determine the effects of *C. cactorum* attack on the survival and growth rate of plants: 78% of the plants were attacked during the study and the overall survival rate was 76%. Plants attacked by *C. cactorum* were more likely to die than those that were not, and a plant's odds of surviving decreased as attack frequency increased. However, plants that survived the six-year period showed, on average, positive growth and there was no significant difference in growth rates between surviving attacked and unattacked plants (Jezorek et al., 2012).

In summary, the introduction of this species to the Caribbean was unjustified, even given the information and perspective of the period. The decision seems to have been taken with no regard other than possible economic gains for a few island planters whose pastures had become overrun with several *Opuntia* species. The knowledge that high cactus densities were linked to overgrazing was not used to solve the problem through better pasture management. Also ignored were potential economic losses from attack of the moth on production of spineless *Opuntia* for fodder or human food (nopales) and the likely damage to native *Opuntia*. Enough information was available at the time of release that the person(s) responsible for the release could have or should have known that the moth had a broad host range within *Opuntia*. No written records were found within CAB records for the West Indian station providing any insights into how the decision to make this release was taken or if it was subject to any outside review. This case argues strongly for more consultation before releases are made, so that countries with distinct interests (especially Mexico) from those planning the release (the British colonial board for the West Indies) can make their concerns known.

Spotted-wing drosophila, in Hawaii, a hypothetical case (Rank 5: unacceptably high risk)

This case, biological control of the invasive spotted-wing drosophila (*Drosophila suzukii* Matsumura) in Hawaii, is hypothetical. This drosophila species is an important pest of cherries and other fruit crops in its native range in Japan (Kanzawa, 1936). By 1980, this species had reached Hawaii, where it became a dominant drosophila species in riparian habitats (Kido et al., 1996). Later, the spotted-wing drosophila extended its invasion to North America (California, 2008) and Europe (Italy, 2009), attacking many soft fruits (Hauser, 2011; Walsh et al., 2011). Damage to these diverse crops stimulated work on its control, with options including pesticides, traps, and local native natural enemies. Natural enemies present in the invaded area so far have had little effect on spotted-wing drosophila populations. In Europe, of the five main drosophila parasitoids, only two pupal parasitoids with wide host ranges develop on *D. suzukii*. Two specialized larval parasitoids examined were unable to develop, presumably because of a strong immune response from the host (Chabert et al., 2012). In Catalunia (Spain), four Hemiptera predator species were found in infested fruit samples: *Orius laevigatus* (Fieber), *Cardiastethus nazarenus* Reuter, *C. fasciventris* (Garbillietti), and *Dicyphus tamaninii* Wagner (Arnó et al.,

2012), although these seem to have little chance of being specialists even at the level of Drosophilidae.

In the spotted-wing drosophila's native range, its larvae are parasitized by a diapriid in the genus *Phaenopria* (Kanzawa, 1939) and a species of *Ganaspis* (Hymenoptera: Figitidae) (Kasuya et al., 2013). While none of these parasitoids show particular promise for control of spotted-wing drosophila given the importance of the species as a pest in Japan, it is possible that a wider, more intensive search might locate better agents. The question would then arise: Should they be introduced in Hawaii or anywhere else? The greatest risk is to Hawaiian drosophila but even an introduction in Europe or North America might lead to the agent invading Hawaii on its own, using the same pathways that that pest used to spread so widely.

Reasons to reject any proposed classical biological control program against spotted-wing drosophila include:

1 The Hawaiian radiation of *Drosophila* species is one of the most species-rich evolutionary radiations in the world and is of great scientific interest (Kambysellis et al., 1995). This group, because of its high species diversity, marked sexual dimorphism and complex mating behavior, host-plant specificity, and the well-known chronology of the Hawaiian Archipelago's formation, is an excellent model system for evolutionary studies (O'Grady et al., 2011). Preservation of this radiation is thus important as a *supporting ecosystem service* (*sensu* Charles and Dukes, 2007).

2 The size of the radiation is immense, likely 1000 species or more, many of which are still being discovered and described (e.g., Magnacca and O'Grady, 2008). Therefore, attempts to assess the host range of a candidate parasitoid would be quite difficult.

3 Many members of the group are rare species with very small home ranges (e.g., Kambysellis et al., 2000), whose diets would likely either be unknown or hard to reproduce in the laboratory, making it very difficult to create the laboratory colonies needed for host-range testing.

4 Even if the above issues and technical problems could be addressed, other approaches such as suitable pesticides or traps are likely to be more tractable.

5 The potential risks, which would likely damage relations with conservation biologists, would simply not be worth taking.

Acknowledgments

We thank Charlotte Causton and David Wagner for reviews of Chapter 4.

References

Acharya, C. B. 2010. Forest Invasive Plant Management in the Northeastern U.S. M.S. Thesis, Cornell University, Ithaca, New York, USA.

Ahlers, D. and D. Moore. 2009. A review of vegetation and hydrologic parameters associated with the southwestern willow flycatcher – 2002 to 2008, Elephante Butte Reservoir Delta, New Mexico. USDI-BOR Technical Service Center, Denver, USA. Available from: https://www.ntis.gov/Search/Home/titleDetail/?abbr=PB2009111575 [Accessed January 2016].

Ailstock, M. S., C. M. Norman, and P. J. Bushmann. 2001. Common reed *Phragmites australis*: control and effects upon biodiversity in freshwater nontidal wetlands. *Restoration Ecology* 9: 49–59.

Annecke, D. P., W. A. Burger, and H. Coetzee. 1976. Pest status of *Cactoblastis cactorum* (Berg) (Lepidoptera: Phycitidae) and *Dactylopius opuntiae* (Cockerell) (Coccoidea: Dactylopiidae) in spineless opuntia plantations in South Africa. *Journal of the Entomological Society of Southern Africa* 39: 111–116.

Anon. 1984. Ropewick applicators allow selective control of Phragmites. *Bureau of Sugar Experiment Stations Bulletin* no. 6, p. 10.

Arnett, A. E. and S. M. Louda. 2002. Re-test of *Rhinocyllus conicus* host specificity, and the prediction of ecological risk in biological control. *Biological Conservation* 106: 251–257.

Arnó, J., J. Riudavets, and R. Gabarra. 2012. Survey of host plants and natural enemies of *Drosophila suzukii* in an area of strawberry production in Catalonia (northeast Spain). *IOBC/WPRS Bulletin* 80: 29–34.

Balciunas, J. 2007. *Lixus cardui*, a biological control agent for scotch thistle (*Onopordum acanthium*): safe for Australia but not for the USA? *Biological Control* 41: 134–141.

Balciunas, J. K., D. W. Burrows, and M. F. Purcell. 1994. Field and laboratory host ranges of the Australian weevil *Oxyops vitiosa* (Coleoptera: Curculionidae), a potential biological control agent for the paperbark tree, *Melaleuca quinquenervia*. *Biological Control* 4: 351–360.

Baum, B. R. 1967. Introduced and naturalized tamarisks in the United States and Canada (Tamaricaceae). *Baileya* 15: 19–25.

Bean, D. W., D. J. Kazmer, K. Gardner, et al. 2013. Molecular genetic and hybridization studies of *Diorhabda* spp. released for biological control of *Tamarix*. *Invasive Plant Management* 6: 1–15.

Beck, R. A. 1971. Phragmites control for urban, industrial and wildlife needs. *Proceedings of the Northeastern Weed Science Society* 25: 89–90.

Belokobylskij, S. A., G. I Yurchenko, J. S. Strazanac, et al. 2012. A new emerald ash borer (Coleoptera: Buprestidae) parasitoid species of *Spathius* Nees (Hymenoptera: Braconidae: Doryctinae) from the Russian Far East and South Korea. *Annals of the Entomological Society of America* 105: 165–178.

Bennett, F. D. 1971. Some recent successes in the field of biological control in the West Indies. *Revista Peruana Entomologia* 14(2): 369–373.

Bennett, C. A. and R. W. Pemberton. 2008. *Neomusotima fuscolinealis* (Lepidoptera: Pyralidae) is an unsuitable biological control agent of *Lygodium japonicum*. *Florida Entomologist* 91: 26–29.

Bevill, R. L., S. M. Louda, and L. M. Stanforth. 1999. Protection from natural enemies in managing rare plant species. *Conservation Biology* 13: 1323–1331.

Blossey, B. and J. McCauley. 2000. A plan for developing biological control of *Phragmites australis* in North America. *The Wetlands Journal* 12: 23–28.

Blossey, B., V. Nuzzo, H. Hinz, and E. Gerber. 2001. Developing biological control of *Alliaria petiolata* (M. Bieb.) Cavara and Grande (garlic mustard). *Natural Areas Journal* 21: 357–367.

Boldt, P. E. and J. A. Jackman. 1993. Establishment of *Rhinocyllus conicus* Froelich on *Carduus macrocephalus* in Texas. *Southwestern Entomologist* 18: 173–181.

Boldt, P. E. and L. T. Kok. 1982. Bibliography of *Rhinocyllus conicus* Froel. (Coleoptera: Curculionidae), an introduced weevil for biological control of *Carduus* and *Silybum* thistles. *Bulletin of the Entomological Society of America* 28 (4): 355–358.

Bowles, M. L., K. A. Jacobs, and J. L. Mengler. 2007. Long-term changes in an oak forest's woody understory and herb layer with repeated burning. *Journal of the Torrey Botanical Society* 134: 223–237.

Buckingham, G. R. 2004. Paper-bark tree alters habitats in Florida, p. 172. *In*: Wittenberg R. and M. J. W. Cock (eds.). *Invasive Alien Species: A Toolkit of Best Prevention and Management Practices*. CAB International. Wallingford, UK. 228 pp.

Buntin, G. D., R. D. Hudson, and T. R. Murphy.1993. Estabishment of *Rhinocyllus conicus* (Coleoptera: Curculionidae) in Georgia for control of musk thistle. *Journal of Entomological Science* 28: 213–217.

Busch, D. E. and S. D. Smith. 1995. Mechanisms associated with decline of woody species in riparian ecosystems of the southwestern U.S. *Ecological Monographs* 65: 347–370.

Carpenter, J. E., K. A. Bloem, and S. Bloem, S. 2001. Applications of F$_1$ sterility for research and management of *Cactoblastis cactorum* (Lepidoptera: Pyralidae). *Florida Entomologist* 84: 531–536.

Castellano, S. M. and D. L. Gorchov. 2012. Reduced ectomycorrhizae on oak near invasive garlic mustard. *Northeastern Naturalist* 19: 1–24.

Center, T. D., P. D. Pratt, P. W. Tipping, et al. 2007. Initial impacts and field validation of host range for *Boreioglycaspis melaleucae* Moore (Hemiptera: Psyllidae), a biological control agent of the invasive tree *Melaleuca quinquenervia* Blake (Myrtales:

Mytaceae: Leptospermoideae). *Environmental Entomology* **36**: 569–576.

Center, T. D., P. D. Pratt, P. W. Tipping, et al. 2008. Biological control of *Melaleuca quinquenervia*: goal-based assessment of success, pp. 655–664. *In*: Julien, M. H., R. Sforza, M. C. Bon, et al. (eds.). *Proceedings of the XII International Symposium on Biological Control of Weeds, La Grande Motte, France, 22–27 April, 2007.* CAB International. Wallingford, U.K.

Center, T. D., M. F. Purcell, P. D. Pratt, et al. 2012. Biological control of *Melaleuca quinquenervia*: an Everglades invader. *Biological Control* **57**: 151–165.

Chabert, S., R. Allemand, M. Poyet, P. Eslin, and P. Gibert. 2012. Ability of European parasitoids (Hymenoptera) to control a new invasive Asiatic pest, *Drosophila suzukii*. *Biological Control* **63**: 40–47.

Charles, H. and J. S. Dukes. 2007. Impacts of invasive species on ecosystem services, pp. 217–233. *In*: Nentwig, W. (ed.). *Biological Invasions. Ecological Studies*, vol. **193**. Springer-Verlag. Berlin, Germany.

Christensen, E. M. 1962. The rate of naturalization of *Tamarix* in Utah. *American Midland Naturalist* **68**: 51–57.

Cohan, D. R., B. W. Anderson, and R. D. Omart. 1979. Avian population responses to salt cedar along the lower Colorado River, pp. 371–382. *In*: Johnson, R. R. and J. F. McCormick (eds.). *National Symposium on Strategies for Protection and Management of Floodplain Wetlands and other Riparian Ecosystems, 11–13 December, 1978, Pine Mountain, Georgia.* General Technical Report WO-12. USDA Forest Service. Available from: https://archive.org/details/CAT89908686 [Accessed January 2016].

Cole, R. J., C. M. Litton, M. J. Koontz, and R. K. Loh. 2012. Vegetation recovery 16 years after feral pig removal from a wet Hawaiian forest. *Biotropica* **44**: 463–471.

Côté, S. D., T. P. Rooney, J.-P. Tremblay, et al. 2004. Ecological impacts of deer overabundance. *Annual Review of Ecology and Systematics* **35**: 113–147.

Crins, W. L. 1989. The Tamaricaceae in the southeastern United States. *Journal of the Arnold Arboretum* **70**: 403–425.

Cripps, M. G., A. Gassmann, S. V. Fowler, et al. 2011. Classical biological control of *Cirsium arvense*: Lessons from the past. *Biological Control* **57**: 165–174.

Dávalos, A. and B. Blossey. 2004. Influence of the invasive herb garlic mustard (*Alliaria petiolata*) on ground beetle (Coleoptera: Carabidae) assemblages. *Environmental Entomology* **33**: 564–576.

Dávalos, A., E. Simpson, V. Nuzzo, and B. Blossey. 2015. Non-consumptive effects of native deer on introduced earthworm abundance. *Ecosystems* **18**: 1029–1042.

Davis, A. S., D. A. Landis, V. Nuzzo, et al. 2006. Demographic models inform selection of biological control agents for garlic mustard (*Alliaria petiolata*). *Ecological Applications* **16**: 2399–2410.

Decary, R. 1965. Some spreading or noxious plants of Madagascar. *Journal d'Agriculture Tropicale* **12** (6/7/8): 343–350.

Delnatte, C. and J. Y. Meyer. 2012. Plant introduction, naturalization, and invasion in French Guiana (South America). *Biological Invasions* **14**: 915–927.

DeLoach, C. J., D. Gerling, L. Fornasari, et al. 1996. Biological control programme against saltcedar (*Tamarix* spp.) in the United States of America: progress and problems, pp. 253–260. *In*: Moran, V. C. and J. H. Hoffmann (eds.). *Proceedings of the 9th International Symposium on Biological Control of Weeds, Stellenbosch, South Africa, 19–26 January, 1996.* University of Rondebosch and the University of Cape Town. Cape Town, South Africa.

DeLoach, C. J., R. I. Carruthers, J. Lovich, et al. 2000. Ecological interactions in the biological control of saltcedar (*Tamarix* spp.) in the U.S.: Toward a new understanding, pp. 819–874. *In*: Spencer, N. R. (ed.). *Proceedings of the X International Symposium on Biological Control of Weeds, July 1999, Bozeman, Montana.* Montana State University. Bozeman, Montana, USA.

DePringer-Levin, M. E., T. A. Grant III, and C. Dawson. 2010. Impacts of the introduced biological control agent, *Rhinocyllus conicus* (Coleoptera: Curculionidae), on the seed production and population dynamics of *Cirsium ownbeyi* (Asteraceae), a rare, native thistle. *Biological Control* **55**: 79–84.

Desrochers, A. M., J. F. Bain, and S. I. Warwick. 1988. A biosystematics study of the *Carduus nutans* complex in Canada. *Canadian Journal of Botany* **66**: 1621–1631.

Dhileepan, K., M. Treviño, G. P. Donnelly, and S. Raghu. 2005. Risk to non-target plants from *Charidotis auroguttata* (Chrysomelidae: Coleoptera), a potential biological control agent for cat's claw creeper, *Macfadyena unguis-cati* (Bignoiaceae), in Australia. *Biological Control* **32**: 450–460.

Dickel, T. S. 1991. *Cactoblastis cactorum* in Florida (Lepidoptera: Pyralidae: Phycitinae). *Tropical Lepidoptera* **2**(2): 117–118.

Dobson, A. and B. Blossey. 2015. Earthworm invasion, white-tailed deer and seedling establishment in deciduous forests of north-eastern North America. *Journal of Ecology* **103**: 153–164.

Dodd, A. P. 1936. Host restriction and host preference, more particularly among cactus insects. *Proceedings of the Entomological Society of Queensland* **1936**, pp. 1–7.

Duan, J. J., M. D. Ulyshen, L. S. Bauer, et al. 2010. Measuring the impact of biotic factors on populations of immature emerald ash borer (Coleoptera: Buprestidae). *Environmental Entomology* **39**: 1513–1522.

Duan, J. J., L. S. Bauer, M. D. Ulyshen, et al. 2011. Development of methods for the field evaluation of *Oobius agrili* (Hymenoptera: Encyrtidae) in North America, a newly introduced egg parasitoid of the emerald ash borer (Coleoptera: Buprestidae). *Biological Control* **56**: 170–174.

Duan, J. J., G. Yurchenko, and R. Fuester. 2012a. Occurrence of emerald ash borer (Coleoptera: Buprestidae) and biotic factors affecting its immature stages in the Russian Far East. *Environmental Entomology* **41**: 245–254.

Duan, J. J., L. S. Bauer, K. J. Abell, and R. G. Van Driesche. 2012b. Population responses of hymenopteran parasitoids to the emerald ash borer (Coleoptera: Buprestidae) in recently invaded areas in north central United States. *Biological Control* **57**: 199–209.

Duan, J. J., L. Bauer, K. J. Abell, and R. Van Driesche. 2012c. Population responses of hymenoperan parasitoids to the emerald as borer (Coleoptera: Buprestidae) in recently invaded areas in Michigan. *Biological Control* **57**: 199–209.

Duan, J. J., L. Bauer, K. J. Abell, et al. 2013. Establishment and abundance of *Tetrastichus planipennisi* (Hymenoptera: Eulophidae) in Michigan. Potential for success in classical biological control of the invasive emerald ash borer (Coleoptera: Buprestidae). *Journal of Economic Entomology* **106**: 1145–1154.

Duan, J. J., K. J. Abell, L. S. Bauer, et al. 2014. Natural enemies implicated in the regulation of an invasive pest: a life table analysis of the population dynamics of the emerald ash borer. *Agricultural and Forest Entomology* **79**: 36–42. doi: 10.1111/afe.12070

Duan, J. J., L. S. Bauer, K. J. Abell, et al. 2015. Population dynamics of an invasive forest insect and associated natural enemies in the aftermath of invasion: implications for biological control. *Journal of Applied Ecology* **52**: 1246–1254.

Dudley, T. L. and D.W. Bean. 2012. Tamarisk biological control, endangered species risk, and resolution of conflict through riparian restoration. *Biological Control* **57**: 331–347.

Dudley, T. L. and C. J. DeLoach. 2004. Saltcedar (*Tamarix* spp.), endangered species, and biological control – can they mix? *Weed Technology* **18**: 1542–1551.

Dudley, T. L. and D. J. Kazmer. 2005. Field assessment of the risk posed by *Diorhabda elongata*, a biological control agent for control of saltcedar (*Tamarix* spp.), to a nontarget plant, *Frankenia salina*. *Biological Control* **35**: 265–275.

Dudley, T. L., C. J. DeLoach, H. Lovich, and R. I. Carruthers. 2000. Saltcedar invasion of western riparian areas: impacts and new prospects for control, pp. 345–381. *In: Transactions of the 65th North American Wildlife and Natural Resources Conference, March 2000, Chicago, Illinois.* Wildlife Management Institute, Washington, DC.

Drus, G. M., T. L. Dudley, M. L. Brooks, and J. R. Matchett. 2012. Influence of biological control on fire intensity profiles of tamarisk (*Tamarix ramosissima* Lebed.). *International Journal of Wildland Fire* **22**: 446–458.

Emerald Ash Borer Information, 2013. Emerald ash borer information published online at http://emeraldashborer.info/ [Accessed January 2016].

Everitt, B. L. 1980. Ecology of saltcedar: A plea for research. *Environmental Geology* **3**: 77–84.

Fell, P. E., R. S. Warren, J. K. Light, et al. 2003. Comparison of fish and macroinvertebrate use of *Typha angustifolia*, *Phragmites australis*, and treated *Phragmites* marshes along the lower Connecticut River. *Estuaries* **26** (2B): 534–551.

Finch, D. M., S. I. Rothstein, J. C. Boren, et al. 2002. Final recovery plan: southwestern willow flycatcher (*Empidonax traillii extimus*). *Albuquerque, New Mexico: Region 2*, U.S Fish and Wildlife Service, 10 pp.

Florens, F. B. V., J. R. Mauremootoo, S. V. Fowler, et al. 2010. Recovery of indigenous butterfly community following control of invasive alien plants in a tropical island's wet forests. *Biodiversity and Conservation* **19**: 3835–3848.

Fullaway, D. T. 1958. Biological control of *Opuntia megacantha* and *Lantana camara* in Hawaii, pp. 549–552. *In*: Becker, E. C. (ed.). *Proceedings of the 10th International Congress of Entomology, Montreal 1956*. [Publisher not identified.]

Garcia-Tuduri, J. C., L. F. Martorell, and S. Medina Gaud. 1971. Geographical distribution and host plants of the cactus moth, *Cactoblastis cactorum* (Berg) in Puerto Rico and the United States Virgin Islands. *Journal of Agriculture of the University of Puerto Rico* **55** (1): 130–134

Gassmann, A. and S. M. Louda. 2001. *Rhinocyllus conicus*: initial evaluation and subsequent ecological impacts in North America, pp. 147–183. *In*: Wajnberg, E., J. K. Scott, and P. C. Quimby (eds.). *Evaluating Indirect Ecological Effects of Biological Control*. CAB International Publishing. Wallingford, UK.

Gerber, E., H. L. Hinz, and B. Blossey. 2007. Interaction of specialist root and shoot herbivores of *Alliaria petiolata* and their impact on plant performance and reproduction. *Ecological Entomology* **32**: 357–365.

Gerber, E., H. L. Hinz, and B. Blossey. 2008. Pre-release impact assessment of two stem-boring weevils proposed as biological control agents for *Alliaria petiolata*. *Biological Control* **45**: 360–367.

Gould, J. 2007. Proposed release of three parasitoids for biological control of the emerald ash borer (*Agrilus planipennis*) in the continental United States. *Environmental Assessment, April 2, 2007*. USDA APHIS. Available from: http://www.emeraldashborer.info/files/07-060-1%20ea_leah_environassess.pdf [Accessed January 2016].

Grant, J. and P. L. Lambdin. 1993. Release of plant-feeding weevils for biological control of musk thistle in Tennessee. *Tennessee Farm and Home Science* **165**: 26–29.

Gratton, C. and R. F. Denno. 2005. Restoration of arthropod assemblages in a *Spartina* salt marsh following removal of the invasive plant *Phragmites australis*. *Restoration Ecology* **13**: 358–372.

Haack, R. A., E. Jendek, H.-P. Liu, et al. 2002. The emerald ash borer: a new exotic pest in North America. *Newsletter of the Michigan Entomological Society* **47**(3–4): 1–5.

Hale, A. N., S. J. Tonsor, and S. Kalisz. 2011. Testing the mutualism disruption hypothesis: physiological mechanisms for invasion of intact perennial plant communities. *Ecosphere* **2** (10) art110.

Hauser, M. 2011. A historic account of the invasion of *Drosophila suzukii* (Matsumura) (Diptera: Drosophilidae) in the continental United States, with remarks on their identification. *Pest Management Science* **67**: 1352–1357.

Heard, T. A., C. W. O'Brien, I. W. Forno, and J. A. Burcher. 1998. *Chalcodermus persimilis* O'Brien n. sp. (Coleoptera: Curculionidae): description, biology, host range, and suitability for biological control of *Mimosa pigra* L. (Mimosaceae). *Transactions of the American Entomological Society* **124** (1):1–11.

Hernández, L. R. and T. C. Emmel. 1993. *Cactoblastis cactorum* in Cuba (Lepidoptera: Pyralidae: Phycitinae). *Tropical Lepidoptera* **4** (1): 45–46.

Herr, J. C. 2000. Evaluating non-target effects: the thistle story, pp. 12–17. In: Hoddle, M. (ed.). *California Conference on Biological Control II. The Historic Mission Inn, Riverside, California, 11–12 July, 2000.* Center for Biological Control, College of Natural Resources, University of California, Berkeley, California.

Herr, J. C., R. I. Carruthers, D. W. Bean, et al. 2009. Host preference between saltcedar (*Tamarix* spp.) and native non-target *Frankenia* spp. within the *Diorhabda elongata* species complex (Coleoptera: Chrysomelidae). *Biological Control* **51**: 337–345.

Hight, S. D., J. E. Carpenter, K. A. Bloem, et al. 2002. Expanding geographical range of *Cactoblastis cactorum* (Lepidoptera: Pyralidae) in North America. *Florida Entomologist* **85**: 527–529.

Hight, S. D., J. E. Carpenter, S. Bloem, and K. S. Bloem. 2005. Developing a sterile insect release program for *Cactoblastis cactorum* (Berg) (Lepidoptera: Pyralidae): effective overflooding ratios and release-recapture field studies. *Environmental Entomology* **34**: 850–856.

Hudgeons, J. L., A. E. Knutson, K. M. Heinz, et al. 2007. Defoliation by introduced *Diorhabda elongata* leaf beetles (Coleoptera: Chrysomelidae) reduces carbohydrate reserves and regrowth of *Tamarix* (Tamaricaceae). *Biological Control* **43**: 213–221.

Hughes, R. F. and J. S. Denslow. 2005. Invasion by a N2-fixing tree alters function and structure in wet lowland forests of Hawaii. *Ecological Applications* **15**: 1615–1628.

Jezorek, H. A., P. D. Stiling, and J. E. Carpenter, J. E. 2010. Targets of an invasive species: oviposition preference and larval performance of *Cactoblastis cactorum* (Lepidoptera: Pyralidae) on 14 North American opuntioid cacti. *Environmental Entomology* **39**: 1884–1892.

Jezorek, H., A. J. Baker, and P. Stiling. 2012. Effects of *Cactoblastis cactorum* on the survival and growth of North American *Opuntia*. *Biological Invasions* **14**: 2355–2367.

Johnson, D. M. and P. D. Stiling. 1996. Host specificity of *Cactoblastis cactorum* (Lepidoptera: Pyralidae), an exotic *Opuntia*-feeding moth, in Florida. *Environmental Entomology* **25**: 743–748.

Julian, P., II, E. M. Everham III, and M. B. Main. 2012. Influence of a large-scale removal of an invasive plant (*Melaleuca quinquenervia*) on home-range size and habitat selection by female Florida panthers (*Puma concolor coryi*) within big Cypress National Preserve, Florida. *Southeastern Naturalist* **11**: 337–348.

Kalisz, S., R. Spigler, and C. Horvitz. 2014. In a long-term experimental demography study, excluding ungulates reversed invader's explosive population growth rate and restored natives. *Proceedings of the National Academy of Sciences of the United States of America* **111**: 4501–4506.

Kambysellis, M. P., K-F. Ho, E. M. Craddock, et al. 1995. Pattern of ecological shifts in the diversification of Hawaiian *Drosophila* inferred from a molecular phylogeny. *Current Biology* **5**: 1129–1139.

Kambysellis, M. P., E. M. Craddock, S. L. Montgomery, et al. 2000. Ecology and evolution of *Drosophila ambochila*, a rare picture-winged species endemic to the Wai'anae range of O'ahu, Hawaiian Islands. *Pacific Science* **54**: 169–181.

Kanzawa, T. 1936. Studies on *Drosophila suzukii* Mats. *Journal of Plant Protection* **23**: 66–70; 127–132; 183–191.

Kanzawa, T. 1939. Studies on *Drosophila suzukii* Mats. Yamanashi Agricultural Experiment Station. Yamanashi, Japan. 49 pp.

Kasuya, N., H. Mitsui, S. Ideo, et al. 2013. Ecological, morphological and molecular studies on *Ganaspis* individuals (Hymenoptera: Figitidae) attacking *Drosophila suzukii* (Diptera: Drosophilidae). *Applied Entomology and Zoology* **48**(1): 87–92.

Kido, M. H., A. Asquith, and R. I. Vargas. 1996. Nontarget insect attraction to methyl eugenol traps used in male annihilation of the oriental fruit fly (Diptera: Tephritidae) in riparian Hawaiian stream habitat. *Environmental Entomology* **25**: 1279–1289.

Kok, L. T. 1976. Biological control of *Carduus* thistles in the northeastern U.S.A., pp. 101–104. In: Freeman, T. E. (ed.). *Proceedings of the IV International Symposium on Biological Control of Weeds, Gainesville, Florida.* Center for Environmental Programs, Institute of Food and Agricultural Sciences, University of Florida, Gainesville, Florida, USA.

Kok, L. T., T. J. McAvoy, and W. T. Mays. 2004. Biological control of *Carduus* thistles in Virginia – a long-term perspective, three decades after the release of the exotic weevils, pp. 554–558. In: Cullen, J. M., D. T. Briese, D. J. Kriticos, et al. (eds.). *Proceedings of the XI International Symposium on Biological Control of Weeds, Canberra, Australia, 27 April–2 May, 2003.* CSIRO. Canberra, Australia.

Kulesza, A. E., J. R. Holomuzki, and D. M. Klarer. 2008. Benthic community structure in stands of *Typha angustifolia* and herbicide-treated and untreated *Phragmites australis*. *Wetlands* **28**: 40–56.

Lankau, R. 2010. Soil microbial communities alter allelopathic competition between *Alliaria petiolata* and a native species. *Biological Invasions* **12**: 2059–2068.

Lankau, R., V. Nuzzo, G. Spyreas, and D. Davis. 2009. Evolution limits ameliorate the negative impact of an invasive plant. *Proceedings of the National Academy of Sciences* **106**: 15363.

Laroche, F. B. 1998. Managing melaleuca (*Melaleuca quinquenervia*) in the Everglades. *Weed Technology* **12**: 726–732.

Lewis, P. A., C. J. DeLoach, J. C. Herr, et al. 2003. Assessment of risk to native *Frankenia* shrubs from an Asian leaf beetle, *Diorhabda elongata deserticola* (Coleoptera: Chrysomelidae), introduced for biological contro of saltcedars (*Tamarix* spp.) in the western United States. *Biological Control* **27**: 148–166.

Liu, H-P., L. S. Bauer, R. Gao, et al. 2003. Exploratory survey for the emerald ash borer, *Agrilus planipennis* (Coleoptera: Buprestidae), and its natural enemies in China. *The Great Lakes Entomologist* **36**: 191–204.

Liu, H., L. S. Bauer, D. L. Miller, et al. 2007. Seasonal abundance of *Agrilus planipennis* (Coleoptera: Buprestidae) and its natural enemies *Oobius agrili* (Hymenoptera: Encyrtidae) and *Tetrastichus planipennisi* (Hymenoptera: Eulophidae) in China. *Biological Control* **42**: 61–71.

Longland, W. S. and T. L. Dudley. 2008. Effects of a biological control agent on the use of saltcedar habitat by passerine birds. *Great Basin Birds* **10**: 21–26.

Louda, S. M. 1998. Population growth of *Rhinocyllus conicus* (Coleoptera: Curculionidae) on two species of native thistles in Prairie. *Environmental Entomology* **27**: 834–841.

Louda, S. M. 2000. *Rhinocyllus conicus* – insights to improve predictability and minimize risk of biological control of weeds, pp. 187–193. *In*: Spencer, N. R. (ed.). *Proceedings of the X International Symposium on Biological Control of Weeds, Bozeman, Montana, 4–14 July, 1999*. Montana State University. Bozeman, Montana.

Louda, S. M., D. Kendall, J. Connor, and D. Simberloff. 1997. Ecological effects of an insect introduced for biological control of weeds. *Science* **277** (5329): 1088–1090.

Louda, S. M., A. E. Arnett, T. A. Rand, and F. L. Russell. 2003. Invasiveness of some biological control insects and adequacy of their ecological risk assessment and regulation. *Conservation Biology* **17**: 73–82.

Louda, S. M., T. A. Rand, A. E. Arnett, et al. 2005a. Evaluation of ecological risk to populations of a threatened plant from an invasive biological control insect. *Ecological Applications* **15**: 234–249.

Louda, S. M., T. A. Rand, F. L. Russell, and A. E. Arnett. 2005b. Assessment of ecological risks in weed biological control: input from retrospective ecological analyses. *Biological Control* **35**: 253–264.

Louda, S. M., T. A. Rand, A. A. R. Kula, et al. 2011. Priority resource access mediates competitive intensity between an invasive weevil and native floral herbivores. *Biological Invasions* **13**: 2233–2248.

Mafokoane, L. D., H. G. Zimmermann, and M. P. Hill. 2007. Development of *Cactoblastis cactorum* (Berg) (Lepidoptera: Pyralidae) on six North American *Opuntia* species. *African Entomology* **15**(2): 295–299.

Magnacca, K. N. and P. M. O'Grady. 2008. Revision of the 'nudidrosophila' and 'ateledrosophila' species groups of Hawaiian *Drosophila* (Diptera: Drosophilidae), with descriptions of twenty-two new species. *Systematic Entomology* **33**: 395–428.

Marks, M., B. Lapin, and J. Randall. 1994. *Phragmites australis* (*P. communis*): threats, management and monitoring. *Natural Areas Journal* **14**: 285–294.

Martin, L. J. and B. Blossey. 2013. The runaway weed: costs and failures of *Phragmites australis* management in the USA. *Estuaries and Coasts* **36**: 626–632.

Martin, M. R., P. W. Tipping, K. R. Reddy, et al. 2010. Interactions of biological and herbicidal management of *Melaleuca quinquenervia* with fire: consequences for ecosystem services. *Biological Control* **54**: 307–315.

McCarty, M. K. 1978. The genus *Carduus* in the United States, pp. 7–10. *In*: Frick, K. E. (ed.). *Biological Control of Thistles in the Genus Carduus in the United States. A Progress Report*. U.S. Department of Agriculture. Stoneville, Mississippi.

McCullough, D. G. 2015. Other options for emerald ash borer management: eradication and chemical control, pp. 75–82. *In*: Van Driesche, R. G. and R. Reardon (eds.). *The Biology and Control of Emerald Ash Borer*. FHTET 2014-09. USDA Forest Service. Morgantown, West Virginia, USA.

Milbrath, L. R. and C. J. DeLoach. 2006a. Host specificity of different populations of the leaf beetle *Diorhabda elongata* (Coleoptera: Chrysomelidae), a biological control agent of saltcedar (*Tamarix* spp.). *Biological Control* **36**: 32–48.

Milbrath, L. R. and C. J. DeLoach. 2006b. Acceptability and suitability of athel, *Tamarix aphylla*, to the leaf beetle *Diorhabda elongata* (Coleoptera: Chrysomelidae), a biological control agent of saltcedar (*Tamarix* spp.). *Biological Control* **35**: 1379–1389.

Moore, R. J. and C. Frankton. 1969. Cyto-taxonomy of some *Cirsium* species of the eastern United States, with a key to eastern species. *Canadian Journal of Botany* **47**: 1257–1275.

Moore, R. J. and C. Frankton. 1974. *The Thistles of Canada*. Monograph Research Branch, Canada Department of Agriculture No. 10. Ottawa, Canada. 112 pp.

Moran, P. J. 2010. Lack of establishment of the Mediterranean tamarisk beetle, *Diorhabda elongata* (Coleoptera: Chrysomelidae), on athel (*Tamarix aphylla*) (Tamaricaceae) in south Texas. *Southwestern Entomologist* **35**: 129–145.

Moran, P. J., C. J. DeLoach, and A. E. Knutson. 2009. Leaf beetles lasso tamarisk without hurting the relatives in Texas and southwestern USA. *Biological Control News and Information* **30**(3): 50N–52N.

Moreira, I., A. Monteiro, and E. Sousa. 1999. Chemical control of common reed (*Phragmites australis*) by foliar herbicides under different spray conditions. *In*: Caffrey, J. M., P. R. F. Barrett, M. T. Ferreira, et al. (eds.). *Biology, Ecology and Management of Aquatic Plants. Proceedings of the 10th International Symposium on Aquatic Weeds*, European Weed Research Society. *Hydrobiologia* **415**: 299–304.

Motooka, P., G. Nagai, and L. Ching. 1983. Cut-surface application of glyphosate to control tropical brush species. Abstracts, 1983 Meeting of the Weed Science Society of America, p. 96.

Mozdzer, T. J., C. J. Hutto, P. A. Clarke, and D. P. Field. 2008. Efficacy of imazapyr and glyphosate in the control of non-native *Phragmites australis*. *Restoration Ecology* **16**: 221–224.

Nuzzo, V. A. 1991. Experimental control of garlic mustard *Alliaria petiolata* (Bieb.) Cavara and Grande in northern Illinois using fire, herbicide and cutting. *Natural Areas Journal* **11**: 158–167.

Nuzzo, V. 1999. Invasion pattern of herb garlic mustard (*Alliaria petiolata*) in high quality forests. *Biological Invasions* **1**: 169–179.

Nuzzo, V. A. 2006. *Ground layer vegetation and deer herbivory in the Big Woods and open and exclosed plots at the Fermi National*

Accelerator Laboratory, Batavia, Illinois. Report to the Fermi National Accelerator Laboratory, Batavia, Illinois. Richford, New York, USA.

Nuzzo, V. A., J. C. Maerz, and B. Blossey. 2009. Earthworm invasion as the driving force behind plant invasion and community change in northeastern North American forests. *Conservation Biology* **23**: 966–974.

Oelrichs, P. B., J. K. MacLeod, A. A. Seawright, and P. B. Grace. 2001. Isolation and identification of the toxic peptides from *Lophyrotoma zonalis* (Pergidae) sawfly larvae. *Toxicon* **39**: 1933–1936.

OEQC. 2011. 2011-11-08-FEA-Biological control-Strawberry-Guava. Available from: http://oeqc.doh.hawaii.gov/Shared%20Documents/EA_and_EIS_Online_Library/Statewide/2010s/2011-11-08-FEA-Biological control-Strawberry-Guava.pdf [Accessed January 2016].

O'Grady, P. M., R. T. Lapoint, J. Bonacum, et al. 2011. Phylogenetic and ecological relationships of the Hawaiian *Drosophila* inferred by mitochondrial DNA analysis. *Molecular Phylogenetics and Evolution* **58**: 244–256.

Pattison, R. R., C. M. D'Antonio, T. L. Dudley, et al. 2010. Early impacts of biological control on canopy cover and water use of the invasive saltcedar tree (*Tamarix* spp.) in western Nevada, USA. *Oecologia* **165**: 605–616.

Pemberton, R. 1995. *Cactoblastis cactorum* (Lepidoptera: Pyralidae) in the United States: an immigrant biological control agent or an introduction of the nursery industry? *American Entomologist* **41** (4): 230–232.

Pemberton, R. W. and H. A. Cordo. 2001. Potential and risks of biological control of *Cactoblastis cactorum* (Lepidoptera: Pyralidae) in North America. *Florida Entomologist* **84**: 513–526.

Pemberton, R. W. and H. Liu. 2007. Control and persistence of native *Opuntia* on Nevis and St. Kitts 50 years after the introduction of *Cactoblastis cactorum*. *Biological Control* **41**: 272–282.

Pettey, F. W. 1939. The protection of spineless cactus plantations against *Cactoblastis*. *Farming in South Africa* (No. 2). Pretoria, South Africa, 3 pp.

Pratt, P. D., V. Quevedo, L. Bernier, et al. 2005. Invasions of Puerto Rican wetlands by the Australian tree *Melaleuca quinquenervia*. *Caribbean Journal of Science* **41**: 42–54.

Pratt, P. D., M. B. Rayamajhi, L. Bernier, and T. D. Center. 2006. Geographic range expansion of *Boreioglycapis melaleucae* (Hemiptera: Psyllidae) to Puerto Rico. *Florida Entomologist* **89**: 529–533.

Pratt, P. D., M. B. Rayamajhi, T. D. Center, et al. 2009. The ecological host range of an intentionally introduced herbivore: a comparison of predicted versus actual host use. *Biological Control* **49**: 146–153.

Purcell, M. F. and J. A. Goolsby. 2005. Herbivorous insects associated with the paperbark *Melaleuca quinquenervia* and its allies: VI. Pergidae (Hymenoptera). *Australian Entomologist* **32**: 37–48.

Rand, T. A. and S. M. Louda. 2004. Exotic weed invasion increases the susceptibility of native plants to attack by a biological control herbivore. *Ecology* **85**: 1548–1554.

Rand, T. A. and S. M. Louda. 2006. Invasive insect abundance varies across the biographic distribution of a native host plant. *Ecological Applications* **16**: 877–890.

Rand, T. A. and S. M. Louda. 2012. Exotic weevil invasion increases floral herbivore community density, function, and impact on a native plant. *Oikos* **121**: 85–94.

Rand, T. A., F. L. Russell, and S. M. Louda. 2004. Local- vs. landscape-scale indirect effects of an invasive weed on native plants. *Weed Technology* **18** (suppl.): 1250–1254.

Rayamajhi, M. B., P. D. Pratt, T. D. Center, et al. 2009. Decline in exotic tree density facilitates increased plant diversity: the experience from *Melaleuca quinquenervia* invaded wetlands. *Wetlands Ecology and Management* **17**: 455–467.

Rayamajhi, M. B., P. D. Pratt, T. D. Center, and T. K. Van. 2010. Insects and a pathogen suppress *Melaleuca quinquenervia* cut-stump regrowth in Florida. *Biological Control* **53**: 1–8.

Rees, N. E. 1978. Interactions of *Rhinocyllus conicus* and thistles in the Gallatin Valley, pp. 31–38. *In*: Frick, K. E. (ed.). *Biological Control of Thistles in the Genus Carduus in the United States. A Progress Report.* U.S. Department of Agriculture, Stoneville, Mississippi, USA.

Riemer, D. N. 1976. Long-term effects of glyphosate applications to phragmites. *Journal of Aquatic Plant Management* **14**: 39–43.

Robinson, T. W. 1965. Introduction, spread, and areal extent of saltcedar (*Tamarix*) in the western states. *Report No 491-A*. U. S. Geological Survey. Washington, DC.

Rockermann, P. 2011. *Implications for Invasion by Emerald Ash Borer in New York: Ash Abundance in Riparian Areas and Moth Assemblages in Upland and Wetland Forests with High and Low Ash Densities.* M.S. Thesis. State University of New York, College of Environmental Science and Forestry, Syracuse, New York, USA.

Roduner, M., G. Cuperus, P. Mulder, et al. 2003. Successful biological control of the musk thistle in Oklahoma using the musk thistle head weevil and the rosette weevil. *American Entomologist* **49** (2): 112–120.

Rose, K. E., S. M. Louda, and M. Rees. 2005. Demographic and evolutionary impacts of native and invasive insect herbivores on *Cirsium canescens*. *Ecology* **86**: 453–465.

Russell, F. L. and S. M. Louda. 2004. Phenological synchrony affects interaction strength of an exotic weevil with Platte thistle, a native host plant. *Oecologia* **139**: 525–534.

Russell, F. L., K. E. Rose, and S. M. Louda. 2010. Seed availability and insect herbivory limit recruitment and adult density of native tall thistle. *Ecology* **91**: 3081–3093.

Saltonstall, K. 2002. Cryptic invasion by a non-native genotype of the common reed, *Phragmites australis*, into North America. *Proceedings of the National Academy of Science USA* **99**: 2445–2449.

Saltonstall, K., P. M. Peterson, and R. J. Soreng. 2004. Recognition of *Phragmites australis* subsp. *americanus* (Poaceae: Arundinoideae) in North America: evidence from morphological and genetic analyses. *SIDA* **21**: 683–692.

Sauer, S. A. and K. L. Bradley. 2008. First record for the biological control agent *Rhinocyllus conicus* (Coleoptera: Curculionidae) in a threatened native thistle, *Cirsium hillii* (Asteraceae), in Wisconsin, U.S.A. *Entomological News* **119**: 90–95.

Schroder, D. 1980. The biological control of thistles. *Biological Control News and Information* **1** (1): 9–26.

Serbesoff-King, K. 2003. Melaleuca in Florida: a literature review on the taxonomy, distribution, biology, ecology, economic importance and control measures. *Journal of Aquatic Plant Management* **41**: 98–112.

Shafroth, P. B., V. B. Beauchamp, M. K. Briggs, et al. 2008. Planning riparian restoration in the context of *Tamarix* control in western North America. *Restoration Ecology* **16**: 97–112.

Sher, A. and M. F. Quigley (eds.). 2013. *Tamarix: A Case Study of Ecological Change in the American West*. Oxford University Press. New York.

Silvers, C. S., P. D. Pratt, A. P. Ferriter, and T. D. Center. 2007. T.A.M.E. Melaleuca: a regional approach for suppressing one of Florida's worst weeds. *Journal of Aquatic Plant Management* **45**: 1–8.

Sime, K. 2002. Chemical defense of *Battus philenor* larvae against attack by the parasitoid *Trogus pennator*. *Ecological Entomology* **27**: 337–345.

Simmonds, F. J. and F. D. Bennett. 1966. Biological control of *Opuntia* spp. by *Cactoblastis cactorum* in the Leeward Islands (West Indies). *Entomophaga* **11**: 183–189.

Slaughter, B. S., W. W. Hochstedler, D. L. Gorchov, and A. M. Carlson. 2007. Response of *Alliaria petiolata* (garlic mustard) to five years of fall herbicide application in a southern Ohio deciduous forest. *Journal of the Torrey Botanical Society* **134**: 18–26.

Soberon, J., J. Golubov, and J. Sarukhán. 2001. The importance of *Opuntia* in Mexico and routes of invasion and impact of *Cactoblastis cactorum* (Lepidoptera: Pyralidae). *Florida Entomologist* **84**: 486–492.

Sogge, M. K., S. J. Sferra, and E. H. Paxton. 2008. *Tamarix* as habitat for birds: implications for riparian restoration in the southwestern United States. *Restoration Ecology* **16**: 146–154.

Stinson, K., S. Kaufman, L. Durbin, and F. Lowenstein. 2007. Impacts of garlic mustard invasion on a forest understory community. *Northeastern Naturalist* **14**: 73–88.

Stocker, R. K. and D. R. Sanders, Sr. 1981. Chemical control of *Melaleuca quinquenervia*, pp. 129–134. *In*: Anon. *Proceedings of Melaleuca Symposium. September 23–24, 1980*. Florida Division of Forestry. Tallahassee, Florida, USA.

Stronge, D. C., R. A. Fordham, and E.O. Minot. 1997. The foraging ecology of feral goats *Capra hircus* in the Mahoenui giant weta reserve, southern King Country, New Zealand. *New Zealand Journal of Ecology* **21**: 81–88.

Sun, H. B., A. Brown, J. Coppen, and P. Steblein. 2007. Response of *Phragmites* to environmental parameters associated with treatments. *Wetlands Ecology and Management* **15**: 63–79.

Surles, W. W. and L. T. Kok. 1978. *Carduus* thistle seed destruction by *Rhinocyllus conicus*. *Weed Science* **26**: 264–269.

Tessele, B., J. S. Brum, A. L. Schild, et al. 2012. Sawfly larval poisoning in cattle: report on new outbreaks and brief review of the literature. *Pesquisa Veterinária Brasileira* **32**: 1095–1102.

Tipping, P. W., M. R. Martin, K. R. Nimmo, et al. 2009. Invasion of a West Everglades wetland by *Melaleuca quinquenervia* countered by classical biological control. *Biological Control* **48**: 73–78.

Tipping, P. W., M. R. Martin, R. Pierce, et al. 2012. Post-biological control invasion trajectory for *Melaleuca quinquenervia* in a seasonally inundated wetland. *Biological Control* **60**: 163–168.

Turner, C. E., R. W. Pemberton, and S. S. Rosenthal. 1987. Host utilization of native *Cirsium* thistles (Asteraceae) by the introduced weevil *Rhinocyllus conicus* (Coleoptera: Curculionidae) in California. *Environmental Entomology* **16**: 111–115.

Tracy, J. L. and C. J. DeLoach. 1999: Biological control of saltcedar in the United States: Progress and projected ecological effects, pp. 111–154. *In*: Bell, C.E. (ed.). *Proceedings of the Arundo and Saltcedar Workshop, Arundo and Saltcedar: The Deadly Duo. 17 June 1998. Ontario, California*. University of California Cooperative Extension Service, Holtville, California. Available from: http://www.cal-ipc.org/symposia/archive/pdf/Arundo_Saltcedar1998_1-71.pdf [Accessed January 2016].

Unruh, T. R. and R. D. Goeden. 1987. Electrophoresis helps to identify which race of the introduced weevil, *Rhinocyllus conicus* (Coleoptera: Curculionidae), has transferred to two native southern California thistles. *Environmental Entomology* **16**: 979–983.

USDA PLANTS Database 2013. http://plants.usda.gov/java/. United States Department of Agriculture, Nature Resource and Conservation Services, Washington, D.C. [Accessed January 2016].

Vigueras G. and A. L. Portillo. 2001. Uses of *Opuntia* species and the potential impact of *Cactoblastis cactorum* (Lepidoptera: Pyralidae) in Mexico. *Florida Entomologist* **84**: 493–498.

Wagner, D. 2007. Emerald ash borer threatens ash-feeding Lepidoptera. *News of the Lepidopterists' Society* **49**(1): 10–11.

Wagner, D. L. and K. J. Todd. 2015. Conservation implications and ecological impacts of the emerald ash borer in North America, pp. 15–62. *In*: Van Driesche, R. G. and R. Reardon (eds.). *Biology and Control of Emerald Ash Borer*. Technical Bulletin FHTET 2014-09, USDA Forest Service. Morgantown, West Virginia, USA.

Walsh, D. B., M. P. Bolda, R. E. Goodhue, et al. 2011. *Drosophila suzukii* (Diptera: Drosophilidae): invasive pest of ripening soft fruit expanding its geographic range and damage potential. *Journal of Integrated Pest Management* **2** (1): G1–G7.

Wang, X-Y., Z-Q. Yang, J. R. Gould, et al. 2010. Host-seeking behavior and *Spathius agrili* Yang (Hymenoptera: Braconidae), a parasitoid of the emerald ash borer. *Biological Control* **52**: 24–29.

Warner, K. D. and F. Kinslow. 2013. Manipulating risk communication: value predispositions shape public understandings of invasive species science in Hawaii. *Public Understanding of Science* **22** (2): 203–218.

Watts, M. S., D. J. Kriticos, and L. K. Manning. 2009. The current and future potential distribution of *Melaleuca quinquenervia*. *Weed Research* **49**: 381–390.

Wessels, F. J., J. P. Cuda, M. T. Johnson, and J. H. Pedrosa-Macedo. 2007. Host specificity of *Tectococcus ovatus* (Hemiptera: Eriococcidae), a potential biological control agent of the invasive strawberry guava, *Psidium cattleianum* (Myrtales: Myrtaceae), in Florida. *Biological Control* **52**: 439–449.

West, C. J. 2003. Eradication of alien plants on Raoul Island, Kermadec Islands, New Zealand, pp. 365–373. *In*: Veitch, C. R. and M. N. Clout (eds.). *Turning the Tide: The Eradication of Invasive Species: Proceedings of the International Conference on Eradication of Island Invasives*. Occasional Paper No. 27. IUCN (International Union for Conservation of Nature), Gland, Switzerland.

Wheeler, G. S. 2005. Maintenance of a narrow host range by *Oxyops vitiosa*: a biological control agent of *Melaleuca quinquenervia*. *Biochemical Systematics and Ecology* **33**: 365–383.

Wiggins, G. J., J. F. Grant, P. L. Lambdin, et al. 2010. Host utilization of field-caged native and introduced thistle species by *Rhinocyllus conicus*. *Environmental Entomology* **39**: 1858–1865.

Wixted, K. L. and J. B. McGraw. 2010. Competitive and allelopathic effects of garlic mustard (*Alliaria petiolata*) on American ginseng (*Panax quinquefolius*). *Plant Ecology* **208**: 347–357.

Wikler, C., J. H. Pedrosa-Macedo, M. D. Vitorino, et al. 2000. Strawberry guava (*Psidium cattleianum*) – prospects for biological control, pp. 659–665. *In*: Spencer, N. R. (ed.). *Proceedings of the X International Symposium on Biological Control of Weeds, Bozeman, Montana, USA, 4–14 July, 1999*. Montana State University. Bozeman, Montana, USA.

Wilson, F. 1960. *A Review of the Biological Control of Insects and Weeds in Australia and Australian New Guinea*. Technical Communication No.1, Commonwealth Institute of Biological Control, Ottawa, Canada, CAB, Farnham Royal, England, see pp. 51–56.

Wolfe, B. E., V. L. Rodgers, K. A. Stinson, and A. Pringle. 2008. The invasive plant *Alliaria petiolata* (garlic mustard) inhibits ectomycorrhizal fungi in its introduced range. *Journal of Ecology* **96**: 777–783.

Woodall, S. L. 1982. Herbicide tests for control of Brazilian-pepper and melaleuca in Florida. *Research Note SE-314*, Southeastern Forest Experiment Station, USDA Forest Service. 10 pp.

Wu, H., X-Y. Wang, M-L. Li, Z-Q. et al. 2008. Biology and mass rearing of *Sclerodermus pupariae* Yang et Yao (Hymenoptera: Bethylidae), an important ectoparasitoid of the emerald ash borer, *Agrilus planipennisi* (Coleoptera: Buprestidae) in China. *Acta Entomologica Sinica* **51**(1): 46–54.

Yahaya, I. 2003. Forest ecosystems of the Comoros: biodiversity, principle [sic] threats, prospect for improvement – the case of Kartala forest on Ngazidja island, pp. 105–118. *In*: Mauremootoo, J. R. (ed.). *Proceedings of the Regional Workshop on Invasive Alien Species and Terrestrial Ecosystem Rehabilitation in Western Indian Ocean Island States, Seychelles, 13–17 October 2003: Sharing Experience, Identifying Priorities and Defining Joint Action*. International Union for Conservation of Nature (IUCN). Gland, Switzerland.

Yang, Z-Q., J. S. Strazanac, P. M. Marsh, et al. 2005. First recorded parasitoid from China of *Agrilus planipennis*: A new species of *Spathius* (Hymenoptera: Braconidae: Doryctinae). *Annals of the Entomological Society of America* **98**: 636–642.

Yang, Z-Q., J. S. Strazanac, Y-X. Yao, and Y-X Wang. 2006. A new species of emerald ash borer parasitoid from China belonging to the genus *Tetrastichus* Haliday (Hymenoptera: Eulophidae) parasitizing emerald ash borer from China. *Proceedings of the Entomological Society of Washington* **108**: 550–558.

Yang, Z-Q., X-Y. Wang, J. R. Gould, and H. Wu. 2008. Host specificity of *Spathius agrili* Yang (Hymenoptera: Braconidae), an important parasitoid of the emerald ash borer. *Biological Control* **47**: 216–221.

Yang, Z-Q., X-Y. Wang, J. Y-X Yao, et al. 2012. New species of *Sclerodermus* (Hymenoptera: Bethylidae) parasitizing *Agrilus planipennis* (Coleoptera: Buprestidae) From China, with a key to Chinese species in the genus. *Annals of Entomological Society of America* **105**: 619–627.

Yobo, K. S., M. D. Laing, W. A. Palmer, and R. G. Shivas. 2009. Evaluation of *Ustilago sporoboli-indici* as a classical biological control agent for invasive *Sporobolus* grasses in Australia. *Biological Control* **50**: 7–12.

Zhang, Y-Z., D-W. Huang, T-H. Zhao, et al. 2005. Two new egg parasitoids (Hymenoptera: Encyrtidae) of economic importance from China. *Phytoparasitica* **33**: 253–260.

Zimmermann, H. G., V. C. Moran, and J. H. Hoffmann. 2000. The renowned cactus moth, *Cactoblastis cactorum*: its natural history and threat to native *Opuntia* floras in Mexico and the United States of America. *Diversity and Distributions* **6**(5): 259–269.

Zimmerman, N., R. F. Hughes, S. Cordell, P. Hart, et al. 2008. Patterns of primary succession of native and introduced plants in lowland wet forests in eastern Hawai'i. *Biotropica* **40**: 277–284.

Zwölfer, H. and M. Preiss. 1983. Host selection and oviposition behavior in West-European ecotypes of *Rhinocyllus conicus* Froel. (Col., Curculionidae). *Zweitschrift für Angewandte Entomologie* **95**: 113–122.

CHAPTER 5

Benefit–risk assessment of biological control in wildlands

Roy G. Van Driesche[1] and Daniel Simberloff[2]

[1] Department of Environmental Conservation, University of Massachusetts, USA
[2] Department of Ecology & Evolutionary Biology, University of Tennessee, USA

Who gets to decide which species are targeted

Whether an introduced species in wildlands (lands supporting functioning native ecosystems, excluding farms, pastures, tree plantations, and developed land) should be targeted for management at all, by biological control or any other means, is itself a decision that should engage many stakeholders. (See Chapters 12 and 14 for discussions on conflict management and the influence of economics on target selection.) Parties with an interest in such discussions typically include wildlands managers, policymakers, and landowners in the vicinity of the target area. Many members of the general public have a substantial stake in the outcome. Certain invasive plants, for instance, are considered desirable, and great animosity can be engendered by attempts to remove them, particularly when particular individuals feel they have had little input in the decision-making. For instance, eucalyptus trees have long been a cultural icon in California, appreciated for their aesthetic qualities and shade (Nuñez and Simberloff, 2005). When the US National Park Service planned in 1979 to remove eucalyptus trees from Angel Island in the Golden Gate National Recreation Area on ecological grounds, they were delayed for five years by the members of POET (Preserve Our Eucalyptus Trees) who attacked Park Service employees as "plant racists" and "plant Nazis" (Williams, 2002). In the Chicago area, a plan to remove introduced European common buckthorn (*Rhamnus cathartica* L.) and glossy buckthorn (*R. frangula* L.) as part of a restoration of 3000 ha of exotic-dominated woodlands and shrublands to native oak savanna and prairie spawned enormous controversy, on the grounds that the buckthorn was more aesthetically appealing than prairie, provided shade, and served as good deer habitat (Shore, 1997; Helford, 2000). In Florida, invasive "Australian pine" (*Casuarina* spp.) damages native vegetation and hinders sea turtle nesting (Schmitz et al., 1997), but projects by federal, state, and municipal authorities to remove it frequently encounter opposition on the grounds that it provides shade (e.g., Gasco, 1995). In Hawaii, strawberry guava (*Psidium cattleianum* Sabine) has replaced expanses of native trees and supports fruit flies that are agricultural pests, but a 2008 proposal by the USDA to control it generated opposition from Save Our Strawberry Guava, an organization of citizens who like the fruit and find the tree attractive (Tummons, 2008) (see Chapter 12).

Some invasive animals also have defenders. Management of feral pigs in Hawaii has long been a source of dispute between conservationists and hunters, including native Hawaiians (Burdick, 2005). Conservationists are concerned about the massive erosion caused by rooting by pigs, the vectoring of avian malaria by introduced mosquitoes that breed in wallows, and dispersal by pigs of seeds of ecologically damaging non-native plants such as strawberry guava and firebush (*Morella faya* [Ait.] Wilbur). Hunters, however, like to hunt pigs, as did the ancestors of today's native Hawaiians who brought pigs with them when they colonized the islands. Hunters opposed to pig control have even formed an unusual alliance with animal rights advocates, some of whom object to deliberate

Integrating Biological Control into Conservation Practice, First Edition. Edited by Roy G. Van Driesche, Daniel Simberloff, Bernd Blossey, Charlotte Causton, Mark S. Hoddle, David L. Wagner, Christian O. Marks, Kevin M. Heinz, and Keith D. Warner.
© 2016 John Wiley & Sons, Ltd. Published 2016 by John Wiley & Sons, Ltd.

killing of all sentient vertebrates and all of whom reject the most effective control method for pigs in Hawaii, snaring, as brutal and inhumane (Knickerbocker, 1994). The development of principles (such as Public Trust Thinking, see Chapters 8 and 15) that reflect the public interest and provide guidance on desirable outcomes and processes would strengthen the ability to make such decisions.

The decision to begin to manage an invasive species is not a trivial matter. Authors agree that it is essential that all stakeholders are able to participate – the operative word, in fact, is "participatory" (Light, 2000; Gobster and Barro, 2000; Evans et al., 2008; Boudjelas, 2009). In some instances, such as eradication of the northern pike (*Esox lucius* L.) in Lake Davis in California, fostering participation by previously excluded stakeholders led to a satisfactory resolution of a controversial management program (Elmendorf et al., 2005). Other times, when utterly different worldviews and values are represented among stakeholders, it is not clear how resolution can ever be achieved. An example might be the antipathy of some animal rights advocates towards any killing of a sentient being (Simberloff, 2012a). Evans et al. (2008) advocate "participative adaptive management" in the management of introduced aquatic plants, with stakeholders involved at every step of the planning and implementation of a program. Maris and Béchet (2010) argue that much more may be needed than simply stakeholder involvement, and that stakeholder involvement can even freeze antagonistic positions and prevent resolution of a controversy. Instead, they call for all parties to grant the legitimacy of one another's ethical commitments and to begin a process termed "adjustive management" that will lead to a resolution in spite of what appear to be irreconcilable moral absolutes. The key, in their view, is that initial restriction of the discussion to stakeholders, who are aiming to maximize satisfaction of their own interests, automatically hardens all positions and works against any evolution of viewpoints. Instead, philosophers or ethicists should be engaged at the outset, and the initial exploration should be of the core values ostensibly underpinning the conflicting views on a particular issue, rather than on the facts and complexities of the issue itself. Exemplifying the process with a hypothetical attempt to resolve the heated controversy about the ongoing French eradication campaign against the sacred ibis (*Threskiornis aethiopicus* Latham) in natural areas (Clergeau et al., 2005), they

suggest that profound consideration of these values may allow a progressive evolution of views on the specific case at hand and, ultimately, a program of actions acceptable to all. They envision this evolution of viewpoints as running in parallel with the "adaptive management" of actions on the ground.

Once a decision has been made that a species should be managed, who should decide what method is to be used? Many persons would see themselves as having a stake in the method, including some not engaged in the initial decision that the species should be managed. Certain approaches (e.g., aerial spraying of herbicides for weeds or insecticides for insects, use of poison baits or traps for mammals) automatically attract attention from particular groups of citizens. Aerial spraying, for example, draws complaints ranging from non-target impacts, including human health concerns (Telg and Dufresne, 2001), to damage to car paint (Getz, 1989). Possible non-target impacts of classical biological control may concern many people, and the facts that established biological control agents probably cannot be removed and that they can spread autonomously impose a particular duty to allow all stakeholders to have input into a decision to employ biological control. In addition to wildlands managers and policymakers, the biological control professionals engaged in the particular project under discussion, and landowners in the vicinity of the target area, stakeholders may include members of the public who have particular concerns about non-target impacts. For instance, native plant enthusiasts might worry about the possibility that a phytophagous agent would attack a rare native plant, or a conservationist could fear that a predatory beetle will attack native insects of conservation concern. If the affected wildlands include land in the public domain, various other members of the public, such as outdoors enthusiasts, hunters, or fishermen, may see themselves as having a stake in the decision. As with the matter of whether a population should be managed in the first place, participation of all stakeholders in the decision about what management method to use is a necessary but not sufficient condition. The adjustive management approach advocated by Maris and Béchet (2010), in which philosophers and ethicists join with stakeholders at the outset to try to reveal the real ethical commitments and underlying values, may aid in resolving conflicts over methods as well as over targets.

Evidence needed to justify use of biological control

For a biological control project to be justified for use in wildlands, it must first be established that the target should be managed. For wildlands, this would generally entail demonstrating that the target species is harming native species or impairing ecosystem function (although one can imagine other reasons, such as the possibility that wildlands are harboring an invasive species that affects public health). Simply observing a great abundance of some invasive species, even if that abundance were accompanied by decreasing abundance of one or more native species or deterioration of some ecosystem process, would not suffice to conclude that the invasive species is actually causing harm. MacDougall and Turkington (2005) have argued that plant invasions are more often the consequence of some other environmental change (usually some anthropogenic disturbance like grazing or agriculture) that is the ultimate cause of the harm, at least in degraded ecosystems (see Chapter 2 for passengers vs. drivers discussion). In such cases, targeting an introduced species for management may be the wrong approach, and instead one should ameliorate, if possible, whatever factor is driving the invasion. However, Martin et al. (2009) have pointed to many examples of physically undisturbed plant communities that nevertheless became substantially invaded by some introduced species. Regardless, MacDougall and Turkington (2005)'s point is well taken that one cannot simply assume that massive presence of an invasive species is the ultimate cause of some ecological harm and removing that invasive species will remove the harm. Conversely, even if an invader is shown to be a passenger rather than a driver of an environmental change, this does not automatically mean that managing it is a worthless endeavor. Some species that begin as passengers may eventually become drivers if not contained. For instance, the marine alga *Caulerpa racemosa* (Forsskål) J. Agardh in the Mediterranean Sea commences invasion as a passenger of human disturbances to reef environments, but after accumulating sediment it becomes an algal turf that subsequently transforms the entire reef ecosystem (Klein and Verlaque, 2008).

Of course the laborious demonstration that an introduced species is actually harming native species or ecosystems need not be undertaken at every site where that

species is a suspect. Although introduced species may behave differently in different settings, for a well known species with a record of demonstrated harm in other sites it would be foolish not to consider its record elsewhere when deciding whether to target a new occurrence for management. An additional consideration, for any apparently harmless introduced population, is the fact that some introduced species remain restricted in distribution and ecologically innocuous for an extended period, even several decades, before suddenly exploding across the landscape and wreaking substantial environmental damage (Crooks, 2005). Giant reed (*Arundo donax* L.) in California (Dudley, 2000) and Brazilian pepper (*Schinus terebinthifolius* Raddi) in Florida (Ewel, 1986) followed this trajectory. The causes of certain of these time lags and the reason they end are understood in some cases and remain mysterious in others, but the fact of the phenomenon is not in doubt. The possibility that the status of a currently unproblematic introduction could change should therefore be considered in decisions about whether to attempt management of an invader.

Once it is decided that an invasion warrants management, all possible management methods and combinations, not only biological control, should be considered, at least initially, and one factor that must weigh in the choice of the method to pursue is the likelihood that successful control will be achieved. If no method has a good chance of success, it may well be valid to decide not to attempt management unless the perceived cost of the invasion is so great that even a low-probability effort is worth trying. It must also be determined that management of the target species by a particular method will not simply result in its replacement by another invasive species or, if that might occur, that it could it be prevented by some additional management, such as addition of native plant propagules to restored areas early in the restoration. Replanting of native willows and cottonwoods within enclosures (to prevent loss of seedlings to grazing) in areas where saltcedar (*Tamarix*) is being defoliated by biological control agents (Dudley, pers. comm.) illustrates the need for planting and protecting propagules of native plants to achieve restoration goals (here, protection of endangered flycatchers).

Most management methods fall into one of three categories – mechanical, biological, and chemical control (Simberloff, 2002; Clout and Williams, 2009) – although a variety of other methods have been successfully employed in particular cases, such as the use of sterile males (Vreysen et al., 2007), pheromones to disrupt mating (e.g., Witzgall et al., 2008), or ecosystem management (Simberloff, 2002) (see Chapter 3). Each of these methods has succeeded in at least some instances and failed in others, so there is obviously no automatic presumption that any one of them should be the default choice in all situations. In any particular case, the factors in Table 5.1 should be considered.

In any specific case, the answers to the questions in Table 5.1 should guide the decision on what method, if any, to use in a management project. Certain sorts of issues will arise repeatedly with respect to particular methods. For chemical control, matters of cost, non-target impacts, and the possibility of evolution of resistance will always be concerns. In general, herbicides and pesticides are expensive, require repeated application, and are potentially polluting. For wildlands, which, unlike agriculture, are not primarily producing marketed products whose sale allows costs to be passed on to consumers, treatment expenses become prohibitive as the

Table 5.1 Factors to be considered in choosing a management method for an invasive population.

1 Feasibility: can the method actually be implemented on the ground or in the water?
  a Does the requisite technology exist to use it in this specific case?
  b Can individuals with the necessary expertise in the method be engaged?
  c What is the estimated cost of the project, and can resources be assembled to meet this cost?
  d Can the requisite facilities and equipment be marshaled?
2 What is the likelihood of success?
  a What would costs of failure be?
  b Would successful removal of the target result in invasion by another invasive species?
3 What are the types and likely extents of possible non-target impacts?
  a How can the likelihood of non-target impacts be estimated?
  b Are anticipated non-target impacts acceptable to all relevant stakeholders?
  c Can potential non-target impacts be mitigated?
4 What objections will arise to this method?
  a Can these objections be lessened or accommodated?
  b Can individuals who object prevent the project from succeeding?

area treated expands. Target species frequently develop resistance to herbicides and pesticides. For instance, *Hydrilla verticillata* (L. f.) Royle, long controlled adequately by fluridone in small- to medium-sized lakes in Florida, has now evolved a fluridone-resistant strain (Puri et al., 2007). The advent of resistance means that, if the herbicide or pesticide can be used at all, it must be used in increasing concentrations, thus increasing the cost and the possibility of non-target impacts. Although non-target impacts of many herbicides and other pesticides are far lower than half a century ago in the days of 2,4,5-T, parathion, and DDT, particularly if they are used according to directions and for situations for which they are registered, this is still an issue (e.g., increased trematode infection in amphibians in waters contaminated by atrazine, which raises snail density [Rohr et al., 2008]). In any event, absence of substantial non-target impact would have to be demonstrated in each particular wildland setting before pesticides could be used safely over large natural areas (see further arguments in Chapter 8). Also, the public in many parts of the United States, since the 1960s, has become reflexively suspicious of pesticides (including those intended to protect human health) even when there is no direct evidence of adverse effects from use, a phenomenon termed "chemophobia" (Williams, 1997). Both chemophobia and the combined problems of cost and resistance suggest that chemical control is more likely to be useful in eradication programs rather than long-term maintenance management, and indeed several successful eradications of invasive plants and animals have relied on chemicals (Simberloff, 2009). Herbicides are routinely used in some preserves to suppress patches of invasive plants. This is, however, often done by staff, out of view of the general public.

Mechanical control, including use of such simple hand tools as trowels for weed-pulling or heavy mechanized equipment modified to remove vegetation, has achieved notable successes against some invasive plants and animals (Simberloff, 2009; Pickart, 2013). Mechanical control has often been combined with chemical control in particular projects. An issue is often the cost of labor, but the availability of free labor such as prison work crews or, in some nations, citizen volunteers can render feasible what would have been a prohibitively expensive project. In Kentucky, for example, musk thistle (*Carduus nutans* L.) has for over a decade been well controlled in most state parks and preserves by supervised prisoners convicted of traffic violations (J. Bender, pers. comm.). Volunteer or convict labor must be heavily supervised (for instance, to avoid removing the wrong species), and the likely availability of both the source of labor and support of such supervision over the long term must be assessed. Of course, trampling by laborers or disturbance caused by the movement and operation of heavy machinery may cause various non-target impacts. For instance, many invasive plant species are favored by soil disturbance (Davis et al., 2000), and it is possible that mechanical removal of one invader would facilitate its replacement by another invasive species.

As part of a surge of interest in the 1980s and 1990s in ecosystem management as a general approach to resource management (Keystone Center, 1993), the idea was proposed that managing ecosystem processes, particularly maintaining them as close as possible to natural processes, would inhibit new invasions and help to manage old ones. Louda (2000), for example, has argued that good pasture management and prevention of overgrazing can minimize invasion by musk thistle. Fire regime adjustments through use of prescribed burns are routinely used in management programs for *Melaleuca quinquenervia* (Cav.) S. T. Blake, *S. terebinthifolius*, *Casuarina* spp., and *Lygodium microphyllum* (Cav.) R. Br. (F. Laroche, pers. comm.), though for none of them is fire the main tool, and seed and seedling suppression via biological control of melaleuca (Center et al., 2012) and potential for biological control of *Lygodium* (Boughton and Pemberton, 2009) are likely to change the management needs for those species, especially melaleuca. There is little doubt that certain natural fire regimes prevent establishment of a number of non-native plants. For instance, the remaining remnants of the fire disclimax longleaf pine-wiregrass community that formerly dominated much of the southeastern United States have notably few introduced plants, despite the presence nearby of many ornamental plants that have become invasive in other settings. This situation could change, however, when cogon grass (*Imperata cylindrica*) [L.] P. Beauv.) reaches these remnants, as it seems adapted to the regime of frequent growing-season fires that maintains the system (Schmitz et al., 1997).

For biological control, a first question to ask is whether the same species has already been targeted by a biological control program, and what the outcome of that program has been. If particular biological control agents have already been successfully used, this would give increased confidence in their probability of success in the new setting and would also eliminate or greatly lessen the development costs for a new project. However, success against the same target elsewhere is not a guarantee that a biological control agent will succeed in a new setting, and misplaced confidence may, in some cases, lead to a worsened problem. For instance, Koster's curse (*Clidemia hirta* [L.] D. Don), discovered on Oahu in 1941, was for at least a decade restricted to a small area and perhaps could have been eradicated by chemical and mechanical means (Mack and Lonsdale, 2002). However, because it had been controlled on Fiji by an introduced thrips, *Liothrips urichi* Karny, authorities believed it could be managed on Oahu by the same means if it should start to spread (Pemberton, 1957). In Hawaii, however, high rates of predation on this thrips by the invasive ant *Pheidole megacephala* (Fabricius) occurred and may have contributed to the ineffectiveness of the thrips in Hawaii (Reimer, 1988), where it failed to control the plant, which now occupies ca. 100,000 ha throughout the Hawaiian Islands (Smith, 1992).

If no successful biological control project has been undertaken against a proposed target, a critical concern is the likely expense and length of time needed to mount such a project, and the likelihood that a successful agent will be found. One can imagine the possibility of a stopgap measure employing chemical or mechanical control during development of biological control. Such is currently the case in the Galápagos Islands, where a recently invasive parasitic fly, *Philornis downsi* Dodge & Aitken (Diptera: Muscidae), poses major new threats to native passerine birds, such as warbler finches. To contain the immediate threat for the most endangered species, nests are treated with pesticides in combination with an attempt to develop effective fly traps. These are viewed as potentially temporary measures that might be superseded by introduction of larval and pupal parasitoids of the fly, pending favorable results from longer-term research on the fly's potential biological control (Causton, pers. comm.).

Historically, disastrous non-target impacts are known from some early biological control activities that employed vertebrates with a generalized diet and were carried out with little scientific oversight or basis; in fact seven species on the widely cited list of 100 of the world's worst invaders (Lowe et al., 2001) were originally introduced for biological control purposes

(Simberloff, 2012b). Among these were introductions in the 1700s of the myna (*Acridotheres tristis* [L.] and in the 1800s of the small Indian mongoose (*Herpestes javanicus* É. Geoffroy Saint-Hilaire ssp. *auropunctatus*), stoats (*Mustela erminea* L.), and western mosquitofish (*Gambusia affinis* [S. F. Baird & Girard]). In the 1980s, further damage was done by the introduction of two generalist snail predators to the Pacific Islands to control the giant African snail (*Achatina fulica* Bowditch). The first species was the New Guinea flatworm *Platydemus manokwari* De Beauchamp (which had spread widely in the Pacific on its own), which was also deliberately introduced to Guam and the Okinawa Islands, where native snail populations were affected (Yamaura, 2008). The second snail predator was the rosy wolf snail, *Euglandina rosea* (Férussac), which was moved from Florida throughout the Pacific. It had major non-target effects on native snails, even causing local extinctions (Murray et al., 1988). Why generalist predators able to cause such obvious harm to native species should have been used in this way at that time (when knowledge was available on the dangers of such actions) is unclear, but likely the decisions were based on the limited expertise applied in the decision-making process. These cases argue for collective decision-making to allow more expertise to be brought to bear. Similarly, the mosquito fishes *G. affinis* and *Gambusia holbrooki* Girard are still being distributed by various mosquito control agencies into areas where they are not native and where they compete with local native fish with similar feeding habits (Courtenay and Meffe, 1989; Pyke, 2008).

Since the 1980s, however, many biological control scientists and many groups affected by invasive species have become very concerned to avoid non-target impacts (Pemberton, 2000; Simberloff, 2012b), and the knowledge of how to do so has been greatly expanded by research and changes in societal attitudes. For biological control of plants, modified versions of the centrifugal phylogeny method (Wapshere, 1974) (which uses plant relatedness to predict risk) have minimized the probability of non-target impacts, and with current procedures such risks are likely to be small and to be confined to plants in the same genus as the target weed or other plants accepted during host-range evaluation (Pemberton, 2000). For biological control of insects, the centrifugal phylogeny method is of less use (Simberloff, 2012b), largely because insects generally lack taxonomically linked secondary defense compounds, which are

often the basis for the relationship between plant taxonomy and diets of their specialized insect herbivores. Rather, for parasitoids of pest insects, host ranges are frequently shaped by how the biologies of the host and parasitoid interact. For many species, an early filter is parasitoid response to odors associated with the host's body, its host plant, or habitat. For internal parasitoids, a second filter is whether or not the parasitoid progeny can defeat the host's efforts to kill them by encapsulation (Van Driesche et al., 2008).

Certain other generic aspects of classical biological control provide important advantages that chemical and mechanical control often lack. When a biological agent succeeds, it does not have to be periodically reapplied because the biological control agent and target will cycle in perpetuity in homeostatic fashion. Although cases of evolution of target pest populations toward enhanced resistance to biological control agents, or naturally occurring parasitoids, have been reported (e.g., Muldrew, 1953; Dubuffet et al., 2007), such cases seem to be rare, but may simply be understudied. Biological control also does not entail the substantial physical disturbance that often accompanies mechanical control, so the facilitation of other invasions is probably not an issue; however, in eutrophic aquatic systems, suppression of the dominant invasive plant through biological control releases nutrients that are then available for other ecologically similar invasive plants found in the habitat.

Because the underlying principle of biological control entails the continued persistence of the pest population (albeit in low numbers) as well as that of the control agent, a decision to attempt to eradicate a pest population is not compatible with a decision to attempt to use biological control. Expensive large-scale eradication campaigns that failed but had major non-target impacts have received enormous publicity, such as those against white pine blister rust (*Cronartium ribicola* J. C. Fisch.) (Maloy, 1997), European barberry (*Berberis vulgaris* L.) (Campbell and Long, 2001), the gypsy moth (*Lymantria dispar* [L.]) (Spear, 2005), and the red imported fire ant (*Solenopsis invicta* Buren) (Buhs, 2004). Such cases have fostered great pessimism; Dahlsten (1986), for instance, focusing on insect management, thought eradication of a widespread invasive insect population would rarely if ever be feasible. However, just as biological control technology has advanced enormously in the last few decades, so have various approaches to eradication (Veitch et al., 2011). Along with many failed eradication campaigns, there are an increasing number of successes, especially of vertebrates and ants on islands, including some against targets dispersed over wide areas (Simberloff, 2009). Aquatic invasions and most insect invasions have remained particularly refractory. However, once a decision is made that a species should be targeted for management, the possibility of eradication should also be considered, along with the various technologies for long-term maintenance management, and the same considerations raised in Table 5.1 for maintenance management should be raised for eradication.

Even aside from the inherent incompatibility of biological control and eradication, other cases exist in which biological control cannot be employed alongside other management techniques. For instance, if an insect pest is targeted for management by biological control (say by a parasitoid or insect predator), the simultaneous use of an insecticide would not be possible unless the chemical were specific to the pest or applied in a selective time or manner. While this has been done often in agriculture, it has rarely been done for pests of wildlands, apart from the use of biopesticides such as *Bacillus thuringiensis* Berliner or lepidopteran viruses, both of which are intrinsically compatible with conservation of introduced parasitoids of targeted species. However, some management campaigns can profitably employ biological control with chemical and other methods. For example, the campaign to control melaleuca (*M. quinquenervia*) in Florida (see Chapter 4), which has approximately halved the coverage of this major invader, employed chemical, mechanical, and biological control as well as prescribed burns (F. Laroche, pers. comm.). In this instance, the biological control agents attack the reproductive ability of the tree, reducing seeding and lowering survival of seedlings or stump sprouts (Center et al., 2012). Thus the biological agents have effectively neutered the plant population in most areas and should prevent its further spread by diminishing the supply of propagules. Cutting or herbiciding can be used effectively to clear stands of mature plants where this is desired. This has been done on most public lands of high conservation value. Stands on private lands, no longer a source of seeds, are now less of a threat and therefore not as high a concern for management.

Finally, how to manage a targeted pest should consider developments from different directions than those traditionally employed. For example, the idea of using genetically based autocidal methods, raised in the 1960s and 1970s but discarded in the absence of the ability to find and select for appropriate genes, in the age of genetic manipulation and transgenes has returned as a potentially feasible method, with various schemes under consideration and field tests already underway for pest mosquitoes and fish (Thresher et al., 2014).

What principles and processes should guide decision-making?

Openness
A basic principle guiding biological control projects should be free and open presentation of information, particularly concerning the science involved, at each stage of a project. Above we have described the process of designating a species as suitable for management and of determining that biological control is the preferred method. After these steps, further public scrutiny should come at two additional points: (1) agent selection for release and (2) post-release discussion of results. At each of these points, broad-based public discussion by panels of experts should occur, as described below (see also Chapter 15). Information concerning the issues debated and the discussions held should be posted on websites that allow other parties to register comments, to allow for as broad a public participation as possible (see concluding remarks at end of book).

Insulation from urgency
In some invasions, damage increases rapidly and the invader spreads quickly. The social pressure to "do something" in such situations may become intense. For these cases, initial efforts probably should focus on control measures whose effects end when management ends, such as chemical or mechanical control. Biological control, however, should not be adopted in haste, as its effects typically cannot be reversed. Consequently, the normal decision-making process should not be abridged. Rather, in each case, sufficient time must be taken to ensure that the invader needs to be suppressed, that biological control is potentially the most effective approach, and that the agent to be introduced would be safe if released at the target location.

Separation of functions

To avoid conflict of interest, decision-making for approvals versus activities at the various stages of a biological control project should be kept separate. Currently, three stages (agent approval, conduct of project, and area-wide implementation) can be defined in US-based projects. However, the system could be improved by formalizing two additional stages: (1) the decision that an invasive species needs to be managed and that biological control is the best approach and (2) the post-project assessment of final outcomes.

Potential conflicts are best understood by carefully considering the functions of each of the above stages.

1 *Target and control method selection*. Currently, in the United States, the determination that an invader must be managed is an informal judgment that emerges from the actions and convictions of those affected by the consequences of the invasion, who then advocate for a program for its control. Need is judged, if at all, only in the sense that potential funders are either persuaded to provide financial aid or not. A more formal process would reduce the advocacy role of affected parties, who would also be required to submit data justifying the need for control. Similarly, the selection of biological control as the optimal tool would best be made by a committee specialized in making such choices, rather than the scientists who would be in charged with carrying out such a project. Slightly different approaches are used in other countries commonly using biological control (see Chapter 11).

2 *Conduct of project*. Another distinction that could reduce conflicts would be, as in New Zealand, to separate conduct of a project from advocacy for it, with public agencies rather than individual scientists applying for project and agent approvals. This would end the phenomenon of having the scientists wishing to do the work also being the parties petitioning for approvals from regulatory agencies (see also Chapters 11 and 15).

3 *Approval of release of agents*. The current arrangement in the United States, in which APHIS (in USDA) approves permits for release of agents for projects that APHIS itself is carrying out, should end. One solution would be for APHIS to divest itself of its action units that carry out biological control projects and move such groups into other agencies. Alternatively, power to issue permits for new biological control agents

could be taken from APHIS and placed in another group or with a special independent committee, which APHIS might be part of but not control. How this conflict is avoided in other countries that conduct many biological control projects is discussed in Chapter 11.

4 *Post-release evaluation of agent safety and project success*. In the United States, neither post-release evaluations of agents nor global post-project assessments of outcomes are required, but they should be. If a system to do so is created, evaluation of the data on outcomes should be presented to a group that is different from the one doing the actual work to ensure objectivity. Ideally, the same government committees reviewing petitions to release agents used in projects might assess the post-release impacts of those agents, as they would be the best informed concerning both the potential risks and benefits of the agents, as known before their release. Global assessments of projects would require wider expertise. Such assessments might be done by *ad hoc* external review committees assembled to review just a single project, either at the project's completion or at critical points of reassessment where new directions are needed. (See also Chapter 15 for more on this discussion.)

5 *Area-wide implementation*. Once a project has demonstrated success at the local research-plot level, a need exists to release agents or take other restoration actions over the whole of the affected landscape. As noted above, the agency charged with such work should not itself be in charge of regulatory compliance and permits for the project.

Risk assessment of biological control agents

Biological control agents, those used for either the protection of agriculture or, as in this book, the restoration of natural systems or species damaged by invasive insects or plants, are not a unique category of invasive species with special risks, but rather nest within the broader categories of invasive species, which have many sources and different characteristics. Horticulture, agriculture, the pet industry, and general trade of goods are major reasons for invasions of insects and plants to new areas. The insects and plant pathogens used as biological control agents are not more risky than these broad

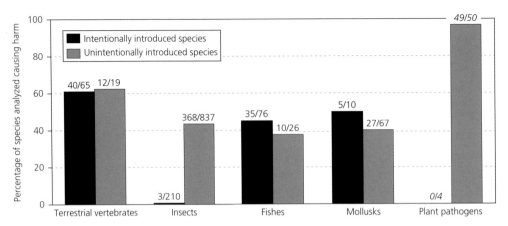

Figure 5.1 Differences by taxon in rates at which intentionally introduced and accidentally introduced species become pests suggest that, for terrestrial vertebrates, fish, and mollusks, deliberate introduction does not lower risk. In contrast, for insects and plant pathogens, categories for which biological control introductions would dominate the intentionally released groups, risk is highly reduced over random introduction. This reduction suggests that biological agent selection seeks safe agents and avoids release of risky species, which is not the case on average for intentionally released species in other taxa (Office of Technology Assessment, 1993).

categories, but rather less so because exotic insects and plants (excepting food crops and some horticultural plants) in general often have characteristics predisposing toward invasion, such as high vigor and lack of pests, while biological control agents are nowadays chosen for both their likely harm to the target invader and safety to non-target species. Historical rates of the percentages of invasive species by taxonomic group show that deliberately introduced insects (i.e., biological control agents) that successfully establish become damaging much less frequently than do accidentally introduced insects, or than either accidentally or deliberately introduced species in other groups (fish, vertebrates, mollusks, pathogens) (Figure 5.1) (Office of Technology Assessment, 1993).

Among introduced species, biological control agents are the only group consistently subject to pre-release risk assessment, resulting in a relatively high level of specificity (Figure 5.2). In other taxa, such risk assessments are done in some cases, but because they are usually not required, only a small percentage of introductions are assessed for risk of becoming a damaging invasive. Programs in which risk was assessed for species other than biological control agents include assessment of the potential impacts of lodgepole pine (*Pinus contorta* Douglas) before its widespread planting in Sweden (Andersson et al., 1999) and assessment of the potential effects on Chesapeake Bay of the planned introduction of the suminoe oyster (*Crassostrea ariakensis* [Fujita]) (United States National Research

Council, 2004). Most deliberate introductions for purposes other than biological control in the United States and most other countries as well, excepting New Zealand and Australia, are undertaken either as a matter of right (i.e., horticultural introductions made by commercial firms for economic benefits, which in the United States must only be shown not to carry pests and not to be on a short prohibited list) or as a matter of government policy to pursue benefits in areas such as agriculture, aquaculture, or sport fisheries, but without any formal effort to predict invasiveness, or to assess risks to the environment should they become invasive. Some countries (New Zealand, Australia) have now added such risk evaluations to the process of permitting most new organisms, across all categories. The evaluation of biological control agents, however, pioneered consideration of invasiveness and associated risks in many countries, and risk assessment was implemented much earlier. Chapter 11 explores how the formal assessment of biological control agents developed and how it differs among countries and between herbivorous (weed control) and carnivorous (insect control) agents. The countries that have imported the largest number of biological control agents include the United States, South Africa, Australia, Canada, and other Commonwealth countries (through CABI Bioscience in the UK), New Zealand, and several European countries (for insect biological control only until recently). These are compared in detail in Chapter 11.

Figure 5.2 Biological control is a method of invasive species management that generally shows high levels of target specificity. Shown here are two brown clumps of purple loosestrife (*Lythrum salicaria*) selectively defoliated by *Galerucella* leaf beetles introduced to North America, while all remaining wetland vegetation remains untouched despite food limitation for developing larvae. Photo credit: Bernd Blossey. (*See Plate 1 for the color representation of this figure.*)

Differences in risk assessment of herbivorous and carnivorous biological control agents

One difference that emerged early in evaluation of potential risks of biological control agents, in nearly all participating countries, was a higher level of concern to ensure protection of non-target plants, with little or no concern until the 1990s for protection of non-target insects. In large measure this difference developed because invasive herbivorous insects were seen as a risk to economic plants and herbivorous biological control agents were regulated to reduce that risk. By contrast, for biological control of insects, few insects had market value or excited public concern, so the threat of non-target impacts was considered unimportant until the 1970s, when some insects were added to lists developed under laws to protect endangered species. Thus, in the first major weed biological control project, for control of

invasive *Opuntia* cacti in Australia in the 1920s, important local crops underwent extensive testing, quite independently of the knowledge that cactus-feeding insects could not feed on crops such as wheat or apples. Nevertheless, the negative results of such tests helped convince political figures charged with protecting agriculture that such biological control importations did not threaten economic values.

Concern for plants over time expanded from a focus on crops and ornamentals to consideration of rare native plants, and finally to all native plants. Expansion of protection for rare plants developed because of heightened public concern for biodiversity and the rise of the conservation movement, which in various nations resulted in laws protecting endangered species. Concurrent with this expansion of focus was a shift in the strategy of estimating host ranges, from list-ticking (proving safety to particular species selected by various criteria) to estimating fundamental host ranges (van Klinken and Heard, 2000; Sheppard et al., 2005). Estimating such host ranges was based on the notion that for specialized herbivorous insects host ranges were largely determined by secondary plant defensive chemistry, which often paralleled plant taxonomy (Wapshere, 1974). In contrast, such relationships tend to be weaker between carnivorous insects and the species they attack. However, for carnivorous insects, predators and parasitoids need to be distinguished as categories and specialization attributes sought in the biology of each proposed agent, particularly focusing on how they find hosts or prey or their host plants and how internal parasitoids defeat host defenses and regulate host physiological attributes to favor parasitoid progeny. While some relationships exist within parasitoid groups as to the types of hosts they attack, these seem more determined by host biology than host chemistry.

As a consequence of these differences in historical focus (types of concerns) and biology (importance of secondary defense compounds or lack thereof), the approaches taken to define host ranges in herbivorous and carnivorous insects have developed along rather different lines.

How risks of newly proposed agents are evaluated, and more specifically how herbivorous vs. carnivorous agents are handled, varies among countries, and such differences are discussed in Chapter 11. Here we look at the specific case of regulation in the United States and contrast approaches used for herbivorous and carnivorous biological control agents, asking how the process might be improved.

The US review system – a discussion

In the United States, herbivorous biological control agents are directly regulated by USDA APHIS (Animal and Plant, Health Inspection Service), but APHIS decisions take into account a recommendation from an interagency group (TAG), a committee with representation from federal agencies with interests in weed biological control, including the USDA Natural Resources Conservation Service, four US Department of the Interior (USDI) agencies (Fish and Wildlife Service, the National Park Service, the Bureau of Land Management, and Bureau of Indian Affairs), the US Environmental Protection Agency (EPA), and the Department of Defense, among others.

This committee reviews non-target species test lists submitted by researchers (who are the petitioners) and later the data from the tests themselves. Petitions for release of new species cover several issues, including the importance of the impacts of the invasive species, non-target species or communities that potentially might be affected by release of the agent, and the likelihood of success in suppressing the target. Petitions also address obligations under a variety of fesderal statutes, including laws for the protection of wildlife (Lacey Act) and the environment (NEPA). TAG recommendations are not binding on APHIS. APHIS can take a contrary view if it so chooses.

Several potential improvements to the TAG process are possible. First, as currently defined, the TAG is comprised of members representing the interests of various federal agencies who discuss petitions and consider reviews of petitions submitted by outside independent scientists chosen by the committee. Non-federal scientists do not participate directly as TAG committee members. Better TAG committees could be formed by choosing members based on expertise, selected from universities, agencies, or organizations (such as native plant societies) rather than being selected to serve as representatives of federal agencies. A second improvement would be to make the TAG decision binding rather than advisory to APHIS. Finally the consultation process with the federal Fish and Wildlife Service needs to be better defined and streamlined, as this has proved to be the point where the review process has faltered in recent years.

Petitions for release of carnivorous insect biological control agents (parasitoids and predators) are not subjected to the TAG process in the strict sense. Rather, *ad hoc* review panels are created by APHIS, whose members submit comments individually on petitions. While less formally organized than TAG deliberations on herbivorous insects, this informal process for carnivorous agents is able to draw expertise from both university and government scientists.

Such review systems could be improved in a variety of ways. A logical improvement might be to bring post-project outcome reviews to the same groups carrying out pre-project risk evaluations. Completing the circle in this way would make more direct and explicit the link between post-release monitoring results and pre-release risk predictions. It would also improve the knowledge of the committee members, making their future risk assessment judgments progressively more realistic and accurate. For a complete hypothetical description of an optimal system, see Chapter 15.

International coordination of reviews of release petitions

Attention to differences among countries in their perspective on potential risks of release of new agents can improve outcomes. In North America, coordination among Canada, the United States, and Mexico is formally instituted during petition review through NAPPO (North American Plant Protection Organization). A more complicated geographical area is the Floridian region of the United States and the Caribbean, for which the large number of countries and lack of regional reviewing authority pose significant potential risk of invasions of one nation by species released by another nation. Other geographic areas of natural joint decision-making would include the Pacific Island nations, the major ecological blocks of South or Central America, New Zealand, Australia, and potentially the European Union. Issues that should be addressed by such regional bodies include (1) adjustment of test species lists for host-range estimation that represent the region rather than just the importing country and (2) identification of new ecological or economic interests as the geographic range expands for which the release of the agent is being planned or its spread is anticipated. Several examples follow that illustrate the need for such consultation or planning.

1 *Saltcedar*. Release of Mediterranean tamarisk beetles (*Diarhabda elongata* [Brullé]) (Coleoptera: Chrysomelidae) in the southwestern United States for control of invasive saltcedar species (*Tamarix* spp.) was done in consultation with Mexico, in part because one exotic non-target species of tamarix (athel) (*Tamarix aphylla* [L.] Karsten) has economic value as an urban shade tree in Mexico but not in the United States. Consequently, specific consideration of the likelihood and magnitude of risk to this shade tree had to be evaluated (Moran, 2010), and a mutually agreeable decision taken based on projected impacts.

2 *Hill raspberry in the Galápagos*. In the Galápagos, an invasive raspberry (*Rubus niveus* Thunb.) threatens biodiversity of *Scalesia* forests (Rentería et al., 2012) and other moist upland plant communities. Biological control of the species is under consideration (St. Quinton et al., 2011), most likely with rust fungi. However, in selecting a specific rust species or strain,

consideration must be given not just to the islands, but also to the related plants of mainland Ecuador and surrounding Andean areas, because it is very likely that fungal spores would be spread from the islands to the mainland by the high volume of tourist and commercial traffic (Gardener and Grenier, 2011). This likelihood forces explicit evaluation of potential risks to other exotic *Rubus* species that may be more highly valued for fruit on the mainland than on the islands, or native *Rubus* that might be put at risk.

3 *Melaleuca in the Caribbean and South America*. The biological control of melaleuca (*M. quinquenervia*) in Florida (Center et al., 2012) provided the opportunity for the agents used to spread to areas outside the planned project area in Florida because of Miami's role as a plane and boat hub (Pratt and Center, 2012), with the result that the melaleuca weevil, *Oxyops vitiosa* Pascoe, was detected in parts of the Bahamas and the melaleuca psyllid, *Boreioglycaspis melaleucae* Moore, in Puerto Rico and Los Angeles, all locations where these species were not released. Project managers, however, in originally selecting these agents, anticipated possible spread to the nearby islands (Pratt and Center, 2012) and determined that genus-level specificity would provide adequate protection for non-target plants in these additional areas. While desirable, a formal mechanism for vetting releases within this region, with many small countries with limited scientific expertise in biological control, is lacking and may be impractical.

Opportunities probably do exist for multi-country coordination that should be encouraged. The Pacific island nations, for example, should develop such region-wide review processes given the rapidity with which invasive species tend to spread throughout the region (e.g., Petit et al., 2009). Such coordination might be led by some of the major biological control practitioners in the region, such as ACIAR (Australia), CABI (UK), or the University of Hawaii (USA).

Politically, how such multi-country decision-making might function requires careful consideration. For example, the difficulty in the European Union (28 countries) of setting new regulations for commercial activities within the zone illustrates the slowness of tackling new issues and in some cases the inability to make decisions under policy structures in which consensus must be reached. Use of a consensus model, in which veto powers are invested in every participating country, would be impractical in large groupings, as most projects would likely result in some countries perceiving risks in any project. In fact, 15 of the member states do not include certain large groups of species when listing which introduced species are invasive, different member states use different risk assessment procedures for species introductions, and some member states (e.g., Bulgaria and the Czech Republic) do not use formal risk assessment for this purpose (European Commission and BIO Intelligence Service, 2011).

Trouble spots: bad practice
Within-country redistribution of naturally arriving agents

The current regulatory framework for control of importations acts at international borders. This creates a loophole allowing redistribution of agents that are not permitted for first-time release if they arrive naturally first. If a country making an initial release of a biological control agent considers only its own non-target species, native species in adjacent countries that are subsequently invaded by the agent through natural spread may be put in risk. An example of this process includes the redistribution of the weevil *Larinus planus* (F.), which apparently arrived autonomously in the northeastern United States before 1968. It was subsequently studied as a potential biological control agent for use in Alberta, Canada, against Canada thistle (*Cirsium arvense* [L.] Scopoli) (McClay, 1990). There it was ultimately permitted and released (McClay et al., 2001), with subsequent spread into the north central United States. There, weed control officials subsequently redistributed it at the local level even though it had never been reviewed or permitted as a biological control agent in the United States. This weevil now attacks and threatens rare species of thistles in the United States (Louda and O'Brien, 2002; Havens et al., 2012). Redistribution of such adventitious species that invade autonomously or are studied and permitted for release in an adjacent country should be prohibited from active release in a secondarily invaded country, unless their potential safety to native plants is assessed.

Deliberate spread of species never permitted as biological control agents

Some species that were never properly permitted anywhere as biological control agents may be redistributed by third parties for biological control purposes. This practice easily creates problems, both biologically from the consequences that follow and politically, because it blurs the line between controlled use of screened agents and uncontrolled use of species that function as biological control agents, but without safety reviews. Species used in this manner may be (1) native species or (2) adventive (self-invading) species. Examples include the pathogen (and associated vectoring mite) causing rose rosette disease in invasive multiflora rose (*Rosa multiflora* Thunb.), which is a native virus associated with *Rosa woodsii* Lindl. in Colorado (Amrine, 2002). When the Asian multiflora rose invasion spread to Colorado, it was attacked by this native rose virus. Strong temptation exists to hasten the spread of this virus into areas of the eastern United States with dense multiflora rose stands. However, this virus is not an approved weed biological control agent and would not be approved as such because its host range includes native and commercial roses (Amrine, 2002). Similar concerns arise over a native North America aquatic weevil, *Euhrychiopsis lecontei* Dietz, which has expanded its host range to include invasive Eurasian water milfoil (*Myriophyllum spicatum* L.) (Roley and Newman, 2006). Native from at least Vermont to Minnesota, this weevil in some instances appears to contribute to suppression of mats of this invasive weed (Creed and Sheldon, 1993). Interest exists in spreading it to other invaded water bodies (Sheldon and O'Bryan, 1996), but again, it has never been formally evaluated for use as a federally permitted biological control agent, and so its safety with respect to native related plants is incompletely understood.

Cases of unsafe biological control introductions

A few poorly chosen biological control agents have been widely discussed (Strong, 1997; Louda et al., 1997; Boettner et al., 2000; Stiling and Moon, 2001; Kellogg et al., 2003; Selfridge et al., 2006; Hernández et al., 2007; Koch and Galvin, 2008). Though a very small percentage of all agents released historically, these cases have helped shape perception of biological control practice in the eyes of a generation of conservation biologists. Also, these cases are instructive because the characteristics of such past mistakes must be understood so that appropriate corrections to practice can be incorporated into new projects. While various such cases could be identified, at or near the top of most such lists might be (1) the introduction of *Cactoblastis cactorum* (Bergroth) in the Caribbean in the 1950s, (2) the release in North America of the oligophagous thistle-feeding weevil *Rhinocyllus*

Figure 5.3 Larvae of *Cactoblastis cactorum*, an herbivore that attacks various native and pest *Opuntia* cacti. Photo courtesy of Ignacio Baez, USDA Agricultural Research Service, Bugwood.org. (*See Plate 2 for the color representation of this figure.*)

conicus Frölich in the 1960s, (3) selection of a polyphagous tachinid parasitoid *Compsilura concinnata* (Meigen) for release in Massachusetts in 1905, and (4) release of coccinellids such as *Harmonia axyridis* (Pallas) in North America in the 1960s to 1990s.

1 Cactoblastis cactorum *in the Caribbean in the 1950s.* The introduction of the cactus-feeding pyralid moth *C. cactorum* (Figure 5.3) into the Caribbean in the 1950s (Simmonds and Bennett, 1966) (see Chapter 4), following its safe and effective use against invasive prickly pear cacti in Australia in the 1920s to 1930s (Dodd, 1940), shows that safety depends on the location. With no native cacti outside of the Americas, release of this moth in Australia was completely safe, even without host-range testing. Conversely, release of the same moth in the Caribbean in the 1950s without clear demonstration of near species-level host specificity was highly irresponsible because many native species exist in the same genus as the principal target pest (*Opuntia triacantha* [Willd.] Sweet). Before such a release, it would have been necessary to prove through host-range testing that this moth was species specific (which it is not). This project is an example of neglect of the duty to protect native plants. The consequences of this failure to consider the interests of native cacti include damage to a rare native cactus (Florida semaphore cactus, *Consolea corallicola* Small) (Stiling and Moon, 2001), in addition to potential

damage, with eventual spread, to perhaps many species of cacti in Mexico and the southwestern United States (Pemberton and Liu, 2007), as well as likely economic damage to commercial cactus production in Mexico for nopales and cochineal insects. Also, *Opuntia* cacti are a keystone group in both Chihuahuan and Sonoran deserts, where they are a source of shelter, pollen, food, and water for many animal species. While eradication of the moth with sterile insect releases may be feasible (Hight et al., 2005), funding for such a project has been rejected. Biological control of the moth itself, with introduced parasitoids from its native range, may also be possible. Final effects on native cacti of the invasion of the United States and Mexico of *C. cactorum* are unlikely to be fully known for decades. However, native *Opuntia* cacti and their relatives on St. Kitts and Nevis are still present 50 years after *C. cactorum* release (Pemberton and Liu, 2007), suggesting that whatever non-target effects there may have been at that location did not lead to species extinction. Population density changes induced by the moth are difficult to evaluate quantitatively for lack of data before the moth was released.

2 *The thistle feeding weevil* Rhinocyllus conicus *in the 1960s.* In North America, several invasive species of European thistles became pasture weeds, leading to the introduction in the 1960s and 1970s of the thistle-feeding weevil *R. conicus* (Figure 5.4) (see Chapter 4),

Figure 5.4 *Rhinocyllus conicus*, a polyphagous thistle-head feeding insect that is well known to attack a range of native thistles. Photo courtesy of Loke Kok, Bugwood.org. (*See Plate 3 for the color representation of this figure.*)

first in Canada (Anon., 1972 [notes 1968 release in Craik, Saskatoon] and then the United States (Hawkes, 1972 [notes a 1971 release in California]). Studies of the host range of this species showed it fed on thistle heads of European species in three genera (*Carduus*, *Cirsium*, and *Silybum*); few thistles native to North America were included in testing, but those that were included were attacked (Gassmann and Louda, 2001). The estimated host range thus logically includes many North American species in these genera. Because this introduction preceded laws protecting rare native plants in the United States, and because thistles in the 1960s were popularly viewed as pests despite their real ecological value, the decision was made to release this weevil based on the ideas that (1) it was acceptable to suppress all thistles and (2) some level of adult preference by the weevil (in choice test conditions) would focus most of the agent's impact on nodding thistle, *Carduus nutans* L., and several other European species of pasture-infesting thistles in these genera. While this weevil did suppress the target thistles in many areas (e.g., Kok and Surles, 1975 [Virginia]; Rees, 1975 [Montana]; Horber, 1980 [Kansas]), neither assumption about effects on non-target species proved true. Over time isolated populations of many native North American thistles became infested (Turner et al., 1987; Strong, 1997; Louda et al., 1997, 1998; Sauer and Bradley,

2008), although, for many species, use was at minor levels not expected to reduce populations (Herr, 2000). Some species, however, suffered density reductions owing to loss of seeds to *R. conicus* feeding (Russell and Louda, 2004; Russell et al., 2007; DePrenger-Levin et al., 2010).

3 *Polyphagous tachinid*, Compsilura concinnata, *in Massachusetts in 1905*. To combat the forest defoliator gypsy moth, *Lymantria dispar* (L.) (Lepidoptera: Erebidae, Lymantriinae) in Massachusetts, many parasitoids and predators were introduced in the early 1900s (Fuester et al., 2014). Among these was a highly polyphagous tachinid fly, *C. concinnata*, which at the time was recognized as attacking more than 50 insect species (MacClaine, 1916; Culver, 1919). The rationale for this action was that native insects were not valued (except those with economic uses), and polyphagy in biological control agents was valued as allowing higher densities of the agents to be sustained when density of the target pest was low, especially for species like gypsy moth that experienced periodic outbreaks. Furthermore, while some tachinids listed in the literature as having broad host ranges are actually groups of more specialized species (Smith et al., 2007), there appears to be little in the biology of *C. concinnata* narrowing the field host range, as it appears to hunt visually for larvae on a wide range of deciduous plants (trees and herbaceous plants)

Figure 5.5 Adult polyphemus moth, *Antheraea polyphemus* (Cramer), one of the giant silkmoths strongly attacked by the tachinid *Compsilura concinnata*. Photo courtesy of Lacy L. Hyche, Auburn University, Bugwood.org. (*See Plate 4 for the color representation of this figure.*)

without limitations imposed by attraction to volatiles from particular kinds of plants. The most important damage from this introduction has been to silk moths (Figure 5.5) and buck moths (both Saturnidae), and parasitism appears to be very high in some areas (e.g., Massachusetts [Boettner and Elkinton, 2000]), but low to moderate in other locations (e.g., New York [Selfridge et al., 2007] and Virginia [Kellogg et al., 2003]). It is possible that the level of impact in the northeastern United States may decline now that gypsy moth outbreaks there either no longer occur or are rare, owing to another biological control agent, the fungus *Entomophaga maimaiga* Humber, Shimazu, and R. S. Soper, as fly parasitism rates were found to be lower where pathogen infection rates were high (Hajek and Tobin, 2011).

4 *Harmonia axyridis in North America in the 1960s to 1990s.* Classical biological control of insects began with the successful use of a highly specialized ladybird (*Rodolia cardinalis* [Mulsant]) in California (USA) in the 1880s to suppress a polyphagous scale (*Icerya purchasi* Maskell) attacking citrus (Caltagirone and Doutt, 1989). In 1990–2010, this same agent later protected native plants being severely damaged by the same scale on the Galàpagos Islands (Calderón Alvarez et al., 2012; Hoddle et al., 2013). However, most ladybirds are not specialized feeders and most do not regulate their prey populations. Nevertheless, the early stunning success with one specialized ladybird led biological control scientists to seek to repeat themselves, introducing many species of ladybirds into North America (Harmon et al., 2007). This

phenomenon was termed "ladybird madness" by Caltagirone (1989). Few introduced species established and none had much effect on their target hosts. Among ladybirds, efficacy is linked to the relative generation times of the predator and its prey ("generation time ratio [GTR]" is the ratio of the length of the predator generation to that of the prey), as this affects the ability of the predator population to undergo a numerical response to the prey (Kindlmann and Dixon, 1999). Aphid-feeding species are particularly unlikely to regulate their prey (e.g., Kindlmann et al., 2005), while mite and scale feeders may be both more specialized and have generation times more nearly equal to that of their prey, making them more effective (Kindlmann and Dixon, 1999). The enthusiasm for ladybirds, however, has not been excised from US biological control culture and as late as the 1960s (Shands et al., 1972; *Coccinella septempunctata* L.) and 1970s (Tedders and Schaefer, 1994; *Harmonia axyridis* [Pallas]) ladybird introductions continued, leading to introduction and redistribution of aggressive, generalist European coccinellids. These releases, especially for *C. septempunctata*, were unfocused in that there was no single target prey species, but rather the focal target was the vague category of "aphids," which obviously includes many non-pest species. These large, aggressive ladybirds subsequently displaced many native coccinellids that formerly had been common in North American agricultural fields (Harmon et al., 2007; Koch and Galvin, 2008; Hesler et al., 2009). While most of these species appear to have survived, their densities in agricultural habitats have plummeted, and a few species became almost unrecoverable (Losey et al., 2007). Competitive displacement caused by greater prey consumption by *H. axyrdis* is particularly severe (Leppanen, 2012), and this Asian species has also spread to other continents (Katsoyannos et al., 1997) owing to its adoption for a period of time by the commercial insectary business for use in greenhouses (e.g., Sighinolfi et al., 2008). Subsequent world-level invasiveness was increased through the creation of an invasive strain owing to hybridization between strains that do not coexist in the Asian native range of the species, following introductions and spread that occurred in North America (Lombaert et al., 2010). This invasive strain then reached South Africa, South America, and Europe from North America, by use in greenhouses or unknown means.

Analysis of these cases points to obvious areas where practice was deficient. Most of these points are now understood and have been incorporated into current agent selection processes. However, there would be considerable value in continuing to review and track modern cases of other non-target effects as they are recognized, and for this review to be done by some authority such as APHIS or TAG that could assemble and pass on institutional memory of such cases. This body of information would provide increasingly accurate advice on approval of future agents.

Post-release risk evaluation

Overview

Post-release studies of impacts on both the target pest and non-target species have been inadequate in many projects, although sometimes procedures were excellent (e.g., *Nezara viridula* [L.] in Australia [Sands and Coombs, 1999; Coombs, 2002]); cottony cushion scale, *I. purchasi*, in the Galápagos [Calderón Alvarez et al., 2012; Hoddle et al., 2013]; melaleuca in Florida [Rayamajhi et al., 2009]; miconia (*Miconia calvescens* DC) in Tahiti [Meyer et al., 2012]; and mist flower, *Ageratina riparia* (Regel) R. King and H. Robinson, in New Zealand [Barton et al., 2007]). In other cases, outcomes have been reconstructed well after original projects ceased activity (e.g., Hawaiian koa bug, *Coleotichus blackburniae* White [Johnson et al., 2005]; *Opuntia* cacti in the Caribbean islands [Pemberton and Liu, 2007]; and browntail moth, *Euproctis chrysorrhoea* [L.] in New England [Elkinton et al., 2006]. Post-release monitoring of outcomes of biological control is becoming a more common practice in current projects. Such studies should be obligatory (as proposed in our code of best practices in Chapter 15) and should include assessments of the released agent's effects on both the target species and on non-target species, with post-release studies examining any species identified in pre-release assessments as potentially being at risk. Also, agencies sponsoring restoration through biological control programs should issue global outcome reports on projects as they are concluded. These would likely take the form of agency reports, as individual research articles will have presented original findings during the projects. The latter, however, will of necessity not present the grand overview, because they are produced as

work proceeds and individual steps are completed or findings arise. For projects overseen by government review committees (e.g., TAG [in the United States, for weed biological control projects] or equivalent) this final report from a project, summarizing effects on the pest, non-target species potentially at risk, and recovery of the affected native species or ecosystem, should be submitted to the agency originally authorizing the project. This process will provide essential feedback on how decisions taken by the committee produced favorable or other outcomes and would lead to improved functioning of such bodies in their future work. Finally, in our view, the national interest in this area would be well served by the creation of one or several national centers devoted to invasive species management, including in natural areas (Schmitz and Simberloff, 2001). Such centers would provide concentrations of expertise in the many disciplines required by invasive species suppression, including biological control. Such centers might be modeled after the invasive aquatic weeds laboratory in Ft. Lauderdale, Florida, the Centers for Disease Control, or the National Interagency Fire Center in Idaho.

General review of risks of past introductions

Comprehensive reviews of biological control introductions as a whole (Lynch and Thomas, 2000; Pemberton, 2000; Suckling and Sforza, 2014) suggest that the rate of risk to non-target species of new introductions of biological control agents declined substantially over the twentieth century. A similar analysis for Florida only found that, for the period 1899–2007, 60 agents were established in the state for classical biological control and that, of these, 10 (17%) may have produced some changes in non-target species populations, and 4 (7%) were likely to have had substantial effects on non-target species (Frank and McCoy, 2007). During the period 1900–2000, as reviewed by Lynch and Thomas (2000), the risk of attack on non-target species (of any magnitude) by an agent introduced against an invasive insect declined from about 3% for agents introduced before 1950, to about 2% in the 1950s, 1960s, and 1970s, and 1% in the 1980s and 1990s, the last decades covered in the review. For insect biological control agents, Lynch and Thomas (2000) provide a table of known examples (through 2000) of non-target effects from parasitoids and predaceous insects.

For biological control agents directed against invasive plants, most species targeted were invasive in natural systems (as opposed to agricultural fields), even though most early project goals were economic (e.g., to reduce losses to grazing or water use values etc.). While some groups of insects introduced before 1970 were later found to be insufficiently specific and to feed on native plants (see table in Pemberton [2000]), these were usually in the same genus as the target pest. Of 117 agents released against invasive plants in the continental United States, Hawaii, and the Caribbean, 15 fed on native plants (13%), and 12 of the attacked plants were congeneric with the target species. Thus only one case of unforeseen attack (0.8%) occurred among 117 released agents (Pemberton, 2000), suggesting the host-specificity testing methods in use predicted species at risk of agent feeding reasonably well. Thus with attention being paid to native plants, even for groups like thistles, most species of which were not traditionally seen as desirable, agents are now selected in such a way as to provide a high degree of safety to native species. Indeed, most of the famous cases of significant non-target attack on plants date from releases made in the 1920–70 period (Pemberton, 2000), including *R. conicus* (attacking native thistles, see references listed in Pemberton [2000]) and *C. cactorum* (*Opuntia* cacti in the Caribbean and southeastern United States).

Since 2000, post-release experimentation to test host-range predictions made during quarantine testing of weed biological control agents has become increasingly common. Knowledge of the agent's non-target plant interactions is consequently expanding significantly. For a number of systems, post-release experiments found no impacts on non-target species predicted to be potential hosts of agents: rubber vine (*Cryptostegia grandiflora* [Roxb. ex R. Br.] R. Br.) (McFadyen et al., 2002), toadflax (*Linaria* spp.) (Breiter and Seastedt, 2007), and melaleuca (*M. quinquenervia*) (Pratt et al., 2009). In other cases, sustained feeding on non-target species at varying levels was detected in some cases, such as projects directed against the invasive species *Mimosa pigra* L. (Taylor et al., 2007), lantana (*Lantana camara* L.) (Snow and Dhileepan, 2008), houndstongue (*Cynoglossum officinale* L.) (Andreas et al., 2008), leafy spurge (*Euphorbia estula* L.) (Baker and Webber, 2008; but cf. Wacker and Butler, 2006), and old man's beard (*Clematis vitalba* L.) (Paynter et al., 2008). Interestingly, Baker and Webber

(2008) found that while the flea beetles released against the invasive *E. esula* did consume native *Euphorbia robusta* Engelmann, the native plant's population actually increased significantly owing to the greater importance of reduced competition from the invasive leafy spurge as the invasive plant came under biological control. Also, Paynter et al. (2008) found that oviposition by the introduced leafmining agromyzid fly *Phytomyza vitalbae* Kaltenbach on the native species *Clematis foetida* Rauol depended on flies having first fed on the invasive species of *Clematis*, with non-target effects decreasing with increasing distance from weed infestations.

For releases of parasitoids and predators against invasive insects, the same general trend appears to be occurring, but with a delay of 20 years. Thus some of the most polyphagous of introduced parasitoids (e.g., *C. concinnata* [Boettner et al., 2000; Kellogg et al., 2003; Selfridge et al., 2007]) were introduced more than a century ago, when little or no value was placed on native insects as wildlife. However, introductions of some generalist predators such as certain ladybird beetles (Tedders and Schaefer, 1994; Harmon et al., 2007 and references therein) that were later associated with declines of native ladybirds (via indirect competition effects, see Chapter 8) continued into the 1980s. In the United States, New Zealand, and Australia, efforts since the 1990s to identify more specialized parasitoids and predators have been greatly strengthened. While selection procedures for predators and parasitoids are not yet on a par with those for herbivorous insects, risks associated with such introductions appear to be much lower than in the past.

Improving post-release monitoring

In order for any practice to improve, actual results after implementation must be measured and compared to predicted results. When biological control is part of restoration programs, predictions about potential non-target risks may have been generated during the host-range testing phase before release of the biological control agent. If the host-range testing program or other information about the system suggests that there is some limited risk to certain closely related species under some circumstances, post-release studies to measure the degree of such effects offer an opportunity to learn if such predictions of limited impact actually hold up. Some examples follow from real projects to illustrate how this might be done.

In the case of biological control of cottony cushion scale (*I. purchasi*) on the Galápagos by introduction of the specialized ladybird *R. cardinalis*, it was predicted based on quarantine studies that no native insects would be attacked. This prediction was subsequently further examined, after release, by means of direct field observations of naturally occurring non-target insect colonies and by staged encounters inside a large field cage in which various native or exotic non-target insects were exposed to the ladybird under semi-natural conditions so that the beetle frequently encountered non-target aphids, scales, mites, and mealybugs. Also in this same case, colonies of such non-target species were examined in the field to see if they had been foraged on by the ladybird. In aggregate some 30 hours of cage studies and 12 hours of field observations found no evidence of non-host attack. In field cages of 351 encounters, 166 were with cottony cushion scale stages and of these 53 (32%) led to attacks, while 185 were with non-target insects (27 with aphids, 113 with scales, 37 with mealybugs, and 8 with mites) and none led to attacks (Hoddle et al., 2013).

For releases of biological control agents against the invasive riparian plant saltcedar (*Tamarix* spp.) in the western United States, two predicted potential impacts were examined in post-release studies. The first of these came from host-range tests that showed that native plants in the genus *Frankenia* might be at risk from feeding by saltcedar beetles (*Diorhabda* spp.). While some oviposition and feeding had occurred in no-choice laboratory tests, these were interpreted as unlikely to occur in the field. To assess if that interpretation was accurate, potted *Frankenia* plants (primarily *F. salina* [Molina] I. M. Johnston) were exposed in the field near stands of saltcedar that were being defoliated by the saltcedar beetle (Dudley and Kazmer, 2005), when large numbers of starving beetles would be present. If *Frankenia* were attractive to the beetle, the large local population would be expected to colonize the plant and feed on it. Actual effects on *Frankenia* plants were minor (fewer than four plants with leaf damage, no oviposition, and no larval development and no dieback of damaged *F. salina* stems). *Frankenia* plants continued growing once tamarix beetle populations subsided.

For the saltcedar project, a second effect of potential concern was also predicted, that an introduced species of *Tamarix*, athel (*T. aphylla*), used in northern Mexico as a shade tree, might be attacked. In the laboratory, this was a suitable host (Milbrath and DeLoach, 2006). It was predicted that some attack would occur from the beetle during large defoliation events. Post-release monitoring detected such attack but also found that damage to athel was moderate and beetle populations did not persist on the plant (Moran, 2010). These examples illustrate the potential to tie post-release monitoring or field experiments to specific predictions made in the pre-release study of a biological control agent. Of course, substantial post-release monitoring would also be likely to detect idiosyncratic indirect impacts that might never have been predicted even by thorough pre-release studies, such as the increase in populations of deer mice (*Peromyscus maniculatus* [Wagner]), subsidized by the introduction of the tephritid gall flies *Urophora affinis* (Frauenfeld) and *Urophora quadrifasciata* (Meigen) to control spotted knapweed (*Centaurea stoebe* L. [formerly known as *C. maculosa* Lam.]) (Pearson and Callaway, 2003; Ortega et al., 2004). Such post-release studies are highly desirable and recommended in our code of best practices in which we describe our view of an ideally conducted use of biological control (Chapter 15).

Conclusions

Given the likelihood that future years will bring additional high-impact invasive plants and insects, the need for biological control programs to suppress them will remain. To make such programs safer, they should be tightly integrated into restoration ecology and conservation biology to ensure that projects have a good perspective on the biodiversity of the communities affected by the invasive species. Laws and practices will shape how biological control targets and agents are selected in particular countries. As discussed above, such laws and practices evolve over time. Here we have made some suggestions concerning what principles might guide the development of better review systems.

Acknowledgments

We thank Christian Marks, Bernd Blossey, David Wagner, and Peter Stiling for reviews of Chapter 5.

References

Amrine, J. W. Jr. 2002. Multiflora rose, pp. 265–292. *In*: R. Van Driesche, B. Blossey, M. Hoddle, et al. 2002. *Biological Control of Invasive Plants in the Eastern United States*. FHTET-2002-04. USDA Forest Service. Morgantown, West Virginia, USA. 413 pp.

Andersson, B., O. Engelmark, O. Rosvall, and K. Sjöberg. 1999. Environmental impact analysis (EIA) concerning lodgepole pine forestry in Sweden. Forestry Research Institute of Sweden. Uppsala, Sweden.

Andreas, J. E., M. Schwarzländer, and R. de Clerck-Floate. 2008. The occurrence and potential relevance of post-release nontarget attack by *Moguones cruciger*, a biological control agent for *Cynoglossum officinale* in Canada. *Biological Control* **46**: 304–311.

Anon. 1972. Biological control of weeds. Research Branch Report #192 T-2002-04. Canada Department of Agriculture, 1971. Ottawa: Canada Department of Agriculture. 413 pp.

Baker, J. L. and N. A. P. Webber. 2008. Feeding impacts of a leafy spurge (*Euphorbia esula*) biological control agent on a native plant, *Euphorbia robusta*. *Invasive Plant Science and Management* **1**: 26–30.

Barton, J., S.V. Fowler, A. F. Gianotti, et al. 2007. Successful biological control of mist flower (*Ageratina riparia*) in New Zealand: agent establishment, impact and benefits to the native flora. *Biological Control* **40**: 370–385.

Boettner, G. H., J. S. Elkinton, and C. J. Boettner. 2000. Effects of a biological control introduction on three nontarget native species of saturniid moths. *Conservation Biology* **14**: 1798–1806.

Boudjelas, S. 2009. Public participation in invasive species management, pp. 93–107. *In*: Clout, M. N. and P. A. Williams (eds.). *Invasive Species Management. A Handbook of Principles and Techniques*. Oxford University Press. Oxford, UK.

Boughton, A. J. and R.W. Pemberton. 2009. Establishment of an imported natural enemy, *Neomusotima conspurcatalis* (Lepidoptera: Crambidae) against an invasive weed, Old World climbing fern, *Lygodium microphyllum*, in Florida. *Biological Control Science and Technology* **19**: 769–772.

Breiter, N. C. and T. R. Seastedt. 2007. Postrelease evaluation of *Mecinus janthinus* host specificity, a biological control agent of invasive toadflax (*Linaria* spp.). *Weed Science* **55**: 164–168.

Buhs, J. B. 2004. *The Fire Ant Wars*. University of Chicago Press. Chicago, Illinois, USA.

Burdick, A. 2005. *Out of Eden: An Odyssey of Ecological Invasion*. Farrar, Straus and Giroux. New York.

Calderón Alvarez, C., C. E. Causton, M. S. Hoddle, et al. 2012. Monitoring the effects of *Rodolia cardinalis* on *Icerya purchasi* populations on the Galápagos Islands. *Biological Control* **57**: 167–179.

Caltagirone, L. E. and R. L. Doutt. 1989. The history of the vedalia beetle importation to California and its impact on the development of biological control. *Annual Review of Entomology* **34**: 1–16.

Campbell, C. L. and D. L. Long. 2001. The campaign to eradicate the common barberry in the United States, pp. 16–43. *In*: Peterson, P. D. (ed.). *Stem Rust of Wheat: From Ancient Enemy to Modern Foe*. APS Press. St. Paul, Minnesota, USA.

Center, T. D., M. F. Purcell, P. D. Pratt, et al. 2012. Biological control of *Melaleuca quinquenervia*: an Everglades invader. *Biological Control* **57**: 151–165.

Clergeau, P., P. Yésou, and C. Chadenas. 2005. Ibis sacré *Threskiornis aethiopicus*. État actuel et impact potentiel des populations introduites en France métropolitaine. Institut National de la Recherche Agronomique. Rennes, France.

Clout, M. N. and P. A. Williams (eds.). 2009. *Invasive Species Management. A Handbook of Principles and Techniques*. Oxford University Press. Oxford, UK.

Coombs, M. 2002. Post-release evaluation of *Trichopoda giacomellii* (Diptera: Tachinidae) for efficacy and non-target effects, pp. 399–406. *In*: Van Driesche, R. G. (ed.). *Proceedings of the 1st International Symposium on Biological Control of Arthropods, Honolulu, Hawaii, 14–18 January 2002*. FHTET 03-05. USDA Forest Service. Morgantown, West Virginia.

Courtenay, W. R., Jr. and G. K. Meffe. 1989. Small fishes in strange places: a review of introduced poeciliids, pp. 319–221. *In*: Meffe, G. K. and F. F. Snelson, Jr. (eds). *Ecology and Evolution of Livebearing Fishes (Poeciliidae)*. Prentice-Hall. Englewood Cliffs, New Jersey, USA.

Creed, R. P., Jr. and S. P. Sheldon. 1993. The effect of feeding by a North American weevil, *Euhrychiopsis lecontei*, on Eurasian watermilfoil (*Myriophyllum spicatum*). *Aquatic Botany* **45**: 245–256.

Crooks, J. A. 2005. Lag times and exotic species: the ecology and management of biological invasions in slow-motion. *Écoscience* **12**: 316–329.

Culver, J. A. 1919. Study of *Compsilura concinnata*, an imported tachinid parasite of the gipsy moth and the brown-tail moth. Bulletin 766. United States Department of Agriculture, Washington, DC. 27 pp.

Dahlsten, D. L. 1986. Control of invaders, pp. 275–302. *In*: Mooney, H. A. and J. A. Drake (eds.). *Ecology of Biological Invasions of North America and Hawaii*. Springer-Verlag. New York.

Davis, M. A., J. P. Grime, and K. Thompson. 2000. Fluctuating resources in plant communities: a general theory of invasibility. *Journal of Ecology* **88**: 528–534.

DePrenger-Levin, M. E., T. A. Grant, III, and C. Dawson, 2010. Impacts of the introduced biological control agent *Rhinocyllus conicus* (Coleoptera: Curculionidae) on the seed production and population dynamics of *Cirsium ownbeyi* (Asteraceae), a rare, native thistle. *Biological Control* **55**: 79–84.

Dodd, A. P. 1940. *The Biological Control Campaign against Prickly Pear*. Commonwealth Prickly Pear Board, Brisbane, Australia. 177 pp.

Dubuffet, A., S. Dupas, F. Frey, et al. 2007. Genetic interactions between the parasitoid wasp *Leptopilina boulardi* and its *Drosophila* hosts. *Heredity* **98**: 21–27.

Dudley, T. L. 2000. *Arundo donax* L., pp. 53–58. *In*: Bossard, C. C., J. M. Randall, and M. C. Hoshovsky (eds.). *Invasive Plants of California's Wildlands*. University of California Press. Berkeley, California, USA.

Dudley, T. L. and D. J. Kazmer. 2005. Field assessment of the risk posed by *Diorhabda elongata*, a biological control agent for control of saltcedar (*Tamarix* spp.), to a nontarget plant, *Frankenia salina*. *Biological Control* **35**: 265–275.

Elkinton, J. S., D. Parry, and G. H. Boettner. 2006. Implicating an introduced generalist parasitoid in the invasive browntail moth's enigmatic demise. *Ecology* **87**: 2664–2672.

Elmendorf, S., J. Byrnes, A. Wright, et al. 2005. *Fear and Fishing in Lake Davis (DVD)*. Flag in the Ground Productions. Davis, California, USA.

European Commission and BIO Intelligence Service. 2011. A comparative assessment of existing policies on invasive species in the EU member states and in selected OECD countries. BIO Intelligence Service. Paris.

Evans, J. M., A. C. Wilkie, and J. Burkhardt. 2008. Adaptive management of nonnative species: moving beyond the "either-or" through experimental pluralism. *Journal of Agricultural and Environmental Ethics* **21**: 521–539.

Ewel, J. J. 1986. Invasibility: lessons from south Florida, pp. 214–230. *In*: Mooney, H. A. and J. A. Drake (eds.). *Ecology of Biological Invasions of North America and Hawaii*. Springer-Verlag. New York.

Frank, J. H. and E. D. McCoy. 2007. The risk of classical biological control in Florida. *Biological Control* **41**: 151–174.

Fuester, R. W., A. E. Hajek, J. S. Elkinton, and P. W. Schaefer. 2014. Gypsy moth (*Lymantria dispar* L.) (Lepidoptera: Erebidae: Lymantriinae), pp. 49–82. *In*: Van Driesche, R. G. and R. Reardon (eds.). *The Use of Classical Biological Control to Preserve Forests in North America*. FHTET- 2013-02. USDA Forest Service, Morgantown, West Virginia, USA. Available from: http://www.fs.fed.us/foresthealth/technology/pub_titles.shtml [Accessed January 2016].

Gasco, D. 1995. Keep up efforts to save Australian pines on A1A. *Palm Beach Post May* **23**: 10.

Gassmann, A. and S. M. Louda. 2001. *Rhinocyllus conicus*: initial evaluation and subsequent impacts in North America, pp. 147–183. *In*: Wajnberg, E., J. K. Scott, and P. C. Quimby, (eds.). *Evaluating Indirect Ecological Effects of Biological Control*. CABI Publishing. Wallingford, UK.

Gardener, M. R. and C. Grenier. 2011. Challenges facing the Galápagos Islands, pp. 73–85. *In*: Baldacchino, G. and D. Niles (eds.). *Island Futures: Conservation and Development across the Asia-Pacific Region*. Springer. Tokyo.

Getz, C. W. 1989. Legal implications of eradication programs, pp. 66–73. *In*: Dahlsten, D. L. and R. Garcia (eds.). *Eradication of Exotic Pests*. Yale University Press. Yale, Connecticut, USA.

Gobster, P. H. and S. C. Barro. 2000. Negotiating nature: making restoration happen in an urban park context, pp. 185–208. *In*: Gobster, P. H. and R. B. Hull (eds.). *Restoring Nature: Perspectives from the Social Sciences and Humanities*. Island Press. Washington, DC.

Hajek, A. E. and P. C. Tobin. 2011. Introduced pathogens follow the invasion front of a spreading alien host. *Journal of Animal Ecology* **80**: 1217–1226.

Harmon, J. P., E. Stephens, and J. Losey. 2007. The decline of native coccinellids (Coleoptera: Coccinellidae) in the United States and Canada. *Journal of Insect Conservation* **11**: 85–94.

Hawkes, R. B., L. A. Andres, and P. H Dunn. 1972. Seed weevil released to control milk thistle *California Agriculture* **26**(12): 14.

Havens, K., C. L. Jolls, J. E. Marik, et al. 2012. Effects of a non-native biological control weevil, *Larinus planus*, and other emerging threats on populations of the federally threatened Pitcher's thistle, *Cirsium pitcheri*. *Biological Conservation* **155**: 202–211.

Helford, R. M. 2000. Constructing nature as constructing science: expertise, activist science, and public conflict in the Chicago Wilderness, pp. 119–142. *In*: Gobster, P. H. and R. B. Hull (eds.). *Restoring Nature: Perspectives from the Social Sciences and Humanities*. Island Press, Washington, DC.

Hernández, J., H. Sánchez, A. Bello, and G. González. 2007. Preventive programme against the cactus moth *Cactoblastis cactorum* in Mexico, pp. 345–350. *In*: Vreysen, M. J. B., A. S. Robinson, and J. Hendrichs (eds.). *Area-Wide Control of Insect Pests: From Research to Field Implementation*. Plant Health General Directorate, SENASICA SAGARPA, Coyoacán México, D.F. 04100, Mexico. Springer. Dordrecht, The Netherlands.

Herr, J. C. 2000. Evaluating non-target effects: the thistle story, pp. 12–17. *In*: Hoddle, M. S. (ed.). *California Conference on Biological Control II. The Historic Mission Inn Riverside, California, USA, 11–12 July, 2000*. Center for Biological Control, University of California, Berkeley, California, USA.

Hesler, L. S., M. A. Catangui, J. E. Losey, et al. 2009. Recent records of *Adalia bipunctata* (L.), *Coccinella transversoguttata richardsoni* Brown, and *Coccinella novemnotata* Herbst (Coleoptera: Coccinellidae) from South Dakota and Nebraska. *Coleopterists Bulletin* **63**: 475–484.

Hight, S. D., J. E. Carpenter, S. Bloem, and K. S. Bloem. 2005. Developing a sterile insect release program for *Cactoblastis cactorum* (Berg) (Lepidoptera: Pyralidae): effective overflooding ratios and release-recapture field studies. *Environmental Entomology* **34**: 850–856.

Hoddle, M. S., C. Crespo Ramírez, C. D. Hoddle, et al. 2013. Post release evaluation of *Rodolia cardinalis* (Coleoptera: Coccinellidae) for control of *Icerya purchasi* (Hemiptera: Monophlebidae) in the Galápagos Islands. *Biological Control* **67**: 262–274.

Horber, E. 1980. Accelerating the biological suppression of the muskthistle (*Carduus nutans* L.). *Journal of the Kansas Entomological Society* **53**: 606.

Johnson, M. T., P. A. Follett, A. D. Taylor, and V. P. Jones. 2005. Impacts of biological control and invasive species on a non-target native Hawaiian insect. *Oecologia* **142**: 529–540.

Katsoyannos, P., D. C. Kontodimas, G. J. Stathas, and C. T. Tsartsalis. 1997. Establishment of *Harmonia axyridis* on citrus and some data on its phenology in Greece. *Phytoparasitica* **25**: 183–191.

Kellogg, S. K., L. S. Fink, and L. P. Brower. 2003. Parasitism of native luna moths, *Actias luna* (L.) (Lepidoptera: Saturniidae) by the introduced *Compsilura concinnata* (Meigen) (Diptera: Tachinidae) in central Virginia, and their hyperparasitism by trigonalid wasps (Hymenoptera: Trigonalidae). *Environmental Entomology* **32**: 1019–1027.

Keystone Center. 1993. *National Ecosystem Management. Forum Meeting Summary*. Keystone Center, Keystone, Colorado, USA.

Kindlmann, P. and A. F. G. Dixon. 1999. Generation time ratios – determinants of prey abundance in insect predator-prey interactions. *Biological Control* **16**: 133–138.

Kindlmann, P., H. Yasuda, Y. Kajita, and A. F. G. Dixon. 2005. Field test of the effectiveness of ladybirds in controlling aphids, pp. 441–447. *In*: Hoddle, M. (ed.). *Second International Symposium on Biological Control of Arthropods, Davos, Switzerland, 12–16 September, 2005*. FHTET-2005-08. USDA Forest Service, Morgantown, West Virginia, USA.

Klein, J. and M. Verlaque. 2008. The *Caulerpa racemosa* invasion: a critical review. *Marine Pollution Bulletin* **56**: 205–225.

Knickerbocker, B. 1994. Snaring of Hawaii's feral pigs angers animal activists. *Christian Science Monitor*, 6 January **6**: 7.

Koch, R. L. and T. L. Galvan. 2008. Bad side of a good beetle: the North American experience with *Harmonia axyridis*. *Biological Control* **53**: 23–35.

Kok, L. T. and W. W. Surles. 1975. Successful biological control of musk thistle by an introduced weevil, *Rhinocyllus conicus*. *Environmental Entomology* **4**: 1025–1027.

Leppanen, C., A. Alyokhin, and S. Gross. 2012. Competition for aphid prey between different lady beetle species in a laboratory arena. *Psyche: A Journal of Entomology*: Article ID 890327.

Light, A. 2000. Restoration, the value of participation, and the risks of professionalization, pp. 163–181. *In*: Gobster, P. H. and R. B. Hull (eds.). *Restoring Nature: Perspectives from the Social Sciences and Humanities*. Island Press. Washington, DC.

Lombaert, E., T. Guillemaud, J.-M. Cornuet, et al. 2010. Bridgehead effect in the worldwide invasion of the biological control harlequin ladybird. *PLoS One* **5**(3): e9743. doi:10.1371/journal.pone.0009743.

Losey, J. E., J. E. Perlman, and E. R. Hoebeke. 2007. Citizen scientist rediscovers rare nine-spotted lady beetle, *Coccinella novemnotata*, in eastern North America. *Journal of Insect Conservation* **11**: 415–417.

Louda, S. M. 1998. Population growth of *Rhinocyllus conicus* (Coleoptera: Curculionidae) on two species of native thistles in Prairie. *Environmental Entomology* **27**: 834–841.

Louda, S. M. 2000. Negative ecological effects of the musk thistle biological control agent, *Rhinocyllus conicus*, pp. 215–243. *In*: Follett, P. A. and J. J. Duan (eds.). *Non-target Effects of Biological Control*. Kluwer. Boston, Massachusetts, USA.

Louda, S. M. and C. W. O'Brien. 2002. Unexpected ecological effects of distributing the exotic weevil, *Larinus planus* (F.),

for the biological control of Canada thistle. *Conservation Biology* **16**: 717–727.

Louda, S. M., D. Kendall, J. Connor, and D. Simberloff. 1997. Ecological effects of an insect introduced for the biological control of weeds. *Science* **277**(5329): 1088–1090.

Lowe, S., M. Browne, and S. Boudjelas. 2001. *100 of the World's Worst Invasive Alien Species. A Selection from the Global Invasive Species Database*. IUCN-ISSG. Auckland, New Zealand.

Lynch, L. D. and M. B. Thomas. 2000. Nontarget effects in the biological control of insects with insects, nematodes, and microbial agents: the evidence. *Biological Control News and Information* **21**(4): 117N–130N.

MacClaine, L. S. 1916. Rearing the parasites of the brown-tail moth in New England for colonization in Canada. *Agricultural Gazette* **3**(1): 22–25.

MacDougall, A. S. and R. Turkington. 2005. Are invasive species the drivers or passengers of change in degraded ecosystems? *Ecology* **86**: 42–55.

Mack, R. N. and W. M. Lonsdale. 2002. Controlling invasive plants: hard-won lessons from continents and islands, pp. 164–172. *In*: Veitch, D. and M. Clout (eds.). *Turning the Tide: Eradication of Invasive Species*. Invasive Species Specialist Group of the World Conservation Union (IUCN). Auckland, New Zealand.

Maloy, O. C. 1997. White pine blister rust control in North America: a case history. *Annual Review of Phytopathology* **35**: 87–109.

Maris, V. and A. Béchet. 2010. From adaptive management to adjustive management: a pragmatic account of biodiversity values. *Conservation Biology* **24**: 966–973.

Martin, P. H., C. D. Canham, and P. L. Marks. 2009. Why forests appear resistant to exotic plant invasions: intentional introductions, stand dynamics, and the role of shade tolerance. *Frontiers in Ecology and the Environment* **7**: 142–149.

McClay, A. S. 1990. The potential of *Larinus planus* (Coleoptera: Curculionidae), an accidentally introduced insect in North America, for biological control of *Cirsium arvense* (Compositae), pp. 173–179. *In*: Anon. *Proceedings of the VIII International Symposium on Biological Control of Weeds*. Istituto Sperimentale per la Vegetale, Ministero dell'Agricoltura e delle Foreste. Rome, Italy.

McClay, A. S., R. S. Bourchier, R. A. Butts, and D. P. Peschken. 2001. *Cirsium arvense* (L.) Scopoli, Canada thistle (Asteraceae), pp. 318–330. *In*: Mason, P. G. and J. T. Huber (eds.). *Biological Control Programmes in Canada, 1981–2000*. CABI Publishing. Wallingford, UK.

McFadyen, R. E. C., M. Vitelli, and C. Setter. 2002. Host specificity of the rubber vine moth, *Euclasta whalleyi* Popescu-Gorj and Constantinescu (Lepidoptera: Crambidae: Pyraustinae): field host-range compared to that predicted by laboratory tests. *Australian Journal of Entomology* **41**: 321–323.

Meyer, J.-Y., M. Fourdrigniez, and R. Taputuarai, R. 2012. Restoring habitat for native and endemic plants through the introduction of a fungal pathogen to control the alien invasive tree *Miconia calvescens* in the island of Tahiti. *Biological Control* **57**: 191–198.

Milbrath, L. R. and C. J. Deloach. 2006. Acceptability and suitability of athel, *Tamarix aphylla*, to the leaf beetle *Diorhabda elongata* (Coleoptera: Chrysomelidae), a biological control agent of saltcedar (*Tamarix* spp.). *Environmental Entomology* **35**: 1379–1389.

Moran, P. J. 2010. Lack of establishment of the Mediterranean tamarisk beetle *Diorhabda elongata* (Coleoptera: Chrysomelidae) on athel (*Tamarix aphylla*) (Tamaricaceae) in south Texas. *Southwestern Entomologist* **35**: 129–145.

Muldrew, J. A. 1953. The natural immunity of the larch sawfly (*Pristiphora erichsonii* [Htg.]) to the introduced parasite (*Mesoleius tenthredinis* Morley), in Manitoba and Saskatchewan. *Canadian Journal of Zoology* **31**: 313–332.

Murray, J., E. Murray, M. S. Johnson, and B. Clarke. 1988. The extinction of *Partula* on Moorea. *Pacific Science* **42**: 150–153.

Nuñez, M. A. and D. Simberloff. 2005. Invasive species and the cultural keystone concept. *Ecology and Society* **10**(1): r4. Available from: http://www.ecologyandsociety.org/vol10/iss1/resp4/ [Accessed January 2016].

Office of Technology Assessment (U.S. Congress). 1993. *Harmful Non-Indigenous Species in the United States*. OTA-F-565. U.S. Government Printing Office. Washington, DC. Available from: http://www.anstaskforce.gov/Documents/OTA_Report_1993.pdf [Accessed January 2016].

Ortega Y., D. E. Pearson, and K. S. McKelvey. 2004. Effects of biological control agents and exotic plant invasion on deer mouse populations. *Ecological Applications* **14**: 241–253.

Paynter, Q., N. Martin, J. Berry, et al. 2008. Non-target impacts of *Phytomyza vitalbae*, a biological control agent of the European weed *Clematis vitalba* in New Zealand. *Biological Control* **44**: 248–258.

Pearson, D. E. and R. M. Callaway. 2003. Indirect effects of host specific biological control agents. *Trends in Ecology and Evolution* **18**: 456–461.

Pemberton, C. E. 1957. Progress in the biological control of undesirable plants in Hawaii, pp. 124–126. *In*: Anon (Secretariat of Congress) (ed.). *Proceedings of the Ninth Pacific Science Congress. Volume 9*. Department of Science. Bangkok, Thailand.

Pemberton, R. W. 2000. Predictable risk to native plants in weed biological control. *Oecologia* **125**: 489–494.

Pemberton, R. W. and H. Liu. 2007. Control and persistence of native *Opuntia* on Nevis and St. Kitts 50 years after the introduction of *Cactoblastis cactorum*. *Biological Control* **41**: 272–282.

Petit, J. N., M. S. Hoddle, J. Grandgirard, et al. 2009. Successful spread of a biological control agent reveals a biosecurity failure: elucidating long distance invasion pathways for *Gonatocerus ashmeadi* in French Polynesia. *Biological Control* **54**: 485–495.

Pickart, A. J. 2013. Dune restoration over two decades at Lanphere and Ma-le'l Dunes in northern California, pp. 159–171. *In*: Martínez, M. L., J. B. Gallego-Fernández, and P. A. Hesp (eds.). *Restoration of Coastal Dunes*. Springer. Berlin, Germany.

Pratt, P. D. and T. D. Center. 2012. Biological control without borders: the unintended spread of introduced weed biological control agents. *Biological Control* **57**: 319–329.

Pratt, P. D., M. B. Rayamajhi, T. D. Center, et al. 2009. The ecological host range of an intentionally introduced herbivore: a comparison of predicted versus actual host use. *Biological Control* **49**: 146–153.

Puri, A., G. E. MacDonald, and W. T. Haller. 2007. Stability of fluridone-resistant hydrilla (*Hydrilla verticillata*) biotypes over time. *Weed Science* **55**: 12–15.

Pyke, G. H. 2008. Plague minnow or mosquito fish? A review of the biology and impacts of introduced *Gambusia* species. *Annual Review of Ecology, Evolution, and Systematics* **39**: 171–191.

Rayamajhi, M., P. Pratt, T. Center, et al. 2009. Decline in exotic tree density facilitates increased plant diversity: the experience from *Melaleuca quinquenervia* invaded wetlands. *Wetlands Ecology and Management* **17**: 455–467.

Reimer, N. J. 1988. Predation on *Liothrips urichi* Karny (Thysanoptera: Phlaeothripidae): a case of biotic interference. *Environmental Entomology* **17**: 132–134.

Rentería, J. L., M. R. Gardener, F. D. Panetta, et al. 2012. Possible impacts of the invasive plant *Rubus niveus* on the native vegetation of the *Scalesia* forest in the Galapagos Islands. *PLoS One* **7** (10): e48106.

Rees, N. E. 1977. Impact of *Rhinocyllus conicus* on thistles in southwestern Montana. *Environmental Entomology* **6**: 839–842.

Rohr, J. R., A. M. Schotthoefer, T. R. Raffel, et al. 2008. Agrochemicals increase trematode infections in a declining amphibian species. *Nature (London)* **455**(7217): 1235–1239.

Roley, S. S. and R. M. Newman. 2006. Developmental performance of the milfoil weevil, *Euhrychiopsis lecontei* (Coleoptera: Curculionidae), on northern watermilfoil, Eurasian watermilfoil, and hybrid (northern × Eurasian) watermilfoil. *Environmental Entomology* **35**: 121–126.

Russell, F. L. and S. M. Louda. 2004. Phenological synchrony affects interaction strength of an exotic weevil with Platte thistle, a native host plant. *Oecologia* **139**: 525–534.

Russell, F. L., S. M. Louda, T. A. Rand, and S. D. Kachman. 2007. Variation in herbivore-mediated indirect effects of an invasive plant on a native plant. *Ecology* **88**: 413–423.

Sands, D. P. A. and M. T. Coombs. 1999. Evaluation of the Argentinian parasitoid, *Trichopoda giacomellii* (Diptera: Tachinidae), for biological control of *Nezara viridula* (Hemiptera: Pentatomidae) in Australia. *Biological Control* **15**: 19–24.

Sauer, S. A. and K. L. Bradley, 2008. First record for the biological control agent *Rhinocyllus conicus* (Coleoptera: Curculionidae) in a threatened native thistle, *Cirsium hillii* (Asteraceae), in Wisconsin, USA. *Entomological News* **119**: 90–95.

Schmitz, D. C. and D. Simberloff. 2001. Needed: a national center for biological invasions. *Issues in Science and Technology* **17**(4): 57–62.

Schmitz, D. C., D. Simberloff, R. H. Hofstetter, et al. 1997. The ecological impact of nonindigenous plants, pp. 39–61. *In*: Simberloff, D., D. C. Schmitz, and T. C. Brown (eds.). *Strangers in Paradise: Impact and Management of Nonindigenous Species in Florida*. Island Press. Washington, DC.

Selfridge, J. A., D. Parry, and G. H. Boettner. 2007. Parasitism of barrens buck moth, *Hemileuca maia* Drury, in early and late successional pine barrens habitats. *Journal of the Lepidopterists' Society* **61**: 213–221.

Shands, W. A., G. W. Simpson, and R. H. Storch. 1972. Insect predators for controlling aphids on potatoes. Part 9. Winter survival of *Coccinella* species in field cages over grassland in northeastern Maine. *Journal of Economic Entomology* **65**: 1392–1396.

Sheldon, S. P. and L. M. O'Bryan. 1996. Life history of the weevil *Euhrychiopsis lecontei*, a potential biological control agent of Eurasian watermilfoil. *Entomological News* **107**: 16–22.

Sheppard, A. W., R. D. van Klinken, and T. A. Heard. 2005. Scientific advances in the analysis of direct risks of weed biological control agents to nontarget plants. Series Title: Special issue: Science and decision making in biological control of weeds: benefits and risks of biological control. *Biological Control* **35**: 215–226.

Sighinolfi, L., G. Febvay, M. L. Dindo, et al. 2008. Biological and biochemical characteristics for quality control of *Harmonia axyridis* (Pallas) (Coleoptera: Coccinellidae) reared on a liver-based diet. *Archives of Insect Biochemistry and Physiology* **68**: 26–39.

Shore, D. 1997. Controversy erupts over restoration in Chicago area. *Restoration and Management Notes* **15**: 25–31.

Simberloff, D. 2002. Managing established populations of alien species, pp. 269–278. *In*: Claudi, R., P. Nantel, and E. Muckle-Jeffs (eds.). *Alien Invaders in Canada's Waters, Wetlands, and Forests*. Natural Resources Canada, Canadian Forest Service. Ottawa, Ontario, Canada.

Simberloff, D. 2009. We can eliminate invasions or live with them. Successful management projects. *Biological Invasions* **11**: 149–157.

Simberloff, D. 2012a. Nature, natives, nativism, and management: worldviews underlying controversies in invasion biology. *Environmental Ethics* **34**: 5–25.

Simberloff, D. 2012b. Risks of biological control for conservation purposes. *Biological Control* **57**: 263–276.

Simmonds, F. J. and F. D. Bennett. 1966. Biological control of *Opuntia* spp. by *Cactoblastis cactorum* in the Leeward Islands (West Indies). *Entomophaga* **11**: 183–189.

Smith, C. W. 1992. Distribution, status, phenology, rate of spread, and management of *Clidemia* in Hawai'i, pp. 241–253. *In*: Stone, C. P., C. W. Smith, and J. T. Tunison (eds.). *Alien Plant Invasions in Native Ecosystems of Hawaii*. University of Hawaii Cooperative National Park Resources Studies Unit. Honolulu, Hawaii, USA.

Smith, M. A., D. M. Wood, D. H. Janzen, et al. 2007. DNA barcodes affirm that 16 species of apparently generalist tropical parasitoid flies (Diptera: Tachinidae) are not all generalists. *Proceedings of the National Academy of Sciences of the United States of America* **104**(12): 4967–4972.

Snow, E. L. and K. Dhileepan. 2008. The suitability of non-target native mangroves for the survival and development of the lantana bug *Aconophora compressa*, an introduced weed biological control agent. *Biological Control* **53**: 699–707.

Spear, R. J. 2005. *The Great Gypsy Moth War*. University of Massachusetts Press. Amherst, Massachusetts, USA.

Stiling, P. and D. C. Moon. 2001. Protecting rare Florida cacti from attack by the exotic cactus moth, *Cactoblastis cactorum* (Lepidoptera: Pyralidae). *Florida Entomologist* **84**: 506–509.

St. Quinton, J. M., M. F. Fay, M. Ingrouille, and J. Faull. 2011. Characterisation of *Rubus niveus*: a prerequisite to its biological control in oceanic islands. *Biological Control Science and Technology* **21**: 733–752.

Strong, D. R. 1997. Fear no weevil? *Science (Washington)* **277** (5329): 1058–1059.

Suckling, D. M. and R. F. H. Sforza. 2014. What magnitude are observed non-target impacts from weed biological control? *PLoS One* **9** (1) e84847.

Taylor, D. B. J., T. A. Heard, Q. Paynter, and H. Spafford. 2007. Nontarget effects of a weed biological control agent on a native plant in Northern Australia. *Biological Control* **42**: 25–33.

Tedders, W. L. and P. W. Schaefer. 1994. Release and establishment of *Harmonia axyridis* (Coleoptera: Coccinellidae) in the southeastern United States. *Entomological News* **105**: 228–243.

Telg, R. and M. G. Dufresne. 2001. Agricultural communications efforts during Florida's medfly infestations of 1997 and 1998. *Journal of Applied Communications* **85**: 7–22.

Thresher, R. E., K. Hayes, N. J. Bax, et al. 2014. Genetic control of invasive fish: technological options and its role in integrated pest management. *Biological Invasions* **16**: 1201–1216.

Tummons, P. 2008. Haiwaii biological control controversy. *Environment Hawaii* **19**: 1.

Turner, C. E., R. W. Pemberton, and S. S. Rosenthal. 1987. Host utilization of native *Cirsium* thistles (Asteraceae) by the introduced weevil *Rhinocyllus conicus* (Coleoptera: Curculionidae) in California. *Environmental Entomology* **16**: 111–115.

United States National Research Council. 2004. *Nonnative Oysters in the Chesapeake Bay*. National Academy Press. Washington, DC.

Van Driesche, R. G., M. Hoddle, and T. Center. 2008. *Control of Pests and Weeds by Natural Enemies: An Introduction to Biological Control*. Blackwell. Oxford, UK.

van Klinken and T. A. Heard. 2000. Estimating fundamental host range: a host-specificity study of a potential biological control agent for *Prosopis* species (Leguminosae). *Biological Control Science and Technology* **10**: 331–342.

Veitch, C. R., M. N. Clout, and D. R. Towns (eds.). 2011. *Island Invasives: Eradication and Management. Proceedings of the International Conference on Island Invasives*. International Union for the Conservation of Nature, Gland, Switzerland and Centre for Biodiversity and Biosecurity, Auckland, New Zealand.

Vreysen, M. J. B., A. S. Robinson, and J. Hendrichs (eds.). 2007. *Area-wide Control of Insect Pests*. Springer. Dordrecht, The Netherlands.

Wacker, S. D. and J. L. Butler. 2006. Potential impact of two *Aphthona* spp. on a native nontarget *Euphorbia* species. *Rangeland Ecology and Management* **59**: 468–474.

Wapshere, A. J. 1974. A strategy for evaluating the safety of organisms for biological weed control. *Annals of Applied Biology* **77**: 201–211.

Williams, T. 1997. Killer weeds. *Audubon* **99**(2): 24–31.

Williams, T. 2002. America's largest weed. *Audubon* **104**(1): 24–31.

Witzgall, P., L. Stelinski, L. Gut, and D. Thomson. 2008. Codling moth management and chemical ecology. *Annual Review of Entomology* **53**: 503–522.

Yamaura, Y. 2008. Potential impacts of the invasive flatworm *Platydemus manokwari* on arboreal snails. *Biological Invasions* **11**: 737–742.

CHAPTER 6
Systematics and biological control

Jeremy C. Andersen[1] and David L. Wagner[2]

[1]*Department of Environmental Science, Policy, & Management, University of California, USA*
[2]*Department of Ecology & Evolutionary Biology, University of Connecticut, USA*

"The beginning of wisdom is to call things by their proper name."

Chinese proverb often attributed to Confucius

Introduction

There are few fields of study in ecology where the correct identification of an organism, whether it is to the species or population level, has so great an effect on ecosystem functioning and services as in the study of the biological control of pest species. Taxonomic identifications can be the difference between life and death for both humans and native organisms when food supplies or ecosystems are being threatened by invasive species (Wagner and Van Driesche, 2010) (see Box 6.1). To counter threats in both agricultural and wildlands, biological control programs have often been used to greatly reduce the abundance of invasive species and their detrimental effects (Van Driesche et al., 2010). If conservation biologists are to safely add biological control to their toolkit of management strategies (as is the thesis of this book), then they must be confident that the risks associated with the implementation of a biological control program have been clearly identified and evaluated (see Simberloff, 2012 and Chapter 5). We believe that one of the most important risk factors – and thankfully one that can easily be addressed – is the correct identification of target pests and their candidate biological control agents. In this chapter, we will first review how incorrect

taxonomic determinations can greatly affect the speed, safety, and success of biological control efforts (e.g., Bin et al., 2012). Then, we will highlight the critical nature of voucher materials and finish by reviewing molecular methods available for identification of biological control agents and their target pests.

Identification's effect on biological control service

Identification of target pests and their natural enemies plays an important role in the safety (both perceived and actual) and success of biological control services (Miller and Rossman, 1995). In particular, misidentifications (e.g., Pemberton, 2003) and failure to identify cryptic species complexes (e.g., DeBach, 1959) have delayed the discovery of biological control agents and implementation of control programs. Misidentifications have also resulted in the introduction of the wrong biological control agents. Examples presented in this chapter represent just a fraction of the cases where taxonomic issues proved problematic. Moreover, given both the nature and frequency of taxonomic problems that have surfaced to date, one can be certain that molecular reassessments (e.g., Hoddle et al., 2015) will long continue to reveal additional instances where misidentifications were responsible for delays in establishments, resulted in program failures, or proved detrimental to non-target species.

Integrating Biological Control into Conservation Practice, First Edition. Edited by Roy G. Van Driesche, Daniel Simberloff, Bernd Blossey, Charlotte Causton, Mark S. Hoddle, David L. Wagner, Christian O. Marks, Kevin M. Heinz, and Keith D. Warner.
© 2016 John Wiley & Sons, Ltd. Published 2016 by John Wiley & Sons, Ltd.

Misidentification of target pests

Pest misidentifications frequently occur when the target pest is a member of a cryptic species complex. An important example is the case of the biological control of the cassava mealybug, *Phenacoccus manihoti* Matile-Ferrero (Hemiptera: Pseudococcidae), a devastating pest of cassava (manioc) that is invasive in central Africa (Nwanze et al., 1979). When this pest first appeared in Africa in the 1970s its identity was unknown (Matile-Ferrero, 1977), although it was hypothesized to be of South American origin. A subsequent search for natural enemies in South America was initially misled due to the collection of a morphologically similar and previously undescribed mealybug feeding on cassava, *Phenacoccus herreni* Cox & Williams. However, once Cox and Williams (1981) examined specimens from Africa and South America, they found that individuals separated into two species, *P. manihoti* and *P. herreni*, and that these differed in their modes of reproduction. Subsequently, workers were able to rear a parasitic wasp from *P. manihoti* collected in subtropical South America (*P. herreni* being fully tropical) that effectively reduced pest densities of the cassava mealybug in Africa.

A second heralded example of a misidentified wildlands pest is the floating fern, *Salvinia molesta* D. S. Mitchell (Salviniales, Salviniaceae), one of the world's top 100 invasive species (Luque et al., 2014). This invasive aquatic weed is found in over 30 countries and greatly degrades water quality and aquatic habitats (Julien et al., 2002). When this invasive plant was first observed outside of its native range it was assumed to be *Salvinia auriculata* Aubl. (Room, 1990). Three potential biological control agents (the weevil *Cyrtobagous singularis* Hust., the crambid moth *Samea multiplicalis* [Guenée], and the acridid grasshopper *Paulinia acuminata* [Deg.]) were collected from *S. auriculata*'s native range in Brazil (Room, 1990). However, when they were released into the wild, herbivory on floating fern was only modest, insufficient to reduce populations of the weed (Julien, 1987). Further review of herbarium specimens in Brazil revealed that the invasive populations belonged to a different *Salvinia* species, giant salvinia (subsequently described as *S. molesta*), with its own herbivore fauna. Upon this discovery, a congener of the weevil *C. singularis*, *C. salviniae* Calder & Sands, was collected and released, which proved enormously effective in control of this floating fern (Room et al., 1981; Thomas and Room, 1986).

Misdirection of candidate natural enemy searches resulting from misidentifications continues today, but the frequency of such mistakes is being reduced by the use of more exacting taxonomic methods (see the section titled "Molecular methods for identification" for more examples). In the Florida Everglades, an invasive "lobate lac" scale insect believed to be *Paratachardina lobata lobata* (Chamberlin) (Hemiptera: Kerriidae) was found damaging native hardwood trees and was subsequently targeted for biological control (Pemberton, 2003). Consequently, natural enemy surveys were conducted in India and adjacent areas where *P. lobata* is native. Parasitoids collected from *P. lobata*, however, failed to reproduce successfully on scales from Florida (Schroer and Pemberton, 2007). Upon further investigation of specimens from Florida, Kondo and Gullan (2007) determined that the scale represented an undescribed species, which they named *P. pseudolobata*. Moreover, the discovery that the invasive individuals belonged to a different taxon catalyzed renewed efforts to locate the origin of the Floridian entity and secure more efficacious biological control agents.

Misidentification of target pests sometimes becomes problematic when unrecognized intraspecific differences are not considered during the selection (and release) of biological control agents. Brazilian peppertree, *Schinus terebinthifolius* Raddi (Sapindales, Anacardiaceae), is a widely distributed invasive plant (Cuda et al., 2006), and recent analyses have revealed that populations in Florida include at least two genetically distinct entities (Williams et al., 2005, 2007) (Figure 6.1). Subsequently, it was found that the candidate biological control agent *Pseudophilothrips ichini* (Hood) (Thysanoptera: Phlaeothripidae) also includes either cryptic species (Manrique et al., 2008) or ecotypes (Manrique et al., 2014), and that these species/ecotypes differ with regard to their virulence across the different populations and hybrids of Brazilian peppertree (Manrique et al., 2008).

Misidentification of natural enemies

Misidentifications have led to the rearing and release of unsuitable parasitoids across the globe (e.g., the wrong species of *Trichogramma* [Hymenoptera, Trichogrammatidae] for the biological control of *Diatraea saccharalis* [Fabricus] [Lepidoptera, Pyralidae] and the wrong species of *Prospaltella* [now *Encarsia*] [Hymenoptera: Aphelinidae] for the biological control of *Quadraspidiotus pernicious* [Comstock] [Hemiptera,

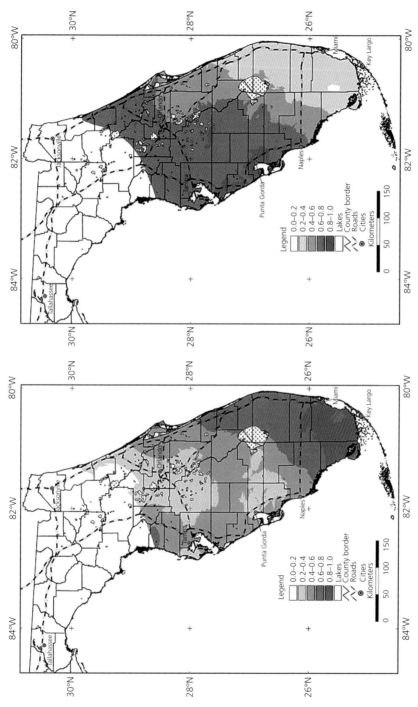

Figure 6.1 Haplotype data can be used to identify geographic distributions of invasive organisms. Williams et al. (2007) mapped the probability of having the eastern (left) and western (right) haplotypes of *Schinus terebinthifolius* in Florida. The identification of these genetic and geographic differences also led to the determination that different haplotypes vary in their susceptibility to a candidate biological control agent, *Pseudophilothrips ichnini*, which itself appears to represent a cryptic-species complex (Manrique et al., 2008). Williams et al. 2007. Reproduced with permission from Nature Publishing Group.

Diaspididae], reviewed in Bin et al., 2012). The taxonomic uncertainties surrounding the 33 predators released to control the balsam woolly adelgid (*Adelges piceae* [Ratzeburg]) (Hemiptera: Adelgidae) in Canada and the United States are legion and the vouchering so inadequate that there is little hope of ever knowing what exactly was released (Montgomery and Havill, 2014; see Chapter 10). The biological control of Florida and California red scales (*Chrysomphalus aonidum* [L.] and *Aonidiella aurantii* [Maskell], both Hemiptera, Diaspididae) was delayed by the presence of cryptic species of parasitoid wasps in the genus *Aphytis* (Hymenoptera: Aphelinidae) (DeBach, 1960), or more correctly stated, by the inability of workers to recognize the taxonomic complexity of *Aphytis*. In Hong Kong, where *Aphytis lingnanensis* Compere, 1955 was collected, it appeared to be the dominant parasitoid of *C. aonidum* (DeBach, 1959), and it was introduced successfully to control *C. aonidum* in Israel (DeBach, 1960). However, in the United States it only attacked *A. aurantii* and ignored *C. aonidum* (Flanders, 1956). Upon closer examination of morphological characters and through using mating trials, it was discovered that the field-collected individuals of *A. lingnanensis* in Hong Kong included two closely related taxa that were both host specific (DeBach, 1960). The two species were subsequently isolated and have been introduced for the successful biological control of both *C. aonidum* and *A. aurantii* (DeBach, 1960).

A more recent example of the challenges presented by cryptic species is the biological control of the Cuban laurel thrips, *Gynaikothrips ficorum* (Marchal) (Thysanoptera: Phaleothripidae), an important pest of *Ficus* (Rosales, Moraceae) (Denmark et al., 2014). Cuban laurel thrips is preyed upon by the minute pirate bug *Montandoniola moraguesi* (Puton) (Hemiptera: Anthocoridae) (Pluot-Sigwalt et al., 2009), and while *M. moraguesi* was successfully introduced to Florida, attempts to establish the species in California and Texas failed (Dobbs and Boyd, 2006). Pluot-Sigwalt et al. (2009) conducted a detailed morphological survey of individuals of *M. moraguesi* across its range and identified as many as four distinct cryptic species, with each pirate bug inhabiting a distinct geographic region (Pluot-Sigwalt et al., 2009). Perhaps some failures to establish *M. moraguesi* in other parts of the United States will be found to relate to introductions of climatically mismatched cryptic species.

Misidentifications can also handicap biological control efforts when a candidate agent is believed to already be established in the target region. For example, the fly *Delia radicum* (L.) (Diptera: Anthomyiidae) is an important pest of canola crops in Canada, causing up to $100 million in damage annually (Hemachandra, 2004). The staphylinid beetle *Aleochara bipustulata* (L.) is a candidate biological control agent for *D. radicum*, but published records suggested that the beetle was already present in Canada (Hemachandra et al., 2005). However, after examining museum specimens, Hemachandra et al. (2005) determined that individuals previously reported as *A. bipustulata* were misidentified and instead represented *Aleochara verna* Say. Currently, studies are underway to examine the potential value of introducing *A. bipustulata* in Canada (Holliday et al., 2012).

Similarly, the invasive weevil *Ceutorhynchus obstrictus* Marsham is an important pest of many Brassicaceae, including canola and rapeseed crops (Buntin et al., 1995). When this weevil was first noticed in the western United States, more than 80 years ago, surveys were conducted to locate natural enemies in its native and introduced ranges (Baker, 1936; Gahan, 1941; Doucette, 1944, 1948; Hanson et al., 1948). In Europe, two chalcidoid wasps (*Mesopolobus morys* [Walker] and *Trichomalus fasciatus* [Thomson]) parasitized 50–90% of *Ceutorhynchus* larvae (Buntin, 1998; Murchie and Williams, 1998). Initial surveys in North America reported both wasp species to be present – though field parasitism rates were less than 15% (Dosdall et al., 2009). When Gibson et al. (2005) examined adult specimens reared from *C. obstrictus* from Europe and North America, they found that the North American chalcidoids had been misidentified and that neither of the previously reported European parasitoids had yet established in western North America. There is now considerable interest in introducing both *M. morys* and *T. fasciatus* to North America (Gillespie et al., 2006).

Using classifications to make biological predictions

Classifications can play an important role in facilitating decisions that influence the success of biological control programs. Natural classifications that reflect underlying relatedness help to identify the region of origin of an organism (e.g., Qin and Gullan, 1998; Gwiazdowski et al., 2006) and allow predictions about host ranges (e.g., Sheppard et al., 2005). Host-range predictions have traditionally been dependent on existing classifications and phylogenies (see Morrison [2005] or Vandamme

[2009] for a review of phylogenetic methods). Whether inferring regions of origins or host ranges, approaches based on phylogenies are, however, only as good as the available classifications or phylogenies (Simberloff, 2012), and thus they should be used with caution where taxa of concern have not yet been the subject of modern systematic analyses (Messing, 2001).

Using phylogenies to predict the native range

In biological control settings, phylogenies can be used to identify the likely origins of both pests and control agents. For example, a cladogram revealed the likely region of origin of the Chinese wax scale, *Ceroplastes sinensis* Del Guercio (Hemiptera: Coccidae), an important pest of citrus crops. By the 1990s this scale pest was found worldwide and there existed considerable uncertainty over its origin (Qin and Gullan 1998). To determine the

scale's origin, Qin et al. (1994) constructed a phylogeny for wax scales based on 57 morphological characters. They then mapped the range for each of the 73 scales onto their cladogram (Figure 6.2). Surprisingly, they found that *C. sinensis* belonged to a clade whose other members were from South and Central America. Upon further review of the literature, the authors concluded that *C. sinensis* likely originated in Argentina; a subsequent field expedition to Argentina located *C. sinensis* and several of its parasitoids that were then introduced to other continents (including China) for control of the Chinese wax scale (Qin and Gullan, 1998). The above case is emblematic of the need to consult (or in this case, to construct) a phylogeny, because common sense would tell one that *C. sinensis* had originated in China, given that *sinensis* in Latin means "of China!"

Phylogenies and/or phylogeographic structure can also be useful when it is unclear whether a target pest

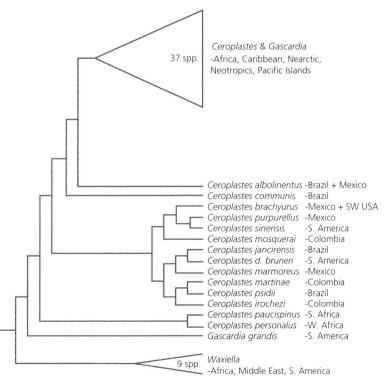

Figure 6.2 Cladogram for wax scales. The Chinese wax scale *Ceroplastes sinensis* was a cosmopolitan pest of citrus at the time it was described by Del Guercio in 1900 (his holotype was collected in Italy). As indicated by its name, Del Guercio assumed the scale to be of Asian origin, but a cladistic analysis by Qin et al. (1994) based on 57 morphological characters, suggested otherwise. In Qin et al.'s (1998) tree, *C. sinensis* grouped in a clade that contained nine Neotropical wax scales. Subsequent studies suggested the species was native to southern South America, and an expedition to Argentina resulted in the species' rediscovery as well as parasitoids that were successfully introduced elsewhere. Qin and Gullan 1995. Reproduced with permission from John Wiley & Sons.

is a native or introduced species. Hemlock woolly adelgid, *Adelges tsugae* Annand (Hemiptera: Adelgidae), is an important pest of hemlock trees (Pinales, Pinaceae) in the eastern United States (McClure, 1987), and for several decades the source of the eastern US introduction was unknown (Havill et al., 2006). After collecting samples from China, Taiwan, Japan, and western and eastern North America, Havill et al. (2006) constructed a woolly adelgid phylogeny based on mitochondrial sequence data. Their results suggested that the populations in the eastern United States had been introduced from Japan, and that they were genetically distinct from genotypes that had established in western North America.

Population genetic tools can be valuable when it is suspected that multiple species or genetically distinct lineages have been introduced and the origins of the introductions need to be determined. Such is the case with the biological control program for invasive saltcedar, *Tamarix* spp. (Caryophyllales, Tamaricaceae), in the western United States (Gaskin and Schaal, 2002; Williams et al., 2014). Using an intron from a nuclear gene amplified from *Tamarix* collected throughout saltcedar's native and introduced ranges, and then reconstructing the relationships among individuals using a network analysis, Gaskin and Schaal (2002) were able to identify the regions of origin for two species of *Tamarix* found in the United States (*T. ramosissima* Ledeb. from central Asia, and *T. chinensis* Lour. from eastern China) and produce evidence of high levels of hybridization between the two species. Subsequently, multiple ecotypes of the leaf beetle *Diorhabda elongata* (Brullé) (Coleoptera: Chrysomelidae) were introduced to the western United States from populations collected throughout central Asia and eastern China (DeLoach et al., 2004). As might have been predicted, recent work has revealed the various ecotypes of this beetle to be members of a cryptic species complex (Tracy and Robbins, 2009), with each lineage having distinct geographic preferences. Collectively the set of *Diorhabda* are having considerable impact on the fitness of invasive saltcedar in the American West and Southwest.

Using phylogenies to predict the region of origin can be particularly effective in matching invasive pests with potential candidate biological control agents. Goolsby et al. (2006) used phylogenetic reconstructions to evaluate the genetic diversity and region of origin of the climbing fern *Lygodium microphyllum* (Cavanilles)

(Pteridophyta: Lygodiaceae), an invasive plant in the Florida Everglades, as well as a candidate biological control agent, the mite *Floracarus perrepae* Knihinicki and Boczek (Acari: Eriophyidae). Based on their inferred phylogenies, invasive populations of climbing fern most closely matched those populations collected in northern Australia and southern Asia, and populations of the mite collected in this same region showed greatest virulence (Goolsby et al., 2006).

Using phylogenies and classifications to predict host ranges

Related herbivores often have shared host preferences (Futuyma, 1976; Farrell, 2001; Novotny et al., 2002; Novotny and Basset, 2005). Consequently, knowledge of phylogenetic relatedness can be used to predict the host range of a biological control agent or target pest (Sheppard et al., 2005). However, predicting potential host ranges should not be based solely on phylogenetic relationships because numerous ecological (Hoffmeister, 1992; Stireman and Singer, 2003) and behavioral (König et al., 2015) factors also influence host associations. When explicit phylogenies are unavailable, biological classifications can be harnessed to select sets of species appropriate for testing. One method, known as the "Step-Function Method" and implemented by USDA APHIS, requires extensive testing of susceptibility of non-target plant species in the same genus as those that are known to support pathogens or pests of concern and decreasing amounts of testing as one moves from congeners to members of the same tribe, family, and so forth. A similar approach is known as the "Centrifugal Phylogeny Approach" of Wapshere (1974).

To examine the effects of phylogeny on predicting susceptibility, Gilbert and Webb (2007) conducted experiments in Panama at nursery and natural forest study sites. At each location they cultured pathogens, preserved voucher specimens, created inoculations from their cultures, and then tested resident plant species for susceptibility. Using the angiosperm phylogeny of Davies et al. (2004) and logistic regression, Gilbert and Webb (2007) found that phylogenetic distance was highly correlated with pathogen susceptibility. The authors cautioned that using taxonomic cutoffs such as genus or family designations could lead to errors in estimating susceptibility, and that approaches based on phylogenetic distance are inherently more accurate (Figure 6.3).

Box 6.1 Taxonomy and ecological risk assessments

Taxonomy should play a central role in environmental risk assessments associated with biological control programs. Records of species occurrences and host associations are only as accurate as the taxonomic names of the species involved. Poor taxonomy will result in errors of inclusion (Type I) and of exclusion (Type II). In the first case, resources and efforts might be wasted on taxa that are not imperiled. In the second case, the failure to recognize a taxon (that is or could be imperiled as a consequence of decisions made concerning a biological control program) presents a more serious problem. In deciding when to release (and when not to release) a biological control agent into wildlands, one consideration should be the number and nature of the organisms potentially threatened by the established (exotic) pest. For example, when examining control options for the emerald ash borer, *Agrilus planipennis* Fairmaire (Coleoptera: Buprestidae), it is important to assess the potential ecological consequences of the functional loss of ash from North America's woodlands and forests and to weigh the potential costs of releasing exotic wasps versus a decision not to do so – both actions will have environmental consequences beyond the demographic effects on the target pest.

A recent assessment of the herbivores threatened by emerald ash borer (Gandhi and Herms, 2010) highlights the importance of accounting for both types of error. Among their list of imperiled ash specialists were invalid taxa (mostly junior synonyms), taxa that are less specialized in diet breadth than previously thought (Type I errors), ash specialists treated as generalists (Type II errors), exotic or extralimital species, and species otherwise erroneously assessed as being imperiled. Forty-five insect herbivores not mentioned in their assessment were found to be threatened entities of conservation significance, including high-risk taxa that are globally imperiled (Wagner and Todd, 2015). Most of these errors traced back to dated literature. Errors of exclusion were discovered mostly through correspondence with taxonomic authorities who had special knowledge of the current literature, access to burgeoning genetic data, or were aware of other information relevant to the biosystematics of their taxon. In ecological risk assessments, the evolutionary fates of non-target species are at stake. Getting the taxonomy right matters. We urge workers to be cautious when using older literature, secondary resources, and names applied to poorly studied taxa, and to embrace modern resources, techniques, and especially molecular tools to verify identifications. As much as possible, workers should seek out active systematists familiar with relevant taxa, as they often have unpublished data and special knowledge that might otherwise be overlooked (Wagner and Todd, 2016).

Figures for Box 6.1 Taxonomic errors can contaminate efforts to evaluate the risks associated with a biological control action (or a decision not to take action). *Manduca brontes* (left) (photo courtesy of Andy Warren) has been treated as a species at risk from the loss of ash, but the species is largely an extralimital, temporary breeder in the continental United States and published reports of the caterpillar feeding on ash were based on no-choice laboratory-reared caterpillars; temporary populations in Florida are believed to use yellow trumpetbush, *Tacoma stans* (L.). Grant's Hercules beetle (*Dynastes granti*) is the largest beetle in the North America with some males approaching 80 mm in length (right) (photo courtesy of Margarethe Brummermann). It and three other rhinoceros beetles appear to be dependent on ash either as larvae (*Xyloryctes*) or as adults (*Dynastes*) (Wagner and Todd, 2015, 2016). Previous risk assessments for EAB in North America overlooked these charismatic beetles. See text.

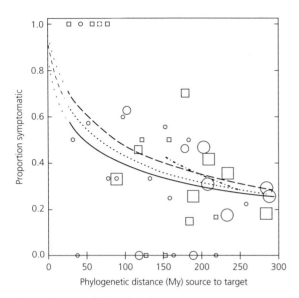

Figure 6.3 Susceptibility of tropical angiosperms to pathogenic fungi correlates with phylogenetic distance. Such relationships across trophic levels demonstrate the value of using phylogenies and classifications to design non-target/specificity studies that can be used to gauge potential non-target impacts that could accrue from the release of a biological control agent. Reproduced with permission from Gilbert and Webb (2007).

In addition, while well-resolved phylogenies provide optimal data, sound genetic distance measures and sound biological classifications can still play important roles in estimating diet breaths and non-target effects. While evaluating *Colletotrichum gloeosporioides* (Penz) (Glomerellales, Glomerallaceae) as a potential biological control agent of Russian thistle, *Salsola tragus* (L.) (Caryophyllales, Amaranthaceae), Berner et al. (2009a,b) constructed a list of potential non-target hosts. Included on this list, however, were many species that were difficult to obtain. The authors obtained DNA sequence data and constructed a distance matrix. This allowed them to predict the susceptibility of 33 species (plus the 59 they had available for testing) based on inferred evolutionary relationships. They concluded that while *C. gloeosporioides* is mainly confined to hosts closely related to *S. tragus*, it also has the potential to infect more distantly related hosts including several species considered endangered in their native ranges (Berner et al., 2009b).

Environmental niche modeling (ENM)

Climate matching allows agents to be sourced from areas with similar conditions to the area of intended release. This process led to the successful biological

control, for example, of the walnut aphid, *Chromaphis juglandicola* (Kaltenbach) (Hemiptera: Aphididae), by a heat-tolerant strain of the braconid *Trioxys pallidus* Halliday (Hymenoptera: Aphidiidae) collected from Iran that was well suited to the hot, dry summers of California's Central Valley (van den Bosch et al., 1970; Frazer and van den Bosch, 1972). More recently, studies have taken advantage of climate-modeling software for identifying the suitability of novel habitats for candidate agents based on the species' niche in its native range (e.g., Lozier and Mills, 2011). Just as for methods using phylogenies to predict host ranges or origins, niche models require accurate circumscription of species-level taxa or, in some cases, population segments.

For the biological control of yellow starthistle, *Centaurea solstitialis* L. (Asteraceae), an invasive weed that causes extensive damage to rangeland and natural areas in the western United States, several biological control agents have been introduced (Jetter et al., 2003). One agent, the rust fungus *Puccinia jaceae* G. H. Otth var. *solstitialis* (Pucciniales, Pucciniaceae), has successfully established in several US locations (reviewed in Fisher et al. [2011]), but did not do so in California. Climate differences between source locations in Eurasia and release locations in California were suspected to play a role in the failure of this control agent to establish. Fisher et al. (2011) performed climate comparison analyses with the software CLIMEX (Sutherst et al., 1999) and found that populations from northern Africa, Spain, and locations along the Mediterranean Sea were best matched for coastal regions of California. These populations, however, are not well matched for the climate of California's Central Valley; therefore, additional source locations would be needed to provide biological control services across the whole invaded range of *C. solstitialis*.

When doing climate analyses, taking phylogeographic structure into account can facilitate biological control efforts. Lozier and Mills (2009) (Figure 6.4) analyzed molecular data collected from populations of *Aphidius transcaspicus* Telenga (Hymenoptera: Braconidae), a potential biological control agent for the mealy plum aphid, *Hyalopterus pruni* (Geoffroy) (Hemiptera: Aphididae), to estimate the distributions of genetically distinct lineages and then to identify key climatic variables associated with each lineage using the ENM software MaxEnt (Phillips et al., 2006; Phillips and Dudik, 2008). Similarly, Manrique et al. (2014) accounted for genetic differences between

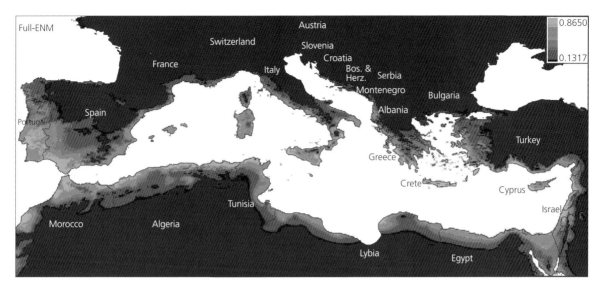

Figure 6.4 Ecological niche modeling (ENM) can be integrated with the results from molecular analyses to identify variation in environmental factors influencing the distributions of candidate biological control agents. Using ENM software, Lozier and Mills (2009) identified different distributions for evolutionary significant units (ESUs) of the parasitoid *Aphidius transcaspicus*. They then used these results to identify climatic variables that might restrict each of these ESUs and may influence their establishment as biological control agents. Lozier and Mills 2009. Used under CC-BY-4.0, http://creativecommons.org/licenses/by/4.0/. (*See Plate 5 for the color representation of this figure.*)

two populations of *P. ichini* during their ENM analyses and found that only one of the two proposed populations would be able to establish in the United States based on their models (Manrique et al., 2014).

The importance of voucher specimens

Voucher specimens that are well preserved, labeled, and deposited in public institutions become a scientific legacy and allow for repeatability and scientific inquiry (Huber, 1998). They anchor the results of scientific studies. To quote Anna Behrensmeyer (Curator of Vertebrate Paleontology at the United States National Museum-Smithsonian Institution), "We must have the objects [vouchers] themselves to serve as the factual basis for knowledge, the final arbitrator in matters of contested identity or meaning, the 'ground truth' that underlies our understanding of the world." The many cases of mistaken identifications (and associated costs and risks) in the examples cited previously underscore their value and relevance. Modern biological control programs conducted without adequate vouchering are reckless and bad science.

Vouchers across all disciplines must be well labeled with precise locality data, collection date, and the identification so far as known. Collections from different source populations or markedly different dates should also be included among the set of vouchered specimens. Subsequent sampling from laboratory cultures, especially at points in time when field releases are made, is also recommended. Accurate labeling and related referencing of specimens deposited in museums and or databases are essential. Voucher specimens labeled only with a researcher's or study's code number are vulnerable as few such vouchers ever find their way into permanent, curated collections and their utility rarely extends beyond the tenure of the investigator or laboratory.

Barratt et al. (2010) dedicated a section of their review of risk assesment in biological control programs to the importance of vouchering. The most obvious and widespread applications of voucher specimens are for verifying and correcting the names of either the target pests or that of natural enemies (e.g., see Berry [2003]) or determining which species have established when multiple agents have been introduced (e.g., Hardy and Delfinado, 1980). Voucher specimens have particular value in their role in allowing workers to reconstruct points of origin, timing, and geographic spread of biological invasions. Specimens in collections also add locales that can parameterize ENM analyses (Aragón et al., 2013). Andersen and Mills (2012) were able to screen previously vouchered museum specimens to correctly identify a European braconid reared from a population of the invasive light brown apple moth, *Epiphyas postvittana* (Walker) (Lepidoptera: Tortricidae), in California that had never before been recorded from North America (see Box 6.2: Using non-invasive DNA extraction techniques with vouchers). Rapidly evolving nucleotides isolated from voucher material could have substantial value for exploring post-introduction evolution of a pest or its biological control agents. Each voucher specimen deposited contributes to scientific infrastructure. Future possibilities will be limited only by the lack of preserved specimens in collections.

Some governments (e.g., New Zealand) require specimen vouchers for introductions (Barratt et al., 2010). It is lamentable – given the importance of voucher specimens for future research and the recognition of their importance for the safety and sustainability of biological control services (McCoy and Frank, 2010) – that voucher specimens are not required by all governments (e.g., the United States) for biological control introductions (Marsico et al., 2010). Until legislation can be passed, such as that in place in New Zealand, it strikes us as prudent for journals (reviewers and editors) publishing the results of biological control work, and especially those involving releases, to be more vigilant about requiring authors to deposit vouchers and requesting documentation that voucher material has been received before publication. Too often workers with good intentions and time constraints fall short of providing adequately preserved and labeled voucher material. One way forward would be for journals as well as funding and regulatory agencies to require that properly labeled voucher specimens be deposited in public institutions, in the same way that many journals and agencies now require deposition of genetic data in GenBank.

Advances in DNA extraction and amplification methodologies are greatly increasing the value and utility of museum collections, including but not limited to entomological collections (e.g., Heintzman et al., 2014) and herbaria (e.g., Besnard et al., 2014). As sequence databases (e.g., GenBank and the Barcode of Life) become more densely populated, it will be possible to unambiguously establish identities from older specimens (i.e., degraded, shorter sequence fragments) in museums (e.g., Tin et al., 2014). Likewise, new techniques are allowing workers to amplify sequences from smaller samples or tissue fragments. Given recent methodological advances and those on the horizon, virtually every insect voucher specimen in museums will have value as a repository of historical genetic data and help to put museums back at the center of biodiversity research.

Teams of workers from the Barcode of Life project have visited several North American and European museums to sample thousands of insect species (Hebert et al., 2013). Typically, a voucher specimen's CO1 is sequenced, all associated label information is databased, and the individual is imaged, and within months all information from a museum "blitz" is made publicly available online (Hebert et al., 2013). For both nuclear and mitochondrial sampling, it is commonplace to remove a single leg from the voucher (insect) specimen, although advances in non-destructive sampling are making it possible to extract DNA from the body while preserving the intact integument for morphological analysis (Gilbert et al., 2007; Porco et al., 2010; Andersen and Mills, 2012; Tin et al., 2014: Box 6.2).

What constitutes a voucher specimen is changing. Historically, vouchering has required the preservation of an intact, well-labeled specimen – usually a dry specimen in a museum drawer or, for soft-bodied taxa, a specimen in 70% ethanol. We still advocate for such specimens as they represent a time-anchored, irreplaceable resource of considerable future value: not only is the animal's whole body preserved, but also its surface debris (such as pollen grains), its internal morphology, its tissues, and its microbial fauna, which at some point in the future could be further analyzed. For some purposes, photographs can serve as vouchers. Better and more economical cameras are making it possible for workers to get exceptional macrophotographic image vouchers. Soon it will be possible to use high-end CT scanners and focus-stacking techniques to acquire superb digital imagery that will allow for detailed study of the internal and external morphology of insect specimens. Photographic vouchers lack the scientific value of specimen vouchers, but do have great utility and are easily stored, curated, reproduced, shared, and made public on databases such as MorphoBank (http://www.morphobank.org/).

Box 6.2 Using non-invasive DNA extraction techniques with voucher specimens

Non-invasive DNA extraction techniques are allowing genomic material to be collected from museum specimens without physically damaging them. Recently, Andersen and Mills (2012) non-destructively extracted DNA from both freshly collected and museum specimens to confirm the identification of a parasitoid in the genus *Meteorus* (Hymenoptera: Braconidae) found attacking the invasive light brown apple moth in California. Based on morphological comparisons, the specimens appeared to be *M. trachynotus* Viereck, a native parasitoid known to attack leaf rollers. However, all individuals reared from field populations of light brown apple moth were

female, while *M. trachynotus* populations are sexual, with males and females (Bürgi and Mills, 2013). Andersen and Mills (2012) compared nuclear and mitochondrial sequences from specimens at the University of California Berkeley to specimens reared from recent larval collections of the moth as well as published records in GenBank and found that the field-collected specimens of the wasp were *Meteorus ictericus* (Nees), a species native to Europe, which had never been recorded from North America. Moreover, *M. ictericus* is proving to be an important biological control agent for light brown apple moth (Bürgi and Mills, 2013).

 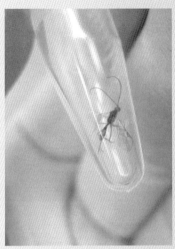 

Figures for Box 6.2 Improvements in non-destructive DNA extraction methods are allowing museum specimens to be integrated into molecular analyses with little to no morphological damage to the specimens. Andersen and Mills (2012) extracted DNA from individual museum specimens of *Meteorus* sp. On the left is a specimen prior to DNA extraction, in the center the specimen placed in a 1.5 ml microcentrifuge tube, and on the right after articulation following DNA extraction. While not visible in the photo, post-extraction specimens appeared slightly translucent compared to pre-extraction specimens. Photos by Jeremy C. Andersen.

Wherever possible, vouchers suitable for genetic analyses should be given priority. Ethanol solutions of 95–100% will preserve genomic information for decades and beyond, especially if held in an ultracold freezer. Dry storage and new alternatives to alcohol (e.g., RNALater) are also possible, especially when samples are kept frozen, ideally at -80 °C. However, most cryo-collections are dependent on grant funds, personnel, functioning equipment, and continuous electrical power; most are idiosyncratically curated; and, as noted above, few will survive beyond the working life of the laboratory's lead scientist. The Monell Cryo Collection at the American

Museum of Natural History is a notable exception given that it has dedicated funding, multiple fail-safe mechanisms, and a full-time curator.

Molecular technologies have made it possible to voucher genetic data in GenBank and other genomic databases. These various approaches, however, are not either/or options: modern vouchering can and should involve the deposition of an actual specimen in a public institution, posting or publication of one or more publicly viewable photographs, cryo-preservation, and the uploading of diagnostic sequence data into a genetic database.

Given the quintessential role of voucher specimens as anchors for biological studies, their increasing value as genetic repositories, the potential of misidentifications to plague biological control studies, and risks to non-target taxa, there is much to be gained from demanding more rigorous standards for the deposition of voucher materials in museums and public institutions where specimens or data can be adequately housed and studied long into the future. Active endorsement and participation in vouchering by biological control workers would raise the scientific value of the institutional collections where material has been deposited and, in so doing, help justify continued investment in the infrastructure of natural history collections, which face increasing costs associated with storage and maintenance of materials (Abrol, 2013) and reductions in staff (Sluys, 2013). The availability of voucher specimens can facilitate their use in integrative taxonomic studies, which use both morphological and molecular approaches to study the biosystematics of a taxon (Polaszek et al., 2004; Okassa et al., 2011; Gebiola et al., 2012; Pina et al., 2012).

Molecular methods

Historically, biological control programs have relied on morphological characters, host associations, and behavioral and life history data to make identifications. More recently, due to the many taxonomic problems presented by minute, cryptic, or otherwise taxonomically intractable taxa, molecular techniques are increasingly being used to sort out species- and population-level differences in both pests and their natural enemy complexes. Molecular approaches allow researchers to collect genetic data that can be used to verify identifications, reconstruct phylogenies, circumscribe subspecific population segments, confirm hybridization, quantify levels of gene flow, and more (see review by Heraty [2004]). Genetic data can be especially valuable in cases that have confounded classical taxonomic methods, such as to associate disparate life stages (especially where taxonomic resources are nascent or non-existent) or sexually dimorphic taxa and to identify body fragments or damaged specimens. Because some methods of genetic analysis require very little starting material, they are well suited for studies of minute organisms or old museum specimens (Hebert *et al.*,

2013; CBOL, 2014). Several reviews of molecular methods and their applications for taxonomists and biological control practitioners have been published over the past decade (e.g., Greenstone, 2006; Stouthamer, 2006; Gaskin et al., 2011; Roderick and Navajas, 2012). Below we provide an overview of the most common techniques, as well as examples of biological control programs that have successfully employed each method.

DNA "barcodes"
Background

Many workers are now routinely using molecular methods to identify specimens by comparing diagnostic sequence data from a target specimen to sequences in a reference database or to additional (library) sequences generated by the researcher. Principal genomic databases include GenBank in the United States (Benson et al., 2013), EMBL in Europe (Kanz et al., 2005), DDBJ in Japan (Kodama et al., 2012), or the Barcode of Life Database (BOLD) in Canada (Ratnasingham and Hebert, 2007). The first three databases share information on a daily basis so that DNA sequences deposited in one can be accessed through queries to any of the three. Likewise, queries initiated through the BOLD Systems v3 database automatically include sequence matches from each of the previously mentioned databases (however queries initiated through GenBank, EMBL, and DDBJ do not necessarily include sequence matches in the BOLD database). As would be expected, existing genetic databases tend to be reliable for well-known taxa but increasingly less reliable for obscure, exotic, and poorly known organisms. For example, Tixier et al. (2011) compared sequences they generated from predatory mites in the family Phytoseiidae (Acari: Mesostigmata), many of which are biological control agents, to those stored in GenBank and found frequent examples of misidentified species in the publicly available genomic databases.

In contrast to GenBank, EMBL, and DDJJ, where sequences from any genomic region can be deposited, BOLD Systems is decidedly focused on the mitochondrial locus cytochrome-oxidase 1 (referred to as CO1, COI, or COX1) (CBOL, 2014) for animal taxa or bcL, matK, trnH-psbA, and ITS for plant taxa (Smithsonian National Museum of Natural History, 2014). These "DNA barcodes" have revealed the existence of hundreds of cryptic species (e.g., Hebert et al. [2004], Smith et al. [2007] and see Smith et al. [2013] for taxa relevant to biological control settings) as well as demographic variation in pest species (Frewin et al., 2014). Biological control programs have also commonly used DNA barcoding (e.g., Davis et al., 2011; Zhang et al., 2011). For example, Cristofaro et al. (2013) used CO1 to confirm the identity of field-collected weevil larvae that would have been impossible to identify otherwise. The weevil *Ceratapion basicorne* (Illiger) (Coleoptera: Apionidae) is a candidate biological control agent for yellow starthistle (*Centaurea solstitialis* L.) in the western United States (Smith, 2007); however, no-choice laboratory experiments demonstrated that larvae can develop on safflower (Clement et al., 1989; Smith, 2007). Due to the economic importance of safflower, Cristofaro et al. (2013) conducted field experiments in Turkey where *C. basicorne* is native. Safflower blocks were planted next to yellow starthistle, and immature weevil larvae were identified with barcoding. Cristofaro et al. found that only 15 of the over 1000 safflower plants sampled were infected by *Ceratapion* weevils, and none were *C. basicorne*.

BOLD Systems also has developed many software tools that readily display identification trees and provide other metrics for users. One particularly powerful new feature of BOLD Systems v3 warrants mention: BOLD algorithmically identifies cohesive, species-level clusters (bins) and assigns these a unique Barcode Index Number (BIN), whether or not the genetic cluster has an associated name or aligns with existing taxonomy (Ratnasingham and Hebert, 2013). On average, bins differ by about ~2% uncorrected genetic similarity (across the 658-bp sequence), but BOLD and its curators can integrate additional algorithms, genetic data, organismal data, taxonomic authority opinion, and other information to refine BIN numbers so they coordinate with other knowledge. Sequences that are grouped by genetic similarity through the BIN algorithm can become curated into "digital unit trays," becoming comparative tools for taxonomists and common-source sets for DNA-based identification by non-specialists (Gwiazdowski et al., 2015). Each week, all BOLD Systems sequences (2.5 million plus) are reanalyzed and rebinned as a consequence of the addition of new sequences. An especially valuable aspect of BOLD's BIN system, is that it allows all submitted sequence (and aforementioned metadata) to be linked and permanently curated (Ratnasingham and Hebert, 2013),

regardless of misidentifications and name changes, or in the case of some environmental data where submitted sequences arrive without names.

Advantages and limitations

DNA barcoding is a powerful tool for identifying specimens to species and is often capable of distinguishing intraspecific lineages. It is relatively inexpensive and widely available as most universities have associated biotech facilities that can generate barcode data. BOLD Systems can also be contracted to gather sequence data from biological tissues, and the organization has set in place a rigid set of protocols in an effort to raise the rigor of their genetic records (i.e., minimize the number of misidentifications and, more importantly, make it possible for future workers to provide names and correct erroneous records). Considerable information must be supplied before specimens or tissue samples are processed (e.g., associated collection data, image of each specimen, a unique specimen number, the repository for voucher material, the taxonomic authority for named material, permitting documentation, etc.). Furthermore, all such metadata must be successfully uploaded into the BOLD database before tissues are mailed to the sequencing facility in Canada. This metadata is invaluable to future scientists, and its importance is highlighted in the section titled "The importance of voucher specimens."

For biological control practitioners, there are more than half a million insect barcodes in the BOLD database with more than 2000 insect sequences added each week. The taxonomic breadth and depth of BOLD make it the best global reference database for ascertaining the identity of unknown insect specimens, and its value is further bolstered by the wealth of metadata (including voucher image) supplied with each specimen. In addition, the BOLD system has developed software tools that readily display identification trees and provide other metrics for users.

Barcoding, however, like all methods has disadvantages. Chief among these is that CO1 (the barcoding region in animals) is a single, non-recombining, and strictly maternally inherited gene that represents a relatively small region of the mitochondrial genome. Closely related or cryptic species may not have accumulated enough fixed differences in this gene region to be recognized as separate entities. Hybridization, ancient or recent, as well as incomplete lineage sorting, confounds results and produces misleading relationships that do not reflect species boundaries or histories. Amplification of pseudogenes (e.g., non-functioning nuclear copies of mitochondrial genes) or nuclear-mitochondrial genes ("numts") can also result in erroneous and misleading results (Thalmann et al., 2004; Buhay, 2009; Moulton et al., 2010). Finally, an emphasis on arbitrary predetermined level (often circa 2%) of genetic differentiation between species may also lead to inaccurate species-level delimitations (Meier et al., 2006, 2008). Consequently, CO1 is not a reliable marker for delimiting species in the absence of other data (e.g., Heraty et al. [2007]). In addition, while BOLD has rigorous quality standards, it also communicates with other less intensively curated databases, which can result in the matching of sequences with mislabeled or misidentified sequences. Finally, for understudied or highly diverse taxa the scientific names assigned in any genetic databases should be treated as hypotheses. Despite these concerns, given DNA barcoding's overall utility, minimal costs, ready availability, and potential to inform, we recommend it as an important component for all biological control programs.

Overview of fragment length analyses

In contrast to DNA barcoding, which directly compares nucleotide data, several approaches exist that examine variation in the fragment lengths of targeted loci to provide species- and population-level identifications. A major advantage of these approaches is that because sequencing is not required, costs can be greatly reduced. In addition, analyses of fragment lengths generally target nuclear loci, allowing gene flow, introgression, and hybridization to be measured. Hybridization has been identified as an important force in evolution (Mallet, 2005) and is applicable to many aspects of conservation biology (Allendorf et al., 2001). Hybridization phenomena of relevance to biological control programs include hybrid "breakdown" (i.e., reduced fitness of hybrids versus their parents) for biological control agents and target pests (Messing and AliNiazee, 1988; Blair et al., 2008; Bean et al., 2013; Williams et al., 2014) and the hybridization between native species and introduced biological control agents (Havill et al., 2012).

Below we highlight two approaches that have been useful in biological control studies, but both techniques are being replaced by more informative methods. One

approach randomly amplifies DNA (RAPD) fragments across the genome with short primers (8 to 12 base pairs) (Borowsky et al., 1995). (See Black et al. [1992], Polaszek et al. [1992], Edwards and Hoy [1993, 1995], or Kirk et al. [2000] for examples in biological control settings.) However, because amplification results often vary across PCR conditions within and between laboratories (Gallego and Martinez, 1997; Jones et al., 1997), the use of RAPD markers has fallen out of favor. Another approach compares differences in fragment lengths of DNA digested with restriction enzymes (restriction fragment length polymorphisms [RFLP]). For example, Hajek et al. (1990) identified which introduction of *Entomophaga maimaiga* Humber et al. (Entomophthorales, Entomophthoraceae) was responsible for declines in gypsy moth, *Lymantria dispar* (L.) (Lepidoptera: Erebidae), populations in the northeastern United States (see Elkinton and Liebhold [1990] for a review) with RFLP analysis. Similarly, Soper (2012) digested the barcode region of CO1 into species-specific size fragments to differentiate native and introduced species of *Lathrolestes* (Hymenoptera: Ichneumonidae) that are biological control agents for the ambermarked birch leafminer, *Profenusa thomsoni* (Konow) (Hymenoptera: Tenthredinidae). Hirsch et al. (2010) digested fragments of COII to diagnose larvae of 23 different species of *Otiorhynchus* (Curculionidae), an important group of root-feeding weevil larvae that are pests of horticultural crops worldwide (Moorhouse et al., 1992). However, this technique is labor intensive, often requires large amounts of genomic material, and in many cases is being replaced by techniques that yield sequence data (Charlesworth and Charlesworth, 2010).

Amplified fragment length polymorphisms (AFLP)
Background

One well-established technique, at least for plants, bacteria, and fungi (Bensch and Akesson, 2005), combines aspects of both RAPD and RFLP approaches to rapidly produce data about the presence or absence of fragments of DNA, or "amplified fragment length polymorphisms" (AFLPs) (Vos et al., 1995). These markers have been harnessed for studies of invasive plants (Gaskin et al., 2009, 2013a, b, 2014; Chandi et al., 2013), pest insects (Bray et al., 2011; Trissi et al., 2013), nematodes (Dillon et al., 2008), and in biological control settings to quantify genetic variation in populations (e.g., Piyaboon et al., 2014), compare variation between laboratory and field populations (e.g., Kneeland et al., 2012), and characterize hybridization rates (Szűcs et al., 2011).

Rauth et al. (2011) examined pre-release population genetics of the candidate biological control agent *Ceutorhynchus scrobicollis* Nerensheimer and Wagner (Coleoptera: Curculionidae) in Europe with AFLPs. Their survey suggested that between 25–30 individuals from a given source population would capture nearly all the native genetic variation in Germany, and that populations from Greece, Romania, and elsewhere in southern Europe would be needed if additional genetic variability was deemed desirable for releases. AFLPs have also been used in studies of biopesticides, in particular to determine which strain might be the most effective at controlling a target pest (Kim et al., 2008; Larralde-Corona et al., 2008; Franklin et al., 2010; Kupper et al., 2010). Recently, however, concerns have been raised about the need to report error rates associated with AFLP data (Crawford et al., 2012).

Advantages and limitations

The primary advantage of AFLPs is that the start-up time and costs to develop a large number of loci are relatively small (Bensch and Akesson, 2005). AFLPs are particularly well suited to taxonomic groups for which molecular resources (and associated genetic databases) are lacking. The primary drawback to AFLP markers is that results are scored as presence and absence data (usually scored as 1 or 0), and thus heterozygosity is estimated rather than observed. AFLPs are considered to be "dominant" markers (Bench and Akesson, 2005), as compared to microsatellites and single-nucleotide polymorphisms (for which heterozygosity is directly observed and which are considered "co-dominant" markers) (Bhargava and Fuentes, 2010). As a consequence 4–10 times as many dominant markers are needed as co-dominant markers to obtain the same statistical power (Mariette et al., 2002). However, because it is relatively easy to develop AFLP markers, obtaining the additional loci is not an insurmountable barrier.

Microsatellites
Background

One of the most common methods for examining genetic variation within species, microsatellite analysis compares the numbers of repetitions in a targeted section of non-coding DNA. These repeats are variously referred to as microsatellites, simple sequence repeats (SSRs), or simple tandem repeats (STRs). The repeat nature of these regions and their lack of coding functions make them prone to DNA copying errors that generally follow a simple step-wise addition or deletion pattern. This results in a high degree of polymorphism among individuals, as well as the advantage that the step-wise pattern allows for relationships among individuals to be estimated, although (due to mechanisms that are not entirely understood) multiple sets of repeats can be deleted or added in a single generation (Ellegren, 2004). Due to their extensive use, their mutational dynamics have been well studied (Bhargava and Fuentes, 2010). For a review of microsatellite markers in ecological settings, see Selkoe and Toonen (2006).

In a biological control context, microsatellite markers have been employed to evaluate populations of candidate natural enemies (Lozier et al., 2008), to examine the effective population sizes of agents following introductions (Hufbauer et al., 2004), to determine the region of origins of invasive pests (Tsutsui et al., 2001), to disentangle complicated invasion pathways (e.g., the "bridgehead" invasion pattern of the ladybird beetle *Harmonia axyridis* [Pallas]) (Lombaert et al., 2010), and to study hybridization – either between pest species in their native range (Lozier et al., 2007), between invasive species (Williams et al., 2005, 2014), or between native and introduced biological control agents (Havill et al., 2012).

Advantages and limitations

Microsatellites show high degrees of polymorphism with many alleles per locus and thus are well suited to distinguish populations and closely related species. In addition, because these markers are co-dominant, heterozygosity is directly observed and does not need to be estimated. This ensures less error in population genetic analyses. In addition, as the number of polymorphic microsatellite markers increases, the power available to assign individuals to populations increases. With enough polymorphic markers, individuals can also be assigned to kinship groups. Although no biological control study has yet used kinship analysis, it is important for the study of endangered species (King et al., 2005a), other species of conservation concern (King et al., 2005b; Crawford et al., 2008), and invasive/parasitic species (Alderson et al., 1999; King et al., 2011). However, as stated previously, due to deviations from a step-wise addition and deletion of repetitive repeats, modeling evolutionary relationships using microsatellite markers can lead to erroneous or misleading results, especially when investigations examine many species (Ellegren, 2004). One important concern with using microsatellite markers is that the fragment lengths may appear identical, while the sequences themselves show divergence (i.e., homoplasy) (Estoup et al., 2002; Selkoe and Toonen, 2006). The greatest limitation of microsatellite markers has been the time and cost associated with the development of markers. In addition, because they are located in rapidly mutating regions of the genome, for many taxonomic groups, markers must be developed for each species individually. However, advances in next-generation sequencing techniques have greatly reduced the time and cost associated with the development of such markers.

Methods of classification using genomic methods

Background

The genomic revolution brought about by the accessibility of next-generation sequencing (NGS) technologies has increased the speed and reduced the costs associated with the development of molecular resources for non-model organisms (Davey et al., 2011). With little modification to laboratory personnel or practices, the same techniques can be used to sequence and assemble a genome and to identify regions of the target genome that are under selection (Hohenlohe et al., 2010; Chutimanitsakun et al., 2011; Pfender et al., 2011; Peterson et al., 2012), or to conduct fine-scale population genetic surveys where hundreds of individuals are assayed/sequenced in a single analysis (Baird et al., 2008; Hohenlohe et al., 2010). New methods are being developed rapidly (see Buermans and den Dunnen [2014] for a recent review). NGS methods can broadly be grouped into two categories, those that target and amplify a region of interest (e.g., Tringe et al., 2005) and those that sample randomly from across the genome (e.g., Hohenlohe et al., 2010; Peterson et al., 2012). Because these technologies are new, there are few examples in the literature from biological control settings; however, we know of many studies underway where biological control practitioners are embracing NGS techniques.

One of the most obvious ways in which NGS technologies can be helpful for biological control programs is for the rapid identification and development of molecular markers for both phylogenetic (Kuhn et al., 2013) and microsatellite analysis (Abdelkrim et al., 2009; Mlonyeni et al., 2011; Zalapa et al., 2012; Andersen and Mills, 2014; Khidr et al., 2014; Marcari et al., 2014). NGS technologies are first used to sequence data from individuals or populations of interest and then are used to find markers for subsequent studies or analyses. This two-step approach may also be applicable when specimens are physically small (and thus extractable DNA yields are low) or continuous monitoring is needed. In such instances, using techniques that examine fragment length variation may be more appropriate for the needs of the researcher. However, the real promise of genomic techniques is the ability to simultaneously develop markers while identifying individuals (Baird et al., 2008; Hohenlohe et al., 2010; Peterson et al., 2012). Currently, the most commonly used approaches use restriction enzymes to digest the DNA extracted from a specimen and then sequence the regions following the cut site. These approaches are commonly referred to as some variation of RAD-seq (e.g., RAD-tag, or GBS) (Baird et al., 2008; Hohenlohe et al., 2010; Narum et al., 2013; Peterson et al., 2012) and are currently being used to reveal the population structure of individuals in studies of plants (Chutimanitsakun et al., 2011; Pfender et al., 2011; Deschamps et al., 2012; Scaglione et al., 2012; Pegadaraju et al., 2013; Vandepitte et al., 2013; Deokar et al., 2014), endangered species (Hohenlohe et al., 2011; Sharma et al., 2012), and other non-model organisms (Baird et al., 2008; Hohenlohe et al., 2010; Davey et al., 2011; Ekblom and Galindo, 2011; Peterson et al., 2012; Bourgeois et al., 2013; Fauer and Joly, 2015; Macher et al., 2015; Pante et al., 2015).

Genomic techniques will be especially beneficial for biological control when there is a need to identify bacterial or symbiont diversity (Hail et al., 2011; Russell et al., 2013) or study viral biological control agents (Valles et al., 2012). NGS techniques can help characterize the effect of a biological control program on non-target organisms. For example, Terhonen et al. (2013) used NGS sequencing to examine differences in fungal diversity between sites where a fungal biological control agent had been applied versus control sites.

Advantages and limitations

Genomic approaches have the advantage of producing thousands of genetic markers per individual, which can then be directly analyzed or form the basis of subsequent genotyping efforts using any of the above-mentioned techniques. The limitations of this approach are the facilities required and the costs per individual. In addition to standard molecular laboratories, this approach requires access to a core facility with NGS equipment able to provide sequences (increasingly available at universities or commercial companies), and a computer cluster for data analysis. These limitations may make genomic techniques a less realistic option for some researchers; however, as techniques become more affordable and accessible, NGS methods will increasingly become the standard for studies of within- and between-species variation for biological control programs.

Conclusions

Good science requires good taxonomy. Getting the taxonomy right must be regarded as an essential component of all biological control efforts. Only in this way can knowledge be correctly stored and reliably retrieved, our models be accurately parameterized, and our identification tools and genetic databases be trusted. The inefficiencies, costs, and risks of getting it wrong greatly outweigh the time and effort necessary to secure correct identifications and uphold the integrity of a biological control effort. The elevated non-target risks alone make it imperative that we do our best to get the taxonomy right (see Box 6.1: Taxonomy and ecological risk assessments). Properly curated and accurately labeled collections, better microscopy, better trained scientists, and more modern methodologies are helping to improve existing taxonomy for most groups of economic or ecological importance. In particular, widely available, increasingly powerful, and lower-cost molecular methods are making it easy and inexpensive to specify the genetic identity of species, races, biotypes, and other infra- or subspecific entities. Likewise our ability to generate and harness the information content of phylogenies is becoming routine, with direct applications to biological control efforts.

The scientific opportunities from increased use of genomics techniques provide a new perspective on the potential value of specimen vouchers. Biological control releases without adequate attention to essential taxonomy and vouchering is bad science. Biological control practitioners need to embrace vouchering with the same vigor that molecular biologists (and their journals, outside reviewers, and granting agencies) have pushed for the requirement of sequence data deposition in GenBank. When working with species in groups that are poorly known taxonomically, it is essential that adequately prepared voucher material be deposited in a public institution so that the specimens can be re-examined as knowledge of the group improves. New and future genomic methods in particular will be able to obtain extraordinary amounts of genetic information from dry museum specimens. Future possibilities will be limited only by the lack of preserved specimens in collections.

Acknowledgments

We thank Linda Bürgi, Roger Gwiazdowski, Ronald Labbé, Jane O'Donnell, Linda Nolan, Katie Todd, Elizabeth Wade, and Roy Van Driesche for their helpful comments and suggestions on earlier drafts of this chapter. John Heraty and Jim Woolley provided thoughtful reviews that substantially improved our manuscript. We also thank Virge Kask for her help with Figure 6.2 and Margarethe Brummermann and Andy Warren for graciously offering to share photos for the chapter. This work was supported in part by a Robert van den Bosch Memorial Fellowship (to JCA) and a USDA COOP #14-CA-11420004-138 (to DLW).

References

Abdelkrim, J., B. C. Robertson, J. L. Stanton, and N. J. Gemmell. 2009. Fast, cost-effective development of species-specific microsatellite markers by genomic sequencing. *BioTechniques* **46**: 185–192.

Abrol, D. P. 2013. Future of taxonomy in the 21st century – whither or wither. *Current Science* **104**: 12 1594–1594.

Alderson, G. W., H. L. Gibbs, and S. G. Sealy. 1999. Parentage and kinship studies in an obligate brood parasitic bird, the brown-headed cowbird (*Molothrus ater*), using microsatellite DNA markers. *Journal of Heredity* **90**: 182–190.

Allendorf, F. W., R. F. Leary, P. Spruell, and J. K. Wenburg. 2001. The problem with hybrids: setting conservation guidelines. *Trends in Ecology & Evolution* **16**: 11 613–622.

Andersen, J. C. and N. J. Mills. 2012. DNA extraction from museum specimens of parasitic Hymenoptera. *PLoS One* **7**: 10 e45549.

Andersen, J. C. and N. J. Mills. 2014. iMSAT: a novel approach to the development of microsatellite loci using barcoded Illumina libraries. *BMC Genomics* **15**(858): 1–9.

Aragón, P., M. M. Coca-Abia, V. Llorente, and J. M. Lobo. 2013. Estimation of climatic favourable areas for locust outbreaks in Spain: integrating species' presence records and spatial information on outbreaks. *Journal of Applied Entomology* **137**: 610–623.

Baird, N. A., P. D. Etter, T. S. Atwood, et al. 2008. Rapid SNP discovery and genetic mapping using sequenced RAD markers. *PLoS One* **3**(10): e3376.

Baker, W. W. 1936. Notes on the European weevil, *Ceutorhynchus assimilis* Payk., recently found in the State of Washington. *The Canadian Entomologist* **68**(9): 191–193.

Barratt B. I. P., F. G. Howarth, T. M. Withers, et al. 2010. Progress in risk assessment for classical biological control. *Biological Control* **52**: 245–254.

Bean, D. W., D. J. Kazmer, K. Gardner, et al. 2013. Molecular genetic and hybridization studies of *Diorhabda* spp. released for biological control of *Tamarix*. *Invasive Plant Science and Management* **6**: 1–15.

Bensch, S. and M. Akesson. 2005. Ten years of AFLP in ecology and evolution: why so few animals? *Molecular Ecology* **14**: 962–1083

Benson, D. A., M. Cavanaugh, K. Clark, et al. 2013. GenBank. *Nucleic Acids Research* **41**: D1 D36–D42.

Berner, D. K., W. L. Bruckart, C. A. Cavin, and J. L. Michael. 2009a. Mixed model analysis combining disease ratings and DNA sequences to determine host range of *Uromyces salsolae* for biological control of Russian thistle. *Biological Control* **49**: 68–76.

Berner, D. K., W. L. Bruckart, C. A. Cavin, et al. 2009b. Best linear unbiased prediction of host-range of the facultative parasite *Colletotrichum gloeosporioides* f. sp. *salsolae*, a potential biological control agent of Russian thistle. *Biological Control* **51**: 158–168.

Berry, J. A. 2003. *Neopolycystus insectifurax* Girault (Hymenoptera: Pteromalidae) is established in New Zealand, but how did it get here? *New Zealand Entomologist* **26**: 113–114.

Besnard, G., P.-A. Christin, P.-J. G. Male, et al. 2014. From museums to genomics: old herbarium specimens shed light on a C3 to C4 transition. *Journal of Experimental Botany* **65**: 6711–6721.

Bhargava, A. and F. F. Fuentes. 2010. Mutational dynamics of microsatellites. *Molecular Biotechnology* **44**: 250–266.

Bin, F., P. F. Roversi, and J. C. van Lenteren. 2012. Erroneous host identification frustrates systematics and delays implementation of biological control. *Redia-Giornale Di Zoologia* **95**: 83–88.

Black, W. C. I., N. M. DuTeau, G. J. Puterka, et al. 1992. Use of the random amplified polymorphic DNA polymerase chain reaction (RAPD-PCR) to detect DNA polymorphisms in aphids (Homoptera: Aphididae). *Bulletin of Entomological Research* **82**: 151–159.

Blair, A. C., U. Schaffner, P. Haefliger, et al. 2008. How do biological control and hybridization affect enemy escape? *Biological Control* **46**: 358–370.

Borowsky, R. L., M. McClelland, R. Cheng, and J. Welsh. 1995. Arbitrarily primed DNA-fingerprinting for phylogenetic reconstruction in vertebrates – the *Xiphophorus* model. *Molecular Biology and Evolution* **12**: 1022–1032.

Bourgeois, Y. X. C., E. Lhuillier, T. Cezard, et al. 2013. Mass production of SNP markers in a nonmodel passerine bird through RAD sequencing and contig mapping to the zebra finch genome. *Molecular Ecology Resources* **13**: 899–907.

Bray, A. M., L. S. Bauer, T. M. Poland, et al. 2011. Genetic analysis of emerald ash borer (*Agrilus planipennis* Fairmaire) populations in Asia and North America. *Biological Invasions* **13**: 2869–2887.

Buhay, J. E. 2009. "COI-like" sequences are becoming problematic in molecular systematics and DNA barcoding studies. *Journal of Crustacean Biology* **29**: 96–110.

Bürgi, L. P. and N. J. Mills. 2013. Developmental strategy and life history traits of *Meteorus ictericus*, a successful resident parasitoid of the exotic light brown apple moth in California. *Biological Control* **66**: 173–182.

Buermans, H. P. J. and J. T. den Dunnen. 2014. Next generation sequencing technology: advances and applications. *Biochemica et Biophysica Acta – Molecular Basis of Disease* **1842**: 10 1932–1941.

Buntin, G. D. 1998. Cabbage seedpod weevil (*Ceutorhynchus assimilis*, Paykull) management by trap cropping and its effect on parasitism by *Trichomalus perfectus* (Walker) in oilseed rape. *Crop Protection* **17**: 299–305.

Buntin, G. D., J. P. McCaffrey, P. L. Raymer, and J. Romero. 1995. Quality and germination of rapeseed and canola seed damaged by adult cabbage seedpod weevil, *Ceutorhynchus assimilis* (Paykull) Coleoptera, Curculionidae. *Canadian Journal of Plant Science* **75**: 539–541.

CBOL. 2014. Barcode of Life: standards and guidelines. Available from: http://www.barcodeoflife.org/content/resources/standards-and-guidelines. [Accessed January 2016].

Chandi, A., S. R. Milla-Lewis, D. L. Jordan,et al. 2013. Use of AFLP markers to assess genetic diversity in Palmer amaranth (*Amaranthus palmeri*) populations from North Carolina and Georgia. *Weed Science* **61**: 136–145.

Charlesworth, B. and D. Charlesworth. 2010. *Elements of Evolutionary Genetics*. Roberts and Company Publishers. Greenwood Village, Colorado, USA. 734 pp.

Chutimanitsakun, Y., R. W. Nipper, A. Cuesta-Marcos, et al. 2011. Construction and application for QTL analysis of a restriction site associated DNA (RAD) linkage map in barley. *BMC Genomics* **12**: 1–13.

Clement, S. L., M. A. Alonsozarazaga, T. Mimmocchi, and M. Cristofaro. 1989. Life-history and host range of *Ceratapion basicorne* (Coleoptera: Apionidae) with notes on other weevil associates (Apioninae) of yellow starthistle in Italy and Greece. *Annals of the Entomological Society of America* **82**: 741–747.

Cox, J. M. and D. J. Williams. 1981. An account of cassava mealybugs (Hemiptera: Pseudococcidae) with a description of a new species. *Bulletin of Entomological Research* **71**: 247–258.

Crawford, J. C., Z. Liu, T. A. Nelson, et al. 2008. Microsatellite analysis of mating and kinship in beavers (*Castor canadensis*). *Journal of Mammalogy* **89**: 575–581.

Crawford, L. A., D. Koscinski, and N. Keyghobadi. 2012. A call for more transparent reporting of error rates: the quality of AFLP data in ecological and evolutionary research. *Molecular Ecology* **24**: 5911–5917.

Cristofaro, M., A. De Biase, and L. Smith. 2013. Field release of a prospective biological control agent of weeds, *Ceratapion basicorne*, to evaluate potential risk to a nontarget crop. *Biological Control* **64**: 305–314.

Cuda, J. P., A. P. Ferriter, V. Manrique, and J. C. Medal. 2006. Interagency Brazilian Peppertree (*Schinus terebinthifolius*) Management Plan for Florida 2nd Edition: Recommendations from the Brazilian Peppertree Task Force Florida Exotic Pest Plant Council. April, 2006. Florida Exotic Pest Plant Council. Available from: http://www.fleppc.org/manage_plans/bpmanagplan06.pdf [Accessed January 2016].

Davey, J. W., P. A. Hohenlohe, P. D. Etter, et al. 2011. Genome-wide genetic marker discovery and genotyping using next-generation sequencing. *Nature Reviews Genetics* **12**: 499–510.

Davies, T. J., T. G. Barraclough, M. W. Chase, et al. 2004. Darwin's abominable mystery: Insights from a supertree of the angiosperms. *Proceedings of the National Academy of Sciences of the United States of America* **101**: 1904–1909.

Davis, G. A., N. P. Havill, Z. N. Adelman, et al. 2011. DNA barcodes and molecular diagnostics to distinguish an introduced and native *Laricobius* (Coleoptera: Derodontidae) species in eastern North America. *Biological Control* **58**: 53–59.

DeBach, P. 1959. New species and strains of *Aphytis* (Hymenoptera: Eulophidae) parasitic on the California red scale, *Aonidiella aurantii* (Mask.), in the Orient. *Annals of the Entomological Society of America* **52**: 354–362.

DeBach, P. 1960. The importance of taxonomy to biological control as illustrated by the cryptic history of *Aphytis holoxanthus* n. sp. (Hymenoptera: Aphelinidae), a parasite of *Chrysomphalus aonidum*, and *Aphytis coheni* n. sp., a parasite of *Aonidiella aurantii*. *Annals of the Entomological Society of America* **53**: 701–705.

DeLoach, C. J., R. Carruthers, T. Dudley, et al. 2004. First results for control of saltcedar (*Tamarix* spp.) in the open field in the western United States, pp. 505–513. *In:* Cullen, J. (ed.). *Eleventh International Symposium on Biological Control of Weeds*. CSIRO Entomology. Canberra, Australia.

Denmark, H. A., T. R. Fasulo, and J. E. Funderburk. 2014. Cuban laurel thrips, *Gynaikothrips ficorum* (Marchal) (Insecta: Thysanoptera: Phlaeothripidae). *Institute of Food and Agricultural Sciences Extension* EENY-324, pp. 1–5. Available from: http://manatee.ifas.ufl.edu/lawn_and_garden/master-gardener/gardening-manatee-style/c/cuban-laurel-thrips.pdf [Accessed January 2016].

Deokar, A. A., L. Ramsay, A. G. Sharpe, et al. 2014. Genome wide SNP identification in chickpea for use in development of a high density genetic map and improvement of chickpea reference genome assembly. *BMC Genomics* **15**: 1–19.

Deschamps, S., K. Nannapaneni, Y. Zhang, and K. Hayes. 2012. Local assemblies of paired-end reduced representation libraries sequenced with the illumina genome analyzer in maize. *International Journal of Plant Genomics* **2012**: (360598) 1–8.

Dillon, A. B., A. N. Rolston, C. V. Meade, et al. 2008. Establishment, persistence, and introgression of entomopathogenic nematodes in a forest ecosystem. *Ecological Applications* **18**: 735–747.

Dobbs, T. T. and D. W. Boyd. 2006. Status and distribution of *Montandoniola moraguesi* (Hemiptera: Anthocoridae) in the continental United States. *Florida Entomologist* **89**: 41–46.

Dosdall, L. M., G. A. P. Gibson, O. O. Olfert, and P. G. Mason. 2009. Responses of Chalcidoidea (Hymenoptera) parasitoids to invasion of the cabbage seedpod weevil (Coleoptera: Curculionidae) in western Canada. *Biological Invasions* **11**: 109–125.

Doucette, C. F. 1944. The cabbage seedpod weevil, *Ceutorhynchus assimilis* (Payk.). *Bulletin of the Washington Agricultural Experiment Station* **455**: 123–125.

Doucette, C. F. 1948. Field parasitization and larval mortality of the cabbage seedpod weevil. *Journal of Economic Entomology* **41**: 763–765.

Edwards, O. R. and M. A. Hoy. 1993. Polymorphism in 2 parasitoids detected using random amplified polymorphic DNA-polymerase chain-reaction. *Biological Control* **3**: 243–257.

Edwards, O. R. and M. A. Hoy. 1995. Random amplified polymorphic DNA markers to monitor laboratory-selected, pesticide-resistant *Trioxys pallidus* (Hymenoptera: Aphidiidae) after release into three California walnut orchards. *Environmental Entomology* **24**: 487–496.

Ekblom, R. and J. Galindo. 2011. Applications of next generation sequencing in molecular ecology of non-model organisms. *Heredity* **107**: 1–15.

Elkinton, J. S. and A. M. Liebhold. 1990. Population dynamics of gypsy moth in North America. *Annual Review of Entomology* **35**: 571–596.

Ellegren, H. 2004. Microsatellites: simple sequences with complex evolution. *Nature Reviews Genetics* **5**: 435–445.

Estoup, A., P. Jarne, and J. M. Cornuet. 2002. Homoplasy and mutation model at microsatellite loci and their consequences for population genetics analysis. *Molecular Ecology* **11**: 1591–1604.

Farrell, B. D. 2001. Evolutionary assembly of the milkweed fauna: cytochrome oxidase I and the age of *Tetraopes* beetles. *Molecular Phylogenetics and Evolution* **18**: 467–478.

Fauer, D. and D. Joly. 2015. Next-generation sequencing as a powerful motor for advances in the biological and environmental sciences. *Genetica* **143**: 129–132.

Fisher, A. J., L. Smith, and D. M. Woods. 2011. Climatic analysis to determine where to collect and release *Puccinia jaceae* var. *solstitialis* for biological control of yellow starthistle. *Biocontrol Science and Technology* **21**: 333–351.

Franklin, M. T., C. E. Ritland, and J. H. Myers. 2010. Spatial and temporal changes in genetic structure of greenhouse and field populations of cabbage looper, *Trichoplusia ni*. *Molecular Ecology* **19**: 112–1133.

Frazer, B. D. and R. van den Bosch. 1972. Biological control of the walnut aphid in California; the interrelationship of the aphid and its parasite. *Environmental Entomology* **2**: 561–568.

Flanders, S. E. 1956. Hymenopterous parasites of three species of oriental scale insects. *Bollettino del Laboratorio di Zoologia Generale e Agraria, Portici* **33**: 10–28.

Frewin, A. J., C. Scott-Dupree, G. Murphy, and R. Hanner. 2014. Demographic trends in mixed *Bemisia tabaci* (Hemiptera: Aleyrodidae) cryptic species populations in commercial poinsettia under biological control- and insecticide-based management. *Journal of Economic Entomology* **107**: 1150–1155.

Futuyma, D. J. 1976. Food plant specialization and environmental predictability in Lepidoptera. *The American Naturalist* **110**: 285–292.

Gahan, A. B. 1941. A revision of the parasitic wasps of the genus *Necremnus* Thomson (Eulophidae; Hymenoptera). *Journal of the Washington Academy of Science* **31**: 196–203.

Gallego, F. J. and I. Martinez. 1997. Method to improve reliability of random-amplified polymorphic DNA markers. *Biotechniques* **23**: 663–664.

Gandhi, K. J. K. and D. A. Herms. 2010. North American arthropods at risk due to widespread *Fraxinus* mortality caused by the alien emerald ash borer. *Biological Invasions* **12**: 1839–1846.

Gaskin, J. F. and B. A. Schaal. 2002. Hybrid *Tamarix* widespread in U.S. invasion and undetected in native Asian range. *Proceedings of the National Academy of Sciences of the United States of America* **99**: 11256–11259.

Gaskin, J. F., G. S. Wheeler, M. F. Purcell, and G. S. Taylor. 2009. Molecular evidence of hybridization in Florida's sheoak (*Casuarina* spp.) invasion. *Molecular Ecology* **18**: 3216–3226.

Gaskin, J. F., M.-C. Bon, M. J. W. Cock, et al. 2011. Applying molecular-based approaches to classical biological control of weeds. *Biological Control* **58**: 1–21.

Gaskin, J. F., M. Schwarzlaender, H. L. Hinz, et al. 2013a. Genetic identity and diversity of perennial pepperweed (*Lepidium latifolium*) in its native and invaded ranges. *Invasive Plant Science and Management* **6**: 268–280.

Gaskin, J. F., M. Schwarzlaender, C. L. Kinter, et al. 2013b. Propagule pressure, genetic structure, and geographic origins of *Chondrilla juncea* (Asteraceae): an apomictic invader on three continents. *American Journal of Botany* **100**: 1871–1882.

Gaskin, J. F., M. Schwarzlaender, F. S. Grevstad, et al. 2014. Extreme differences in population structure and genetic diversity for three invasive congeners: knotweeds in western North America. *Biological Invasions* **16**: 2127–2136.

Gebiola, M., J. Gomez-Zurita, M. M. Monti, et al. 2012. Integration of molecular, ecological, morphological and endosymbiont data for species delimitation within the *Pnigalio soemius* complex (Hymenoptera: Eulophidae). *Molecular Ecology* **21**: 1190–1208.

Gibson, G. A. P., H. Baur, B. Ulmer, et al. 2005. On the misidentification of chalcid (Hymenoptera: Chalcidoidea) parasitolds of the cabbage seedpod weevil (Coleoptera: Curculionidae) in North America. *The Canadian Entomologist* **137**: 381–403.

Gilbert, G. S. and C. O. Webb. 2007. Phylogenetic signal in plant pathogen-host range. *Proceedings of the National Academy of Sciences of the United States of America* **104**: 4979–4983.

Gilbert, M. T. P., W. Moore, L. Melchior, and M. Worobey. 2007. DNA extraction from dry museum beetles without conferring external morphological damage. *PLoS One* **2**(3): e272.

Gillespie, D. R., P. G. Mason, L. M. Dosdall, et al. 2006. Importance of long-term research in classical biological control: an analytical review of a release against the cabbage seedpod weevil in North America. *Journal of Applied Entomology* **130**: 401–409.

Goolsby, J. A., P. J. De Barro, J. R. Makinson, et al. 2006. Matching the origin of an invasive weed for selection of a herbivore haplotype for a biological control programme. *Molecular Ecology* **15**: 287–297.

Greenstone, M. H. 2006. Molecular methods for assessing insect parasitism. *Bulletin of Entomological Research* **96**: 1–13.

Gwiazdowski, R. A., R. G. Van Driesche, A. Desnoyers, et al. 2006. Possible geographic origin of beech scale, *Cryptococcus fagisuga* (Hemiptera: Eriococcidae), an invasive pest in North America. *Biological Control* **39**: 9–18.

Gwiazdowski, R. A., R. G. Foottit, E. L. Maw, and P. D. N. Hebert. 2015. The Hemiptera (Insecta) of Canada: constructing a reference library of DNA barcodes. *PLoS One* **10**: e0125635.

Hail, D., I. Lauziere, S. E. Dowd, and B. Bextine. 2011. Culture independent survey of the microbiota of the glassy-winged sharpshooter (*Homalodisca vitripennis*) using 454 pyrosequencing. *Environmental Entomology* **40**: 23–29.

Hajek, A. E., R. A. Humber, J. S. Elkinton, et al. 1990. Allozyme and restriction fragment length polymorphism analyses confirm *Entomophaga maimaiga* responsible for 1989 epizootics in North American gypsy moth populations. *Proceedings of the National Academy of Sciences of the United States of America* **87**: 6979–6982.

Hanson, A. J., E. C. Carlson, E. P. Breakey, and R. L. Webster. 1948. Biology of the cabbage seedpod weevil in Northwestern Washington. *Bulletin of the Washington State Agricultural Experiment Station* **498**: 1–15.

Hardy, D. E. and M. D. Delfinado. 1980. *Insects of Hawaii. Volume 13. Diptera: Cyclorrhapha III*. Series Schizophora Section Acalypterae, Exclusive of family Drosophilidae. University Press of Hawaii. Honolulu.

Havill, N. P., M. E. Montgomery, G. Yu, et al. 2006. Mitochondrial DNA from hemlock woolly adelgid (Hemiptera: Adelgidae) suggests cryptic speciation and pinpoints the source of introduction to eastern North America. *Annals of the Entomological Society of America* **99**: 195–203.

Havill, N. P., G. Davis, D. L. Mausel, et al. 2012. Hybridization between a native and introduced predator of Adelgidae: an unintended result of classical biological control. *Biological Control* **63**: 359–369.

Hebert, P. D. N., E. H. Penton, J. M. Burns, et al. 2004. Ten species in one: DNA barcoding reveals cryptic species in the neotropical skipper butterfly *Astraptes fulgerator*. *Proceedings of the National Academy of Sciences of the United States of America* **101**: 14812–14817.

Hebert, P. D. N., J. R. deWaard, E. v. Zakharov, et al. 2013. A DNA 'barcode blitz': rapid digitization and sequencing of a natural history collection. *PLoS One* **8**(7): e68535.

Hemachandra, K. S. 2004. *Parasitoids of Delia radicum (Diptera: Anthomyiidae) in Canola: Assessment of Potential Agents for Classical Biological Control*. Ph.D. Dissertation. University of Manitoba, Winnipeg.

Hemachandra, K. S., N. J. Holliday, J. Klimaszewski, et al. 2005. Erroneous records of *Aleochara bipustulata* from North America: an assessment of the evidence. *The Canadian Entomologist* **137**: 182–187.

Heintzman, P. D., S. A. Elias, K. Moore, et al. 2014. Characterizing DNA preservation in degraded specimens of *Amara alpina* (Carabidae: Coleoptera). *Molecular Ecology Resources* **14**: 606–615.

Heraty, J. 2004. Molecular systematics and biological control, pp. 39–71. *In*: L. E. Ehler, R. Sforza, and T. Mateille (eds.). *Genetics, Evolution and Biological Control*. CABI Publishers. Wallingford, UK.

Heraty, J. M., J. B. Woolley, K. R. Hopper, et al. 2007. Molecular phylogenetics and reproductive incompatibility in a complex of cryptic species of aphid parasitoids. *Molecular Phylogenetics and Evolution* **45**: 480–493.

Hirsch, J., P. Sprick, and A. Reineke. 2010. Molecular identification of larval stages of *Otiorhynchus* (Coleoptera: Curculionidae) species based on polymerase chain reaction-restriction fragment length polymorphism analysis. *Journal of Economic Entomology* **103**: 898–907.

Hoddle, M. S., K. Warner, J. Steggal, and K. M. Jetter. 2015. Classical biological control of invasive legacy crop pests: new technologies offer opportunities to revist old pest problems in perennial tree crops. *Insects* **6**: 13–37.

Hoffmeister, T. 1992. Factors determining the structure and diversity of parasitoid complexes in tephritid fruit flies. *Oecologia* **89**: 288–297.

Hohenlohe, P. A., S. Bassham, P. D. Etter, et al. 2010. Population genomics of parallel adaptation in threespine stickleback using sequenced RAD tags. *PLoS Genetics* **6**(2): e1000862.

Hohenlohe, P. A., S. J. Amish, J. M. Catchen, et al. 2011. Next-generation RAD sequencing identifies thousands of SNPs for assessing hybridization between rainbow and westslope cutthroat trout. *Molecular Ecology Resources* **11**: (S1) 117–122.

Holliday, N. J., U. Kuhlmann, L. Andreassen, and J. Du. 2012. Classical biological control of root maggots in canola with *Aleochara bipustulata*. *Final Project Report: Canola Agronomic Research Program (CARP)*: 1–14.

Huber, J. T. 1998. The importance of voucher specimens, with pratical guidelines for preserving specimens of the major invertebrate phyla for identification. *Journal of Natural History* **32**: 367–385.

Hufbauer, R. A., S. M. Bogdanowicz, and R. G. Harrison. 2004. The population genetics of a biological control introduction: mitochondrial DNA and microsatellie variation in native and introduced populations of *Aphidus ervi*, a parisitoid wasp. *Molecular Ecology* **13**: 337–348.

Jetter, K. M., J. M. DiTomaso, D. J. Drake, et al. 2003. Biological control of yellow starthistle, pp. 225–241. *In*: Sumner, D. A. and F. H. Buck (eds.). *Exotic Pests and Diseases: Biology and Economics for Biosecurity*. Iowa State Press. Ames, Iowa, USA.

Jones, C. J., K. J. Edwards, S. Castaglione, et al. 1997. Reproducibility testing of RAPD, AFLP and SSR markers in plants by a network of European laboratories. *Molecular Breeding* **3**: 381–390.

Julien, M., T. Center, and P. Tipping. 2002. Floating fern (*Salvinia*), pp. 17–32. *In*: Van Driesche, R., B. Blossey, M. Hoddle, et al. (eds.). *Biological Control of Invasive Plants in the Eastern United States*. FHTET-2002-04. USDA, Forest Service. Morgantown, West Virginia, USA.

Julien, M. H. 1987. *Biological Control of Weeds: A World Catalogue of Agents and Their Target Weeds*. CAB International. Wallingford, Oxon, UK.

Kanz, C., P. Aldebert, N. Althorpe, et al. 2005. The EMBL nucleotide sequence database. *Nucleic Acids Research* **33**: SI D29–D33.

Khidr, S. K., I. C. W. Hardy, T. Zaviezo, and S. Mayes. 2014. Development of microsatellite markers and detection of genetic variation between *Goniozus* wasp populations. *Journal of Insect Science* **14**: 1–17.

Kim, J. S., S. C. Shin, J. Y. Roh, et al. 2008. Identificaiton of an entomopathogenic fungus, *Beauveria bassiana* SFB-205 toxic to the green peach aphid, *Myzus persicae*. *International Journal of Industrial Entomology* **17**: 211–215.

King, T. L., M. S. Eackles, A. P. Henderson, et al. 2005a. Microsatellite DNA markers for delineating population structure and kinship among the endangered Kirtland's warbler (*Dendroica kirtlandii*). *Molecular Ecology Notes* **5**: 569–571.

King, T. L., M. S. Eackles, and B. H. Letcher. 2005b. Microsatellite DNA markers for the study of Atlantic salmon (*Salmo salar*) kinship, population structure, and mixed-fishery analyses. *Molecular Ecology Notes* **5**: 130–132.

King, T. L., M. S. Eackles, and D. C. Chapman. 2011. Tools for assessing kinship, population structure, phylogeography, and interspecific hybridization in Asian carps invasive to the Mississippi River, USA: isolation and characterization of novel tetranucleotide microsatellite DNA loci in silver carp *Hypophthalmichthys molitrix*. *Conservation Genetics Resources* **3**: 397–401.

Kirk, A. A., L. A. Lacey, J. K. Brown, et al. 2000. Variation in the *Bemisia tabaci* s. l. species complex (Hemiptera: Aleyrodidae) and its natural enemies leading to successful biological control of *Bemisia* biotype B in the USA. *Bulletin of Entomological Research* **90**: 317–327.

Kneeland, K., T. A. Coudron, E. Lindroth, et al. 2012. Genetic variation in field and laboratory populations of the spined soldier bug, *Podisus maculiventris*. *Entomologia Experimentalis et Applicata* **143**: 120–126.

Kodama, Y., J. Mashima, E. Kaminuma, et al. 2012. The DNA Data Bank of Japan launches a new resource, the DDBJ Omics Archive of functional genomics experiments. *Nucleic Acids Research* **40**: D1 D38–D42.

Kondo, T. and P. J. Gullan. 2007. Taxonomic review of the lac insect genus *Paratachardina* Balachowsky (Hemiptera: Cocdea: Kerridae), with a revised key to genera of Kerriidae and description of two new species. *Zootaxa* **1617**: 1–41.

König, K., E. Krimmer, S. Brose, et al. 2015. Does early learning drive ecological divergence during speciation processes in parasitoid wasps? *Proceedings of the Royal Society B Biological Sciences* **282**: 20141850.

Kuhn, K. L., J. J. Duan, and K. R. Hopper. 2013. Next-generation genome sequencing and assembly provides tools for phylogenetics and identification of closely related species of *Spathius* parasitoids of *Agrilus planipennis* (emerald ash borer). *Biological Control* **66**: 77–82.

Kupper, K. C., E. Wicker, A. P. A. Aukar, et al. 2010. Genetic diversity of *Bacillus subtilis* isolates with potential to biological control of *Colletotrichum acutatum* and *Guignardia citricarpa*. *Summa Phytopathologica* **36**: 195–202.

Larralde-Corona, C. P., M. R. Santiago-Mena, A. M. Sifuentes-Ricon, et al. 2008. Biocontrol potential and polyphasic characterization of novel native *Trichoderma* strains against *Macrophomina phaseolina* isolated from sorghum and common bean. *Applied Microbiology and Biotechnology* **80**: 167–177.

Lombaert, E., T. Guillemaud, J.-M. Cornuet, et al. 2010. Bridgehead effect in the worldwide invasion of the biocontrol harlequin ladybird. *PLoS One* **5**(3): e9743.

Lozier, J. D. and N. J. Mills. 2009. Ecological niche models and coalescent analysis of gene flow support recent allopatric isolation of parasitoid wasp populations in the Mediterranean. *PLoS One* **4**(6): e5901.

Lozier, J. D. and N. J. Mills. 2011. Predicting the potential invasive range of light brown apple moth (*Epiphyas postvittana*) using biologically informed and correlative species distribution models. *Biological Invasions* **13**: 2409–2421.

Lozier, J. D., G. K. Roderick, and N. J. Mills. 2007. Genetic evidence from mitochondrial, nuclear, and endosymbiont markers for the evolution of host plant associated species in the aphid genus *Hyalopterus* (Hemiptera: Aphididae). *Evolution* **61**: 1353–1367.

Lozier, J. D., G. K. Roderick, and N. J. Mills. 2008. Evolutionarily significant units in natural enemies: Identifying regional populations of *Aphidius transcaspicus* (Hymenoptera: Braconidae) for use in biological control of mealy plum aphid. *Biological Control* **46**: 532–541.

Luque, G. M., C. Bellard, C. Bertelsmeier, et al. 2014. The 100th of the world's worst invasive alien species. *Biological Invasions* **16**: 981–985.

Macher, J. N., A. Rozenberg, S. U. Pauls, et al. 2015. Assessing the phylogeographic history of the montane caddisfly *Thremma gallicum* using mitochondrial and restriction-site-associated DNA (RAD) markers. *Ecology and Evolution* **5**: 648–622.

Mallet, J. 2005. Hybridization as an invasion of the genome. *Trends in Ecology & Evolution* **20**: 229–237

Manrique, V., J. P. Cauda, W. A. Overholt, et al. 2008. Effect of host-plant genotypes on the performance of three candidate biological control agents of *Schinus terebinthifolius* in Florida. *Biological Control* **47**: 167–171.

Manrique, V., R. Diaz, L. Erazo, et al. 2014. Comparison of two populations of *Pseudophilothrips ichini* (Thysanoptera: Phlaeothripidae) as candidates for biological control of the invasive weed *Schinus terebinthifolia* (Sapindales: Anacardiaceae). *Biocontrol Science and Technology* **24**: 518–535.

Marcari, V., S. Causse, K. Hoelmer, et al. 2014. Development of microsatellite markers for *Peristenus digoneutis* (Hymenotpera: Braconidae) a key natural enemy of tarnished plant bugs. *Conservation Genetic Resources* **6**: 421–423.

Mariette, S., V. Le Corre, F. Austerlitz, and A. Kremer. 2002. Sampling within the genome for measuring within-population diversity: trade-offs between markers. *Molecular Ecology* **11**: 1145–1156.

Marsico, T. D., J. W. Burt, E. K. Espeland, et al. 2010. Underutilized resources for studying the evolution of invasive species during their introduction, establishment, and lag phases. *Evolutionary Applications* **3**: 203–219.

Matile-Ferrero, D. 1977. A new scale-insect injurious to cassava in Equatorial Africa, *Phenacoccus manihoti* sp. n. (Homoptera: Cocdea, Pseudococcidae). *Annales de la Société Entomologique de France* **13**: 145–152.

McClure, M. S. 1987. Biology and control of hemlock woolly adelgid. *Bulletin of the Connecticut Agricultural Experimental Station* **85**: 1–9.

McCoy, E. D. and J. H. Frank. 2010. How should the risk associated with the introduction of biological control agents be estimated? *Agricultural and Forest Entomology* **12**: 1–8.

Meier, R., K. Shiyang, G. Vaidya, and P. K. L. Ng. 2006. DNA barcoding and taxonomy in Diptera: a tale of high intraspecific variablity and low identification success. *Systematic Biology* **55**: 715–728.

Meier, R., G. Zhang, and F. Ali. 2008. The use of mean instead of smallest interspecific distances exaggerates the size of the "barcoding gap" and leads to misidentification. *Systematic Biology* **57**: 809–813.

Messing, R. H. 2001. Centrifugal phylogeny as a basis for non-target host testing in biological control: is it relevant for parasitoids? *Phytoparasitica* **29**: 187–190.

Messing, R. H. and M. T. AliNiazee. 1988. Hybridization and host suitability of two biotypes of *Trioxys pallidus* (Hymenoptera: Aphidiidae). *Annals of the Entomological Society of America* **81**: 6–9.

Miller, D. R. and A. Y. Rossman. 1995. Systematics, biodiversity, and agriculture – systematic analyses of small but important organisms provide crucial information for improvement of agriculture. *Bioscience* **45**: 680–686.

Mlonyeni, X. O., B. D. Wingfield, M. J. Wingfield, et al. 2011. Extreme homozygosity in Southern Hemisphere populations of *Deladenus siricidicola*, a biological control agent of *Sirex noctilio*. *Biological Control* **59**: 348–353.

Montgomery, M. E. and N. P. Havill. 2014. Balsam woolly adelgid (*Adelges piceae* [Ratzeburg]) (Hemiptera: Adelgidae), pp. 9–19. *In*: Van Driesche, R. G. and R. Reardon (eds.). *The Use of Classical Biological Control to Preserve Forests in North America*. FHTET- 2013-02. USDA Forest Service. Morgantown, West Virginia, USA.

Moorhouse, E. R., A. K. Charnley, and A. T. Gillespie. 1992. A review of the biology and control of the vine weevil, *Otiorhynchus sulcatus* (Coleoptera: Curculionidae). *Annals of Applied Biology* **121**: 431–454.

Morrison, D. A. 2005. Networks in phylogenetic analysis: new tools for population biology. *International Journal for Parasitology* **35**: 567–592.

Moulton, M. J., H. J. Song, and M. F. Whiting. 2010. Assessing the effects of primer specificity on eliminating numt coamplification in DNA barcoding: a case study from Orthoptera (Arthropoda: Insecta). *Molecular Ecology Resources* **10**: 615–627.

Murchie, A. K. and I. H. Williams. 1998. A bibliography of the parasitoids of the cabbage seed weevil (*Ceutorhynchus assimilis* Payk.). *IOBC-WPRS Bulletin* **21**: 163–169.

Narum, S. R., C. A. Buerkle, J. W. Davey, et al. 2013. Genotyping-by-sequencing in ecological and conservation genomics. *Molecular Ecology* **22**: 2841–2847.

Novotny, V. and Y. Basset. 2005. Review – Host specificity of insect herbivores in tropical forests. *Proceedings of the Royal Society B Biological Sciences* **272**: 1083–1090.

Novotny, V., Y. Basset, S. E. Miller, et al. 2002. Low host specificity of herbivorous insects in a tropical forest. *Nature* **416**: 841–844.

Nwanze, K. F., K. Leuschner, and H. C. Ezumah. 1979. Cassava mealybug, *Phenacoccus* sp. in the Republic of Zaire. *Proceedings of the National Academy of Science of the United States of America* **25**: 125–130.

Okassa, M., S. Kreiter, S. Guichou, and M. S. Tixier. 2011. Molecular and morphological boundaries of the predatory mite *Neoseiulus californicus* (McGregor) (Acari: Phytoseiidae). *Biological Journal of the Linnean Society* **104**: 393–406.

Pante, E., J. Abdelkrim, A. Viricel, et al. 2015. Use of RAD sequencing for delimiting species. *Heredity* **114**: 450–459.

Pegadaraju, V., R. Nipper, B. Hulke, et al. 2013. De novo sequencing of sunflower genome for SNP discovery using RAD (Restriction site Associated DNA) approach. *BMC Genomics* **14**: 1–9.

Pemberton, R. W. 2003. Potential for biological control of the lobate lac scale, *Paratachardina lobata lobata* (Hemiptera: Kerriidae). *Florida Entomologist* **86**: 353–360.

Peterson, B. K., J. N. Weber, E. H. Kay, et al. 2012. Double digest RADseq: an inexpensive method for *de novo* SNP discovery and genotyping in model and non-model species. *PLoS One* **7**(5): e37135.

Pfender, W. F., M. C. Saha, E. A. Johnson, and M. B. Slabaugh. 2011. Mapping with RAD (restriction-site associated DNA) markers to rapidly identify QTL for stem rust resistance in *Lolium perenne. Theoretical and Applied Genetics* **122**: 1467–1480.

Phillips, S. J. and M. Dudik. 2008. Modeling of species distributions with MaxEnt: new extensions and a comprehensive evaluation. *Ecography* **31**: 161–175.

Phillips, S. J., R. P. Anderson, and R. E. Schapire. 2006. Maximum entropy modeling of species geographic distributions. *Ecological Modelling* **190**: 231–259.

Pina, T., M. J. Verdu, A. Urbaneja, and B. Sabater-Munoz. 2012. The use of integrative taxonomy in determining species limits in the convergent pupa coloration pattern of *Aphytis* species. *Biological Control* **61**: 64–70.

Piyaboon, O., A. Unartngam, and J. Unartngam. 2014. Effectiveness of *Myrothecium roridum* for controlling water hyacinth and species identification based on molecular data. *African Journal of Microbiology Research* **8**: 1444–1452.

Pluot-Sigwalt, D., J.-C. Streito, and A. Matocq. 2009. Is *Montandoniola moraguesi* (Puton, 1896) a mixture of different species? (Hemiptera: Heteroptera: Anthocoridae). *Zootaxa* **2208**: 25–43.

Polaszek, A., G. A. Evans, and F. D. Bennett. 1992. *Encarsia* parasitoids of *Bemisia tabaci* (Hymenoptera: Aphelinidae, Homoptera, Aleyrodidae) – a preliminary guide to identification. *Bulletin of Entomological Research* **82**: 375–392.

Polaszek, A., S. Manzari, and D. L. J. Quicke. 2004. Morphological and molecular taxonomic analysis of the *Encarsia meritoria* species-complex (Hymenoptera: Aphelinidae), parasitoids of whiteflies (Hemiptera: Aleyrodidae) of economic importance. *Zoologica Scripta* **33**: 403–421.

Porco, D., R. Rougerie, L. Deharveng, and P. Hebert. 2010. Coupling non-destructive DNA extraction and voucher retrieval for small soft-bodied arthropods in a high-throughput context: the example of Collembola. *Molecular Ecology Resources* **10**: 942–945. doi:10.1111/j.1755-0998.2010.2839.x.

Qin, T. K. and P. J. Gullan. 1995. Biogeography of the wax scales (Insecta: Hemiptera: Coccidae: Ceroplastinae). *Journal of Biogeography* **25**: 37–45.

Qin, T. K. and P. J. Gullan. 1998. Systematics as a tool for pest management: a case study using scale insects and mites, pp. 479–488. *In:* Zalucki, M., R. Drew, and G. White (eds.). *Proceedings of the Sixth Australasian Applied Entomological Research Conference, Brisbane, Australia, 29 September–2nd October 1998, vol. 1.* University of Queensland Printery. Brisbane, Australia.

Qin, T. K., P. J. Gullan, G. A. C. Beattie, et al. 1994. The current distribution and geographical origin of the scale insect pest *Ceroplastes sinensis* (Hemiptera: Coccidae). *Bulletin of Entomological Research* **84**: 541–549.

Qin, T. K., P. J. Gullan, and G. A. C. Beattie. 1998. Biogeography of the wax scales (Insecta: Hemiptera: Coccidae: Ceroplastinae). *Journal of Biogeography* **25**: 37–45.

Ratnasingham, S. and P. D. N. Hebert. 2007. BOLD: the barcode of life data system (http://www.barcodinglife.org [Accessed January 2016]). *Molecular Ecology Notes* **7**: 355–364.

Ratnasingham, S. and P. D. N. Hebert. 2013. A DNA-based registry for all animal species: the barcode index number (BIN) system. *PLoS One* **8**(7): e66213.

Rauth, S. J., H. L. Hinz, E. Gerber, and R. A. Hufbauer. 2011. The benefits of pre-release population genetics: a case study using *Ceutorhynchus scrobicollis*, a candidate agent of garlic mustard, *Alliaria petiolata*. *Biological Control* **56**: 67–75.

Roderick, G. K. and M. Navajas. 2012. Evolution and biological control. *Evolutionary Applications* **5**: 419–423.

Room, P. M. 1990. Ecology of a simple plant herbivore system. Biological control of *Salvinia*. *Trends in Ecology & Evolution* **5**: 74–79.

Room, P. M., K. L. S. Harley, I. W. Forno, and D. P. A. Sands. 1981. Successful biological control of the floating weed *Salvinia*. *Nature* **294**: 78–80.

Russell, J. A., S. Weldon, A. H. Smith, et al. 2013. Uncovering symbiont-driven genetic diversity across North American pea aphids. *Molecular Ecology* **22**: 2045–2059.

Scaglione, D., A. Acquadro, E. Portis, et al. 2012. RAD tag sequencing as a source of SNP markers in *Cynara cardunculus* L. *BMC Genomics* **13**: 1–11.

Schroer, S. and R. W. Pemberton. 2007. Host acceptance trials of parasitoids from Indian *Paratachardina lobata* (Hemiptera: Kerriidae) on the invasive lobate lac scale in Florida. *Florida Entomologist* **90**: 545–552.

Selkoe, K. A. and R. J. Toonen. 2006. Microsatellites for ecologists: a practical guide to using and evaluating microsatellite markers. *Ecology Letters* **9**: 615–629.

Sharma, R., B. Goossens, C. Kun-Rodrigues, et al. 2012. Two different high throughput sequencing approaches identify thousands of *de novo* genomic markers for the genetically depleted Bornean elephant. *PLoS One* **7**(11): e49533.

Sheppard, A. W., R. D. van Klinken, and T. A. Heard. 2005. Scientific advances in the analysis of direct risks of weed biological control agents to nontarget plants. *Biological Control* **35**: 215–226.

Simberloff, D. 2012. Risks of biological control for conservation purposes. *Biocontrol* **57**: 263–276.

Sluys, R. 2013. The unappreciated, fundamentally analytical nature of taxonomy and the implications for the inventory of biodiversity. *Biodiversity and Conservation* **22**: 1095–1105.

Smith, L. 2007. Physiological host range of *Ceratapion basicorne*, a prospective biological control agent of *Centaurea solstitialis* (Asteraceae). *Biological Control* **411**: 120–133.

Smith, M. A., D. M. Wood. D. H. Janzen, et al. 2007. DNA barcodes affirm that 16 species of apparently generalist tropical parasitoid flies (Diptera: Tachinidae) are not all generalists. *Proceedings of the National Academy of Science of the United States of America* **104**: 4967–4972.

Smith, M. A., J. L. Fernandez-Triana, E. Eveleigh, et al. 2013. DNA barcoding and the taxonomy of microgastrinae wasps (Hymenoptera: Braconidae): impacts after 8 years and nearly 20,000 sequences. *Molecular Ecology Resources* **13**: 168–176.

Smithsonian National Museum of Natural History. 2014. Plant DNA Barcode Project. Available from: http://botany.si.edu/projects/dnabarcode/ [Accessed January 2016].

Soper, A. 2012. Biological Control of the Ambermarked Birch Leafminer (*Profenusa thomsoni*) in Alaska, pp. 1–128. Ph.D. Dissertation. University of Massachusetts, Amherst.

Stireman, J. O. and M. S. Singer. 2003. What determines host range in parasitoids? An analysis of a tachinid parasitoid community. *Oecologia* **135**: 629–638.

Stouthamer, R. 2006. Molecular methods for the identification of biological control agents at the species and strain level, pp. 187–201. *In:* Bigler, F., D. Babendreier, and U. Kuhlmann (eds.). *Environmental Impact of Invertebrates for Biological Control of Arthropods: Methods and Risk Assessment*. CABI Publishing. Wallingford, UK.

Sutherst, R. W., G. F. Maywald, T. Yonow, and P. M. Stevens. 1999. *CLIMEX: Predicting the Effects of Climate on Plant and Animals*. CSIRO Publishing. Collingwood, Australia.

Szűcs, M., M. Schwarzläender, and J. F. Gaskin. 2011. Reevaluating establishment and potential hybridization of different biotypes of the biological control agent *Longitarsus jacobaeae* using molecular tools. *Biological Control* **58**: 44–52.

Terhonen, E., H. Sun, M. Buee, et al. 2013. Effects of the use of biocontrol agent (*Phlebiopsis gigantea*) on fungal communities on the surface of *Picea abies* stumps. *Forest Ecology and Management* **310**: 428–433.

Thalmann, O., J. Hebler, H. N. Poinar, et al. 2004. Unreliable mtDNA data due to nuclear insertions: a cautionary tale from analysis of humans and other great apes. *Molecular Ecology* **13**: 321–335.

Thomas, P. A. and P. M. Room. 1986. Taxonomy and control of *Salvinia molesta*. *Nature* **320**: 581–584.

Tin, M. M-Y., E. P. Economo, and A. S. Milkheyev. 2014. Sequencing degraded DNA from non-destructively sampled museum specimens for RAD-Tagging and low-coverage shotgun phylogenetics. *PLoS One* **9**(5): e96793. doi:10.1371/journal.pone.0096793.

Tixier, M.-S., F. A. Hernandes, S. Guichou, and S. Kreiter. 2011. The puzzle of DNA sequences of Phytoseiidae (Acari: Mesostigmata) in the public GenBank database. *Invertebrate Systematics* **25**: 389–406.

Tracy, J. T. and T. O. Robbins. 2009. Taxonomic revision and biogeographcy of the *Tamarix*-feeding *Diorhabda elongata* (Brullé, 1983) species group (Coleoptera: Chrysomelidae: Galerucinae: Galerucini) and analysis of their potential in biological control of tamarisk. *Zootaxa* **2101**: 1–152

Tringe, S. G., C. von Mering, A. Kobayashi, et al. 2005. Comparative metagenomics of microbial communities. *Science* **308**: 554–557.

Trissi, A. N., M. El Bouhsini, M. N. Alsalti, et al. 2013. Genetic diversity among summer and winter *Beauveria bassiana* populations as revealed by AFLP analysis. *Journal of Asia-Pacific Entomology* **16**: 269–273.

Tsutsui, N. D., A. V. Suarez, D. A. Holway, and T. J. Case. 2001. Relationships among native and introduced populations of the Argentine ant (*Linepithema humile*) and the source of introduced populations. *Molecular Ecology* **10**: 2151–2161.

Valles, S. M., D. H. Oi, F. Yu, et al. 2012. Metatranscriptomics and Pyrosequencing facilitate discovery of potential viral natural enemies of the invasive Caribbean crazy ant, *Nylanderia pubens*. *PLoS One* **7**(2): e31828.

Vandamme, A. M. 2009. Basic concepts of molecular evolution. *In*: Lemey, P., M. Salemi, and A. M. Vandamme (eds.). *The Phylogenetic Handbook: A Practical Approach to Phylogenetic Analysis and Hypothesis Testing*. Cambridge University Press. Cambridge, UK.

van den Bosch, R., B. D. Frazer, C. S. Davis, et al. 1970. *Trioxys pallidus* – an effective new walnut aphid parasite from Iran. *California Agriculture* **24**: 8–10.

Vandepitte, K., O. Honnay, J. Mergeay, et al. 2013. SNP discovery using paired-end RAD-tag sequencing on pooled genomic DNA of *Sisymbrium austriacum* (Brassicaceae). *Molecular Ecology Resources* **13**: 269–275.

Van Driesche, R. G., R. I. Carruthers, T. Center, et al. 2010. Classical biological control for the protection of natural ecosystems. *Biological Control* **54**: S2–S33.

Vos, P., R. Hogers, M. Bleeker, et al. 1995. AFLP – a new technique for DNA-fingerprinting. *Nucleic Acids Research* **23**: 4407–4414.

Wagner, D. L. and R. G. Van Driesche. 2010. Threats posed to rare or endangered insects by invasions of non-native species. *Annual Review of Entomology* **55**: 547–568.

Wagner, D. L. and K. J. Todd. 2015. Conservation implications and ecological impacts of the emerald ash borer in North America, pp. 15–62. *In*: Van Driesche, R. G. and R. Reardon (eds.). *The Biology and Control of Emerald Ash Borer*. FHTET 2014-09. USDA Forest Service. Morgantown, West Virginia, USA.

Wagner, D. L. and K. J. Todd. 2016. New ecological assessment for the emerald ash borer: a cautionary tale about unvetted host-plant literature. *American Entomologist* **62**: 26–35.

Wapshere, A. J. 1974. Strategy for evaluating safety of organisms for biological weed control. *Annals of Applied Biology* **77**: 201–211.

Williams, D. A., W. A. Overholt, J. P. Cuda, and C. R. Hughes. 2005. Chloroplast and microsatellite DNA diversities reveal the introduction history of Brazilian peppertree (*Schinus terebinthifolius*) in Florida. *Molecular Ecology* **14**: 3643–3656.

Williams, D. A., E. Muchugu, W. A. Overholt, and J. P. Cuda. 2007. Colonization patterns of the invasive Brazilian peppertree, *Schinus terebinthifolius*, in Florida. *Heredity* **98**: 284–293.

Williams, W. I., J. M. Friedman, J. F. Gaskin, and A. P. Norton. 2014. Hybridization of an invasive shrub affects tolerance and resistance to defoliation by a biological control agent. *Evolutionary Applications* **7**: 381–393.

Zalapa, J. E., H. Cuevas, H. Zhu, et al. 2012. Using next-generation sequencing approaches to isolate simple sequence repeat (SSR) loci in the plant sciences. *American Journal of Botany* **99**: 193–208.

Zhang, Y-Z., S-L. Si, J-T. Zheng, et al. 2011. DNA barcoding of endoparasitoid wasps in the genus *Anicetus* reveals high levels of host specificity (Hymenoptera: Encyrtidae). *Biological Control* **58**: 182–191.

CHAPTER 7

Forecasting unintended effects of natural enemies used for classical biological control of invasive species

Mark S. Hoddle

Department of Entomology, University of California, USA

Introduction

Classical biological control is the deliberate introduction and establishment of an organism, a natural enemy, outside its native range to control an invasive pest (e.g., non-native weed, insect, or vertebrate) that is causing significant ecological or economic damage in the invaded area. For a general description of the steps in the process, see Table 7.3. Foreign exploration, searching for and collecting natural enemies from the home range of the pest, is a necessary part of the development of a classical biological control program. It is during this phase of the biological control program that organisms potentially exerting top-down control (i.e., predation, parasitism, or disease) of the target are sought. If such agents can be found, ideally ones that are specialized feeders unlikely to affect non-target species, they are collected and taken into a quarantine laboratory for further study. In quarantine, natural enemies undergo host-range testing, which are studies intended to delineate what other species might be used as food or for reproduction by the natural enemy. Additional impact studies may be run in the native range. Selecting specialized natural enemies that do not feed on species other than the target pest (i.e., do not have multiple food web linkages to other prey in the native system) helps to ensure target specificity, which in turn may prevent the unwanted exploitation of non-target prey in the introduced range. This process, by necessity, attempts to re-establish a small, highly specific and truncated part of a "consumer (predator)–resource (prey)" or herbivore–pest weed community in a novel area, often far from the native range, where the interaction has not historically existed (Hoddle, 2004a, b).

The fundamental paradigm underlying the reconstruction of these *strong* (i.e., inflicting high levels of mortality on pest populations or, for plant pests, high levels of damage, which in either case stabilizes or reduces pest population growth and spread), *host-specific, top-down connections* in biological control is that such specialized natural enemies have the potential to reduce pest populations to less damaging levels. In turn, populations of effective biological control agents usually collapse to low levels in response to reduced prey or host availability. Uncontrolled population growth of an introduced species may have occurred, in part, because of escape from natural enemies that normally regulate population growth in the native range. This concept of unrestrained population growth in the absence of top-down control is referred to as the "enemy release hypothesis" (Keane and Crawley, 2002; Torchin et al., 2003; Prior and Hellmann, 2015), and is one of several hypotheses about what permits abundance increases of introduced species (Green et al., 2011; Perkins and Nowark 2013). Reconstruction of part of a food web through the introduction of a natural enemy differs significantly in approach and outcomes when compared to the control provided by native resident natural enemies (termed "natural control" by biological control workers) (Hawkins et al., 1999). Successful biological control of insects, for example, often results from one natural enemy, often a parasitoid, while successful "natural

Integrating Biological Control into Conservation Practice, First Edition. Edited by Roy G. Van Driesche, Daniel Simberloff, Bernd Blossey, Charlotte Causton, Mark S. Hoddle, David L. Wagner, Christian O. Marks, Kevin M. Heinz, and Keith D. Warner.
© 2016 John Wiley & Sons, Ltd. Published 2016 by John Wiley & Sons, Ltd.

control" is often due to the actions of a guild of natural enemies including parasitoids and generalist predators. The success of a biological control agent is due to the strong linkage it forms with the target pest in a highly simplified food web. In contrast, "natural control" often results from multiple linkages in complex food webs (Hawkins et al., 1999).

Successful pest suppression by natural enemies may allow adversely affected ecosystems to return to conditions similar to those before the invasion event without additional human-mediated management. This is often the desired outcome for a biological control program of a pest of conservation importance, but in some cases the pre-invasion state may itself have been degraded and so biological control may assist in lessening the accumulating effect of negative influences. When the invasive species is a plant that has severely suppressed native plants, some form of ecosystem restoration may be needed to ensure project goals are met (see Chapter 2). Alternatively, natural enemies, if unable to fully suppress the target pest, may make other management technologies more feasible, effective, or economically viable because lower pest densities are now amenable to suppression with other tactics (see Chapter 3).

The deliberate introduction of a novel organism in a new area with the intent of permanent establishment and subsequent widespread self-dispersal carries with it an element of risk – this being the uncertainty about the effects this new organism may have in the receiving environment, especially with respect to non-target species. The intended effect of a natural enemy introduction is for this organism to have strong linkages to the target pest (i.e., high host specificity), preventing or greatly limiting exploitation of other organisms that the biological control agent may encounter. High specificity is a desirable attribute because it greatly reduces the possibility of significant *non-target effects*, which may be defined as unintended or unwanted attacks on non-target species, food web subsidies, or other important ecological community changes resulting from the introduced biological control agent. Non-target attacks and unintended ecological disturbances resulting from the deliberate introduction of biological control agents, especially those on native species of high conservation, scientific, or cultural value, have been documented, and these cases have directly challenged the belief that biological control is usually a safe and environmentally friendly form of pest suppression (Howarth, 1991;

Pearson et al., 2000; Louda et al., 2003). However, despite these concerns, severe unintended ecological effects from biological control agents don't appear to be common and are estimated to be <1% for introduced arthropod natural enemies (van Lenteren et al., 2006a).

Natural enemies with insufficient specificity are able to form linkages to non-target organisms in the new environment where they are released and have not previously existed. Therefore, introduced biological control agents can influence the structure and functioning of ecological systems. Understanding the mechanisms underlying the trophic interactions that shape ecological communities is an important issue in ecology (Ohgushi, 2008), and descriptions of food webs are an attempt to recognize such interactions among community constituents (i.e., plants, their herbivores, and natural enemies and the herbivores they exploit). Food webs (sometimes referred to as food cycles, interaction webs, or indirect-interaction webs if non-consuming interactions are considered) are maps, typically presented as schematic diagrams, that identify connections between organisms (Pimm et al., 1991; Ohgushi, 2008). Food webs function as tools to describe and help us understand how ecological communities are structurally organized (Ohgushi, 2008) and to document the intensity of interactions within the matrix (Memmott, 2000). Adding a natural enemy for the control of an invasive species introduces a new pathway into an existing ecosystem in which the pest is already well established. The issue of interest following the establishment of the natural enemy in this new system is learning how strong the connection to the target is and whether the biological control agent develops ecologically significant, quantifiable non-trophic or trophic linkages to other organisms in this environment. Understanding these food or interaction web effects helps us understand how the natural enemy interacts within its new ecosystem, where it has not existed previously, but where the target may have long been established.

Consequently, this chapter will examine both non-trophic (i.e., non-feeding interactions) and trophic effects (i.e., feeding interactions by herbivores on invasive plants; carnivores [i.e., predators and parasitoids] on invasive insects, and pathogens on plants and insects) that deliberately introduced natural enemies have had following their introduction for the control of an invasive pest.

Forecasting non-trophic effects

Theoretical basis for non-trophic effects

Non-trophic interactions are species relationships that do not involve direct consumer–resource exploitation. Such phenomena are incredibly diverse and widespread and drive species diversity and composition, community patterns, productivity, and persistence (Kéfi et al., 2012). It is argued that non-trophic effects may exceed in importance the most commonly studied form of interspecies interactions, consumer–resource exploitation (Ohgushi, 2008), because indirect non-trophic effects are diffuse and can sometimes have profound and penetrating effects on resource acquisition, reproductive success, and behavior (Kéfi et al., 2012).

Ecosystem engineering

A well-studied example of a non-trophic interaction is habitat modification by a particular species that indirectly affects other species, either positively or negatively. Manipulation of habitat resources is referred to as *ecosystem engineering* (Jones et al., 1994; Ohgushi, 2008; Vieira and Romero, 2013) or *niche construction* (Laland and Boogert, 2010). Ecosystem engineering is a term that describes the modification, maintenance, or creation of habitats by organisms (Laland and Boogert, 2010), and Jones et al. (1994) recognized five types of ecosystem engineering. Niche construction is a concept from evolutionary biology and is the process by which an organism's activities modify its own niche and/or niches of other species thereby directing the forces of natural selection in their own environment (Laland and Boogert, 2010). These two terms, ecosystem engineering and niche construction, can be considered synonymous (Laland and Boogert, 2010). Modeling of habitat modification, ecosystem engineering, and niche construction indicates that non-trophic interactions that result from these processes increase species diversity and biomass, productivity, spatial distribution of species, and resilience to external disturbances (Jones et al., 1994; Kéfi et al., 2012; Vieira and Romero, 2013).

Ecosystem engineers can provide shelter, protection from enemies, or increased food availability (Jones et al., 1994). An example of an ecosystem engineer is a galling herbivore whose feeding modifies host-plant architecture by creating new physical structures, galls (i.e., abnormal outgrowths of plant tissue), on host plants. Galls in turn provide indirectly a resource that

can increase the abundance, richness, and biomass of other guilds (e.g., predators, herbivores, detritivores, or omnivores) that inhabit the galled plant. Depending on the time of year, non-trophic effects mediated by ecological engineering may elicit pronounced temporal responses by affected species (Vieira and Romero, 2013). Alternatively, ecosystem engineering by herbivores may reduce the abundance of natural enemies by modifying habitat so as to remove refugia that predators utilize or altering habitat structure to make it less attractive for foraging parasitoids (Meinke et al., 2009). Weed biological control agents, especially gall formers like tephritid flies, wasps, or rust pathogens, can act as ecosystem engineers by providing resources for inquilines (Veldtman et al., 2011).

Habitat complexity

Other examples of non-trophic interactions include effects of leaf litter abundance on the attack rates of ants foraging for caterpillars – the thicker the litter, the less effective ants are at finding prey (Karban et al., 2013). This example is part of a much wider body of work that clearly demonstrates the importance of increasing structural complexity of habitat on interactions among organisms. Architectural complexity often reduces encounter rates between organisms, especially cursorial predators and prey, because more refuges exist for prey, making it more difficult for predators to find them, and this in turn affects abundance and rarity of interacting species (Langellotto and Denno, 2004). Alternatively, increasing complexity may benefit some predators, like web-spinning spiders, because there is more substrate on which to attach webs, and this can increase spider densities, web diameters, prey capture rates, and reproductive output (Pearson 2009, 2010). Interestingly, while leaf litter production can affect predation rates, generation of this resource, a by-product of feeding by one species of native caterpillar, can benefit other native caterpillar species by providing protection from desiccation and predation, leading to significant population increases (Karban et al., 2012).

Anti-predation behaviors

Detection of signs (or semiochemicals) of predators can modify behaviors of herbivores so that they pre-emptively drop from plants to reduce predation risk (Ramirez et al., 2010). Similar studies have been conducted in highly modified agro-ecosystems and have shown that native predators affect pest behavior in potatoes (Ramirez et al., 2010). One guild of predators (ladybugs) may indirectly provide additional food for predatory ground beetles by causing aphids on radishes to initiate escape behaviors by dropping to the ground where they are attacked by ground beetles (Prasad and Snyder, 2010). Anti-predation behaviors such as resource abandonment are important non-trophic channels through which introduced predators may indirectly protect plants and could be as beneficial to plants as effects of direct attacks on herbivores by natural enemies (Preisser et al., 2005).

Competition

Some forms of competition can mediate non-trophic interactions and these can have strong ecosystem effects. Exploitative competition, for example, occurs when one species removes resources more efficiently than another species, and this resource depletion causes populations of the competing species to decrease. For example, invading plants usurp resources (e.g., space and nutrients), causing decline in native competing plants, which in turn lowers densities of specialized native herbivores associated with the less competitive native plants, as these herbivores usually cannot feed on the invasive weed (Yoshioka et al., 2010). Generalist insect predators, such as the invasive ladybug *Harmonia axyridis* (Pallas) (Coleoptera: Coccinellidae), can displace native predators. *Harmonia axyridis* is a superior competitor for resources, and when niche overlap occurs with native coccinellids, *H. ayxridis* dominates owing to its higher reproductive output (Michaud, 2002) and the ability of its larvae to consume eggs or larvae of competing native coccinellids (Yasuda et al., 2004). These attributes allow generalist invasive insects like *H. axyridis* to reach high densities, leading to competitive displacement of native coccinellids (Crowder and Snyder, 2010).

Altered defenses

Non-trophic interactions can also result from the production of toxins or other compounds by one organism that modify the growth and development of another species (A'Bear et al., 2010). Herbivory, for example, can induce defensive chemical and physical changes in plants, referred to as phenotypic changes. These modifications may affect the feeding of other species of herbivores and their subsequent vulnerability

to parasitism and predation (Poelman et al., 2011). Herbivore-induced changes in plant chemistry may be influenced by whether or not the herbivore is parasitized because parasitoids alter herbivore physiology and this can cause plant responses that differ from those induced by non-parasitized herbivores (Poelman et al., 2011). For example, different species of parasitoids that share a herbivorous host (e.g., caterpillar) on the same host plant (e.g., brassica) can modify host-plant responses (e.g., changes in protective secondary metabolites) induced by the parasitized herbivore, so that induced changes increase the mortality of one of the competing parasitoid species thereby benefiting the other species. Asymmetric outcomes of this nature are examples of interactions among different species of consumers (i.e., parasitoids) that share a common prey (i.e., the same species of caterpillar), mediated by a non-trophic channel (i.e., alteration of plant chemistry through an induced response to herbivory) (Poelman et al., 2011). It is possible that such effects induced by one guild of organisms (e.g., larval parasitoids attacking a herbivorous host) may exert effects on other guilds exploiting non-related prey (e.g., aphid parasitoids) on the same host plant via induced phenotypic changes (e.g., changes in host plant chemistry or morphology) that consequently establish unique interactions with other apparently unrelated community members (Poelman et al., 2011).

The examples given above of non-trophic interactions are astonishingly diverse, and attempts to classify these interactions have been based on fitness outcomes between two interacting species as being either positive (+), negative (−), or neutral (0) (i.e., no positive or negative outcome results from the interaction). Therefore, categories of non-trophic interactions seen in the ecology literature include mutualism (++), commensalism (+0), antagonism (+−), neutralism (00), amensalism (−0), and competition (−−) (Kéfi et al., 2012). If direct predator–prey interactions are included, nine possible binary outcomes are possible (Fath, 2007). Often interactions between species are not binary and may include interactions among several species. Consequently, an attempt has been made to refine these categories to broaden understanding of non-trophic interactions through a functional class concept that assesses the impact of the interaction between species rather than the specific nature of the interaction (Kéfi et al., 2012).

Kéfi et al. (2012) propose three such categories: (1) non-trophic interactions that directly modify feeding and can include plant responses that recruit natural enemies of herbivores, interspecific predator interference, refuge provisioning, or prey escape responses; (2) non-trophic interactions that modify non-feeding attributes of species, which can include ecosystem engineering, allelopathy, competition for space and resources, maintenance of territories, and pollination; and (3) non-trophic interactions that influence flows of matter across system boundaries, which include alarm pheromones triggering escape and avoidance behaviors, seed dispersal, recruitment facilitation, alteration of inputs, and losses of abiotic resources (e.g., nutrient inputs, shading that reduces water loss, or substrate stabilization). Similarly, Olff et al. (2009) have proposed re-organization of food web research to include non-trophic interactions and they identify six main types of interactions that exist within ecosystems and suggest that these interactions be plotted along axes that represent trophic position (vertical axis) and palatability (horizontal axis).

Examples of non-trophic effects from classical biological control

The non-trophic interactions presented above are diverse, and their outcomes are sometimes subtle and not immediately obvious. Consequently, significant field work or complicated laboratory experiments may be needed to disentangle actual causes and effects. From the perspective of this section of the chapter, the key question is: "Are there non-trophic effects caused by deliberately released natural enemies that have resulted in significant unwanted ecological perturbations?" An electronic review of the literature using Web of Knowledge with "non-trophic" and "biological control" as keywords returned fewer than ten publications with these two keywords. None of the returned articles explicitly addressed the non-trophic actions of natural enemies used in classical biological control programs.

There are potentially several reasons for this scarcity of published research on non-trophic interactions as part of classical biological control programs targeting invasive pests when compared to the general ecology literature. First, such interactions may be so difficult to predict *a priori* that they are not considered during the quarantine assessment of the safety of a biological control agent. Furthermore, without any concrete

hypotheses about potential non-trophic risks, there would be little incentive to search for such interactions after the release of a natural enemy with nothing in particular to look for. Second, it may be that non-trophic effects of importance simply do not happen and therefore have not been studied. Third, little research activity on non-trophic effects in biological control programs may be because these systems are too "simple" to have the interesting, complex, non-trophic interactions that ecologists prefer to study. Fourth, it may be that biological control scientists do not perceive non-trophic effects as likely to be significant and therefore they devote limited research funds to monitoring post-release trophic effects of natural enemies on the target pest and related non-target species. Additionally, biological control experts may lack the research skills needed to investigate complex community ecology questions. Fifth, host-range testing for weed and arthropod natural enemies is mostly conducted in quarantine laboratories and the emphasis of this research is on the outcomes of direct consumer–resource interactions. These data are used for the preparation of detailed Environment Assessment Reports for regulatory agencies to review, and they generate concrete hypotheses on risks to non-target species for follow-up field work should agents be released from quarantine. Consequently, predicting non-trophic effects from these studies seems tangential to the underlying concept of successful pest control, which is the reduction in target density with no collateral damage. This fundamental goal drives funding and research for classical biological control projects. Sixth, if non-trophic effects mediated by a deliberately released natural enemy are documented and demonstrated to be important, does this enhance the success of the program if the effect is positive or tarnish the project if the effect is negative, and how would these non-trophic interactions be assessed in relation to the overall outcome (i.e., full, partial, or no suppression of the pest) of the project or in comparison to the "do nothing" option?

The selected examples presented above provide some possible areas where non-trophic effects may be expected to occur, if indeed they exist, for biological control programs. Much work published on non-trophic interactions strongly suggests that the majority are plant-mediated via the activity of herbivores. If this is true, then perhaps the systems that should be first studied for non-trophic effects would be the control of invasive plants with herbivores. Growing evidence strongly suggests that invasive plants often lack rich guilds of native insect herbivores (Tallamy, 2004), and if native herbivores are using invasive weeds, the herbivores tend to be generalists or more specialized species that have evolved on closely related native congeneric species (Tallamy et al., 2010; Burghardt et al., 2010). Given this greatly reduced native biota on invasive plants, especially with respect to native Lepidoptera (Burghardt et al., 2010), these weed biological control systems could be ideal models for studying non-trophic effects of an introduced weed biological control agent on native species because food webs associated with the weed are significantly simplified.

Depending on the system of interest and the biological control agents being used, two non-trophic interactions that could be studied are (1) the effects of specialist co-evolved herbivores (e.g., the biological control agent) on plant phenotype and subsequent flow-on effects on native species feeding on the weed and (2) effects of ecosystem engineers, like gall-forming insects or plant pathogens, on measures of biodiversity on galled and non-galled weeds.

Herbivores can mediate non-trophic interactions via changes in phenotype (e.g., plant chemistry) that can affect competing herbivorous species or parasitoids and predators using these prey (Poelman et al., 2011). Invasive weeds subjected to biological control with herbivorous insects, for example, may have the potential to affect plant chemistry and thus resident herbivores and natural enemy guilds associated with these herbivores. Effects may be subdivided depending on whether the native herbivore is a generalist or a specialist on a congeneric native plant that has adopted the weed as a food resource, and studies could be further refined to assess effects by herbivores if they are parasitized or unparasitized. Ecosystem engineers, such as gall formers, may be pre-disposed to generating non-trophic food web linkages because the galling habitat is conducive to the formation of new parasitoid or predator linkages to the novel herbivore (Paynter et al., 2010) and galls provide shelter for inquilines (i.e., an organism that lives commensally within the gall) (Veldtman et al., 2011).

Because herbivores are at the base of most plant food webs, they comprise essential food resources for parasitoids and predators, including spiders, lizards, amphibians, bats, birds, and rodents, whose abundance could influence higher-order predators that use these lower-order predators for food. Consequently,

non-trophic effects such as changes in plant phenotype or ecological engineering caused by biological control agents may influence the abundance of food that these upper-level consumers require. This could be especially important in vast monotypic weed stands where resources (e.g., food or refugia) are scarce and non-trophic effects may positively or negatively affect these organisms. Many non-trophic effects resulting from the addition of a biological control agent to a community are going to be hard to predict *a priori*. However, some effects may be predictable ahead of time, such as the potential consequences of galling herbivores; others may be predicted from the literature or assessed via experimental manipulation in the country of origin before release (see Paynter et al., 2010). For example, the non-trophic effects of increased litter production from defoliation could be studied by deliberately creating a range of litter densities and assessing effects on key non-target organisms likely to be affected. Habitat modification resulting from widespread death of host plants (a form of ecosystem engineering) could be assessed by creating time-sensitive mortality gradients (needed to mimic the anticipated speed and spread of death of the weed caused by the biological control agent) in replicated weed stands using non-chemical methods like girdling, and outcomes for key species could be monitored (e.g., nesting and fledgling rates by native birds such as the southwestern willow flycatcher, *Empidonax traillii extimus* Phillips, which nests in stands of invasive species of saltcedar shrubs [*Tamarix* spp.] [Shafroth et al., 2005]). Assessing and demonstrating non-trophic effects using small-scale, highly controlled experiments where few variables are manipulated is one thing, extrapolating findings to predict complex community-wide effects is another issue entirely, and such predictions must be considered cautiously in terms of the potential benefit that could result from successful biological control or possible environmental degradation if the invader is left unchecked.

Forecasting unwanted trophic effects

Trophic interactions are species interactions that involve consumer–resource exploitation. They may be direct or indirect. Direct trophic effects occur when the biological control agent consumes a non-target species. Indirect trophic effects occur when the biological control agent's suppression of the target pest affects non-target species that use the controlled invasive species for food or habitat, or when the biological control agent is consumed by non-target species and thus becomes a food subsidy in the local food web.

Given the fact that many indirect interactions (non-trophic [see previous section] and trophic [see next section]) exist in communities, the attempt to simplify our view of food web interactions by grouping species into functional classes (Kéfi et al., 2012) may not greatly aid or improve our understanding of interactions within multi-species communities. This is likely true, even for simplified communities associated with invasive weed or insect pests because of the many ways species can interact and the complex dynamics governing such systems. However, despite the considerable work that is needed to generate food webs (Holt and Hochberg, 2001) and the inherent uncertainty in the validity of the methods used to construct and analyze them (Vasas and Jordán, 2006), these networks nonetheless may be useful for retrospective analyses of natural enemy effects post release, especially in native habitats on non-target species (Memmott, 2000; Henneman and Memmott, 2001), which in turn could assist with predictions of non-target effects for proposed future releases of biological control agents (Memmott, 2000; Veldtman et al., 2011).

Another approach to understanding the community interactions of biological control agents, especially indirect trophic effects, is to use community modules that focus on only a small number of species interactions, perhaps as few as three to six interactions (Holt and Hochberg, 2001). These modules can be thought of as ecological guilds; sympatric species of consumers jointly exploiting common resources (Crowley and Cox, 2011). In the context of biological control, theoretical modeling using community modules suggests one aspect of indirect food web interactions to be highly important: shared predation (Holt and Hochberg, 2001). Shared predation is the situation where a natural enemy can attack two or more species of prey, and this may include (1) herbivores feeding on multiple host plants including the target, (2) parasitoids or predators attacking several hosts including the pest, (3) hyperparasitoids sustained by two or more primary parasitoids, one or more of which were released to control a pest, and (4) pathogens infecting multiple hosts including the target (Holt and Hochberg, 2001). Using a community module

approach, Holt and Hochberg (2001) developed several "rules of thumb" using mathematical models to predict potential unwanted effects from biological control agents that share two or more hosts. First, they suggest high risk to non-target species may exist when biological control agents that are not highly specific are also only marginally effective. This risk to non-target species occurs because target populations are not suppressed and the introduced agent maintains large populations on the target and attacks on non-target species increase as a result. This situation likely existed before the 1990s in the northeastern United States (before collapse of gypsy moth densities because of an introduced fungal pathogen in the 1990s) for the tachinid fly, *Compsilura concinna* (Meigen), a parasitoid of gypsy moth larvae (*Lymantria dispar* [L.]). In field experiments, this fly displayed rapid aggregation to artificial gypsy moth outbreaks, indicating that flies were abundant in the environment even when regional gypsy moth populations were low (Gould et al., 1990; Fergusson et al., 1994). Since this fly is known to reproduce in a variety of native moth species (Boettner et al., 2000), gypsy moth may increase this non-target effect by sustaining higher than expected densities of the fly. Second, highly fecund targets maintain robust populations because they can offset heavy population losses to natural enemy attack through their high reproductive rates, which elevates the natural enemy's population, which in turn leads to increased attacks on non-target species. Third, highly mobile natural enemies that can successfully exploit non-preferred but acceptable non-target species in the absence of preferred target species can pose threats to these less desirable, non-target species. This may occur when natural enemies infiltrate sensitive sites where targets are lacking, but suitable less-preferred non-target species are abundant and consequently exploited. This is particularly relevant if the natural enemies have tribe- or genus-level specificity, for example, and can successfully feed and reproduce on related hosts other than the target (Rand et al., 2004; see Chapter 4 for discussion of thistle biological control agents and their effects on native US thistles). Fourth, specialist natural enemies may increase populations of resident generalist natural enemies when these parasitoids are exploited by intra-guild predators or hyperparasitoids. This causes an unintended increase in population densities of these opportunists, which in turn increases their attacks on non-target species. For example, *Cotesia glomerata* (L.), a parasitoid introduced for the suppression of a pest butterfly on brassica crops, was heavily parasitized by resident hyperparasitoids (Gaines and Kok, 1999). Elevated populations of those hyperparasitoids could in turn have inflicted higher levels of mortality on their other hosts that would not have occurred in the absence of *C. glomerata*. This possibility has not been investigated but would be amenable to manipulative experimentation.

Trophic effects of weed biological control agents

Predicting indirect trophic effects of herbivores

Indirect trophic effects of biological agents occur via an interaction chain whereby the impact of one organism on another is mediated by a third species. Several types of such effects can be envisioned: (1) When natural enemies (e.g., a parasitoid) affect the densities of the target (e.g., herbivores are attacked and killed), the fitness of the plant the herbivores were feeding on may increase. (2) *Ecological replacement* happens when the control of the target pest leads to the loss of some useful function formerly performed by the controlled species, such as providing nesting sites or food for native birds. (3) In other cases, attack on plants my cause *compensatory responses*, which occur when plants respond to herbivory (inflicted here by the biological control agents) with increased growth. Such increased growth by the invasive plant may harm non-target species directly because of the weed's increased biomass, or indirectly by other mechanisms such as production of allelochemicals. (4) Finally, the biological control agent may itself become a food resource for other species, and such food web subsidies may inadvertently increase the fitness of other species (Pearson and Callaway, 2003).

Holt and Hochberg's (2001) first rule of thumb – failure of a natural enemy to reduce pest biomass that in turn supports high-density natural enemy populations as the response to abundant host availability – is an example of a food web subsidy that is an indirect trophic effect of a biological control agent. Food web subsidies have been documented for the biological control of the invasive spotted knapweed, *Centaurea stoebe* L. (formerly known as *C. maculosa* Lam.) by the host-specific, gall-inducing tephritid flies *Urophora affinis* Frauenfeld and *U. quadrifasciata* (Meigen). Galls increased the abundance of native deer mice (*Peromyscus maniculatus* [Wagner]), which use these galling fly larvae for food (Pearson et al.,

2000). Spotted knapweed, native to eastern Europe, is a highly invasive weed in North America and infests millions of hectares of natural areas and rangelands to the detriment of native plant and forage species. This weed is allelopathic, it facilitates mycorrhizal colonization that inhibits growth of other species, it removes nutrients more efficiently than competitors, and it shows effective compensatory growth in response to herbivory (Skurski et al., 2013). In addition to the two galling flies mentioned above, ten additional (for a total of twelve) exotic insects of Eurasian origin were introduced into North America for the biological control of spotted knapweed (Story et al., 2010). The two gall flies lay their eggs in immature knapweed flower heads and their larvae induce gall formation. The intent behind the establishment of these natural enemies was to lower knapweed reproductive fitness through reduced seed production via gall formation. Galls can act as nutrient sinks that may lower resource availability for reproduction elsewhere on the plant.

The high year-round abundance of fly larvae in knapweed galls attracted foraging of native vertebrate and arthropod predators that either attack and open galls for prey (e.g., *P. maniculatus* and the black-capped chickadee, *Parus atricapillus* [L.]) or capture and eat adult gall flies (e.g., orb-weaving spiders) (Pearson et al., 2000). In Pearson et al.'s (2000) study, *Urophora* fly larvae accounted for >80% of mouse diets and, during winter, fly larvae were the most common stomach content recovered from field-captured mice, indicating that this resource might have supported higher mouse population densities than would otherwise have occurred in the absence of this resource. Consequently, resource subsidies could be facilitating greater mouse survival and subsequently higher than normal breeding populations the following year (Ortega et al., 2004). Trapping data also indicated that mouse populations change their foraging strategies and habitat selection in response to the abundance of knapweed galls (Pearson et al., 2000). Food subsidies benefiting native consumers could result in the restructuring of community relationships (Pearson and Callaway, 2003) in complex ways. An unforeseen consequence of subsidizing deer mouse populations is a potential increase in exposure of people to mouse-borne Hantaviruses that cause potentially deadly pulmonary disease (mortality rates are ~20–60% for infected victims [http://www.cdc.gov/hantavirus/surveillance/annual-cases.html]). Human disease risk may be elevated when deer mouse populations are higher because the transmission ecology of this virus is altered owing to greater virus prevalence in the environment. Hantavirus-infected mice may be up to three times more common in plots of knapweed infested with galling flies (Pearson and Callaway, 2006). Removal of gall-fly-infested knapweed is correlated with a decrease in deer mouse population densities, which possibly results from lower food availability rather than a lack of knapweed habitat (Pearson and Fletcher, 2008). A question that flows from the finding that gall flies increase deer mouse densities and Hantavirus prevalence is, "is there evidence to suggest that human Hantavirus infections and mortality have increased as Pearson and Callaway (2006) predicted?" Examination of Center for Disease Control statistics for Hantavirus reports up to July 2013 (http://www.cdc.gov/hantavirus/surveillance/state-of-exposure.html) indicates that most cases of human infections are reported from the southwestern United States, where knapweed is recorded (http://www.nps.gov/plants/alien/map/cest1.htm), but where dry conditions also favor hantavirus transmission. This latter factor may be more important than any link to elevated deer mouse populations caused by resource subsidization from an ineffective weed biological control agent. Indeed, New Mexico, the state with the highest reports of hantavirus infections (91 records, http://www.cdc.gov/hantavirus/surveillance/state-of-exposure.html), apparently lacks infestations of knapweed (http://www.nps.gov/plants/alien/map/cest1.htm). Additionally, humans may not encounter hantavirus-infected mice inhabiting dense knapweed stands because these habitats tend to be rural with relatively low human densities associated with them, and weed-infested areas are likely not habitats that attract the intensive outdoor recreational activities that would expose humans to mouse feces that harbor hantavirus.

The knapweed–gall fly–deer mouse system has been very well studied (it has taken Pearson and colleagues approximately 15 years of field work and experimentation to discover the interrelationships in this system, a remarkable achievement given the vagaries of funding), and convincing data have been produced that strongly suggest that an abundant and ineffective weed biological control agent, the knapweed gall fly, has had a significant indirect non-target effect on a native generalist predator, the deer mouse, as the fly acts as an abundant food source that was previously lacking. However, there is no evidence that increased mouse density has led to higher transmission of hantavirus to humans. Nor is there evidence for higher trophic level perturbations, as might be expected if high mouse densities subsidized higher trophic level predators such as owls and coyotes.

Reduction of recruitment of native grasses caused by asymmetric seed predation by deer mice (i.e., mice feed more intensively on the large seeds of native grasses vs. smaller seeds produced by native forbs) has been demonstrated through perturbation experiments that remove gall fly-infested knapweed via herbicide applications, suggesting that elevated mouse densities could affect the persistence of some native plants sympatric with knapweed–gall fly–deer mouse populations (Pearson and Callaway, 2008). The long-term effects on the survival and stability of native plants sympatric with knapweed infestations, either in the presence or absence of seed predators like deer mice, is less clear. Herbicide applications for control of knapweed were also detrimental to native plant recruitment rates (Pearson and Callaway, 2008) and promoted invasions by other invasive plants (Skurski et al., 2013), indicating there are trade-offs to managing invasive weeds with different tools. Also, in the absence of control measures (biological and chemical), knapweeds have strong suppressive effects (e.g., allellopathy) on native plants (Skurski et al., 2013). Presumably in areas not invaded by knapweed, the dynamics of native plant recruitment are subjected to many factors, including the normal fluctuations of seed predation by mice, whose population densities in turn would be modulated by more typical biotic and abiotic factors.

Predicting *a priori* indirect non-target effects, especially the intensity and importance of these types of food web interactions, is extraordinarily difficult for weed biological control agents because there is a paucity of ecological and biological control theory to support reasonable speculation about potential unintentional outcomes (Pearson and Callaway, 2003; Thomas et al., 2004; Paynter et al., 2010; Fowler et al., 2012). Ecologists (Pearson and Callaway, 2003) and weed biological control practitioners (Fowler et al., 2012) agree that improved theory supported by experimental data or meta-analyses is necessary because biological control is a particularly powerful and cost-effective tool for managing invasive species and should not be discarded because of potential unintended risk to non-target species (Thomas and Willis, 1998). This lack of theory is in part caused by insufficient funding to support this necessary research (Maron et al., 2010).

A rule of thumb that work on the knapweed–galling fly–deer mouse system has provided is that natural enemies with high levels of host specificity must also be sufficiently efficacious so as to reduce densities of the target pest and in turn reduce their own population densities. This density-dependent feedback will minimize risk to non-target species because the natural enemies will no longer be superabundant in the environment and will not be a resource that can be exploited by other organisms, thereby eliminating unwanted trophic linkages (Pearson and Callaway, 2005). However, this does not necessarily mean that "ineffective" agents should be excluded from weed biological control programs. Guilds of agents may be necessary to control weeds, and if effects of guild members are additive or synergistic and strong enough to reduce weed density then the prevalence of all host-specific agents (i.e., weed and natural enemies) in the system will decline, and resources that could possibly trigger indirect secondary effects diminish, thereby lessening unwanted trophic effects. This situation may now be occurring with spotted knapweed following the introduction of a weevil that attacks the roots of this weed, which, together with other introduced knapweed natural enemies, is likely reducing weed densities (Story et al., 2006). The risk with establishing guilds of herbivores is that even guilds may sometimes fail to control the target weed and thus provide even greater novel resources potentially exploitable by resident organisms (Pearson and Callaway, 2003).

Approaches already exist that can lead the way in developing theory and improving understanding of potential indirect non-target effects from biological control agents with the aim of mitigating unwanted effects. The simplest approach is to apply the rule of parsimony: introduce the fewest agents necessary to achieve weed suppression and eliminate redundancy by not introducing multiple agents that perform the same ecological function.

Meta-analyses of datasets can help predict the likelihood of weed natural enemies acquiring their own natural enemies (e.g., parasitoids) from the environment into which they are introduced (Paynter et al., 2010). The habitus of a weed biological control agent (e.g., plant gall makers, defoliators, stem borers, etc.)

may be a useful predictor of species' potential for generating indirect trophic effects (Veldtman et al., 2011). The complexity of the system (e.g., simplified agro-ecosystems vs. complex natural systems that expose natural enemies more frequently to non-target species) into which agents are introduced may also correlate with the likelihood of non-target effects (Hawkins et al., 1999). Analytical food webs can be used as predictive tools for weed biological control (Memmott, 2009), and it has been proposed that use of this technique for generating and testing hypothesis concerning introductions of weed biological control agents should be increased (Fowler et al., 2012). The failure of food web research to generate hypotheses for experimental testing has been criticized previously (Pimm and Kitching, 1988; Strong 1988), as have techniques used for acquiring, assembling, and analyzing data (Paine, 1988). However, there is a small but growing body of literature indicating that food webs can provide insight into biological control systems, and food webs can lead to the development of testable hypotheses regarding indirect non-target effects resulting from introductions of host-specific weed natural enemies (Carvalheiro et al., 2008; Willis and Memmott, 2005).

Predicting direct trophic effects of herbivores

Predicting direct trophic effects of weed natural enemies on non-target species is the best studied and understood aspect of all classical biological control subdisciplines, because eliminating direct attacks on non-target plants (i.e., agricultural crops and native species) has long been an important objective of host-range testing of herbivorous natural enemies. Forecasting direct trophic effects by herbivorous natural enemies is guided by an established paradigm that assumes that most herbivorous insects are specialists and exploit only a few plant lineages for growth and reproduction (Bernays and Graham, 1988). Specialist herbivores have evolved traits that enable them to circumvent particular host plant defenses (often chemical) by developing unique behaviors and physiological adaptations to avoid or overcome these protections (Ehrlich and Raven, 1964). Consequently, specialist herbivores are predicted and have been repeatedly shown to be restricted to plant lineages that deploy defensive plant compounds on which they specialize (Weiblen et al., 2006). Because members of a single plant lineage typically share chemical defenses based on a category of secondary plant compounds (Berenbaum, 1990), an invasive non-native plant, the target weed, is most likely to share its defensive compounds with another species (e.g., native non-target species) if it is closely related to that species (e.g., they are in the same genus) (Pemberton, 2000). An exception to this paradigm could exist if unrelated allopatric plant species have independently evolved similar chemical defenses that their respective specialist herbivore groups both recognize should they become newly associated owing to processes such as invasion that put them into sympatry. This phenomenon could occur if unrelated plants contained chemicals that coincidentally possessed the necessary structures and polarities to activate gustatory receptors that stimulate feeding (Tallamy, 2000). To be adequate food sources for the natural enemy, the non-target plant would also have to sustain larval development, with no loss in adult fitness or other debilitating effects.

Insect specialization on secondary plant compounds has led to the development of predictive models for assessing host risk and herbivore safety. The best known of these models is the centrifugal phylogenetic method (or its modifications) (Wapshere, 1974). This model uses relevant biological and taxonomic information to select non-target plants for host-specificity trials in quarantine (Harris and Zwölfer, 1968), with the goal of eliminating unsuitable agents without unnecessarily rejecting candidates because of laboratory artifacts in the data (Wapshere, 1989). Model predictions for non-target effects (or lack thereof) have been assessed through field surveys following natural enemy releases, and subsequent meta-analyses of available datasets suggest that model predictions are accurate and this approach is robust (McFadyen, 1998; Pemberton, 2000).

The identification and selection of specialist herbivores from a guild of plant feeders associated with the weed in its native range that exhibit narrow host ranges and high host specificity is the goal of any weed biological control program. Determining if a particular candidate biological control agent has these qualities is often done in quarantine, where one measures both the agent's willingness to feed on non-target species predicted to be at highest potential risk based on their relatedness to the target plant and the fitness of resulting adults of the natural enemy and their subsequent progeny. To ascertain the acceptance of target and non-target plant species by a potential weed biological control agent, both feeding and oviposition (i.e., egg laying)

trials are conducted under choice and no-choice conditions. No-choice tests present a single non-target plant species to the herbivore of interest, and the ability of adults and larvae to feed and sustain themselves and, in the case of adults, to oviposit is assessed (Hill, 1999). Exposing different herbivore life stages to non-target species will determine if there is variation in non-target host use by adults and larvae (Edwards, 1999). This no-choice evaluation process simulates what could be expected if a natural enemy invades an area that lacks the target weed but has closely related non-target species. Of critical importance is whether the agent being studied could use these non-target species to survive and reproduce.

Variations on no-choice tests exist. The classic no-choice test is static; that is the herbivore is exposed to just one non-target species for a prescribed amount of time and feeding and oviposition are recorded for that individual on that host plant (Hill, 1999). Sequential no-choice tests involve exposing the herbivore to one non-target species for a prescribed amount of time, then the same individual is moved to a different plant species, and so on, and feeding and oviposition activity is recorded (Edwards, 1999). The key aspect of all no-choice tests, regardless of the style used, is that the agent is exposed only to one species of plant at any given time, and features of interest (e.g., feeding, oviposition, or survival rates) are recorded. A further point of importance is that naïve individuals (i.e., they have had no previous contact with the most preferred host) may accept less preferred plant species that would be rejected by experienced herbivores previously exposed to the target weed.

If oviposition occurs during no-choice tests, the suitability of the non-target species is assessed for larval survivorship and subsequent fitness of any individuals that develop to adulthood. If, in comparison to development on the target, there is significantly reduced fitness on the non-target species, results suggest that the natural enemy is poorly adapted to the novel host and would be unlikely to cause significant direct non-target effects. Conversely, fitness measures equal to, or exceeding, those found when the target weed is eaten would indicate that the non-target species is a very good host for the natural enemy, which is likely to attack it in the field. No-choice tests are extremely conservative; negative results (i.e., no observed attacks) strongly indicate that non-target attacks are unlikely (Hill, 1999).

However, the probability of false positives is high, and attacks on non-target species that would not normally occur in the field may happen in quarantine because of the highly artificial experimental conditions of starvation and close confinement (Edwards, 1999; Hill, 1999; Withers and Browne, 2004).

As a follow up to positive results from no-choice tests, choice-test experiments are recommended to ensure promising candidate agents are not discarded unnecessarily (Hill, 1999). In some instances, the predicted host ranges from quarantine studies have been investigated to determine if they correspond to the realized range in the field. Conclusions reached are that risk assessments based on physiological ranges as determined by host-specificity testing in quarantine are conservative compared to host ranges in the field (Pratt et al., 2009). Choice tests are modifications of no-choice experiments, except the agent now has more than one host plant to choose from in the experimental arena. This set up involves the target (i.e., the weed) plus at least one or more non-target plants (Edwards, 1999). Choice tests are used to gauge the degree of specificity of herbivores to target and non-target plants when there is more than one potential food source available. As with no-choice tests, variations on choice tests exist (Edwards, 1999), and outcomes of choice tests, especially attacks on non-targets species, need follow-up assessment to ensure these have not occurred because of experimental conditions (Edwards, 1999). Sometimes cage-induced aberrant behavior cannot be resolved in the laboratory, and field testing in the natural enemy's country of origin may be necessary (Briese, 1999). Open-field host-specificity testing may offer the most realistic method for assessing host ranges, as natural enemies are not restricted in movement and host selection behaviors are not modified by small experimental cages. Implementation of field studies in the native range depends on the researcher being able to obtain the desired species of plants (from the invaded country) to include in the test, which may be possible (if, as is common, they have been imported previously as ornamentals) or their special importation may be permitted subject to conditions preventing rooting or seeding during tests. Briese (1999) argues that field tests are not alternatives to laboratory no-choice and choice tests, but should be used to complement these traditional assays. Open-field host-selection studies have

been made post release to verify risk to non-target species as predicted by host-range testing (Moran et al., 2009; Cristafaro et al., 2013).

A key question from the review presented above on techniques used to forecast direct trophic interactions with non-target species is this: How many cases have been identified where host-range and host-specificity assessments, as conducted at the time of release, for weed biological control agents were later found to be incorrect and to have resulted in damage to non-target plant populations? Pemberton (2000) reports that just one weed biological control agent, the lantana lace bug, *Teleonemia scrupulosa* Stål (Hemiptera: Tingidae), released for the biological control of *Lantana camara* L. (Verbenaceae) in Hawaii in 1902, has apparently attacked a distantly related native plant in Hawaii, naio, *Myoporum sandwicense* Gray (Myoporaceae). This natural enemy was reportedly released in Hawaii with no host-range testing, and follow-up field surveys have failed to corroborate naio use by this bug where the target (lantana) and naio grow close together (Pemberton, 2000). Analysis of published literature suggests that patterns of non-target plant use, especially native species, can be predicted accurately using existing testing protocols before the introduction of natural enemies (Pemberton, 2000; Barton, 2012). Data on field attacks by 112 insects, 3 fungi, 1 mite, and 1 nematode established for the biological control of weeds in Hawaii, the continental USA, and the Caribbean demonstrate that risk to native non-target flora, as predicted by theory, is to species most closely related (same genus) to the target weed (Pemberton, 2000). About ~50% of biological control projects against weeds with close relatives in the receiving area have resulted in some level of non-target species attack (ranging from very minor to significant) compared to ~4% of projects against weeds without close relatives (Pemberton, 2000). The inability of host-range tests to predict attacks on non-target weed species that are closely related to the target pest has been demonstrated. This failure can be influenced by the geographic source of the natural enemy population (some populations, even though genetic testing indicates that they are the same species, exhibit different host ranges and may need to be subjected to further host-range testing) and the level of synchrony between the preferred stage of the target weed and natural enemy's reproductive period (Paynter et al., 2008).

Pemberton's (2000) meta-analysis provides two useful rules of thumb for minimizing direct non-target effects on plants in the receiving range of a herbivorous natural enemy for the biological control of an invasive weed: (1) select as targets those weeds that have few or no native congeners in the area of introduction and (2) introduce only highly specific agents that attack just the weed and are expected to do the greatest amount of damage. These rules of thumb clearly show why the likelihood of important direct attacks by introduced herbivores of thistles in North America and cacti in the Caribbean (see next section) was obvious before the release of the agents involved. These releases, however, are exceptions, made for political rather than scientific reasons, and do not justify the perception by ecologists that practitioners of weed biological control commonly make releases of natural enemies that have unpredictable population-level consequences.

Examples of direct attacks by weed biological control agents on non-target plants

Two high-profile cases documenting attacks by insufficiently specialized biological control agents on native plants are attacks on indigenous *Cirsium* thistles in North America by the deliberately released thistle head weevil *Rhinocyllus conicus* Frölich (native to Europe) and the threat posed to native *Opuntia* spp. cacti in the United States, Mexico, and the Caribbean from the cactus moth *Cactoblastis cactorum* Bergroth (native to South America but released in the Caribbean). Both of these control agents were known to have only genus-level or higher specificity before their release.

Rhinocyllus conicus (see also Chapter 4) was introduced into North America in 1968 for biological control of invasive musk thistle, *Carduus nutan* L. (family Asteraceae: tribe Cynareae), native to Europe. Host-range testing done at the time indicated that *R. conicus* could attack and develop successfully in the flower heads of thistles in three genera (*Carduus*, *Cirsium*, and *Silybum* [all Cynareae]), but data showing that this weevil has a preference, when given a choice, for *C. nutans*, were used to conclude that native North American *Cirsium* thistles would not be at risk (Zwölfer and Harris, 1984). In hindsight, we can now see that this inference was incorrect and existing data at the time were sufficient to suggest risk, which later proved to be the case (Louda et al., 1997; Gassmann and Louda, 2001; Arnett and Louda, 2002; Cripps et al., 2011). That this outcome reflects the

original host range of the insect was shown by re-testing of populations of *R. conicus* by Arnett and Louda (2002), who found that this biological control agent does exhibit a preference for *C. nutans* over *Cirsium* spp., but that it can develop successfully on *Cirsium* species. This study suggests that there has been no post-release evolution by this weevil increasing its performance and subsequent fitness on non-target *Cirsium* spp. (Arnett and Louda, 2002). Larvae of this weevil feed in thistle flower heads, reducing seed number, and this weevil may have reduced pest musk thistle populations by as much at 94% in some areas (Wiggins et al., 2010). The decision to release *R. conicus* was made in pursuit of these benefits, while making poor assumptions that minimized estimates of risk to native thistles, and this natural enemy, if evaluated today, would likely be rejected (Hoddle, 2004a).

Because of its documented host range, *R. conicus* was released for biological control of the invasive thistle *Cirsium arvense* (L.) Scop. (native to Europe) in New Zealand, a country with no native thistles (Cripps et al., 2011). Given the oligophagous nature of *R. conicus*, it is not surprising that in North America it has fed on least 25 species of *Cirsium*, including state and federally listed endangered species (Wiggins et al., 2010). There is concern that reduced set seed by native thistle species suffering attacks from *R. conicus* may threaten the long-term viability of populations of some of these species because recruitment is seed limited (Louda et al., 1997). However, Pemberton (2000) points out that although these *Cirsium* spp. are within the physiological host range of *R. conicus* (i.e., it can eat them), some thistles may not be in the ecological host range of the weevil because they set flowers at times of the year when female weevils are not reproductively active. Because female weevils oviposit only into flower heads, asynchrony of weevil reproduction and seed head formation determine which thistles are attacked and to what extent.

In Australia, *C. cactorum* provided excellent control of invasive *Opuntia* cacti (native to the Americas) following its release in the 1920s. Soon after the moth's release, feeding larvae cleared dense stands of cacti from millions of hectares of pasture and wilderness (Dodd, 1940). This moth was later released in several African countries for control of weedy *Opuntia* species, and successful control was documented in South Africa (Zimmermann et al., 2001). In 1957, this moth was released in the Caribbean (see also Chapter 4) for control of native weedy *Opuntia* species (Simmonds and Bennett, 1966; Stiling et al., 2004), an area where *C. cactorum* is not native, putting it in very close proximity to highly diverse *Opuntia* floras in the United States and Mexico (Zimmermann et al., 2001). This moth invaded the continental United States in 1987, possibly via the movement of infested cactus plants being exported from the Caribbean (Pemberton, 1995), and it now threatens *Opuntia* biodiversity in North America (Zimmermann et al., 2001). The use of *C. cactorum* for suppression of native and some exotic *Opuntia* in the Caribbean raises two important criticisms of biological control: (1) is the release of exotic natural enemies for suppression of unwanted native species justified? (Lockwood, 1993a, b) and (2) has the spread (natural or human assisted) of a natural enemy been properly forecast to predict movement into high-value areas where it is not wanted? This example of unwanted spread supports rule of thumb three from Holt and Hochberg (2001), that is that highly mobile, polyphagous natural enemies can pose threats to non-target species if they infiltrate areas where the target species is lacking but suitable non-target species are abundant. Interestingly, models have been produced that indicate that non-target native plant species might be driven to extinction by weed biological control agents under these circumstances. Several major risk factors were identified, including that the introduced herbivore causes significant attacks on non-target species that suppress their populations and facilitates competition by other plant species (Chalak et al., 2010). Despite theoretical predictions, field surveys in the Caribbean have demonstrated that native *Opuntia* species have survived more than 50 years of attack by *C. cactorum*, and extinction of native non-target cacti has not occurred (Pemberton and Liu, 2007).

Predicting unwanted trophic effects of carnivorous biological control agents

Significant recent interest in forecasting unwanted trophic effects by deliberately or accidentally introduced predacious and parasitic natural enemies has been generated because of studies that have revealed attacks on native non-target species leading to possible population declines and claims of extinctions (Howarth, 1983, 1991, 2001; Boettner et al., 2000; Obrycki et al., 2000; Louda et al., 2003; Elkinton and Boettner, 2012). In comparison to introductions of weed biological control agents, releases of arthropod natural enemies had been assumed to present minimal risk to non-target species, and host-range testing, at least in the United States, was almost never done before release of agents from quarantine. In fact, unlike specialist weed herbivores, many arthropod natural enemies, including parasitoids, are likely not monophagous (i.e., feeding and reproducing on just one species of host) and are instead more likely to be oligophagous (i.e., having a limited host range, indicating some level of "specialization") or polyphagous species (i.e., broad host range) (Holt and Lawton, 1993). These more "specialized" oligophagous natural enemies can present risks to non-target species and even this level of specialization may not be present in some taxonomic groups considered for use in biological control programs (Barratt et al., 1999).

Growing concern from ecologists and taxonomists over probable non-target effects (Simberloff and Stiling, 1996) and increasing regulatory restrictions on natural enemy importations and releases (Syrett, 2003; Messing and Wright, 2006) prompted close examination of these issues by biological control practitioners (Follett and Duan, 2000; Lockwood et al., 2001; Wajnberg et al., 2001), development of codes of best practice (Schulten, 1997), and consideration of methods to evaluate risk potentially posed by natural enemies to non-target species (Withers et al., 1999; Van Driesche and Reardon, 2004; Bigler et al., 2006). Increased advocacy for use of biological control for suppression of invasive species of conservation importance also resulted (Hoddle and Syrett, 2003; Hoddle, 2004a, b; Van Driesche et al., 2010; De Clercq et al., 2011), with concurrent challenges to proposed benefits of this technology for ecosystem restoration (Simberloff, 2012) and criticism of both sides for lacking adequate data to substantiate claims on the magnitude of non-target effects or lack thereof (Parry, 2009).

Undoubtedly, forecasting indirect and direct trophic effects is a nascent area of classical biological control of arthropods. With respect to measuring direct trophic effects of carnivorous natural enemies, work by practitioners in this area is moving this subdiscipline into closer alignment with good practices developed over decades of experience by weed biological control practitioners. Progress and issues associated with assessing indirect and direct effects of predators and parasitoids will be addressed below.

Predicting indirect trophic effects of carnivores

As discussed above for weed biological control agents (parts of the materials presented in this section are also applicable to weed biological control but have been saved for discussion here), indirect trophic interactions can have important effects on species within communities. Indirect trophic effects mediated by carnivorous natural enemies (e.g., predators and parasitoids) that have frequently been studied include *apparent competition*, *apparent mutualism*, *trophic cascades*, and *direct competition*. As with herbivorous natural enemies, potential unintended population-level effects resulting from introductions of carnivorous natural enemies are extraordinarily difficult to predict and assess *a priori* in quarantine. Necessary experiments simulating conditions of the receiving environment would be impossible to run in the laboratory. Similarly, assessment of indirect trophic effects by means of post-release field studies may also be very difficult given the spatial and temporal scales over which effects would have to be evaluated across generations and varying habitats occupied by non-target species.

Apparent competition, apparent mutualism

Indirect effects of predators and parasitoids can be subtle. For example, a particular prey species (i.e., the target of the biological control program) may attract or retain predators or parasitoids, in particular the introduced biological control agents, and this numerical response may positively alter natural enemy densities, which may indirectly result in increased attacks on other sympatric, non-target prey species. This spillover effect of increasingly abundant natural enemies in response to a common or preferred prey (i.e., the pest) onto a less common or less preferred prey (i.e., a non-target species) is referred to as *apparent competition* (Holt and Lawton, 1993; Prasad and Snyder, 2010), and this phenomenon can occur naturally in many insect assemblages (Holt and Lawton, 1993). In some instances it may even benefit and facilitate the co-existence of competing host species, which can be considered an example of *apparent mutualism* (Tack et al., 2011; van Veen et al., 2006). Sometimes apparent competition can completely exclude the herbivore species least able to sustain its population growth in the presence of a shared parasitoid (Bonsall and Hassell, 1998), and models suggest that the most likely outcome from apparent competition from multiple host use by

polyphagous parasitoids is the exclusion of all but one host species (Holt and Lawton, 1993). This situation is termed *dynamic monophagy*, and the persistence of a polyphagous natural enemy on a single host species gives the appearance, in the absence of its other potential hosts, that it is monophagous (i.e., host specific) (Holt and Lawton, 1993). However, this situation is seldom, if ever, observed to persist in nature as multiple parasitoid species manage to co-exist with numerous host species (Holt and Lawton, 1993). If dynamic monophagy does exist, it could have very important implications when conducting foreign exploration for natural enemies, and the polyphagy of such a natural enemy may or may not be revealed in quarantine, depending on the selection of non-target species for host-range testing. Mathematical models have been used to predict under which circumstances species extinctions are likely to occur under varying levels of apparent competition by biological control agents (Noonberg and Byers, 2005). Consequently, in biological control research investigating indirect non-target effects, interest has focused on potential negative or unwanted effects of apparent competition on non-target species.

The historical native range of a native pierid butterfly, *Pieris oleracea* Harris, has suffered significant reduction in the northeastern United States (Benson et al., 2003a). One of the reasons postulated for this, coupled with other factors such as habitat loss, was parasitism by a biological control agent, the braconid parasitoid *C. glomerata*, introduced for the control of the invasive white cabbage butterfly, *Pieris rapae* L. (Benson et al., 2003a). The pest butterfly, *P. rapae*, has a greater reproductive output than the native *P. oleracea*. Abundant parasitoid reproduction on *P. rapae* allowed for parasitization of *P. oleracea*, which *C. glomerata* attacked at higher rates than the target, *P. rapae*, when both species were exposed as trap hosts at the same sites and under the same conditions (Benson et al., 2003a). Attacks on summer generations of *P. oleracea* in meadows (the preferred foraging habitat for *C. glomerata*) in Massachusetts likely reduced densities of larvae in the over-wintering generation, causing their eventual near extirpation in the state (Benson et al., 2003a). However, further north, in Vermont (USA), *P. oleracea* is predominantly a univoltine forest species, and consequently it suffers significantly less non-target parasitism by *C. glomerata*, allowing *P. oleracea* populations to persist

(Benson et al., 2003a). While the above is an accurate description of the system pre-2000, further study revealed two important new developments. First, *C. glomerata* densities (on both *P. oleracea* and *P. rapae*) have been greatly reduced owing to competitive effects of *Cotesia rubecula* (Marshall), a more specific biological control agent of *P. rapae* introduced to New England in 1988 (Herlihy et al., 2012). Second, *P. oleracea* adults now oviposit on a new (exotic) host plant (cuckoo flower, *Cardamine pratensis* L.), which supports dense populations of the butterfly in some areas of Massachusetts, providing a long-season resource and partial protection from parasitism (Herlihy et al., 2014).

This example of apparent competition (the *P. oleracea–C. glomerata–P. rapae* system in the northeast United States) illustrates an important point: species extinctions over an entire range of a non-target species are unlikely to occur because of indirect trophic interactions. This example demonstrates the importance of spatial and temporal refuges for escaping attacks, a mechanism predicted to sustain co-existence between natural enemies and their prey under conditions of apparent competition (Holt and Lawton, 1993). Holt and Lawton (1993) also suggest the following mechanisms may promote persistent co-existence between natural enemies and their prey despite model predictions of extinction caused by apparent competition: (1) labile host and parasitoid behaviors could reduce the intensity of negative interactions, resulting in greater survival of non-target species (as when generalist parasitoids fail to act in a density-dependent manner and consequently do not have strongly suppressive effects); (2) host switching by parasitoids can reduce pressure on non-target species at certain times when attacks are focused on more preferred prey; (3) spatial heterogeneity of natural enemy and host/prey population densities and distributions may cause variation in rates of non-target attacks; (4) hyperparasitoids, that is parasitoids of the parasitoids attacking the host, may reduce populations of primary parasitoids thereby relieving pressure on target and non-target hosts, or the host (non-target native species) may adopt new host plants that provide refuge from parasitism, or the attacking parasitoid may be displaced from use of the invasive pest as its host by arrival of a more competitive, more specific parasitoid.

Trophic cascades

Cascades are an example of an indirect population-level response that can occur from natural enemy introductions. This type of effect is ideally the goal of any successful classical biological control program targeting a pest arthropod or weed. A trophic cascade occurs when a biological control agent (e.g., a carnivore or herbivore) significantly suppresses the population density of the species it consumes; this in turn removes population regulation (direct or indirect) of organisms dominated by the targeted pest. Spectacular examples of this type of event abound in the classical biological control literature and include the suppression (>95% population reduction) of highly damaging, invasive leaf hopper populations in French Polynesia with an egg parasitoid that in turn improved plant health and likely facilitated the recovery of native spider populations (Suttle and Hoddle, 2006; Grandgirard et al., 2008).

Control of an exotic scale by a highly host-specific coccinellid beetle in the Galápagos Islands (with pest population reductions of 60–98% [Hoddle et al., 2013]) facilitated the recovery of threatened and endangered native plants being attacked by the scale, as well as protecting the plants' associated endemic insects and snails. In this project, one possible negative indirect non-target effect was anticipated: potential harm to native birds that might gorge on the biological control agent, whose populations increased markedly in response to high density prey populations (Lincango et al., 2011). The coccinellids released have protective compounds in their bodies to repel predators that, if ingested in large amounts, might harm birds feeding on them. However, studies with Galápagos birds found that beetle repellency was high enough to limit ingestion to non-damaging levels (Lincango et al., 2011).

Trophic cascades resulting from weed biological control projects have resulted in the elimination of monotypic stands of invasive plants. Reduced abundance of the exotic weed ideally assists the recovery of native plants and the communities associated with them. Meta-analyses suggest plant diversity can increase by as much as 88% because of weed biological control agents that reduce the size, biomass, and reproductive output of their targets (Clewley et al., 2012; see also Chapter 8 for more on evaluation of outcomes of weed biological control projects). These indirect trophic interactions that occur because of trophic cascades are overwhelmingly positive and are the motivation for all biological control projects targeting pest arthropods and weeds in support of conservation efforts (Van Driesche et al., 2010).

Direct competition

Direct competition has been documented for parasitoids attacking fruit flies in Hawaii (Messing and Wang, 2009). A Mediterranean fruit fly (*Ceratitis capitata* [Weidemann]) egg–pupal parasitoid that is superior to a competing larval–pupal species of this pest fly caused the inferior larval parasitoid species to instead focus its attacks on a gall-forming fly, a tephritid (same family as the pest fly) that was released for the biological control of lantana, an invasive weed. Exploitation of this competitor-free space caused an unintended direct trophic non-target impact by an inferior competing fruit fly parasitoid on a biological control agent released to suppress a weed. This example indicates that under conditions of competition between members of a natural enemy guild, one natural enemy may exploit hosts it would likely not use, or not do so as intensively, in the absence of competition (Messing and Wang, 2009).

This outcome supports the rule of thumb of parsimony, as discussed earlier under indirect trophic effects of herbivorous natural enemies. The fewest agents necessary to achieve pest suppression should be introduced, eliminating redundancy by introducing only agents that perform distinct ecological functions. Introduction of multiple agents with the intent of letting nature "sort out" the most effective natural enemies has been likened to a "lottery" or "shotgun" approach (Strong and Pemberton, 2001). Proponents of the "lottery" approach for arthropod biological control (this strategy has not been used in weed biological control) argue that it is impossible to predict ahead of time which natural enemies, if any, will be effective (González and Gilstrap, 1992), so the best approach is to hedge one's bets, release them all, with the anticipatory attitude "may the best natural enemy win." Further, it has been argued that release of generalist natural enemies that can exploit alternate hosts in addition to the target pest, and thus are more likely to establish, is beneficial as it maintains natural enemy numbers at high densities when pest populations are low. This idea accepts non-target effects as necessary and essential because natural enemy populations must be maintained in a constant elevated state of preparation to exert rapid control of pest populations should

they rebound. Past advocacy of the "lottery" approach for natural enemy selections and use of generalist natural enemies for arthropod biological control has been strongly criticized (deservedly so) because of the potential environmental risks posed by such introduction strategies (Strong and Pemberton, 2001).

Despite the acceptance that indirect ecological effects from natural enemy introductions are hard to predict, a framework for investigating top-down and bottom-up effects has been developed. Messing et al. (2006) provide a solid foundation for investigating likely non-target effects while candidate natural enemies are still undergoing quarantine testing. These studies can encompass descriptive studies that use direct observations of predation or parasitism tests in the laboratory in petri dishes or large cages ("ecosystem" size, although size and number of cages will be constrained by space limitations in quarantine), or under field conditions in the home range or in other areas where the natural enemy is already established to provide preliminary assessments of interactions between natural enemies, targets, and victims of interest. Hoddle et al. (2013) assessed host specificity of *Rodolia cardinalis* (Mulsant) (Coleoptera: Coccinellidae) in the Galápagos using field cages stocked with native plants infested with target and non-target prey. Native plants infested with the target pest were compared to un-infested native plants to determine how often *R. cardinalis* foraged in habitats without targets, thereby putting it in proximity to non-target prey. These types of studies are termed "focal studies" (direct observations) and manipulative "large cage" experiments by Messing et al. (2006). Further insight may be derived via "surrogate" experiments to assess the likelihood of apparent competition. Under this scenario, components of the system are manipulated (e.g., victims or natural enemy populations are altered to test different hypotheses) and outcomes are quantified. Statistical analyses may provide insights into factors mediating parasitoid assemblages on non-target species and have potential to identify habitats where target and non-target attacks may occur (Kaufman and Wright, 2011).

Messing et al. (2006) provide additional rules of thumb for considering indirect effects from arthropod natural enemies: (1) select those with the most restricted host range and avoid polyphagous species; (2) avoid species that can act as facultative hyperparasitoids as these are likely to have indirect effects when compared to obligate parasitoids; (3) avoid parasitoid taxa that are most likely to be subjected to high levels of hyperparasitism as this increases the risk of apparent competition on non-target species; and (4) be cautious of oversimplification when designing host-range studies because natural enemies may have diverse life styles that are not understood or recognized, and simply designed host-range tests may not detect these attributes.

Predicting direct trophic effects of carnivores

A significant concern pertaining to direct attacks on non-target species by arthropod natural enemies has been the claim that biological control agents have caused the extinction of non-target species (Howarth, 1983, 1991, 2001). This is undoubtedly the most devastating non-target impact possible, and extinction is an ecological concept that is easily understood by lay people. Boettner et al. (2000) provide very strong evidence that reductions of some once common moth species, including some giant silkworm moths (Saturniidae), in the northeastern United States have resulted from a polyphagous tachinid fly (*C. concinnata*) released in 1906 for the control of the gypsy moth (*L. dispar*), a forest defoliator native to Europe (Boettner et al., 2000; Elkinton and Boettner, 2012). The overall impact of *C. concinnata* on non-target native moth species appears to vary by species and locality (Selfridge et al., 2007). The introduction of *C. concinnata* was also directed against the brown-tail moth, *Euproctis chrysorrhoea* (L.), an invasive pest of equal or greater threat to native forests and human health than the gypsy moth. The brown-tail moth was completely suppressed by *C. concinnata* (Elkinton and Boettner, 2012).

Documenting non-target effects of predacious or parasitic natural enemies on non-target species is the easiest first step in determining whether or not significant direct trophic interactions are likely. The assumption that non-target attacks automatically have severe negative consequences for native species has often been made (Howarth, 1983, 1991, 2001) without supporting evidence (Parry, 2009). Parry (2009) suggests that field-based studies are needed to examine these claims and recommends analytical tools, such as life tables, food web analyses, and manipulative deployment of experimental victim populations, be used to quantify population-level effects on non-target species by introduced natural enemy species. In instances where field work has attempted to corroborate population declines of native species by biological control agents, especially butterflies and moths, field-based evidence for non-target attacks has been shown, but in many instances these attacks, when subjected to analytical techniques to measure impact severity, have not been sufficiently strong to support claims that deliberately introduced natural enemies are responsible for perceived population declines (Hickman, 1997; Benson et al., 2003b; Johnson, et al., 2005; Barron, 2007; Kaufman and Wright, 2009, 2010; King et al., 2010).

Four conclusions appear justified by the studies listed above: (1) introduced arthropod natural enemies do sometimes attack native non-target species; (2) population-level effects at least on some Lepidoptera (the example of *C. concinnata* is excluded here) and pentatomid bugs by introduced natural enemies are not always sufficient to cause population declines; (3) attack rates on non-target species by biological control agents are highly variable, exhibiting strong temporal and spatial effects; and (4) accidentally introduced parasitoids and predators (referred to as adventive species) often cause greater mortality of native species. This latter observation is not surprising because adventive species are frequently polyphagous. The consistency of these findings is especially important because the majority of the studies (Benson 2003b excluded) were conducted on islands (Hawaii and New Zealand), habitats known to harbor native species that tend to be extraordinarily vulnerable to invasive species. In retrospect, some of these non-target attacks were possibly predictable and may likely have been detected in host-range tests if they had been conducted. However, what is much more difficult to predict *a priori* is the environmental range (species, habitats, and area) vulnerable to infiltration by introduced biological control agents and the magnitude of effects over time and location. There are additional systems for which food web analyses have been completed that demonstrate the association of introduced biological control agents with native species in New Zealand (Munro and Henderson, 2002) and Hawaii (Henneman and Memmott, 2001), but whose population-level effects are unknown. Examples like these where associations are already documented deserve greater attention to determine if the tentative conclusions proposed above are generally correct.

Howarth (1991) states that about 100 species (an alarmingly large and convenient number to remember) have been driven to extinction by biological control agents, a list that includes species affected by releases of predacious vertebrates. The most commonly cited example of a purported extinction by an insect introduced for biological control, however, is that of a zygaenid moth, the purple coconut moth, *Levuana iridescens* Bethune-Baker, in Fiji by the tachinid fly *Bessa remota* (Aldrich), introduced from southeast Asia in 1925 (Tothill et al., 1930). Levuana moth was most likely accidentally introduced into Fiji from an unknown area in the late 1870s and thereafter caused immense damage to coconut plantations, the mainstay of Fijian culture and this island nation's economy (Tothill et al., 1930). Despite extensive efforts to locate the home range of the levuana moth, it was never recorded from anywhere else, and it has been repeatedly described, incorrectly, as native to Fiji, despite highly convincing historical evidence that this moth was an invasive species that was not native to Fiji (Tothill et al., 1930; Kuris, 2003; Hoddle, 2006). The release of the fly provided rapid control (within six months of release) of the levuana moth, although periodic outbreaks were recorded in Fiji up to 1956, after which research on coconut pests dwindled because British entomologists were called home as Fiji prepared for independence in 1970. This drawdown of scientific expertise corresponded with a concomitant reduction in publications on coconut pest research and the subsequent cessation of reports of levuana moth outbreaks or reports of its parasitism by *B. remota* (Hoddle, 2006). This led to the erroneous assumption that lack of continued documentation of periodic levuana moth outbreaks demonstrated extinction by the parasitic fly. Two points deserve to be made from this extinction claim and its repetition by others (e.g., Wagner and Van Driesche, 2010) as an example of arthropod natural enemies causing the extinction of native insect species. First, merely accepting the repeated citation of articles that express an author's opinion without supporting evidence is poor scholarship. Second, since extinctions of insects are often hard to prove, claims for or against extinction should be based on actual data, including the extent and correctness of sampling efforts that are made to settle the issue.

Interest in forecasting direct trophic effects by parasitoids and predators on non-target species is the most significant recent development in classical biological control of arthropods. There is increasing use of existing techniques or development of new approaches to assess host and ecological specificity of candidate arthropod natural enemies under consideration for use in classical biological control (Bigler et al., 2006). The motivation for abandoning natural enemy release "lotteries" is to improve the safety and efficacy of arthropod biological control. The challenges of selecting non-target test species and then designing and conducting host-range tests in quarantine are complex. Practical obstacles include our frequently limited understanding of the taxonomy, biology, and rearing of candidate species, and uncertainty as to how best to select a subset of related native species for testing when there are potentially hundreds or thousands of candidate non-target species that could be evaluated (Van Driesche, 2004). This latter issue is significant, because a mature theoretical framework does not exist to provide criteria to guide the selection of non-target test species, especially for parasitoids, whose biology and behavior may not be constrained by phylogenic relationships between the target and close relatives in the intended receiving range (Van Driesche, 2004).

Methods for testing the host range of entomophagous natural enemies vary (Van Driesche and Murray, 2004; van Lenteren et al., 2006b), but they always employ laboratory-based manipulative experimentation to assess attack rates and subsequent fitness (e.g., reproductive output, longevity, etc.) in the presence or absence of target and non-target species in small cages. In small exposure arenas in quarantine, attacks can be recorded that may be unlikely to occur in the field. The artificiality of host-range assessments conducted in small test arenas may exaggerate host-range estimates if experimental conditions alter the behavior of natural enemies (van Lenteren et al., 2006b). The results of these experiments require the application of statistical tests so that conclusions about attack rates are supported with a high level of confidence (Hoffmeister et al., 2006; Berner, 2010). Development of host-range testing to determine the specificity of arthropod natural enemies has followed the centrifugal phylogenetic approach developed by weed biological scientists. Variations on this approach have been developed to account for the possibility that some natural enemies of interest may not be strictly monophagous or that the herbivore's host plant or its habitus (borer, sucker, defoliator, leaf miner, etc.) exert a stronger influence on the parasitoid's choice of hosts than does taxonomic relatedness of a test species to the target pest.

Emphasis has been placed on testing the most closely related non-target species (those in the same genus) (Mason et al., 2011), or if congeners do not exist, non-target species in the same tribe or family are tested (Hoddle and Pandey, 2014). However, given the possible oligophagous nature of some natural enemies undergoing host-range testing, additional evaluation criteria for choosing non-target species may be advisable. In addition to selecting taxonomically related non-target species, Hoddle and Pandey (2014) also tested non-target species inhabiting host plants related to the target's preferred hosts, in case host plant volatiles attracted natural enemies to non-target species (Yong et al., 2007; Yang et al., 2008). The likelihood of encountering non-target species in natural habitats in surrounding areas may occur should the proposed natural enemy migrate out of areas where it is likely to be most active (e.g., agricultural areas). Spillover of organisms from managed (e.g., agro-ecosystems) to unmanaged natural areas is a poorly studied area requiring greater attention (Blitzer et al., 2012).

Kuhlmann et al. (2006) developed a comprehensive set of evaluation criteria with decision trees for determining host ranges of arthropod natural enemies. In addition to taxonomic relatedness, Kuhlmann et al. (2006) recommend that the construction of non-target species lists consider ecological similarities (e.g., target and non-target occupy same habitat) and include safeguards (i.e., consideration of beneficial organisms, such as weed biological control agents, that could be exposed to attack). Filters are built into these test selection categories to refine safety estimation, including spatial, temporal, morphological, and physiological attributes that may affect candidate safety. For example, natural enemy phenology may protect non-target species from biological control agents. Alternatively, climate may preclude natural enemy infiltration into some areas. Use of climate matching or niche modeling software (Baker, 2002) or thematic mapping to predict the potential range of pests and natural enemies (Pilkington and Hoddle, 2006) are assumed to be important tools for assessing spread (and subsequent risk), but data in support of this argument are largely lacking (Phillips et al., 2008).

Physiological attributes of the natural enemy may need to be considered as part of risk assessment, and host-range tests must be designed accordingly. For parasitoids, the biological relationship to the host strongly influences the degree of specialization required for successful parasitism. Two groups of parasitoids have been recognized based on such biology. Koinobionts are internal parasitoids that allow their hosts to continue to develop but as a consequence must defeat the host's immune system, which attempts to kill the parasitoid. Idiobionts are external parasitoids of larvae that paralyze their hosts upon attack and whose larvae then feed externally as predators on the host. These differences in biology result in higher levels of specialization by internal larval parasitoids (i.e., koinobionts) than by idiobionts (i.e., external larval parasitoids).

A second filter considers the rarity of non-target species or the difficulty of rearing sufficient numbers for experiments. In this instance, common congeners should be substituted as surrogates to predict likelihood of attack on rare or endangered species (Kuhlmann et al., 2006). Additionally, Hopper (2001) suggests that understanding the host range in the source region is an additional way to help design studies to predict the potential host range of a natural enemy before introduction in a new area. Retroactive studies assessing host ranges in areas of origin and introduced areas have provided insight into realized non-target host ranges in the field following introductions (Barratt et al., 2010a). This is possible in some instances because of improved understanding of phylogenetic relationships of the target with non-target species, field studies documenting the use of target and non-target species in the native and introduced ranges, and advances in molecular biology for distinguishing biotypes (see Chapter 6). When used together, phylogeny, field surveys, and molecular analyses provide substantial insight when one assesses potential risk to non-target native species and whether natural enemy populations exist as mixtures of biotypes that exhibit different host ranges (Barratt et al., 2010a). The general conclusion reached is that polyphagous natural enemies have potential to cause population reductions of non-target species and that such potential adverse effects can be predicted from host-range testing (Barratt, 2011). Furthermore, difficult to detect trophic interactions can be investigated using molecular tools to detect prey species in the guts of predators or the presence of parasitoids in target and non-target species. These data can be used to help elucidate risk to non-target species and quantify the intensity of impacts should they be detected (Gariepy et al., 2008; Barratt, 2011).

Another largely undeterminable post-release process that could affect target and non-target species is evolution, the ability of a natural enemy (or its prey or host plant in the case of weed agents, which are included in this section where relevant) to change certain traits over time in response to selection pressures (Müller-Schärer et al., 2004). Research focused on determining host range has almost always failed to investigate genetic variation in host use and responses by hosts to natural enemies (Roderick et al., 2012). Post-release natural enemy population sizes, their structure, and the strength of connectivity between them are affected by selection, genetic drift, and inbreeding, all of which may affect fitness, population growth rates, and potential host use patterns. Hufbauer and Roderick (2005) have demonstrated that there are opportunities for natural enemies to undergo rapid microevolutionary changes owing to selection, genetic drift, gene flow, mutation, and possibly hybridization (Roderick et al., 2012; McEvoy et al., 2012; Vorsino et al., 2012), but these processes are typically not evaluated, and their effects are largely unknown (see also Chapter 8). Local adaptations by released natural enemies can cause greater levels of target mortality (Phillips et al., 2008), and greater genetic variability at the time of initial introduction may affect evolutionary trajectories of biological control agents. This possibility needs to be balanced against possible risks to non-target species because subsequent ecological risk caused by post-release evolution is unpredictable (Phillips et al., 2008). Alternatively, understanding genetic factors underlying certain natural enemy attributes, such as emergence times from diapause or onset of reproductive activity, may allow selection and exploitation of natural enemy biotypes with characteristics that minimize their temporal overlap with non-target species (Phillips et al., 2008) or enhances their impact on target species (Bean et al., 2012). Better understanding of evolutionary processes affecting biological control agents has the potential not only to improve establishment rates and efficacy of natural enemies, but also to mitigate potential non-target effects (Vorsino et al., 2012).

Predicting unwanted trophic effects of pathogens used for biological control

Pathogens (i.e., fungi, viruses, bacteria, and nematodes) have been used for classical biological control of arthropod pests and weeds. However, pathogens have not been as widely used as carnivorous or herbivorous natural enemies. This may be due to people's innate aversion to employing agents that cause disease, difficulty in finding organisms causing severe disease outbreaks, requirements for alternate hosts, varying levels of virulence

associated with different pathogen strains, difficulty in identifying species and strains without application of molecular methods, or complex issues associated with taxonomic identification and phylogenetic placement.

At least 131 introductions of entomopathogens have targeted 76 insect and 3 mite species (Hajek et al., 2007), and 28 species of plant pathogenic fungi have targeted 20 weed species (Barton, 2012). For entomopathogens released against arthropods, viruses and fungi appear to have been the most successful agents, and good control of target pests appears to be associated with perennial systems such as forests and tree crops (Hajek et al., 2007; Hajek and Delalibera, 2010).

Interestingly, many pathogen introductions (~36% of documented cases) have targeted native insects (Hajek et al., 2007). This approach has been controversial (Lockwood, 1993a, b), especially in the case of a proposal to suppress native US rangeland grasshoppers that periodically outbreak and cause extensive damage, with novel fungal pathogens (and egg parasitoids) that attack grasshoppers in Australia (Carruthers and Onsager, 1993). The rationale for this approach was that indigenous grasshopper natural enemies were ineffective in suppressing outbreaks, and congeneric pathogen species (and deliberately selected pathotypes of these species) from Australia had high potential to lower grasshopper densities (Carruthers and Onsager, 1993). Laboratory host-range testing indicated non-target attacks were possible when some non-pest, native grasshopper species were challenged with the Australian fungus. These data (with other information) were used to prepare the first US Risk Assessment Report for the introduction of an entomopathogen to be used for "new association" biological control. A permit to release was granted and the fungus was subsequently liberated in the states of North Dakota and Alaska (Carruthers and Onsager, 1993). Lockwood (1993a, b) postulated that successful control of grasshoppers by exotic natural enemies could cause extinctions of native grasshopper species and their specialist natural enemies, deprive other native species of important food sources, and alter plant dynamics and nutrient chains that are influenced by grasshopper herbivory.

Predicting indirect trophic effects of pathogens

Entomopathogens with broad host ranges can potentially engage in a variety of indirect trophic interactions, including apparent competition (Meyling and Hajek, 2010). Food web analyses indicate that insect hosts that share a common entomopathogen, like a fungus, may be subjected to higher than normal levels of infection when a certain host species is present because it is a superior host for the fungus. Under these situations, other species suffer higher than normal infection rates because of inoculum pressure emanating from the more susceptible species. Conservation biological control programs for aphids have attempted to capitalize on this phenomenon when multiple aphid species infest crop plants and suffer attack from a shared fungal pathogen (Meyling and Hajek, 2010). Transmission of entompathogenic fungi can be indirectly enhanced when different arthropod species share the same host plant. For example, herbivores that remove leaf area through feeding indirectly increase transmission rates to susceptible species because exposure to inoculum increases as leaf area diminishes. Additionally, parasitoids searching for hosts that are also vulnerable to fungal infection may increase infection rates as hosts move more often to avoid parasitism and in doing so increase their exposure to inoculum (Baverstock et al., 2008). In some circumstances, feeding damage to host plants by herbivores increases infection rates by entomopathogens because plant volatiles produced in response to herbivory increase the infectivity of disease-causing propagules (Baverstock et al., 2005). Under certain circumstances these indirect effects can cause significantly elevated levels of mortality, and most published studies demonstrating indirect effects have been conducted on pest species. It is unclear how widespread or how significant these types of interactions are on non-target species, especially those in natural areas that may be subjected to epizootics of pathogens with broad host ranges.

Predicting direct trophic effects of pathogens

Meta-analyses of use of entomopathogens for arthropod biological control have failed to document pathogens causing substantial mortality to non-target species, and these biological control agents have not had significant effects on human or animal health (Hajek et al., 2007; O'Callaghan and Brownbridge, 2009), even when generalist pathogens (e.g., the fungi *Beauveria bassiana* [Bals.-Criv.] Vuill. and *Metarhizium anisopliae* [Metchnikoff] Sorokin, or the bacterium *Bacillus thuringiensis* Berlinger) have been used as microbial pesticides. This may be because these formulations tend to have short-lived activity in the field and seldom result in permanent populations that cause periodic epizootics.

This reduces persistence and subsequent exposures to non-target species (Hokkanen and Hajek, 2003; O'Callaghan and Brownbridge, 2009). Meta-analyses for plant pathogens strongly indicate that non-target effects are almost non-existent, and when they exist they are minimal and consistent with host-range predictions (Barton, 2012). However, Lockwood (1993b) contends that lack of evidence for non-target effects does not support the suggestion that they are non-existent because insufficient effort has been devoted to long-term investigations for unintended consequences.

Despite these concerns, few field data indicate that significant widespread non-target effects, either of a direct or indirect nature, result from pathogen releases. This is not always the case when pathogens are applied as biopesticides. With this approach, non-target effects, some persistent, have been detected, as for example when *Bt. kurstaki*, a Lepidoptera-specific bacterium, was applied to forests to control gypsy moth. Non-target Lepidoptera were killed and population-level reductions were observed post-application for at least four years (O'Callaghan and Brownbridge, 2009).

Pathogens vary in key characteristics such as host specificity, mode of action, and persistence, which collectively determine the risk they pose to non-target species. As discussed earlier, a key characteristic of a "safe" natural enemy is its specificity. Baculoviruses have been used extensively for insect biological control with no adverse ecological effects recorded despite repeated widespread use against pests in sensitive natural areas (e.g., gypsy moth in forests in the northeastern United States). Similar levels of high specificity and safety have been observed for the nematode *Beddingia* (*Deladenus*) *siricidicola* (Bedding) for the control of *Sirex noctilio* Fabricius, a pest of conifers. The host-specific fungal pathogen *Entomophaga maimaiga* Humber, Shimazu, and Soper has been extraordinarily effective against gypsy moth with no evidence of any important non-target effects (O'Callaghan and Brownbridge, 2009). Host-range testing of microsporidia pathogenic to gypsy moth under field conditions showed high levels of specificity to the target and very low infection rates in just a few non-target Lepidoptera (Solter et al., 2010).

As with carnivorous and herbivorous natural enemies, pathogens used for insect and weed biological control are subjected to host-range testing to determine specificity. One of the primary research interests is assessment of toxicity or infectiousness to non-target species. In addition to taxonomic relatedness, selection of non-target species must account for potential exposure that could arise from sharing a common habitat. For example, soil-active pathogens may pose unique threats to other soil-inhabiting arthropods because of exposure to the natural enemy, even though the non-target species may not be closely related to the target pest. To assess this risk, earthworms are subjected to exposure tests, as they can ingest pathogens as part of their diet and they may also experience exposure through direct contact. It should be noted, however, that surrogate non-target test species may vary in their susceptibility to control agents and, therefore, may not accurately reveal risk to non-target species of interest (O'Callaghan and Brownbridge, 2009).

For entomopathogens, susceptibility studies are conducted in laboratory assays, while plant pathogens are tested against plants maintained in greenhouses. As seen earlier with carnivorous and herbivorous biological control agents, these types of exposure trials tend to overestimate host ranges, and data from other sources may be needed to ensure that potential agents are not rejected prematurely (O'Callaghan and Brownbridge, 2009; Barton, 2012). Despite the disparity between physiological and realized host ranges, bioassays remain a first and necessary step in determining potential susceptibility of non-target species, and they must be designed to mimic likely exposure scenarios and doses likely to be consumed or contacted under field conditions (O'Callaghan and Brownbridge, 2009).

Sublethal effects are effects resulting from exposures that are not sufficient to kill but are debilitating in some way; these have been assessed for entomopathogens. Sublethal effects caused by microbial agents, like viruses or bacteria, may alter developmental rates, reduce fecundity, or lower pupal weights in Lepidoptera. In the case of slower developmental times, this type of sublethal effect may enhance parasitism rates as vulnerable stages are exposed to parasitoids for longer, and this may help conserve biological control agents in areas at critical times when pests are experiencing high levels of pathogen-induced mortality (Nealis et al., 1992). Sublethal effects could potentially occur in non-target species; laboratory studies suggest that pollen dosed with *B. thuringiensis* toxins can reduce coccinellid feeding rates on pest eggs (Giroux et al., 1994). While it is possible that sublethal effects could exist in susceptible non-target species in field situations, the severity of these effects would likely be transient and probably not ecologically significant, and

they would be difficult to detect and quantify in the field in comparison to direct mortality caused by the pathogen.

Pathogenic organisms can exhibit significant diversity within species, and virulence of particular strains or pathotypes can be highly variable as a consequence. This variability may be either beneficial or detrimental, and it depends on the pathogen's mode of action and the susceptibility of individual target pests, which themselves may exhibit different levels of resistance to infection. Consequently, microevolutionary events can occur as a result of these interactions, and outcomes will be affected by selection pressures, pathogen diversity, rates of gene flow as influenced by population connectivity, and host-plant chemistry that may directly affect plant pathogens or act indirectly on entomopathogens through the target herbivore following consumption of these compounds. Together, under certain circumstances, these mechanisms may act in concert to lead to resistance or potentially to host-range expansion, and these possibilities should be considered to avoid these unwanted outcomes (Cory and Franklin, 2012). However, there have been no reports of host-range evolution occurring or long-term negative effects on non-target hosts by pathogens used in the biological control of arthropods and weeds (Barton, 2012; Cory and Franklin, 2012).

Conclusions

Forecasting unwanted trophic impacts resulting from the intentional introductions of biological control agents is inherently difficult because of the complexities of ecosystem interactions. Steps can be taken to reduce considerably the risks posed to non-target species by natural enemies, such as the use of host-specific natural enemies that cause density-dependent feedback loops, resulting in the reduction of pest populations and, in turn, of their own densities as their food supplies dwindle. Substantial research investigating non-target impacts has greatly improved understanding of ecosystem-level effects that natural enemy introductions have caused and may potentially induce via the development of either direct or indirect trophic effects (e.g., apparent competition or resource subsidization) or non-trophic linkages with new organisms. There is no evidence that natural enemies have caused the extinction of non-target plants or arthropods via direct or indirect trophic interactions (acknowledging the fact that such evidence could be difficult to obtain, even if it happened). A growing body of literature assessing non-target effects has provided insights leading to advances in quarantine host-range testing to assess risk to non-target species (Barratt et al., 2010b). In some countries, especially New Zealand and Australia, greater regulatory oversight has been imposed, with the goals of ensuring the continued safety of natural enemy introductions and addressing the concerns of stakeholders (Barratt et al., 2010b; Barratt, 2011). New analytical tools have been used to assess risks of biological control programs. These new methods include improved behavioral studies and their analyses, molecular approaches, and population-level modeling to optimize agent selection. In combination, these new approaches may enhance quantitative predictions of effects of biological control agents on target and non-target species (Mills and Kean, 2010). However, there is one area that still deserves greater attention: post-release monitoring to verify predictions from host-range tests in quarantine. These studies would quantify the magnitude of attacks on target and non-target species and their population-level effects. Consequently, increased use of retroactive studies of biological control projects is recommended, and field studies can be used to determine the accuracy of conclusions from host-range testing (Delfosse, 2005; Barratt, 2011). Given the totality of completed work and ongoing projects in this area, there is optimism that risks associated with future biological control programs can be better identified and managed (Barratt et al., 2010b), thereby preserving biological control as a powerful tool to reverse the damage caused by invasive species in natural areas.

Acknowledgments

I thank Bernd Blossey, David Wagner, Daniel Simberloff, and Hong Liu for reviews of Chapter 7.

References

A'Bear, A. D., L. Boddy, G. Raspotnig, and T. H. Jones. 2010. Non-trophic effects of oribatid mites on cord-forming basidiomycetes in soil microcosms. *Ecological Entomology* **35**: 477–484.

Arnett, A. E. and S. V. Louda. 2002. Re-test of *Rhinocyllus conicus* host specificity, and the prediction of ecological risk in biological control. *Biological Conservation* **106**: 251–257.

Baker, R. H. A. 2002. Predicting the limits of the potential distribution of alien crop pests, pp. 207–241. *In*: Hallman, G. J. and C. P. Schwalbe (eds.). *Invasive Arthropods in Agriculture – Problems and Solutions*. Science Publishers Inc. Enfield, New Hampshire, USA.

Barron, M. C. 2007. Retrospective modeling indicates minimal impact of non-target parasitism by *Pteromalus puparum* on red admiral butterfly (*Bassaris gonerilla*) abundance. *Biological Control* **41**: 53–63.

Barton, J. 2012. Predictability of pathogen host range in classical biological control of weeds: an update. *Biological Control* **57**: 289–305.

Barratt, B. I. P. 2011. Assessing safety of biological control introductions. *CAB Reviews: Perspectives in Agriculture, Veterinary Science, Nutrition and Natural Resources* **6** (042): 1–12.

Barratt, B. I. P., C. M. Ferguson, M. R. McNeil, and S. L. Goldson. 1999. Parasitoid host specificity testing to predict field host range, pp. 70–92. *In*: Withers, T. M., L. Barton Browne, and J. Stanley (eds.). *Host Specificity Testing in Australasia: Towards Improved Assays for Biological Control*. CRC for Tropical Pest Management. Brisbane, Australia.

Barratt, B. I. P., R. G. Oberprieler, D. M. Barton, et al. 2010a. Could research in the native range, and non-target host range in Australia, have helped predict the host range of the parasitoid *Microctonus aethiopoides* Loan (Hymenoptera: Braconidae), a biological control agent introduced for *Sitona discoideus* Gyllenhal (Coleoptera: Curculionidae) in New Zealand? *Biological Control* **57**: 735–750.

Barratt, B. I. P., F. G Howarth, T. M. Withers, et al. 2010b. Progress in risk assessment for classical biological control. *Biological Control* **52**: 245–254.

Baverstock, J., S. L. Elliot, P. G. Alderson, and J. K. Pell. 2005. Response to the entomopathogenic fungus *Pandora neoaphidis* to aphid-induced plant volatiles. *Journal of Invertebrate Pathology* **89**: 157–164.

Baverstock, J., K. E. Baverstock, S. J. Clark, and J. K. Pell. 2008. Transmission of *Pandora neoaphidis* in the presence of co-occurring arthropods. *Journal of Invertebrate Pathology* **98**: 356–359.

Bean, D. W., P. Dalin, and T. L. Dudley. 2012. Evolution of critical day length for diapause induction enables host range expansion of *Diorhabda carinulata*, a biological control agent against Tamarisk (*Tamarix* spp.). *Evolutionary Applications* **5**: 511–523.

Benson, J., R. G. Van Driesche, A. Pasquale, and J. Elkinton. 2003a. Introduced braconid parasitoids and range reduction of a native butterfly in New England. *Biological Control* **28**: 197–213.

Benson, J., R. G. Van Driesche, A. Pasquale, and J. Elkinton. 2003b. Assessment of risk posed by introduced braconid wasps to *Pieris virginiensis*, a native woodland butterfly in New England. *Biological Control* **26**: 83–93.

Berenbaum, M. R. 1990. Evolution of specialization in insect-umbelifera associations. *Annual Review of Entomology* **35**: 319–343.

Bernays, E. and M. Graham. 1988. On the evolution of host specificity in phytophagous arthropods. *Ecology* **69**: 886–892.

Berner, D. K. 2010. BLUP, a new paradigm in host-range determination. *Biological Control* **53**: 143–152.

Bigler, F., D. Babendreier, and U. Kuhlmann (eds.). 2006. *Environmental Impact of Invertebrates for Biological Control of Arthropods – Methods and Risk Assessment.* CABI Publishing. Wallingford, UK.

Blitzer, E. J., C. F. Dormann, A. Holzschuh, et al. 2012. Spillover of functionally important organisms between managed and natural habitats. *Agriculture, Ecosystems and the Environment* **146**: 34–43.

Boettner, G. H., J. S. Elkinton, and C. J. Boettner. 2000. Effects of a biological control introduction on three nontarget native species of saturniid moths. *Conservation Biology* **14**: 1798–1806.

Bonsall, M. B. and M. P. Hassell. 1998. Population dynamics of apparent competition in a host-parasitoid assemblage. *Journal of Animal Ecology* **67**: 918–929.

Briese, D. T. 1999. Open field host-specificity tests: is "natural" good enough for risk assessment? pp. 44–59. *In*: Withers, T. M., L. Barton Browne, and J. Stanley (eds.). *Host Specificity Testing in Australasia: Towards Improved Assays for Biological Control.* CRC for Tropical Pest Management. Brisbane, Australia.

Burghardt, K. T., D. W. Tallamy, C. Philips, and K. J. Shropshire. 2010. Non-native plants reduce abundance, richness, and host specialization in lepidopteran communities. *Ecosphere* **1**: article 11.

Carruthers, R. I. and J. A. Onsager. 1993. Perspective on the use of exotic natural enemies for biological control of pest grasshoppers (Orthoptera: Acrididae). *Environmental Entomology* **22**: 885–903.

Carvalheiro, L. G., Y. M. Buckley, R. Ventim, et al. 2008. Apparent competition can compromise the safety of highly specific biological control agents. *Ecology Letters* **11**: 690–700.

Chalak, M., L. Hemerick, W. van der Werf, et al. 2010. On the risk of extinction of a wild plant species through the spillover of a biological control agent: analysis of an ecosystem compartment model. *Ecological Modelling* **221**: 1934–1943.

Clewley, G. D., R. Eschen, R. H. Shaw, and D. J. Wright. 2012. The effectiveness of classical biological control of invasive plants. *Journal of Applied Ecology* **49**: 1287–1295.

Cory, J. S. and M. T. Franklin. 2012. Evolution and the microbial control of insects. *Evolutionary Applications* **5**: 455–469.

Cripps, M. G., A. Gassmann, S. V. Fowler, et al. 2011. Classical biological control of *Cirsium arvense*: Lessons from the past. *Biological Control* **57**: 165–174.

Cristafaro, M., A. de Biase, and L. Smith. 2013. Field release of a prospective biological control agent of weeds, *Ceratapion basicorne*, to evaluate risk to a nontarget crop. *Biological Control* **64**: 305–314.

Crowder, D. W. and W. E. Snyder. 2010. Eating their way to the top? Mechanisms underlying the success of invasive insect generalist predators. *Biological Invasions* **12**: 2857–2876.

Crowley, P. H. and J. J. Cox. 2011. Intraguild mutualism. *Trends in Ecology & Evolution* **26**: 627–633.

De Clercq, P., P. G. Mason, and D. Babendreier. 2011. Benefits and risks of exotic biological control agents. *Biological Control* **56**: 681–698.

Delfosse, E. S. 2005. Risk and ethics in biological control. *Biological Control* **35**: 319–329.

Dodd, A. P. 1940. *The Biological Campaign against Prickly Pear.* Commonwealth Prickly Pear Board Bulletin. Brisbane, Australia. 177 pp.

Edwards, P. B. 1999. The use of choice tests in host-specificity testing of herbivorous insects, pp. 35–43. *In*: Withers, T. M., L. Barton Browne, and J. Stanley (eds.). *Host Specificity Testing in Australasia: Towards Improved Assays for Biological Control.* CRC for Tropical Pest Management. Brisbane, Australia.

Elkinton, J. S. and G. H. Boettner. 2012. Benefits and harm caused by the introduced generalist tachinid, *Compsilura concinnata*, in North America. *Biological Control* **57**: 277–288.

Ehrlich, P. R. and P. H. Raven. 1964. Butterflies and plants – a study in coevolution. *Evolution* **18**: 586–608.

Fath, B. D. 2007. Network mutualism: positive community-level relations in ecosystems. *Ecological Modelling* **208**: 56–67.

Fergusson, C. S., J. S. Elkinton, J. R. Gould, and W. E. Wallner. 1994. Population regulation of gypsy moth (Lepidoptera: Lymantriidae) by parasitoids – does spatial density dependence lead to temporal density dependence? *Environmental Entomology* **23**: 1155–1164.

Follett, P. A. and J. J. Duan (eds.). 2000. *Nontarget Effects of Biological Control.* Kluwer Academic Publishers. Boston, Massachusetts, USA.

Fowler, S. V., Q. Paynter, S. Dodd, and R. Groenteman. 2012. How can ecologists help practitioners minimize non-target effects in weed biological control? *Journal of Applied Ecology* **49**: 307–310.

Gaines, D. N. and L. T. Kok. 1999. Impact of hyperparasitoids on *Cotesia glomerata* in southwestern Virginia. *Biological Control* **14**: 19–28.

Gariepy T., U. Kuhlmann, C. Gillot, and M. Erlandson. 2008. A large-scale comparison of conventional and molecular methods for the evaluation of host-parasitoid associations in non-target risk-assessment studies. *Journal of Applied Ecology* **45**: 708–715.

Gassmann, A. and S. M. Louda. 2001. *Rhynocyllus conicus*: initial evaluation and subsequent ecological effects in North America, pp. 147–183. *In*: Wajnberg, E., J. K. Scott, and P. C. Quimby (eds.). *Evaluating Indirect Ecological Effects of Biological Control.* CABI Publishing. Wallingford, UK.

Giroux, S., J. C. Cote, C. Vincent, et al. 1994. Bacteriological insecticide M-one effects on predation efficiency and mortality of adult *Coleomegilla maculata lengi* (Coleoptera: Coccinellidae). *Journal of Economic Entomology* **87**: 39–43.

González, D. and F. E. Gilstrap. 1992. Foreign exploration: assessing and prioritizing natural enemies and consequences of preintroduction studies, pp. 53–70. *In*: Kauffman W. C. and J. E. Nechols (eds.). *Selection Criteria and Ecological*

Consequences of Importing Natural Enemies. Proceedings Thomas Say Publications in Entomology, Entomological Society of America. Lanham Maryland.

Gould, J. R., J. S. Elkinton, and W. E. Wallner. 1990. Density-dependent suppression of experimentally created gypsy moth, *Lymantria dispar* (Lepidoptera: Lymantriidae), populations by natural enemies. *Journal of Animal Ecology* **59**: 213–233.

Grandgirard, J., M. S. Hoddle, J. N. Petit, G. K. Roderick, and N. Davies. 2008. Classical biological control of the glassy-winged sharpshooter, *Homalodisca vitripennis*, by the egg parasitoid *Gonatocerus ashmeadi* in the Society, Marquesas, and Austral Archipelagos of French Polynesia. *Biological Control* **48**: 155–163.

Green, P. T., D. J. O'Dowd, K. L. Abbott, et al. 2011. Invasional meltdown: invader-invader mutualism facilitates a secondary invasion. *Ecology* **92**: 1758–1768

Hajek, A. E. and I. Delalibera, Jr. 2010. Fungal pathogens as classical biological control agents against arthropods. *Biological Control* **55**: 147–158.

Hajek, A. E., M. L. McManus, and I. Delalibera, Jr. 2007. A review of introductions of pathogens and nematodes for classical biological control insects and mites. *Biological Control* **41**: 1–13.

Harris, P. and H. Zwölfer. 1968. Screening of phytophagous insects for biological control of weeds. *The Canadian Entomologist* **100**: 295–303.

Hufbauer, R. A. and G. K. Roderick. 2005. Microevolution in biological control: mechanisms, patterns, and processes. *Biological Control* **35**: 227–239.

Hawkins, B. A., N. J. Mills, M. A. Jervis, and P. W. Price. 1999. Is biological control of insects a natural phenomenon? *Oikos* **86**: 493–506.

Henneman, M. L. and J. Memmott. 2001. Infiltration of a Hawaiian community by introduced biological control agents. *Science* **293**: 1314–1316.

Herlihy, M. V., R. G. Van Driesche, M. R. Abney, et al. 2012. Distribution of *Cotesia rubecula* (Hymenoptera: Braconidae) and its displacement of *Cotesia glomerata* in eastern North America. *Florida Entomologist* **95**: 458–464.

Herlihy, M. V., D. L. Wagner, and R. G. Van Driesche. 2014. Persistence in Massachusetts of the veined white butterfly due to use of the invasive form of cuckoo flower. *Biological Invasions* **16**: 2713–2724.

Hill, R. L. 1999. Minimising uncertainty – in support of no choice tests, pp. 1–10. *In*: Withers, T. M., L. Barton Browne, and J. Stanley (eds.). *Host Specificity Testing in Australasia: Towards Improved Assays for Biological Control*. CRC for Tropical Pest Management. Brisbane, Australia.

Hickman, B. 1997. The Effects of the White Butterfly's *(Pieris rapae)* Introduced Parasitoid *(Pteromalus puparum)* on the Native Yellow Admiral, *Bassaris itea*. M.Sc. Thesis, University of Auckland, New Zealand. 97 pp.

Hoddle, M. S. 2004a. Restoring balance: using exotic species to control invasive exotic species. *Conservation Biology* **18**: 38–49.

Hoddle, M. S. 2004b. Biological control in support of conservation: friend or foe? pp. 202–237. *In*: Gordon, M. S. and S. Bartol (eds.). *Experimental Approaches to Conservation Biology*. University of California Press. Berkeley, California, USA.

Hoddle, M. S. 2006. Historical review of control programs for *Levuana iridescens* (Lepidoptera: Zygaenidae) in Fiji and examination of the possible extinction of this moth by *Bessa remota* (Diptera: Tachinidae). *Pacific Science* **60**: 439–453.

Hoddle, M. S. and R. Pandey. 2014. Host range testing of *Tamarixia radiata* (Hymenoptera: Eulophidae) sourced from the Punjab of Pakistan for classical biological control of *Diaphorina citri* (Hemiptera: Liviidae: Euphyllurinae: Diaphorinini) in California. *Journal of Economic Entomology* **107**: 125–136.

Hoddle, M. S. and P. Syrett. 2003. Realizing the potential of classical biological control, pp. 395–424. *In*: Hallman, G. J. and C. P. Schwalbe (eds.). *Invasive Arthropods in Agriculture – Problems and Solutions*. Science Publishers Inc. Enfield, New Hampshire, USA.

Hoddle, M. S., C. C. Ramirez, C. D. Hoddle, et al. 2013. Post release evaluation of *Rodolia cardinalis* (Coleoptera: Coccinellidae) for control of *Icerya purchasi* (Hemiptera: Monophlebidae) in the Galápagos Islands. *Biological Control* **67**: 262–274.

Hoffmeister, T. S., D. Babendreier, and E. Wajnberg. 2006. Statistical tools to improve the quality of experiments and data analysis for assessing non-target effects, pp. 222–240. *In*: Bigler, F., D. Babendreier, and U. Kuhlmann (eds.). *Biological Control of Arthropods using Invertebrates: Methods for Environmental Risk Assessment*. CABI Publishing. Wallingford, U.K.

Hokkanen, H. M. T. and A. E. Hajek (eds.). 2003. *Environmental Impacts of Microbial Insecticides: Need and Methods for Risk Assessment*. Kluwer Academic Publishers. Dordrecht, The Netherlands.

Holt, R. D. and M. E. Hochberg. 2001. Indirect interactions, community modules and biological control: a theoretical perspective, pp. 13–37. *In*: Wajnberg, E., J. K. Scott, and P. C. Quimby (eds.). *Evaluating Indirect Ecological Effects of Biological Control*. CABI Publishing. Wallingford, UK.

Holt, R. D. and J. H. Lawton. 1993. Apparent competition and enemy-free space in insect host-parasitoid communities. *The American Naturalist* **142**: 623–645.

Hopper, K. R. 2001. Research needs concerning non-target impacts of biological control introductions, pp. 39–56. *In*: Wajnberg, E., J. K. Scott, and P. C. Quimby (eds.). *Evaluating Indirect Ecological Effects of Biological Control*. CABI Publishing. Wallingford, UK.

Howarth, F. G. 1983. Biological control: panacea or Pandora's box? *Proceedings of the Hawaii Entomological Society* **24**: 239–244.

Howarth, F. 1991. Environmental impacts of classical biological control. *Annual Review of Entomology* **36**: 485–509.

Howarth, F. G. 2001. Environmental issues concerning the importation of non-indigenous biological control agents, pp. 70–99. *In*: Lockwood, J. A., F. G. Howarth, and M. F. Purcell (eds.). *Balancing Nature: Assessing the Impact of Importing Non-Native Biological Control Agents (An International Perspective)*. Thomas Say Publications in Entomology, Entomological Society of America. Lanham, Maryland, USA.

Jones, C. G., J. H. Lawton, and M. Shachak. 1994. Organisms as ecosystem engineers. *Oikos* **69**: 373–386.

Johnson, M. T., P. A. Follett, A. D. Taylor, and V. P. Jones. 2005. Impacts of biological control and invasive species on a non-target native Hawaiian insect. *Oecologia* **142**: 529–540.

Karban, R., P. Grof-Tisza, and M. Holyoak. 2012. Facilitation of tiger moths by outbreaking tussock moths that share the same host plants. *Journal of Animal Ecology* **81**: 1095–1102.

Karban, R., T. W Mata, P. Grof-Tisza, et al. 2013. Non-trophic effects of litter reduce ant predation and determine caterpillar survival and distribution. *Oikos* **122**: 1362–1370.

Kaufman, L. V. and M. G. Wright. 2009. The impact of exotic parasitoids on populations of a native Hawaiian moth assessed using life table studies. *Oecologia* **159**: 295–304.

Kaufman, L. V. and M. G. Wright. 2010. Parasitism of a Hawaiian endemic moth by invasive and purposely introduced Hymenoptera species. *Environmental Entomology* **39**: 430–439.

Kaufman, L. V. and M. G. Wright. 2011. Ecological correlates of the non-indigenous parasitoid assemblage associated with a Hawaiian endemic moth. *Oecologia* **166**: 1087–1098.

Keane, R. M. and M. J. Crawley. 2002. Exotic plant invasions and the enemy release hypothesis. *Trends in Ecology & Evolution* **17**: 164–170.

Kéfi, S., E. L. Berlow, E. A. Wieters, et al. 2012. More than a meal – integrating non-feeding interactions into food webs. *Ecology Letters* **15**: 291–300.

King, C. B., W. P. Haines, and D. Rubinoff. 2010. Impacts of invasive parasitoids on declining endemic Hawaiian leaf rolling moths (*Omiodes*: Crambidae) vary among sites and species. *Journal of Applied Ecology* **47**: 299–308.

Kuhlmann, U., U. Schaffner, and P. G. Mason. 2006. Selection of non-target species for host specificity tests, pp. 15–37. *In*: Bigler, F., D. Babendreier, and U. Kuhlmann (eds.). *Biological Control of Arthropods Using Invertebrates: Methods for Environmental Risk Assessment*. CABI Publishing. Wallingford, UK.

Kuris, A. M. 2003. Did biological control cause extinction of the coconut moth, *Levuana iridescens*, in Fiji? *Biological Invasions* **5**: 133–141.

Laland, K. N. and N. J. Boogert. 2010. Niche construction, co-evolution and biodiversity. *Ecological Economics* **69**: 731–736.

Langellotto, G. A. and R. F. Denno. 2004. Responses of invertebrate natural enemies to complex-structured habitats: a meta-analytical synthesis. *Oecologia* **139**: 1–10.

Lincango, P. M., C. E. Causton, C. Alvarez Calderón, and G. Jiménez-Uzcátegui. 2011. Evaluating the safety of *Rodolia cardinalis* to two species of Galápagos finch, *Camarhynchus parvulus* and *Geospiza fuliginosa*. *Biological Control* **56**: 145–149.

Lockwood, J. A. 1993a. Environmental issues involved in biological control of rangeland grasshoppers (Orthoptera: Acrididae) with exotic agents. *Environmental Entomology* **22**: 503–518.

Lockwood, J. A. 1993b. Benefits and costs of controlling rangeland grasshoppers (Orthoptera: Acrididae) with exotic organisms: search for a null hypothesis and regulatory compromise. *Environmental Entomology* **22**: 904–914.

Lockwood, J. A., F. G. Howarth, M. F. Purcell (eds.). 2001. *Balancing Nature: Assessing the Impact of Importing Non-Native Biological Control Agents (An International Perspective)*. Thomas Say Publications in Entomology, Entomological Society of America. Lanham, Maryland, USA.

Louda, S. M., D. Kendall, J. Connor, and D. Simberloff. 1997. Ecological effects of an insect introduced for the biological control of weeds. *Science* **277**: 1088–1090.

Louda, S. M., R. W. Pemberton, M. T. Johnson, and P. A. Follett. 2003. Nontarget effects – the Achilles' heel of biological control? Retrospective analyses to reduce risk associated with biological control introductions. *Annual Review of Entomology* **48**: 365–396.

Maron, J. L., D. E Pearson, S. M. Hovick, and W. P. Carson. 2010. Funding needed for assessments of weed biological control. *Frontiers in Ecology and the Environment* **8**: 122–123.

Mason, P. G., A. B. Broadbent, J. W. Wishcraft, and D. R. Gillespie. 2011. Interpreting host range of *Peristenus digoneutis* and *Peristenus relictus* (Hymenoptera: Braconidae) biological control agents of *Lygus* spp. (Hemiptera: Miridae) in North America. *Biological Control* **57**: 94–102.

McEvoy, P. B., K. M. Higgs, E. M. Coombs, et al. 2012. Evolving while invading: rapid adaptive evolution in juvenile development time for a biological control organism colonizing a high-elevation environment. *Evolutionary Applications* **5**: 524–536.

McFadyen, R. E. C. 1998. Biological control of weeds. *Annual Review of Entomology* **43**: 369–393.

Meinke, L. J., T. W. Sappington, D. W. Onstadt, et al. 2009. Western corn rootworm (*Diabrotica virgifera virgifera* LeConte) population dynamics. *Agricultural and Forest Entomology* **11**: 29–46.

Memmott, J. 2000. Food webs as tool for studying nontarget effects in biological control, pp. 147–163. *In*: Follett, P. A. and J. J. Duan (eds.). *Nontarget Effects of Biological Control*. Kluwer Academic Publishers. Boston, Massachusetts, USA.

Memmott, J. 2009. Food webs: a ladder for picking strawberries or a practical tool for practical problems? *Philosophical Transactions of the Royal Society of London. Series B, Biological Sciences* **364**: 1693–1699.

Messing, R. H. and X-G. Wang. 2009. Competitor-free space mediates non-target impact of an introduced biological control agent. *Ecological Entomology* **34**: 107–113.

Messing, R. H. and M. G. Wright. 2006. Biological control of invasive species: solution or pollution? *Frontiers in Ecology and the Environment* **4**: 132–140.

Messing, R., B. Roitberg, and J. Brodeur. 2006. Measuring and predicting indirect impacts of biological control: competition, displacement and secondary interactions, pp. 64–77. *In*: Bigler, F., D. Babendreier, and U. Kuhlmann (eds.). *Environmental Impact of Invertebrates for Biological Control of Arthropods – Methods and Risk Assessment*. CABI Publishing. Wallingford, UK.

Meyling, N. V. and A. E. Hajek. 2010. Principles from community and metapopulation ecology: application to fungal entompathogens. *Biological Control* **55**: 39–54.

Michaud, J. P. 2002. Invasion of the Florida citrus ecosystem by *Harmonia axyridis* (Coleoptera: Coccinellidae) and asymmetric competition with a native species, *Cycloneda sanguinea*. *Environmental Entomology* **31**: 827–835.

Mills, N. J. and J. M. Kean. 2010. Behavioral studies, molecular approaches, and modeling: methodological contributions to biological control success. *Biological Control* **52**: 255–262.

Moran, P. J., C. J. DeLoach, T. L. Dudley, and J. Sanabria. 2009. Open field host selection and behavior by tamarisk beetles (*Diorhabda* spp.) (Coleoptera: Chrysomelidae) in biological control of exotic saltcedars (*Tamarix* spp.) and risks to nontarget athel (*T. aphylla*) and native *Frankenia* spp. *Biological Control* **50**: 243–261.

Müller-Schärer, H., U. Schaffner, and T. Steinger. 2004. Evolution in invasive plants: implications for biological control. *Trends in Ecology & Evolution* **19**: 417–422.

Munro, V. M. W. and I. M. Henderson. 2002. Nontarget effect of entomophagous biological control: shared parasitism between native lepidopteran parasitoids and the biological control agent *Trigonospila brevifacies* (Diptera: Tachinidae) in forest habitats. *Environmental Entomology* **31**: 388–396.

Nealis, V. G., K. van Frankenhuyzen, and B. L. Cardogan. 1992. Conservation of spruce budworm parasitoids following application of *Bacillus thuringiensis* var. *kurstaki* Berliner. *The Canadian Entomologist* **124**: 1085–1092.

Noonberg, E. G. and J. E. Byers. 2005. More harm than good: when invader vulnerability to predators enhances impact on native species. *Ecology* **86**: 2555–2560.

Obrycki, J. J., N. C. Elliott, and K. L. Giles. 2000. Coccinellid introductions: potential for and evaluation of nontarget effects, pp. 127–145. *In*: Follett, P. A. and J. J. Duan (eds.). *Nontarget Effects of Biological Control*, Kulwer Academic Publishers. Boston, Massachusetts, USA.

O'Callaghan, M. and M. Brownbridge. 2009. Environmental impacts of microbial control agents used for control of invasive species, pp. 305–327. *In*: Hajek, A. E., T. R. Glare, and M. O'Callaghan (eds.). *Use of Microbes for Control and Eradication of Invasive Arthropods*. Springer. Dordrecht, The Netherlands.

Ohgushi, T. 2008. Herbivore-induced indirect interaction webs on terrestrial plants: the importance of non-trophic, indirect, and facilitative interactions. *Entomologia Experimentalis et Applicata* **128**: 217–229.

Olff, H., D. Alonso, M. P. Berg, et al. 2009. Parallel ecological networks in ecosystems. *Philosophical Transactions of the Royal Society Series B* **364**: 1755–1779.

Ortega, Y., D. E Pearson, and K. S. McKelvey. 2004. Effects of biological control agents and exotic plant invasion on deer mouse populations. *Ecological Applications* **14**: 241–253.

Paine, R. T. 1988. Road maps of interactions or grist for theoretical development? *Ecology* **69**: 1648–1654.

Parry, D. 2009. Beyond Pandora's Box: quantitatively evaluating non-target effects of parasitoids in classical biological control. *Biological Invasions* **11**: 47–58.

Paynter, Q., A. H. Gourlay, P. T. Oboyski, et al. 2008. Why did specificity testing fail to predict the field host-range of the gorse pod moth in New Zealand? *Biological Control* **46**: 453–462.

Paynter, Q., S. V. Fowler, A. H. Gourley, et al. 2010. Predicting parasitoid accumulation on biological control agents of weeds. *Journal of Applied Ecology* **47**: 575–582.

Pearson, D. E. 2009. Invasive plant architecture alters trophic interactions by changing predator abundance and behavior. *Oecologia* **159**: 549–558.

Pearson, D. E. 2010. Trait- and density-mediated indirect interactions initiated by an exotic invasive plant autogenic ecosystem engineer. *The American Naturalist* **176**: 394–403.

Pearson, D. E. and R. M. Callaway. 2003. Indirect effects of host specific biological control agents. *Trends in Ecology & Evolution* **18**: 456–461.

Pearson, D. E. and R. M. Callaway. 2005. Indirect nontarget effects of host specific biological control agents: implications for biological control. *Biological Control* **35**: 288–298.

Pearson, D. E. and R. M. Callaway. 2006. Biological control agents elevate hantavirus by subsidizing deer mouse populations. *Ecology Letters* **9**: 443–450.

Pearson, D. E. and R. M. Callaway. 2008. Weed-biological control insects reduce native-plant recruitment through second-order apparent competition. *Ecological Applications* **18**: 1489–1500.

Pearson, D. E. and R. J. Fletcher. 2008. Mitigating exotic impacts: restoring deer mouse populations elevated by an exotic food subsidy. *Ecological Applications* **18**: 321–334.

Pearson, D. E., K. S. McKelvey, and L. E. Ruggiero. 2000. Nontarget effects of an introduced biological control agent on deer mouse ecology. *Oecologia* **122**: 121–128.

Pemberton, R. W. 1995. *Cactoblastis cactorum* (Lepidoptera: Pyralidae) in the United States: an immigrant biological control agent or an introduction of the nursery trade. *American Entomologist* **41**: 230–232.

Pemberton, R. W. 2000. Predictable risk to native plants in weed biological control. *Oecologia* **125**: 489–494.

Pemberton, R. W. and H. Liu. 2007. Control and persistence of native *Opuntia* on Nevis and St. Kitts 50 years after the introduction of *Cactoblastis cactorum*. *Biological Control* **41**: 272–282.

Perkins, L. B. and R. S. Nowark. 2013. Invasional syndromes: hypotheses on relationships among invasive species attributes and characteristics of invaded sites. *Journal of Arid Land* **5**: 275–283.

Phillips, C. B., D. B. Baird, I. I. Iline, et al. 2008. East meets west: adaptive evolution of an insect introduced for biological control. *Journal of Applied Ecology* **45**: 948–956.

Pilkington, L. J. and M. S. Hoddle. 2006. Use of life table statistics and degree-day values to predict the invasion success of *Gonatocerus ashmeadi* (Hymenoptera: Mymaridae), and egg parasitoid of *Homalodisca coagulata* (Hemiptera: Cicadellidae), in California. *Biological Control* **37**: 276–283.

Pimm, S. L. and R. L. Kitching. 1988. Food web patterns: trivial flaws or the basis of an active research program? *Ecology* **69**: 1669–1672.

Pimm, S. L., J. H. Lawton, and J. E. Cohen. 1991. Food web patterns and their consequences. *Nature* **350**: 669–674.

Poelman, E. H., R. Gols, T. A. L. Snoeren, et al. 2011. Indirect plant-mediated interactions among parasitoid larvae. *Ecology Letters* **14**: 670–676.

Prasad, R. P. and W. E. Snyder. 2010. A non-trophic interaction chain links predators in different niches. *Oecologia* **162**: 747–753.

Pratt, P. D., M. B. Rayamajhi, T. D. Center, et al. 2009. The ecological host range of an intentionally introduced herbivore: a comparison of predicted versus actual host use. *Biological Control* **49**: 146–153.

Preisser E. L., D. I. Bolnick, and M. F. Benard. 2005. Scared to death? The effects of intimidation and consumption in predator-prey interactions. *Ecology* **86**: 501–509.

Prior, K. M. and J. J. Hellmann. 2015. Does natural enemy release contribute to the success of invasive species? A review of the enemy release hypothesis, pp. 252–280. *In*: Keller, R. P., M. W. Cadotte, and G. Sandiford (eds.). *Invasvice Species in a Globalized World: Ecological, Social, and Legal Perspectives on Policy*. University of Chicago Press. Chicago, Illinois, USA.

Ramirez, R. A., D. W. Crowder, G. B. Snyder, et al. 2010. Antipredator behavior of Colorado potato beetle larvae differs by instar and attacking predator. *Biological Control* **53**: 230–237.

Rand, T. A. and S. M. Louda. 2004. Exotic weed invasion increases the susceptibility of native plants to attack by a biological control herbivore. *Ecology* **85**: 1548–1554.

Roderick, G., R. Hufbauer, and M. Navajas. 2012. Evolution and biological control. *Evolutionary Applications* **5**: 419–423.

Schulten, G. G. M. 1997. The FAO code of conduct for the import and release of exotic biological control agents. *OEPP/EPPO Bulletin* **27**: 29–36.

Selfridge, J. A., D. Parry, and G. H. Boettner. 2007. Parasitism of barrens buck moth *Hemileuca maia* Drury in early and late successional pine barrens habitats. *Journal of the Lepidopterists Society* **61**: 213–221.

Shafroth, P. B., J. R. Cleverly, T. L. Dudley, et al. 2005. Control of *Tamarix* in the Western United States: Implications for water salvage, wildlife use, and riparian restoration. *Environmental Management* **35**: 231–246.

Skurski, T. C., B. D. Maxwell, and L. J. Rew. 2013. Ecological tradeoffs in non-native plant management. *Biological Conservation* **159**: 292–302.

Simberloff, D. 2012. Risks of biological control for conservation purposes. *Biological Control* **57**: 263–276.

Simberloff, D. and P. Stiling. 1996. How risky is biological control? *Ecology* **77**: 1965–1974.

Simmonds, F. J. and F. D. Bennett. 1966. Biological control of *Opuntia* spp. by *Cactoblastis cactorum* in the Leeward Islands (West Indies). *Entomophaga* **11**: 183–189.

Solter, L. F., D. K. Pilarska, M. L. McManus, et al. 2010. Host specificity testing of microsporidia pathogenic to the gypsy moth, *Lymantria dispar* (L.); Field studies in Slovakia. *Journal of Invertebrate Pathology* **105**: 1–10.

Stiling, P., D. Moon, and D. Gordon. 2004. Endangered cactus restoration: mitigating the non-target effects of a biological control agent (*Cactoblastis cactorum*) in Florida. *Restoration Ecology* **12**: 605–610.

Story, J. M., N. W. Callan, J. G. Corn, and L. J. White. 2006. Decline of spotted knapweed density at two sites in western Montana with large populations of the introduced root weevil, *Cyphocleonus achates* (Fahraeus). *Biological Control* **38**: 227–232

Story, J. M., J. G. Corn, and L. J. White. 2010. Compatibility of seed head biological control agents and mowing for management of spotted knapweed. *Environmental Entomology* **39**: 164–168.

Strong, D. R. 1988. Food web theory: a ladder for picking strawberries? *Ecology* **69**: 1647.

Strong, D. R. and R. W. Pemberton. 2001. Food webs, risks of alien enemies and reform of biological control, pp. 57–79. *In*: Wajnberg, E., J. K. Scott, and P. C. Quimby (eds.). *Evaluating Indirect Ecological Effects of Biological Control*. CABI Publishing. Wallingford, UK.

Suttle, K. B. and M. S. Hoddle. 2006. Engineering enemy free space: an invasive pest that kills its predators. *Biological Invasions* **8**: 639–649.

Syrett, P. 2003. New restraints on biological control, pp. 363–394. *In*: Hallman, G. J. and C. P. Schwalbe (eds.). *Invasive Arthropods in Agriculture – Problems and Solutions*. Science Publishers Inc. Enfield, New Hampshire, USA.

Tack, A. J. M., S. Gripenberg, and T. Roslin. 2011. Can we predict indirect interactions from quantitative food webs? – An experimental approach. *Journal of Animal Ecology* **80**: 108–118.

Tallamy, D. W. 2000. Physiological issues in host range expansion, pp. 11–26. *In*: Van Driesche, R., T. Heard, A. McClay, and R. Reardon (eds.). *Proceedings of Session: Host-Specificity Testing of Exotic Arthropod Biological Control Agents – The Biological Basis for Improvement in Safety*. FHTET 99-1. USDA Forest Service. Morgantown, West Virginia, USA.

Tallamy, D. W. 2004. Do alien plants reduce insect biomass? *Conservation Biology* **18**: 1689–1692.

Tallamy, D. W., M. Ballard, and V. D'Amico. 2010. Can alien plants support generalist insect herbivores? *Biological Invasions* **12**: 2285–2292.

Thomas, M. B. and A. J. Willis. 1998. Biological control – risky but necessary? *Trends in Ecology & Evolution* **13**: 325–329.

Thomas, M. B., P. Casula, and A. Wilby. 2004. Biological control and indirect effects. *Trends in Evolution & Ecology* **19**: 61.

Torchin, M. E., K. D. Lafferty, A. P. Dobson, et al. 2003. Introduced species and their missing parasites. *Nature* **421**: 628–630.

Tothill, J. D., T. H. C. Taylor, and R. W. Paine. 1930. *The Coconut Moth in Fiji: A History of its Control by Means of Parasites*. Imperial Bureau of Entomology. London.

Van Driesche, R. G. 2004. Introduction: predicting host ranges of parasitoids and predacious insects – what are the issues? pp. 1–3. *In*: Van Driesche, R. G. and R. Reardon (eds.). *Assessing Host Ranges for Parasitoids and Predators Used in Classical Biological Control: A Guide to Best Practice*. FHTET-2004-03. USDA Forest Service. Morgantown, West Virginia, USA.

Van Driesche, R. G. and T. J. Murray. 2004. Overview of testing schemes and designs used to estimate host ranges, pp. 68–89. *In*: Van Driesche, R. G. and R. Reardon (eds.). *Assessing Host Ranges for Parasitoids and Predators Used in Classical Biological Control: A Guide to Best Practice*. FHTET-2004-03. USDA Forest Service. Morgantown, West Virginia, USA.

Van Driesche, R. G. and R. Reardon (eds.). *Assessing Host Ranges for Parasitoids and Predators Used in Classical Biological Control: A Guide to Best Practice*. FHTET-2004-03. USDA Forest Service. Morgantown, West Virginia, USA.

Van Driesche, R. G., R. I. Carruthers, T. Center, et al. 2010. Classical biological control for the protection of natural ecosystems. *Biological Control* **54** (Supplement 1): S2–S33.

van Lenteren, J. C., J. Bale, F. Bigler, et al. 2006a. Assessing risks of releasing exotic biological control agents of arthropod pests. *Annual Review of Entomology* **51**: 609–634.

van Lenteren, J. C., M. J. W. Cock, T. S. Hoffmeister, and D. P. A. Sands. 2006b. Host specificity in arthropod biological control, methods for testing and interpretation of data, pp. 38–63. *In*: Bigler, F., D. Babendreier, and U. Kuhlmann (eds.). *Biological Control of Arthropods using Invertebrates: Methods for Environmental Risk Assessment*. CABI Publishing. Wallingford, UK.

van Veen, F. J. F., R. J. Morris, and C. J. Godfray. 2006. Apparent competition, quantitative food webs, and the structure of phytophagous insect communities. *Annual Review of Entomology* **51**: 187–208.

Vasas, V. and F. Jordán. 2006. Topological keystone species in ecological interaction networks: considering link quality and non-trophic effects. *Ecological Modelling* **196**: 365–378.

Veldtman, R., T. F. Lado, A. Botes, et al. 2011. Creating novel food webs on introduced acacias: indirect effects of galling biological control agents. *Diversity and Distributions* **17**: 958–967.

Vieira, C. and G. Q. Romero. 2013. Ecosystem engineers on plants: indirect facilitation of arthropod communities by leaf-rollers at different scales. *Ecology* **94**: 1510–1518.

Vorsino, A. E., A. M. Wieczorek, M. G. Wright, and R. H. Messing. 2012. Using evolutionary tools to facilitate the prediction and prevention of host-based differentiation in biological control: a review and perspective. *Annals of Applied Biology* **160**: 204–216.

Wagner, D. L. and R. G. Van Driesche. 2010. Threats posed to rare or endangered insects by invasions of nonnative species. *Annual Review of Entomology* **55**: 547–568.

Wajnberg, E., J. K. Scott, and P. C. Quimby (eds.). 2001. *Evaluating Indirect Ecological Effects of Biological Control*. CABI Publishing. Wallingford, UK.

Wapshere, A. J. 1974. A strategy for evaluating the safety of organisms for biological weed control. *Annals of Applied Biology* **77**: 201–211.

Wapshere, A. J. 1989. A testing sequence for reducing rejection of potential control agents of weeds. *Annals of Applied Biology* **114**: 515–526.

Weiblen, G. D., C. O. Webb, V. Novotny, et al. 2006. Phylogenetic dispersion of host use in a tropical insect herbivore community. *Ecology* **87**: S62–S75.

Wiggins, G. J., J. F. Grant, P. L. Lambdin, et al. 2010. Host utilization of field caged native and introduced thistle species by *Rhinocyllus conicus*. *Environmental Entomology* **39**: 1858–1865.

Willis, A. J. and J. Memmott. 2005. The potential for indirect effects between a weed, one of its biological control agents, and native herbivores: a food web approach. *Biological Control* **35**: 299–306.

Withers, T. M, L. Barton Browne, and J. Stanley (eds.). 1999. *Host Specificity Testing in Australasia: Towards Improved Assays for Biological Control*. CRC for Tropical Pest Management. Brisbane, Australia.

Withers, T. M. and L. B. Browne. 2004. Behavioral and physiological processes affecting outcomes of host range testing, pp. 40–55. *In*: Van Driesche, R. G. and R. Reardon (eds.). *Assessing Host Ranges for Parasitoids and Predators Used in Classical Biological Control: A Guide to Best Practice*. FHTET-2004-03. USDA Forest Service. Morgantown, West Virginia, USA.

Yasuda, H., E. W. Evans, Y. Kajita, et al. 2004. Asymmetric larval interactions between introduced and indigenous ladybirds in North America. *Oecologia* **141**: 722–731.

Yang, Z-Q., X-Y. Wang, J. R. Gould, and H. Wu. 2008. Host specificity of *Spathius agrili* Yang (Hymenoptera: Braconidae), an important parasitoid of the emerald ash borer. *Biological Control* **47**: 216–221.

Yong, T-H., S. Pitcher, J. Gardner, and M. P. Hoffmann. 2007. Odor specificity testing in the assessment of efficacy of non-target risk for *Trichogramma ostriniae* (Hymenoptera: Trichogrammatidae). *Biological Control Science and Technology* **17**: 135–153.

Yoshioka, A., T. Kadoya, S-I. Suda, and I. Washitani. 2010. Invasion of weeping lovegrass reduces native food and habitat resource for *Eusphingonotus japonicas* (Saussure). *Biological Invasions* **12**: 2789–2796.

Zimmermann, H. G., V. C. Moran, and J. H. Hoffmann. 2001. The renowned cactus moth, *Cactoblastis cactorum* (Lepidoptera: Pyralidae): its natural history and threat to native *Opuntia* floras in Mexico and the United States of America. *Florida Entomologist* **84**: 543–551.

Zwölfer, H. and P. Harris. 1984. Biology and host specificity of *Rhinocyllus conicus* (Fröl.) (Col., Curculionidae), a successful for biological control of the thistle *Carduus nutans* L. *Journal of Applied Entomology (Z. ang. Ent.)* **97**: 36–62.

CHAPTER 8

Measuring and evaluating ecological outcomes of biological control introductions

Bernd Blossey

Department of Natural Resources, Cornell University, USA

Introduction

Did we make a difference? Was use of insects or pathogens for control of problem plants or insects safe and without non-target effects? These are two important questions frequently asked when biological control programs are evaluated (Howarth, 1991; Simberloff and Stiling, 1996a; Pemberton, 2000; Strong and Pemberton, 2000; Louda et al., 2003; Van Driesche et al., 2010; Simberloff, 2012). Providing answers is not only important for scientists to develop theory and improve selection and host-specificity screening procedures, but also for regulatory agencies, land managers, and for the public providing funding. This book advocates for an increased role of biological control in conservation and this chapter describes methods to assess outcomes.

Evaluation of biological control programs is not straightforward. Theory and practice in biological control of arthropods and plants vary substantially from each other and may require fundamentally different assessment protocols. Ecosystems and their networks of relationships and interactions are rarely simple and determination of success depends on data availability and attitudes about what is considered success and acceptable risk. Furthermore, it is essential to have appropriate baselines to determine success or failure.

Non-indigenous species are considered major problems in protected area management (Mack et al., 2000; Simberloff et al., 2013; Foxcroft et al., 2014), but whether these species are always the main drivers of environmental degradation is contested (MacDougall

and Turkington, 2005; Davis et al., 2011). For example, invasion success of some introduced plants is linked to introduced earthworms (Nuzzo et al., 2009) and high native or introduced ungulate (deer, sheep, bovids) populations (Knight et al., 2009; Kalisz et al., 2014), both of which have distinct ecosystem impacts (Bohlen et al., 2004; Côté et al., 2004) (Figure 8.1). Can biological control achieve conservation goals if impacts of other major stressors remain, even if the target is successfully suppressed?

Weed biological control programs achieve partial or complete suppression in roughly a third of programs (Crawley, 1989), but does this track record suggest major success or failure? Even a superficial reading reveals claims of widespread successes by advocates (McFadyen, 1998; van Wilgen et al., 2013) and claims of widespread failures by advocates and critics alike (Louda and Stiling, 2004; Thomas and Reid, 2007; Goldson et al., 2014). Are the few well-publicized examples of direct and indirect non-target effects associated with biological control programs (Howarth, 1991; Louda et al., 1997; Lynch et al., 2001; Elkinton and Boettner, 2012; Suckling and Sforza, 2014) historical legacies of an emerging discipline (Barratt et al., 2006, 2010) or systemic problems continuing today? Do we, after all, need biological control given the inherent risks? Furthermore, the level of risk we are willing to accept apparently varies by control techniques and taxa being targeted (Carson, 1962; Thomas and Willis, 1998; Simberloff, 2012, 2014). I will address these concerns but also refer readers to Chapters 4, 5, 7, 13, and 15 for additional analyses of tradeoffs, risks, ethics, and

Integrating Biological Control into Conservation Practice, First Edition. Edited by Roy G. Van Driesche, Daniel Simberloff, Bernd Blossey, Charlotte Causton, Mark S. Hoddle, David L. Wagner, Christian O. Marks, Kevin M. Heinz, and Keith D. Warner.

© 2016 John Wiley & Sons, Ltd. Published 2016 by John Wiley & Sons, Ltd.

Figure 8.1 Forests in eastern North America have been transformed by excessive browse of white-tailed deer (*Odocoileus virginianus*), eliminating forest understories and tree regeneration (Binghamton University forest preserve in New York State, USA). Photo credit, Bernd Blossey.

philosophical, political, and scientific foundations for future biological control programs.

This chapter is divided into five parts. "Setting the stage" provides a very brief history of biological control referencing important summaries. Biological control is not homogeneous; targeting plants, insects, or pathogens requires distinctly different approaches. Yet critics usually do not make distinctions and lump generalist vertebrates (mongoose, grass carp, etc.) with highly specialized parasitoids or herbivores (Howarth, 1991; Simberloff and Stiling, 1996b). This stymies an important and informed discourse about distinctions and particularities that structure and dictate how different control programs are conducted, and how to reform and improve theory and practice of biological control. I focus on classical biological control (importation of natural enemies from the native to the novel range of non-indigenous species) of plants and insect herbivores using arthropods, which constitute the majority of biological control programs. I will not cover augmentative releases, nematodes, viruses, or other microbial systems, but acknowledge their importance.

The second section, "Defining success in biological control programs," focuses on definitions of success, including definitions that should not be used, and in "Evidence for success or failure" I review what we really know. To evaluate outcomes, records must exist. We are fortunate to have elaborate databases tracking programs targeting insects and plants for over a century. BIOCAT (http://www.cabi.org/projects/project/5273) was developed and maintained to track releases of biological control agents targeting insects. Weed biological control programs and their fate are archived (Winston et al., 2014) and updated by scientists and practitioners from around the world. Both offer rich sources for retrospective analyses.

The section "Population growth rates and demographic modeling" briefly reviews justifications for and key aspects of measuring success in biological control using population growth rates. Biological control programs aim to reduce populations of target species, while maintaining or increasing populations of species considered under threat and those considered non-targets. As such, programs aim to affect population growth rates of species. I propose to use assessments of population growth rates as a core metric in evaluating biological control programs, regardless of whether the target is an insect or a pest plant.

The section "Principles of stewardship" proposes using Public Trust Thinking (Hare and Blossey, 2014) as an ethical and normative grounding of stewardship and conservation. Biological control is one of many approaches we use to manage and define our relationship to other biota. Land managers and biological control scientists do not operate in isolation but are part of a larger social–ecological system that involves stakeholders with different views (often contradictory ones), regulators, policymakers, and the public at large. Trust in decision-making, fairness, and an open, informed, and democratic discourse should be hallmarks of any stewardship approach. This requires collecting and disclosing information about performance of techniques we use to manage species and their habitats, and a willingness to be held accountable. I emphasize that this is not a unique requirement for biological control but should be required of any invasive species management method, (mechanical, physical, chemical, or doing nothing). Persistent calls to conduct pre- and post-release assessments (Blossey, 1999; Delfosse, 2005; Downey, 2014) have not resulted in widespread implementation of such programs (often owing to lack of funding). This suggests a continuation of systemic failures in how we as a society approach conservation and biological control. These failures are not unique to any country. Chapter 15 further elaborates on responsibilities of land managers, land management agencies, scientists and practitioners, regulators, politicians, and the public to enable a bright future for biological control.

Setting the stage

Biological control has a long history, but the modern era started with several major sustained successes. The first targeted the Australian cottony cushion scale, *Icerya purchase* (Maskell), a citrus pest in California by releasing the vedalia beetle, *Rodolia cardinalis* Mulsant, followed later by the control of prickly pear cacti (*Opuntia* spp.) in Australia. These particular successes have been repeated in many countries (DeBach, 1974; Crawley, 1989; Greathead, 1995), and they created a short-lived worldwide euphoria, as many programs targeting other

species failed to achieve similar success. Development of "magic" pesticides after World War II shifted research and implementation priorities away from biological control (DeBach, 1974), but concerns over side-effects (Carson, 1962), emergence of pest resistance, and a general shift towards higher environmental awareness renewed interest in biological control solutions (Harris, 1991). Summary assessments of success rates achieved throughout the history of biological control variously list >150 insects and >40 invasive plants as successfully suppressed in at least one country (Greathead, 1995). Considering just the cost savings and environmental benefits of not using pesticides, the benefits of sustained pest suppression by biological control agents appear enormous and continue to accrue.

However, as with any nascent technology, excitement occasionally overrode better judgment, and in retrospect we can only wonder about the reasons and wisdom of releasing generalist vertebrates (mongoose, cane toads) and predatory snails – now the poster children of non-target effects (Simberloff and Stiling, 1996a). Concerns over potential non-target impacts of biological control agents have been expressed since the inception of programs, but these received scant attention from biological control researchers until recently (Barratt et al., 2006).

Harm, or adverse effect, is a subjective interpretation, context dependent, and changes as societal attitudes shift over time. Past weed and insect biological control programs promoted commercial interests, and success was measured as increased crop harvests, stocking rates, or protected saw timber. Protection of native species was not a priority at that time and resulted in poor safety standards. Today, protection of species with little or no commercial interest has gained widespread attention, resulting in significant reform and improvements, particularly for plants but also for insects (Blossey et al., 2001; Kuhlmann et al., 2006; Hunt et al., 2008).

Considering the diversity of habitats, target pests, and biological control agents, it may not be surprising that no standard definition of success exists. Furthermore, variability in local conditions will lead to a patchwork with failure in one region or habitat and success in another using the same or different agents (Hoddle et al., 2013; Center et al., 2014; Weed and Schwarzländer, 2014), but what constitutes success under these circumstances?

Defining success in biological control programs

Abundance of organisms rises and falls in response to many different factors, most of them subtle and difficult to observe (Tylianakis et al., 2008), except when blatantly obvious (e.g. introduced cats eating flightless birds on islands; tree mortality caused by emerald ash borer, *Agrilus planipennis* Fairmaire). Our inability to use simple observations to recognize mechanisms responsible for changes in species' abundance leaves us vulnerable to guesswork and conflicting opinions. Is the arrival of a non-indigenous species responsible for declines in native species or just coincidence? Is greatly reduced enemy pressure (Keane and Crawley, 2002; Mitchell and Powers, 2003) responsible for the success of an introduced species? If pests decline, was this due to release of biological control agents or to climate change, changes in land use, selection for more resistant genotypes, or host-switching of native pathogens, parasitoids, or predators (Carlsson et al., 2009)? There are no easy answers.

Targeting a species with control efforts, biological or otherwise, assumes the target is the driver of undesirable ecosystem changes. Any management effort, including release of biological control agents, should decrease the competitive ability of problematic species, resulting in population declines and a return to ecologically more desirable conditions. But is it always desirable to return to pre-invasion conditions (Hoddle, 2004)? What if those conditions are not desirable because the targeted species just replaced another undesirable non-indigenous species, or areas were overgrazed or otherwise altered and thus not able to deliver conservation or other benefits?

Management goals are to reduce impacts considered detrimental to species or processes of conservation concern. The origin of a species itself is not a useful criterion and does not equate to negative impacts (Cohen et al., 2012; Martin and Blossey, 2013a). Success in invasive species management can be viewed in terms of biological, ecological, economic, and social/democratic outcomes. I want to emphasize that while this chapter and book are focused on biological control, the principles I outline are applicable to any control method, and accountability to provide measures of success is not uniquely applicable to biological control, but to all management methods.

Biological success

Initial establishment of control agents is an early and essential measure of success, as failure to establish is a widespread problem in both weed and insect biological control programs (Crawley, 1989; Greathead, 1995; McFadyen, 1998; Suckling, 2013). After establishment, agent populations need to grow and affect their hosts. For insect targets this can be measured most directly as mortality (of eggs, larvae, pupae, and adults) and population size (number of individuals per unit area). For plants, direct measures include reductions in survival, recruitment, growth, biomass, seed output, competitive ability, and local abundance. For both insect and plant targets, a desirable but rarely measured effect of control agent release is a reduction in rate of spread (Marchetto et al., 2014). Collectively, these measures can be combined as population growth rates (λ), a fitness measure integrating entire life cycles of organisms over time (Caswell, 2001). Values of $\lambda >1$ indicate increasing populations while values of $\lambda <1$ indicate declining populations.

This method is an essential yet historically largely overlooked tool in both weed and insect biological control, but one with increasing appeal. Clear documentation that control agents are responsible for population declines of insect or plant targets is essential to verify that declines are not caused by other factors, such as climate change or land-use changes. Comparing target population growth rates in the presence and absence of control agent(s) delivers this information. If control agents have become so widespread that presence/absence comparisons are difficult, manipulating access by control agents using mechanical or chemical exclusion can help deduce their effects (DeBach, 1974; McEvoy and Coombs, 1999a).

CONCEPT BOX

Biological success can be measured as (1) initial reduction in population growth rate (λ) of target plants or insects to $\lambda <1$ until populations stabilize at lower abundance with λ fluctuating around 1; (2) a reduction in dispersal rates.

Effects of biological control agents on λ will vary temporally and spatially according to local conditions. Furthermore, biological control does not attempt to eradicate target species. Population growth rates hovering around 1 are therefore acceptable *once target abundance has decreased*. Whether biological success translates into ecological success must be evaluated by assessing potential reductions in impacts on species of conservation concern or, for example in a non-conservation context, agricultural productivity.

Ecological success

Non-indigenous species are considered major conservation problems (Simberloff et al., 2013; Foxcroft et al., 2014), and their simple presence, particularly for plants, is often considered problematic (Denslow and D'Antonio, 2005). The assumption that origin is a surrogate for ecosystem deterioration has come under increasing criticism (Davis et al., 2011; Strayer, 2012). Sophisticated analyses using phylogenetic controls confirm that species should be judged by their traits and impacts, not by their origin (Cohen et al., 2012, 2014; Martin and Blossey, 2013a; Perkins and Nowak, 2013). This recognition does not question or reject the fact that introduced species can be major transformative forces in ecosystems, but native species such as white-tailed deer

(*Odocoileus viginianus*) can have similar transformative effects (Wardle et al., 2001; Côté et al., 2004; Kardol et al., 2014).

The recognition that origin alone does not predict negative ecosystem impacts is extremely relevant when assessing success of biological control programs. Reduction in cover or abundance of a target species *by itself* is not a conservation success (Figure 8.2). We are concerned about *impacts* of non-indigenous species, not their simple presence. Any evaluation concerned with ecological successes of control methods needs to assess outcomes on performance of species assumed to be negatively affected by target non-indigenous species (Figure 8.3). Ideally, more than educated guesses about which species are affected should exist before control programs (biological or otherwise) are initiated.

Stabilization or increases in λ for threatened species

Ideally, we would have obtained quantitative evidence for population declines of native species as a result of the arrival, spread, and abundance increases of

Figure 8.2 Leaf beetles (*Galerucella* spp.), biological control agents released against purple loosestrife (*Lythrum salicaria* L.) often eliminate all annual growth, a biological success (here Montezuma wetlands complex in upstate New York). Only if a diverse native flora and fauna replaces purple loosestrife is this biological success also an ecological success. Photo credit, Bernd Blossey.

Figure 8.3 Showy displays of the native spring ephemeral *Trillium grandiflorum* (Michx.) Salisb. were once common in northeastern US forests. The decline in the species has often been associated with non-native plant invasion, but overabundant native deer are a larger demographic threat. Photo credit, Bernd Blossey.

CONCEPT BOX

Ecological success of biological control programs can be measured as (a) stabilization or increases in λ for species threatened by target plants or insects; OR (b) a shift in ecosystem processes (fire, hydrological regimes, nutrient cycling, erosion, etc.) to more desirable conditions; AND (c) absence of reductions in population growth rates of non-target species to λ ≤1).

---

non-indigenous species. This is true for species at all trophic levels, but we expect more pronounced effects for specialists that depend on particular environments or hosts (specialized native herbivores and their natural enemies). It is these declining species that we should target for demographic assessments to evaluate whether biological success of control agent introductions also provides an ecological success in reversing negative population trends. Unfortunately, we often rely on assumed associations instead of data (Blossey, 1999), making the selection of species to assess somewhat

challenging. Which native species are potentially affected will differ from location to location, and with the identity and functional group of the plant (tree, shrub, herb, grass, etc.) or insect (leaf beetle, bark beetle, open feeder, gall maker, etc.) invader. Effects for herbivorous introduced insects may often be more direct and observable (they may kill plants, and affect species dependent on them), while effects of introduced plants are often indirect (occupancy of space, allelopathy, resource competition, and through shifts in primary producer composition affecting food and shelter for other organisms).

Which species are threatened, and which should be monitored, is affected by logistics and local conservation interests, but principles apply across habitats, taxa, and trophic levels. A comprehensive assessment of all species and their demography is impossible. Instead, we need a careful selection of core taxa based on their recognized ecological importance or conservation status. Selecting both primary producers and species in higher trophic levels is highly recommended. For example, specialized herbivores and their suite of natural enemies are likely more affected than native plants by plant invasions. Suppression of target pests should allow specialized consumers to return or increase as their own host plants or herbivorous insects return, creating an ecological success.

The power of this demographic approach is that effects can be compared over large geographic areas and with different species compositions using the "currency" of population growth rates. But there are important caveats to consider when selecting taxa. For example, despite significant efforts and releases of different control agents (Mausel et al., 2008) hemlock woolly adelgid (*Adelges tsugae* Annand) continues to cause widespread stand declines of hemlock (*Tsuga canadensis* [L.] Carrière) in eastern North America. As hemlock stands decline, other native colonizing species show increased population growth rates. Assuming a potent biological control agent can be found, reduced population growth rates of herbaceous plants and trees currently favored by hemlock declines are expected but should not count as harm to non-target species or an indication of lack of ecological success. Appropriate metrics to select would be the demography of hemlock and hemlock-dependent insects (recruiting trees need decades or longer to mature), followed by birds, or understory plants dependent on the dense shade and cool temperatures of healthy hemlock stands. Furthermore, even declines in *A. tsugae* impact may not result in increased hemlock recruitment at current deer densities (Eschtruth and Battles, 2008), but this should not be counted as a failure of biological control.

A second example illustrates the need for careful species selection when assessing ecological outcomes. Human removal of predatory fish and crustaceans released the herbivorous purple marsh crab *Sesarma reticulatum* (Say) from predatory control, resulting in salt marsh die-back, particularly of cordgrass *Spartina alterniflora* Loisel (Bertness et al., 2014). Invasive *Phragmites australis* (Cav.) Trin. ex Steud., a target of ongoing biological control investigations (Tewksbury et al., 2002), is rapidly expanding in the very same areas. Potential future successful suppression of *P. australis*, a species that may take advantage of ecological opportunities provided by crab herbivory, will not result in recovery of *S. alterniflora* unless the trophic structure changes. Selecting *S. reticulatum* as an indicator of ecological success for biological control of *P. australis* would be a mistake, as would be selecting *S. alterniflora* in areas with high crab predation pressure. But these examples highlight the need for improved ecological and demographic understanding of both impacts of introduced species and efforts to ameliorate them.

Shifts in ecosystem processes

Non-indigenous species may be favored by anthropogenic alterations of habitats (Stromberg et al., 2007), but introduced species themselves can cause changes in fire regimes (Keeley, 2006), soil nutrient dynamics (Ehrenfeld, 2003), hydrology (LeMaitre et al., 1996), erosion patterns (Greenwood and Kuhn, 2014), and other ecosystem processes (Strayer, 2012). Alterations in ecosystem processes considered detrimental to conservation interests should be alleviated by successful biological control if the target species is the driver of such changes. Whether habitat conditions "recover" needs to be assessed over time. Ecosystems may have shifted to alternative stable states (Folke et al., 2004; Schooler et al., 2011), and removing a stressor may not be sufficient to achieve conservation goals. This is not a failure of biological control agent(s) to provide relief but reflects system complexities.

Absence of non-target effects

Non-target effects have been called the Achilles heel of biological control (Louda et al., 2003), and a heated debate has pitted biological control advocates against ecologists concerned over control agent introductions, potential for host shifts and host-range evolution, and ripple effects in complex systems that are impossible to predict (Howarth, 1991; Simberloff and Stiling, 1996b; Thomas and Willis, 1998; van Lenteren et al., 2003; Hoddle, 2004; Delfosse, 2005; Moran et al., 2005; Dudley and Bean, 2012; van Wilgen et al., 2013). There is no space to review non-target effects fully here, but available evidence can be summarized as:

a Biological control, like any other management technique, carries inherent risks requiring careful choice of targets and treatments.

b Weed biological control has an overwhelmingly positive safety record with few (preventable) mistakes made in the past (Suckling and Sforza, 2014).

c Arthropod introductions for control of arthropods have had significant non-target effects on some non-target species due to releases of generalist predators or parasitoids (Boettner et al., 2000; Bigler et al., 2006; Messing and Wright, 2006; Elkinton and Boettner, 2012).

d Claims that introduced biological control agents targeting arthropods have caused widespread extinctions of non-target species on islands have not held up to scrutiny and appear sensationalized (Johnson et al., 2005; King et al., 2010).

e Both proponents and critics argue from a limited dataset, since there are no routine or standardized assessments of non-target effects (Parry, 2009).

f Because absence of evidence is not evidence for absence of non-target effects, developing and routinely using protocols to assess potential non-target effects is paramount to retain credibility and advance the scientific basis of agent selection and host-specificity testing procedures.

Our best approach to prevent non-target effects is selecting specific control agents. Selection and testing procedures to identify specific agents have a long and credible history resulting in significant improvements in weed biological control (Blossey et al., 2001; Suckling and Sforza, 2014), and their use in insect biological control is rapidly increasing (Haye et al., 2005; Kuhlmann et al., 2006; Hunt et al., 2008; Strauss, 2009). The principles underpinning host specificity testing are to safeguard all native species or valuable introduced species that have attained cultural, ornamental, or agricultural significance. In the overwhelming majority of cases, the ecological significance of the word "safeguard" means that wild *populations* are maintained and do not suffer declines caused by biological control agent introductions. Cosmetic damage or even substantial damage to or death of *individuals* does not necessarily indicate demographic or ecological consequences.

> **CONCEPT BOX**
>
> A non-target effect is defined as suppression of population growth rate of a non-target species to <1 as the result of successful establishment of a biological control agent.

I consider it overdue to shift assessment procedures from damage to individuals to damage to populations – a significant change but a biologically meaningful one. I am not concerned (although regulatory agencies seem to be) by host *use* but by effects on non-target *demography*. This shift to protect populations would harmonize biological control with standards typically used, for

example in the US EPA risk assessment framework for pesticides where "screening assessments are concerned with perpetuation of populations of non-target species" (Jones et al., 2004). The listing of endangered species by the IUCN follows identical standards (IUCN, 2012).

Note that demographic effects can occur through direct (attack of a non-target) and indirect pathways (often called food web effects, see Chapter 7). My definition captures both. Food web effects can manifest themselves through shared predators or parasitoids, regardless of whether control agents actually affect the demography of intended targets (Pearson and Callaway, 2003), but are likely a function of control agent abundance. Studying non-target effects of biological control agents in the manner proposed here should be standardized and, in fact, should be applied to any management method – biological, mechanical, chemical, or physical. Demographic data can be collected for every plant or animal species suspected to be affected, eliminating guesswork about impacts from proponents as well as critics of any management approach. Ideally, such protocols should be in place *before* initial introductions or other management occurs to capture rapid demographic effects on target or non-target species.

There are many remaining questions over which species to target, duration, spatial extent of investigations, funding, and developing technical expertise that need to be addressed, but this is beyond the scope of this chapter (see Chapter 15). I am not asking resource managers to design, collect, analyze, and publish demographic information, but they should collaborate and assist specialists.

Assessing potential non-target effects is the ethical thing to do. Failure to collect demographic information leaves biological control scientists and practitioners vulnerable to accusations of inappropriate conduct and being held responsible for population declines or extinctions, whether these accusations are true or not. Eliminating guesswork through data collection will increase trust and provide a foundation to enlighten management and stewardship.

Economic success

Successful biological control projects tend to be forgotten. As a pest plant or insect disappears, so does the knowledge of what it was like before biological control, for example of alligator weed (*Alternanthera philoxeroides* [Mart.] Griseb.) in the southern United States,

Opuntia spp. in Australia, or cassava mealybug (*Phenacoccus manihoti* Matile-Ferrero) in many African countries (Neuenschwander, 2001; Van Driesche et al., 2010; De Clercq et al., 2011). Tabulating the resulting economic benefits of biological control programs is notoriously difficult, and various methods have been proposed (see Chapter 14). These include cost savings as the need for alternative control measures is reduced by weed biological control programs (Hoffmann, 1995), or retention of ecosystem services (de Lange and van Wilgen, 2010).

A full accounting of economic benefits should include benefits derived from maintaining the attractiveness of areas for recreation or tourism, and elimination or cost reductions of restoration programs for endangered species management. These evaluations require different forms of analyses because direct monetary benefits are difficult to establish (how much is a bird in a binocular worth, or a duck as part of a waterfowl hunters recreational pursuits, or the continued existence of rare species?). Social scientists will need to develop new assessment approaches to help collect and deliver this information, which is an essential element of holistic assessments. Then benefits derived from biological control need to be compared to costs and benefits derived from other forms of attempted ecological restoration. I will not advance a definition here but call on social scientists and economists to develop more sophisticated assessment tools to be able to gauge economic success.

Democratic or societal success

A very old and nearly universal recognition and acceptance of stewardship responsibilities exists among peoples of the world, regardless of their different religious and cultural norms (Weiss, 1989). Modern societies have delegated responsibilities to protect diversity of life on Earth to state and federal agencies, yet their failure to achieve sufficient protection (Hooper et al., 2012) gave rise to local, regional, and global nongovernmental organizations (World Wildlife Fund, The Nature Conservancy, etc.). This commitment to conservation has collided with agricultural and commercial interests and the welfare of indigenous people or traditional users. The harshest critics have charged conservation with creating crimes against nature by criminalizing traditional use (Jacoby, 2001) or creating conservation refugees (Cronon, 1996), yet biodiversity conservation and reducing poverty can be complementary, if carried out respectfully and appropriately (Adams et al., 2004). Scientists and managers implementing control programs need to be aware of potential conflicts that may arise when abundant and widespread introduced species have gained utility and may even represent important sources of livelihood (Dickie et al., 2014).

Furthermore, introduced species, particularly those with high abundance and long residence times, often have become valuable habitat for native species (D'Antonio et al., 2004; Schlaepfer et al., 2011). Examples include use of invasive plants by endangered birds, such as the southwestern willow flycatcher (*Empidonax traillii extimus* Phillips) nesting in tamarisk (*Tamarix* spp.) (Hultine et al., 2010) and California clapper rails (*Rallus longirostris obsoletus* Ridgway) using invasive cordgrass *S. alterniflora* (Lampert et al., 2014). While it is clear that these birds, and other organisms, do not depend upon invasive species, how and where control is attempted must be carefully approached and monitored.

Potential conflicts with other stakeholders representing cultural, commercial, or conservation interests make it imperative to develop, implement, and assess biological control programs in an open, informed, democratic, and respectful way.

> **CONCEPT BOX**
>
> Social success can be defined as a measure of both the public's knowledge of and willingness to accept certain risks associated with biological or any other invasive species management option.

This is in essence a question of data availability, data accessibility, and openness of decision-making processes. Except for New Zealand, none of the leading countries in biological control (United States, Canada, South Africa, Australia) or the European Union have a formal public and participatory framework for proposed biological control introductions (Sheppard et al., 2003; Hunt et al., 2008) (see also Chapter 11). There are numerous examples where conflicts have created stalemates for individual programs (Dudley and Bean, 2012) or put biological control as a management technique at risk (Sheppard et al., 2003) owing to withering criticism (Strong and Pemberton, 2000, 2001). Other chapters (see Chapters 11 and 12) further explore the implications of risk-averse agencies and lack of public support.

Biological control needs to honor and reflect ecological and social responsibilities (see Chapter 15 for further development of these ideas). It is difficult to imagine how biological control will be able to survive as a long-term management approach, despite its promise to deliver biological and ecological successes, without a fundamentally revised policy that increases the social acceptance and standing of the discipline.

In summary, biological, ecological, economic but, one hopes, not societal success will vary across the range of the target and with habitats. Complete biological and ecological success is unlikely and refugia for problem species will remain and should not be construed as failure. Using the proposed demographic measures for both biological and ecological success will also deter managers and ecologists from pursuing a misguided desire to reduce targeted plants and insects to extremely low levels. This attitude may be a "legacy" effect of pesticide use, but in fact it is not a requirement to achieve success. Thresholds for acceptable population abundance of target invasive plants and insects will vary among species, temporarily and spatially. Using a common currency of population growth rates will enable comparisons across taxa and regions with different species assemblages.

Criteria not useful to measure success

I have proposed biological, ecological, and economic measures to determine outcomes of biological control introductions. I have also outlined the need for social scientists to develop metrics to assess societal acceptance. I have not mentioned some metrics that I consider less meaningful or potentially misleading, particularly when assessing ecological success. I will briefly justify their rejection.

1 *Diversity metrics (Shannon–Wiener or Simpson index).* Using typical diversity metrics to assess the success of biological control releases is problematic because: (a) increases in diversity could be made up of native but common species, while biological control was initiated to protect rare species of conservation concerns; (b) the assumed benefit of increased primary producer diversity to improve or stabilize ecosystem function and productivity is well supported in green

food webs, but less pronounced in brown food webs structured by decomposition processes (Cardinale et al., 2011; Martin et al., 2014), and even less is known about diversity effects at higher trophic levels. Furthermore, habitats with low plant species diversity can have exceptionally high overall biodiversity or conservation value, including brackish salt marshes dominated by *Spartina* spp. along the Atlantic Coast of North America and *Phragmites*-dominated shorelines in Europe (Tscharntke, 1992; Gratton and Denno, 2006). Indeed few people would likely object if biological control were to create near monocultures of an endangered plant species, such as a rare orchid.

2 *Subsequent community composition.* Using replacement faunal or floral community composition after biological control as a metric of success or failure is problematic because biological control agents are neither gardeners nor restorationists, nor do they have the power to affect seed dispersal or competitive, symbiotic, or mutualistic relationships among the majority of community members at different trophic levels. Instead, biological control agents provide a window of opportunity for species to re-establish through dispersal from refugia, or via human-aided releases, reseeding or replanting in areas previously occupied by pests. Failure of native species to expand as anticipated, or invasion by other introduced species in areas under successful biological control, suggests that other fundamental forces (propagule pressure, increased nutrient deposition, climate change, or overabundance of native or introduced ungulates) may be structuring replacement communities (Wardle et al., 2001; Côté et al., 2004; Ripple and Beschta, 2007; Langley and Megonigal, 2010; Hamann and Aitken, 2013; Travis et al., 2013; Yelenik and D'Antonio, 2013; Fujita et al., 2014).

3 *Changes in food web structure.* Simply assigning ecological success or failure to control programs based on changes in food web structure is misguided at best. These changes are to be expected and no ecological surprise. Any newly arriving species will result in changes to trophic structures of ecosystems (Lau, 2013), including introduction of biological control agents. Some have suggested that population increases of biological control agents may create undesirable ecological effects (Pearson and Callaway, 2003), while others consider increased trophic connectivity

an ecological benefit (Harris, 1988; Blossey, 2003). The devil is in the details, as assessments of introduced predators and parasitoids targeting herbivores in Hawaii have shown (Henneman and Memmott, 2001; Messing and Wright, 2006). How we value changes in food web structure may depend on which species and processes we choose to assess, but we should realize that "no-change" when releasing biological control agents is an ecological impossibility.

4 *Ecological benchmarks*. Defining success as the degree to which invaded communities return to their pre-invasion state (Hoddle, 2004; Van Driesche et al., 2008) is problematic for multiple reasons: (a) pre-invasion communities could have been dominated by other invasive species with low conservation value; (b) invasion may have occurred as part of successional change (Yurkonis and Meiners, 2004); (c) alterations in species composition (including extinctions), climate conditions, nutrient dynamics, and other forces exclude certain species from a specific place; and (d) alternative stable states may exist that prevent systems from returning to conditions resembling pre-invasion conditions (Folke et al., 2004; Schooler et al., 2011). Furthermore, ecological systems accumulate histories and only move forward, with no ability to move backwards (Jackson and Sax, 2010). Land-use legacies, age of continents, and effects of atmospheric circulation patterns are reflected in soil profiles, as is their biological history (Jablonski, 2008; Jackson, 2013). Any attempt to restore conditions of the past is, by definition, going to fail and thus an inappropriate goal for defining ecological success in biological control programs.

Managers and scientists must articulate desirable conservation goals that are informed by conditions of the past (Jackson, 2007; Jackson and Hobbs, 2009; Higgs et al., 2014) without trying to reconstitute communities that may no longer be able to exist or were themselves transitional (Jackson and Williams, 2004; van der Leeuw et al., 2011). Furthermore, succession and constant species re-arrangements are part of ecological and evolutionary landscapes. Using historical benchmarks to guide species and ecosystem management (for example to pre-European settlement conditions in North America) ignores human footprints evident in megafaunal collapses around the world (Gill et al., 2009, 2012) and ignores the dynamic nature of species composition and ecological assembly rules (Jackson, 2007; Jackson and Hobbs, 2009). Climax communities are products of human imagination; communities instead are constantly changing in a world of permanent fluxes (Wallington et al., 2005; Williams and Jackson, 2007).

Evidence for success or failure

Success in biological control can be thought of as an omnibus index containing biological, ecological, economic, and societal components. I now turn to available evidence to assess what we know about success rates.

Evaluation of biological and ecological success

Weed biological control

Historically, quantitative evaluation of weed biological control programs was considered an unnecessary expense by sponsors and biological control scientists alike (McFadyen, 1998; Blossey, 1999). The availability of historical records has greatly facilitated possibilities for global and regional outcome assessments (Crawley, 1989; Greathead, 1995), but defining success has bedeviled the discipline since inception. Claims are subjective and vary among observers, and ironically some of the most successful programs were also the ones most likely to fail (Crawley, 1989).

Hoffmann (1995) defined success as (a) "complete," when no other control method is required or used, at least in areas where the agent(s) is established; (b) "substantial," where other methods are needed but effort is reduced (e.g., less herbicide or less frequent application); and (c) "negligible," where despite damage inflicted by agents, control of the weed still depends on other control measures. McFadyen's (1998) review found 60% of agents establish, and by Hoffmann's criteria 33% provide some form of control of their target weed. Program success rates (by weed species, not by individual control agent) ranged from 50 to 80% depending on country and accounting method. But there were large gaps with no analyses available at the time for Australia, the United States, and Canada (McFadyen, 1998).

Hoffmann's definitions of success are similar, but not identical to, my definition of biological success, which is only one aspect of a successful control program. Reduced effort in targeting a species with other control measures could also indicate replacement of a species with another more competitive one, successional change, or change caused by other stressors (deer abundance, earthworm invasion etc.). Such indirect and observational measures are difficult to assign to release of control agents and therefore difficult to defend as cause and effect relationships. In a conservation context, such simple measures also do not provide information on whether species of conservation concern are doing better, worse, or just the same.

Recognition of problems associated with subjective visual assessments is not a new insight. Crawley (1989) reviewed 627 weed biological control programs and found that few provided objective measures of weed reductions. Pre-release plant abundance data were rarely collected and little follow-up work was carried out. Crawley argued that reliance on expert opinion and subjective evaluation hinders progress to understand, predict, manage, and facilitate weed biological control systems. He advocated quantitative evaluations to determine whether biological control agents are sufficient in suppressing and controlling problem plants. My recommendations to use demographic modeling echo and expand his call and that of others (McEvoy and Coombs, 2000; Downey, 2011) for quantification in biological control.

Since McFadyen's (1998) review, individual studies and meta-analyses of weed biological control programs have appeared with some frequency. Some provide overviews, including success of select programs targeting problem plants in protected areas (Van Driesche et al., 2010; Van Driesche and Center, 2014), others discuss projects targeting Australian acacias in South Africa (Impson et al., 2011) or New Zealand's programs collectively (Fowler et al., 2000). All report agent impacts on target plants but lack quantitative information on replacement communities. Two exceptions exist, both targeting weeds with pathogens (Barton et al., 2007; Meyer and Fourdrigniez, 2011), but assessments are short term.

Several studies have incorporated demographic modeling into evaluating effects of weed biological control agents (Shea and Kelly, 1998; McEvoy and Coombs, 1999a; Buckley et al., 2005; Jongejans et al., 2008; Marchetto et al., 2014). These are effective in documenting biological success (or lack thereof) for particular

biological control agents or programs. A meta-analysis using Australian weed biological control programs found that most (75%) evaluated effects of biological control agents on individual target plant growth and reproduction, but only two evaluated outcomes on other plants (Thomas and Reid, 2007). The authors found no clear articulation of conservation goals owing to limited understanding of impacts on other biota, and none of the reviewed studies provided information on populations of desirable species, not even when problem plants were considered environmental weeds (Thomas and Reid, 2007).

Most likely the best evidence for success in weed biological control programs comes from a meta-analysis of 61 studies published from 2000 to 2011 that quantified control agent impacts on individual plants, target populations, and non-target vegetation (Clewley et al., 2012). Biological control agents significantly affected plant size, plant mass, seed production, and target plant density (30–50% reduction), but despite a significant increase in reported non-target plant diversity, the identity of replacement species and their origin was unclear (Clewley et al., 2012). Effects on target demography are not reported, which prevents assignments of biological success as defined earlier. The increase in the number of studies using quantifications is encouraging, although ecological success remains poorly addressed (Denslow and D'Antonio, 2005; Stiling and Cornelissen, 2005; Clewley et al., 2012).

Insect biological control

The situation is similar in insect biological control program evaluations, and conduct has not improved significantly – at least not in terms of ecological assessments – since the early reviews (DeBach, 1974; Hall et al., 1980; Greathead, 1995; Barratt et al., 2006), individual examples not withstanding (Hoddle et al., 2013). Hall et al. (1980) concluded that introduced predators and parasites ($n = 2300$ in 600 countries) provided complete biological success in 16% ($n = 602$) and some degree of control in approximately 60% of cases. A 1992 review of the BIOCAT database indicated that of 4769 introductions against 543 target pests, 1445 (30%) resulted in establishment and 421 (30% of those with established agents; 9% overall) resulted in good control (Greathead, 1995).

Forest or agricultural pests were typical insect biological control targets in the past, while invasive species causing conservation problems have been targeted only recently (Hoddle, 2004; Van Driesche et al., 2010; De Clercq et al., 2011). Consequently, historical reviews of success focus on crop yields or economic impacts. Van Driesche et al. (2010) and Stiling and Cornelissen (2005) include reviews of insect biological control successes for species of conservation concern. Effect sizes reported by Clewley et al. (2012) are only about half of those reported by Stiling and Cornelissen (2005). These discrepancies are difficult to explain, except by study selection (date of publication, experimental vs. field data, etc.). For example, studies selected by Stiling and Cornelissen (2005) included species such as *Bacillus thuringiensis* Berliner making it difficult to assess effects without re-analyses. I am not aware of any other recent comprehensive evaluation of success rates in classical insect biological control.

In summary, both weed and insect biological control programs continue to be plagued by a lack of rigorous follow-up assessments, particularly for ecological outcomes. I address the apparent systemic failures to deliver such information in Chapter 15. Ironically, one of the first successful weed biological control programs, targeting Klamath weed (*Hypericum perforatum* L.), remains a showcase for post-release monitoring (Huffaker and Kennett, 1959). The program delivered a biological success in suppressing the target (no demographic assessment was done), but replacement plant communities were dominated by introduced grasses. This outcome provided benefits for ranching interests, but conservation interests were not well served. Unfortunately, this program did not set a precedent and quantification remains elusive.

Threats posed by introduced species have been called a crisis (Wilcove et al., 1998; Vila et al., 2011; Simberloff et al., 2013). However, management practices and policies implemented to reduce impacts rarely provide quantitative information about outcomes (Blossey, 1999; Downey, 2011, 2014). This is not unique to biological control but a problem for invasive species management in general (Kettenring and Adams, 2011). Lack of information about outcomes of management interventions on target and non-target species alike has greatly hindered our ability to address the problem adequately; it may even have prevented recognition of underlying fundamental problems (such as overgrazing).

Ecological success: avoidance of non-target effects

In addition to the obvious goals of suppressing target pests and achieving conservation benefits, biological control programs must minimize effects on non-target species. My definition of ecological success and absence of non-target effects allows feeding on individuals, even occasional death of non-target individuals by biological control agents, as long as this attack does not result in population declines of non-target species. Preventing such impacts requires reliable host-specificity testing to determine the fundamental host range in an attempt to predict realized host ranges after agents have been released. In addition, we are concerned about the potential for biological control agents to evolve, expanding both their fundamental and realized host ranges.

Specialization is the adaptation to a limited number of hosts and host tissues and has evolved to optimize performance on a host (Henri and Van Veen, 2011; Barrett and Heil, 2012). Specialization is not an evolutionary "dead end," and phylogenetic analyses show that transitions from specialist to generalist strategies are common, and that host shifts are frequent and involve genomic plasticity and rapid evolution of mechanisms underlying specialization over evolutionary time in certain lineages (Stireman, 2005; Barrett and Heil, 2012). But despite host shifts over evolutionary time, phylogenetic analyses also show herbivore–host and parasitoid–host relationships are closely linked to host phylogenetic relationships (Henri and Van Veen, 2011; Barrett and Heil, 2012; Desneux et al., 2012; Harvey et al., 2012). While there are other traits, such as species size and abundance, that may determine parasitoid host choice in some groups (Cagnolo et al., 2011; Henri and Van Veen, 2011), evolved biochemical interactions involving secondary compounds and immune response of natural enemies with their plant and insect hosts exert extremely strong selection pressures to exploit host resources optimally (Henri and Van Veen, 2011; Barrett and Heil, 2012).

The phylogenetic relationships of parasitoids and herbivores with their hosts were mostly studied using realized host ranges. At least for weed biological control agents we also have abundant information about their fundamental host ranges because of extensive host-specificity testing. Unfortunately, we do not have the same information for insect biological control agents, but this field is rapidly accumulating evidence that can be used in future retrospective analyses.

Published evidence suggests that host-specificity testing in weed biological control correctly predicts host use (realized host range) (Blossey et al., 2001; Suckling and Sforza, 2014). We have no evidence for evolution of fundamental host ranges (Pemberton, 2000; Arnett and Louda, 2002; van Klinken and Edwards, 2002), although we have evidence for evolution of improved performance (McEvoy et al., 2012). We have evidence that host specificity may vary within morphologically identical species (Sheppard et al., 2005; Paynter et al., 2008) requiring careful selection and testing of subpopulations. We have reports of control agents using non-target plants (Davies and Greathead, 1967; McFadyen, 1998; Fowler et al., 2000; McEvoy and Coombs, 2000; Pemberton, 2000; Blossey et al., 2001; Schooler et al., 2003; Paynter et al., 2004; Andreas et al., 2008; Fowler et al., 2012), but no reports of population demographic consequences, except for *Rhinocyllus conicus* (Frölich) and *Cactoblastis cactorum* (Bergroth) (Louda et al., 1997; Pemberton 2000). Despite an increased focus on detecting potential non-target effects, no new serious cases have emerged in the past two decades.

Some of the most successful and safe biological control agents, such as the chrysomelid beetles *Chrysolina hyperici* Forster and *C. quadrigemina* Suffrian, introduced to New Zealand in 1943 and 1965, respectively, would not have been approved under current regulatory climates (Groenteman et al., 2011). Instead of embracing retrospective analyses and the good safety record of weed biological control introductions, an ever increasing reliance on laboratory host-testing results prevents releases of promising and safe control agents (e.g., Smith, 2012; Cristofaro et al., 2013). This trend is worrisome, as it increases herbicide use by land area managers, resulting in detrimental long-term effects (Rinella et al., 2009; Kettenring and Adams, 2011; Skurski et al., 2013). There is an urgent need for improved risk assessment and decision-making procedures (Groenteman et al., 2011; Fowler et al., 2012) (see also Chapter 15) and a focus on population-level impacts and demography, not the occasional damage to individuals.

Considerable debate about the safety and risk of biological control introductions (Follett and Duan, 2000; Lynch et al., 2001; Wajnberg et al., 2001) has resulted in often voluntary, sometimes mandated, changes in host-specificity testing and risk assessment for entomophagous control agents (van Lenteren et al.,

2003; Haye et al., 2005; Bigler et al., 2006; Kuhlmann et al., 2006; Hunt et al., 2008; van Lenteren et al., 2008; Barratt et al., 2010). A review revealed 87 known cases of non-target effects, of which 17 apparently involved population reductions in non-target species (Lynch and Thomas, 2000). Despite over 5000 introductions of more than 2000 species of arthropod biological control agents over the last 100 years, there are few reports of negative non-target effects except for polyphagous control agents (Boettner et al., 2000; Lynch and Thomas, 2000; Barratt et al., 2006; Elkinton et al., 2006; van Lenteren et al., 2006; Barratt et al., 2010; De Clercq et al., 2011). Post-release studies to evaluate presumed non-target species effects found minor effects of the implicated biological control agents (Johnson et al., 2005; Messing and Wright, 2006; King et al., 2010), contrary to the view that negative consequences of biological control agent releases were widespread.

We have no indication for evolution of host specificity in entomophagous biological control agents, although this is more difficult to investigate than for weed biological control agents, where details about the fundamental host range exist. It therefore appears that concerns over weakening of host–biological control agent relationships may be unfounded. Strong co-evolutionary relationships, or new host associations, evolve over long time periods (thousands or millions of years); we are unable to detect such processes over the 100 years of biological control history. But host–natural enemy re-arrangements are part of the evolutionary theater that has shaped present communities and their interactions (Janz, 2011; Lankau, 2011; Forister et al., 2012). Over these long time periods, both native and accidentally or purposefully introduced species are expected to have the same overall probabilities for host shifts, although rapid evolution in response to introductions and new environmental conditions are likely (Blossey and Nötzold, 1995; Mitchell et al., 2006; Strauss et al., 2006, 2008; Lankau, 2011; Agrawal et al., 2012).

We need to acknowledge that biological control, like every other management technique, is not risk free and that ecological surprises will occur. Any population-level damage that was not expected must be carefully investigated and may result in changes in assessment procedures. We also need to acknowledge that biological control organisms may spread far from their release sites (Petit et al., 2009; Pratt and Center, 2012).

The reality of long-distance dispersal highlights the importance of host-specificity testing to ensure restricted diet breadth of agents.

Evidence for economic success

Economic analyses show enormous benefits, greatly exceeding program costs, often reaching millions of US dollars in annual savings for individual programs alone (Greathead, 1995; Neuenschwander, 2001; Zeddies et al., 2001). Benefit: cost ratios in weed biological control range from 1 : 1 to 112 : 1 (Coombs et al., 2004; Suckling, 2013), and benefits continue to accrue (where they exist). A recent evaluation focusing on delivery of eco-system services (water, biodiversity, and grazing) in South Africa estimated the annual value protected by biological control agents at nearly US$ 20 billion with benefit: cost ratios from 50 : 1 for invasive subtropical shrubs to 3726 : 1 for invasive Australian trees (de Lange and van Wilgen, 2010). I am not aware of any sophisti-cated economic assessments of biological control pro-grams targeting insect pests of conservation concern. Insect biological control has traditionally focused on pests of agriculture or forestry, where programs have produced enormous economic benefits (see also Chapter 14).

Data limitations and lack of methods prevent more sophisticated economic evaluation of biological control programs for conservation purposes, and this field will need development. Social scientists will need to develop new assessment approaches to help collect and deliver this information. Such assessments, once routinely con-ducted for all management techniques (biological, chemical, physical or mechanical), are likely to favor biological control. Other methods often have unrecog-nized detrimental effects and often fail to achieve ecolog-ical success despite enormous expenditure (Kettenring and Adams, 2011; Martin and Blossey, 2013b).

Evidence for societal success

Biological control may wither and die unless public support increases. Yet we have no comprehensive information about acceptance of biological control and potential non-target impacts among land managers and the general public except for individual case studies. For example, wetland managers in the United States, with very few exceptions, would support biological control of introduced *P. australis* as long as no negative population-level consequences occur for the endemic subspecies *Phragmites australis americanus*; yet some would even

accept population-level impact if spread and abundance of invasive genotypes could be reduced (Martin and Blossey, 2013b). Surveys in New Zealand demonstrate that acceptance of biological (and other) control tech-nologies is a function of the demonstrated need for the technology, host specificity, risks to humans and the environment, fairness regarding distribution of costs and benefits, effects on future generations, and credibility of experts (Wilkinson and Fitzgerald, 1997).

While science is unable to provide the fail-proof assurances that some members of the public and regulatory agencies may be looking for, the issue is not whether public concerns are legitimate, rational consid-erations, but how to integrate them into risk analyses and policy decisions. As in any other pest management technology, inherent risks must be acknowledged and openly discussed, but there never will be unanimous support. One person's weed is another person's orna-mental plant, medicinal plant, or food. One person's dreaded pest is another person's biological control agent (Paine et al., 2010). These conflicts are as old as the dis-cipline, often pitting agricultural and horticultural against conservation interests (Barratt et al., 2006).

At issue are deeply rooted questions of fairness (Binmore, 2014). With the exception of New Zealand, no country using classical biological control has a public review process that can be called open, informed, and democratic (Hunt et al., 2008). A decade ago, the decision-making process regarding biological control in the United States was described as "shrouded in mys-tery; at worst, it appears designed to prevent scientific and policy debate of important public issues. It does not invite, and positively prevents, the active discourse and full implementation of the nation's scientific expertise in facilitating the study and evaluation of proposed and approved biological control agents" (Miller and Aplet, 2005). This continued secrecy in decision-making con-tributes to distrust of agencies charged with doing the public's business (Wilkinson and Fitzgerald, 1997). Yet, at the same time, contemporary research shows that when honest attempts are made to provide public information and education programs, acceptance of management needs and approaches increases, and con-structive cooperation is more likely (Wilkinson and Fitzgerald, 1997; ERMA, 2010; Johnson and Horowitz, 2014). Biological control, as any other conservation effort, is rooted not just in ecology but in a social–ecological system (Calhoun et al., 2014).

In summary, we continue to lack meaningful assessments of biological, ecological, economic, and social success of biological control. This critique does not question claims of enormous benefits derived from biological control programs – they no doubt exist (Van Driesche et al., 2010). But lack of clearly documented biological, ecological, and economic success makes biological control vulnerable to criticism and prevents this method from receiving appropriate societal backing. Continued lack of follow-up work suggests a fundamental structural problem. These structural problems are not unique to any particular country, regardless of whether biological control laws exist (Australia), regulations were revised (New Zealand), or approvals and denials of biological control agents continue to be shrouded in secrecy (United States).

Willingness to improve procedures and engage in an open decision-making process by biological control scientists and practitioners is not sufficient alone; it will require an enabling environment (including regulation and financing) that assigns appropriate responsibilities to stakeholders involved in research, implementation, or regulation, as well as the public (see Chapter 15). An important insight is the recognition that responsibility for biological control programs should *not* rest with biological control scientists or practitioners, but with federal and state agencies responsible for stewardship of natural resources.

I will now return to two core concepts that I propose as cornerstones for the reform and upgrading of biological control activities, regardless of whether they target insects or plants. While this book is about biological control, the concepts I articulate below are applicable to all invasive species management. The first is an ecological grounding using population growth rates and demographic modeling (Caswell, 2001) to assess success (biological and ecological). The second is an ethical or normative grounding of conservation and invasive species management using Public Trust Thinking (Hare and Blossey, 2014).

Population growth rates and demographic modeling

Classical biological control of insects and plants aims to affect *populations and abundance* of organisms. In fact, any management strategy (biological, physical,

mechanical, or chemical) aims to change population performance, that is reduce pest populations, increase native species, and avoid population-level non-target effects, and all can and should be evaluated using the same techniques (Figure 8.4). My recommendations are not specific to biological control but raise the bar for all management and for what legitimately can be claimed a success (hectares treated and number of individuals killed is neither sufficient nor useful).

The tools to assess population trajectories appropriately, under various conditions, including the presence or absence of biological control agents of interest, focus on the assessment of survival, mortality, and performance of individuals. Such data are then integrated through the development of demographic models. In the Preface to the second edition of his book on matrix population models, Hal Caswell appropriately wrote:

Figure 8.4 Permanently marked (note numbered metal tag) *Trillium grandiflorum* used to assess effects of deer browse on plant demography. Photo credit, Bernd Blossey.

"Matrix population models – carefully constructed, correctly analyzed, and properly interpreted – provide a theoretical basis for population management. They raise the bar of what constitutes rigorous analysis in conservation biology. The work may appear daunting but population managers need and deserve the sharpest analytical tools available. Their work is too important to settle for less" (Caswell, 2001). The following is not a comprehensive review; technical background information is available elsewhere (Caswell, 2001; Williams et al., 2001; Jongejans et al., 2008; Caswell et al., 2011; Caswell and Salguero-Gomez, 2013).

Development of demographic models typically requires estimating transition probabilities from one life stage to another (called the vital rates). Each model is tailored to the organism in question, reflecting its life history. For plants with clear developmental stages, a stage-based model requires estimation of transition rates from seed, to seedling, to rosette to flowering plant, to seed output, back to seed and seed bank (Shea and Kelly, 1998; Davis et al., 2006). For insects, we measure transitions from eggs to various larval or nymphal stages, pupal survival, overwinter survival, and adult fecundity. For birds or mammals, age-based models are used to capture transitions from birth through juvenile periods to reproductive adult (Caswell, 2001).

Survival of different stages and the probability that an individual will transition from one stage to the next (or produce propagules such as eggs, seeds or babies) will be affected by local biotic and abiotic conditions. Transition probabilities can then be inferred by monitoring marked individuals and their fate, or through experiments under different conditions, for example by manipulating predator access by caging individuals, or manipulating presence/absence of browsers or grazers along altitudinal gradients (Miller et al., 2009), or comparing native and introduced ranges (Shea et al., 2005). Constructing and populating models with data, and analyzing model performance under different scenarios (often referred to as perturbation analysis), allows comparisons of contributions made by different vital rates (probability of transitioning from one stage to another) for overall population growth rates (Caswell, 2000).

Most management problems involve changing vital rates: we assume that introduced species affect vital rates of native species, decreasing fecundity, juvenile survival, and so on through direct (predation) or indirect (shading, allelopathy, resource competition) pathways. We hope to diagnose correctly the causes of population declines, but what is important is not the diagnosis but the remedy we apply (Nichols and Williams, 2006). Biological control of forest invaders may not be the right remedy if overgrazing by deer facilitates plant invasions and threatens native forest understories (Kalisz et al., 2014) – bows and guns are more appropriate. Data collection informed by understanding of the importance of multiple stressors and demographic models will deliver this information.

Demography and matrix population models have been used for decades (Caswell, 2001) and they are now a common tool in population biology (Crone et al., 2011; Caswell and Salguero-Gomez, 2013). Matrix models are used in invasive plant research and in weed biological control (Shea and Kelly, 1998; McEvoy and Coombs, 1999b; Buckley et al., 2005; Davis et al., 2006; DeWalt, 2006; Carvalheiro et al., 2008; Jongejans et al., 2008; Schutzenhofer et al., 2009; Eckberg et al., 2014). Collaborative efforts can aid in data collection and can be an important tool to inform agent selection (Shea and Kelly, 1998; Davis et al., 2006; DeWalt, 2006).

Arguments that such studies are impossible to fund and not a practical solution (Harris, 1991) ignore the enormous waste of resources poured into unsuccessful agents. For example, use of a matrix population modeling approach could have avoided the pointed debate and waste of monetary and scientific resources regarding the number of seed feeders necessary to control knapweeds, *Centaurea stoebe* (Gugler) Hayek and *Centaurea diffusa* Lam., species that are not seed limited in their introduced range. Such knowledge might have prevented non-target effects of the imported seed-head flies on deer mice (Pearson and Callaway, 2003).

Demographic assessments show that impacts of herbivores vary and are context-specific (Miller et al., 2009; Ramula, 2014), often explaining variable success of control agent(s) (Shea and Kelly, 1998; Shea et al., 2005). Such detailed and informed assessments should replace the rather poor and subjective assessment of success using single versus multiple agents in weed biological control (Denoth and Myers, 2005) that does not hold up to scrutiny (Stephens et al., 2013).

Use of demographic models to assess the role of insect biological control agents on their target hosts has a long history, typically referred to as life table analyses (Bellows et al., 1992; Duan et al., 2014). Constraints in these models have long been recognized (Vickery, 1991; Royama, 1996), and the superior analytical capabilities of new population models have largely replaced traditional life table analyses in the recent literature, and these new models themselves are gaining sophistication (de Valpine et al., 2014). In his classic book, DeBach (1974) makes reference to periodic censuses and life tables, but he recommends against such methods on the grounds that they cannot prove cause and effect, and instead he advocates for experimental assessments of control agent effectiveness using exclusion (see Chapter 9). That these two methods do not need to live separate lives has been elegantly shown for tansy ragwort, *Senecio jacobaea* L., biological control (McEvoy and Coombs, 1999b). The reason for the lack of such work in the recent literature is not entirely clear to me, but use of demographic assessments could have prevented the ill-advised addition of more and more coccinellid predators to control pest aphids (Caltagirone, 1989; Obrycki and Kring, 1998; Rosenheim, 1998; Obrycki et al., 2000), or releases of unstudied, often taxonomically unknown or undefined, species of parasitoids, predators, and fungi, with the vast majority not establishing and none providing satisfactory control (Strong and Pemberton, 2001; Havill et al., 2012; Stehn et al., 2013).

Appropriate monitoring and use of modern demographic population models will allow us to assess ecological success using quantifiable information about performance of species of conservation interest. These species triggered initial biological control efforts, and by focusing on their demography we obtain sophisticated information about the levels of pest suppression needed to maintain or increase populations of species considered under threat. Eradication of widespread species, which are typical biological control targets, is impossible and not required to protect native species; control programs should focus on achieving abundance reductions of pests to reduce or eliminate negative impacts. We need such data to make better management decisions. Without such information, land managers and biological control scientists continue to evade questions about whether their efforts actually made a difference.

The use of demographic models may appear daunting (Bellows et al., 1992; Caswell, 2001). I am not suggesting that all land managers and biological control scientists become experts in matrix population models, but I encourage collaboration with specialists. What I consider most important, and under the control of land managers and biological control scientists, is to (1) recognize the importance and utility of demographic models in achieving conservation and management goals, (2) overcome initial hesitancy for something new and seemingly difficult, and (3) embrace opportunities for new collective learning and collaborative approaches. The long-term nature of biological control programs makes them extremely appealing for demographic modelers since they avoid the errors and lack of accuracy that can plague the forecasting ability of models based on short-term data collection (Ellner and Rees, 2007; Crone et al., 2011).

Society is entitled to have questions regarding performance of invasive species management methods answered in order to make informed decisions about priorities, risk acceptance, and funding. Scientists need to enable managers to collect this information by providing appropriate assessment protocols, and society needs to enable these approaches by requiring accountability and funding (see Chapter 15). Conservation and land stewardship of the future can best be described as holistic management through correctly identifying and then managing stressors according to their importance. Land stewardship and conservation are only as good as the land managers' toolbox – correct diagnostic and management tools are essential, including sophisticated demographic models.

Principles of stewardship

Throughout human history, our societies, our values, and our interpretation of our relationship to and responsibility for each other and the environment have been expressed in many different ways (Pungetti et al., 2012). While it is evident that these relationships and their articulations have undergone dramatic transformations, it is also clear that principles underpinning what we now call *stewardship* are as old as human history. Yet how we define our relationships to other species and the environments we share (de Groot et al., 2011; Blossey, 2012) will determine how we choose to address climate change, pollution, endangered species, biological invasions, and ecosystem management.

The last few decades have seen a resurgence of Public Trust Thinking (PTT) (Hare and Blossey, 2014), most noticeably articulated in the Public Trust Doctrine (PTD) that has been expanded and advanced within environmental law (Sax, 1970; Blumm and Wood, 2013). At its core, PTT recognizes the intimate interdependence of people and ecosystems, and seeks to ensure that benefits we derive from ecosystems are available to everyone, including future generations (Weiss, 1989). Readers interested in exploring these ideas in depth can start with Hare and Blossey (2014) and the references therein to cover ideas, implications, and arguments for wider adoption of PTT in natural resource management.

Philosophically, PTT is appealing because it provides a framework for structuring relationships among citizens, their elected governments, and natural resources and the services they provide (Sagarin and Turnipseed, 2012). PTT asserts that a group of elected or appointed trustees holds certain natural resources in trust in the interests of the beneficiaries – all current and future citizens (Turnipseed et al., 2009; Sagarin and Turnipseed, 2012; Blumm and Wood, 2013). Trustees are charged with overseeing trust resources in a manner that ensures long-term viability and does not privilege any individuals, groups, or uses (Sax, 1970). Trustees turn these management obligations over to trust managers (Smith, 2011; Decker et al., 2014). In turn, beneficiaries (the public) are entitled to hold trustees to account if they are deemed to be in abrogation of their obligations (Sax, 1970; Blumm and Wood, 2013).

PTT has inspired laws, policies, and environmental ethics around the world that share fundamental normative aspirations but differ in scope, antecedents, and institutional arrangements (Takacs, 2008; Blumm and Guthrie, 2012). PTT applications range from water to wildlife to the atmosphere, are located in different elements of the law, and apply at different levels of governance in different countries.

In the United States the National Environmental Policy Act (NEPA) of 1969 is a strong articulation of PTT, broadly asserting the purpose of the statute and establishing the authority and responsibility of the federal government. This landmark piece of legislation mandates federal agencies to consider environmental impacts of any major decision they make, including the release of biological control agents. NEPA asserts that it is the duty of government to "fulfill the responsibilities of each generation as trustee of the environment for succeeding generations" (§ 4331(b) (1)), a clear codification of PTT, and similar language can be found in the Clean Water Act (Sagarin and Turnipseed, 2012). India's public trust arrangements are derived from the constitutional right to a healthy environment, extend to all natural resources, and apply to the Union Government (Takacs, 2008; Blumm and Guthrie, 2012). Modern constitutions such as that of Ecuador – where articulations of Buen Vivir are influenced by the Andean philosophy of *sumak kawsay* (Gudynas, 2009) – advance ideas central to PTT.

What is the relationship between PTT and the topic of this book? PTT is not a single, neatly demarcated set of ideas; rather, it is an orientation towards natural resource governance that uses concepts of trusteeship to frame human stewardship. PTT clearly outlines the responsibilities of those charged with overseeing the trust and those managing public trust resources, and it articulates the rights of beneficiaries. Unleashing biological control for the protection of biodiversity involves enormous responsibilities. It is reassuring that review after review finds that biological control using herbivorous insects and pathogens is a mature and responsible enterprise and the policy mistakes of the past introductions of polyphagous species such as *R. conicus* and *C. cactorum* can easily be avoided (Blossey et al., 2001; Suckling and Sforza, 2014). What is less reassuring is that in the vast majority of the weed and insect biological control cases, we have little if any substantial information on ecological benefits and food web effects. Thus, biological control programs, as presently implemented, fall short of PTT principles and our societal responsibilities.

These shortcomings constitute institutional failures and lack of recognition of various responsibilities of all involved, not just biological control scientists. Resource management agencies (federal, state, local, and environmental NGOs) rarely attempt to provide accurate and timely information to beneficiaries about the status of most of the public trust resources they are managing (Jepson, 2005). As a society, we are left with little evidence to judge whether trustees or their professional managers are fulfilling their obligations of protection and stewardship (Parry, 2009). The usual reasons provided for the lack of this information are resource shortages (staff, funding, expertise), yet a more cynical interpretation argues that institutional arrogance and a lack of recognition of responsibilities under PTT obligations (Smith, 2011) ensures that no fundamental critique of the status quo or operational priorities can be articulated by outsiders.

A fundamental change is in order – for all stewardship and conservation involving invasive species management.

Appropriate use of ecologically sophisticated methods and accountability (Jepson, 2005) should be the hallmarks of biological control. PTT can guide development of appropriate structural transformations to achieve these goals (see Chapter 15 for details).

Conclusions

All authors in this book agree that judicious and careful use of biological control is an appropriate, yet underutilized, form of environmental stewardship. Biological control, when successful, is an enormously potent and powerful tool; judicious use is not only prudent but morally required. Other forms of weed control frequently, if not usually, fail to provide long-term suppression of target species. Biological weed control, by contrast, provides, at least in some projects, long-lasting biological success.

Conflicts arise when opinions about the right things to do differ with conflicting stakeholder interests or belief systems. Furthermore, absence of information about risks and outcomes prevents an informed and democratic discourse about the best approaches. Weed and insect biological control programs have fallen short in their duty to deliver appropriate and timely information about program risks and outcomes to the public, and agencies have traditionally failed to engage in open, informed, and fair approaches to decision-making. Placing biological control into a holistic risk-assessment framework that evaluates chemical, mechanical, and physical management alternatives should elevate biological control to a prime method of choice. The track records of alternatives are unimpressive as they fail to provide long-lasting benefits, except when early stages of an invasion, individuals, or very small populations are targeted (Keeley, 2006; Pearson and Callaway, 2008; Rinella et al., 2009; Kettenring and Adams, 2011; Louhaichi et al., 2012; Skurski et al., 2013), while non-target impacts of many pesticides can be widespread and long lasting (Muir and de Wit, 2010).

Secrecy and absence of accountability in natural resource management are not unique to biological control (Smith, 2011), but they constitute impediments to safeguarding a livable environment for native species threatened by different stressors – introduced plants and insects among them. Using readily available and sophisticated methods to assess the importance of different

threats, as well as success or failure of management methods to protect species and ecosystems we care about, is in the best interest of all of us. We should charge the brightest minds with helping us protect and manage our ecosystems. All of us, and future generations, deserve this care and consideration.

Acknowledgments

I thank Andrea Dávalos, Victoria Nuzzo, Darragh Hare, Roy Van Driesche, Mark Hoodle, Dan Simberloff, Andy Shepard, Kevin Heinz, Christian Marks, Dave Wagner, Peter McEvoy, Richard Casagrande, Hariet Hinz, and S. Raghu for reviews or comments on Chapter 8.

References

Adams, W. M., R. Aveling, D. Briockington, et al. 2004. Biodiversity conservation and the eradication of poverty. *Science* **306**: 1146–1149.

Agrawal, A. A., A. P. Hastings, M. T. J. Johnson, et al. 2012. Insect herbivores drive real-time ecological and evolutionary change in plant populations. *Science* **338**: 113–116.

Andreas, J. E., M. Schwarzländer, and R. de Clerck-Floate. 2008. The occurrence and potential relevance of post-release, nontarget attack by *Mogulones cruciger*, a biological control agent for *Cynoglossum officinale* in Canada. *Biological Control* **46**: 304–311.

Arnett, A. E. and S. M. Louda. 2002. Re-test of *Rhinocyllus conicus* host specificity, and the prediction of ecological risk in biological control. *Biological Conservation* **106**: 251–257.

Barratt, B. I. P., B. Blossey, and H. M. T. Hokkanen. 2006. Post-release evaluation of non-target effects of biological control agents. pp. 166–186. *In*: Bigler, F., D. Babendreier, and U. Kuhlmann (eds.). *Environmental Impact of Invertebrates for Biological Control of Arthropods, Methods and Risk Assessment*. CABI Publishing. Cambridge, Massachusetts, USA.

Barratt, B. I. P., F. G. Howarth, T. M. Withers, et al. 2010. Progress in risk assessment for classical biological control. *Biological Control* **52**: 245–254.

Barrett, L. G. and M. Heil. 2012. Unifying concepts and mechanisms in the specificity of plant-enemy interactions. *Trends in Plant Science* **17**: 282–292.

Barton, J., S. V. Fowler, A. F. Gianotti, et al. 2007. Successful biological control of mist flower (*Ageratina riparia*) in New Zealand: agent establishment, impact and benefits to the native flora. *Biological Control* **40**: 370–385.

Bellows, T. S., Jr., R. G. Van Driesche, and J. S. Elkinton. 1992. Life-table construction and analysis in the evaluation of natural enemies. *Annual Review of Entomology* **37**: 587–614.

Bertness, M. D., C. P. Brisson, T. C. Coverdale, et al. 2014. Experimental predator removal causes rapid salt marsh die-off. *Ecology Letters* **17**: 830–835.

Bigler, F., D. Babendreier, and U. Kuhlmann. 2006. *Environmental Impact of Invertebrates for Biological Control of Arthropods, Methods and Risk Assessment*. CABI Publishing. Cambridge, Massachusetts, USA.

Binmore, K. 2014. Bargaining and fairness. *Proceedings of the National Academy of Sciences of the United States of America* **111**: 10785–10788.

Blossey, B. and R. Nötzold. 1995. Evolution of increased competitive ability in invasive nonindigenous plants: a hypothesis. *Journal of Ecology* **83**: 887–889.

Blossey, B. 1999. Before, during, and after: the need for long-term monitoring in invasive plant species management. *Biological Invasions* **1**: 301–311.

Blossey, B. 2003. A framework for evaluating potential ecological effects of implementing biological control of *Phragmites australis*. *Estuaries* **26**: 607–617.

Blossey, B. 2012. The value of nature. *Frontiers in Ecology and the Environment* **10**: 171.

Blossey, B., R. Casagrande, L., Tewksbury, et al. 2001. Nontarget feeding of leaf-beetles introduced to control purple loosestrife (*Lythrum salicaria* L.). *Natural Areas Journal* **21**: 368–377.

Blumm, M. and R. Guthrie. 2012. Internationalizing the public trust doctrine: natural law and constitutional and statutory approaches to fulfilling the Saxion vision. *UC Davis Law Review* **45**: 741–808.

Blumm, M. C. and M. C. Wood. 2013. *The Public Trust Doctrine in Environmental and Natural Resources Law*. Carolina Academic Press. Durham, North Carolina, USA.

Boettner, G. H., J. S. Elkinton, and C. J. Boettner. 2000. Effects of a biological control introduction on three nontarget native species of saturniid moths. *Conservation Biology* **14**: 1798–1806.

Bohlen, P. J., P. M. Groffman, T. J. Fahey, et al. 2004. Ecosystem consequences of exotic earthworm invasion of north temperate forests. *Ecosystems* **7**: 1–12.

Buckley, Y. M., M. Rees, A. W. Sheppard, and M. J. Smyth. 2005. Stable coexistence of an invasive plant and biological control agent: a parameterized coupled plant–herbivore model. *Journal of Applied Ecology* **42**: 70–79.

Cagnolo, L., A. Salvo, and G. Valladares. 2011. Network topology: patterns and mechanisms in plant–herbivore and host–parasitoid food webs. *Journal of Animal Ecology* **80**: 342–351.

Calhoun, A. J. K., J. S. Jansujwicz, K. P. Bell, and M. L. Hunter, Jr. 2014. Improving management of small natural features on private lands by negotiating the science-policy boundary for Maine vernal pools. *Proceedings of the National Academy of Sciences of the United States of America* **111**(30): 11002–11006. doi: 10.1073/pnas.1323606111.

Caltagirone, L. E. 1989. The history of the vedalia beetle importation to California and its impact on the development of biological control. *Annual Review of Entomology* **34**: 1–16.

Cardinale, B. J., K. L. Matulich, D. U. Hooper, et al. 2011. The functional role of producer diversity in ecosystems. *American Journal of Botany* **98**: 572–592.

Carlsson, N. O. L., O. Sarnelle, and D. L. Strayer. 2009. Native predators and exotic prey – an acquired taste? *Frontiers in Ecology and the Environment* **7**: 525–532.

Carson, R. 1962. *Silent Spring*. (2002, 40th anniversary edition.) Houghton Mifflin Company. New York.

Carvalheiro, L. G., Y. M. Buckley, R. Ventim, et al. 2008. Apparent competition can compromise the safety of highly specific biological control agents. *Ecology Letters* **11**: 690–700.

Caswell, H. 2000. Prospective and retrospective perturbation analyses: their roles in conservation biology. *Ecology* **81**: 619–627.

Caswell, H. 2001. *Matrix Population Models: Construction, Analysis, and Interpretation*. Sinauer Associates. Sunderland, Massachusetts, USA.

Caswell, H. and R. Salguero-Gomez. 2013. Age, stage and senescence in plants. *Journal of Ecology* **101**: 585–595.

Caswell, H., M. G. Neubert, and C. M. Hunter. 2011. Demography and dispersal: invasion speeds and sensitivity analysis in periodic and stochastic environments. *Theoretical Ecology* **4**: 407-421.

Center, T. D., F. A. Dray, E. D. Mattison, et al. 2014. Bottom-up effects on top-down regulation of a floating aquatic plant by two weevil species: the context-specific nature of biological control. *Journal of Applied Ecology* **51**: 814–824.

Clewley, G. D., R. Eschen, R. H. Shaw, and D. J. Wright. 2012. The effectiveness of classical biological control of invasive plants. *Journal of Applied Ecology* **49**: 1287–1295.

Cohen, J. S., J. C. Maerz, and B. Blossey. 2012. Traits, not origin, explain impacts of plants on larval amphibians. *Ecological Appplications* **22**: 218–228.

Cohen, J. S., S. K. D. Rainford, and B. Blossey. 2014. Community-weighted mean functional effect traits determine larval amphibian responses to litter mixtures. *Oecologia* **174**: 1359–1366.

Coombs, E., H. Radtke, and T. Nordblom. 2004. Economic benefits of biological control, pp 122–126. *In*: Coombs, E., J. Clark, G. L. Piper, and A. F. Cofrancesco, Jr. (eds.). *Biological Control of Invasive Plants in the United States*. Oregon State University Press. Corvallis, Oregon, USA.

Côté, S. D., T. P. Rooney, J.-P. Tremblay, et al. 2004. Ecological impacts of deer overabundance. *Annual Review of Ecology and Systematics* **35**: 113–147.

Crawley, M. J. 1989. The successes and failures of weed biological control using insects. *Biological Control News and Information* **10**: 213–223.

Cristofaro, M., A. De Biase, and L. Smith. 2013. Field release of a prospective biological control agent of weeds, *Ceratapion basicorne*, to evaluate potential risk to a nontarget crop. *Biological Control* **64**: 305–314.

Crone, E. E., E. S. Menges, M. M. Ellis, et al. 2011. How do plant ecologists use matrix population models? *Ecology Letters* **14**: 1–8.

Cronon, W. 1996. The trouble with wilderness or, getting back to the wrong nature. *Environmental History* **1**: 7–28.

D'Antonio, C. M., N. E. Jackson, C. C. Horvitz, and R. Hedberg. 2004. Invasive plants in wildland ecosystems: merging the study of invasion processes with management needs. *Frontiers in Ecology and the Environment* **2**: 513–521.

Davies, J. C. and D. J. Greathead. 1967. Occurrence of *Teleonemia scrupulosa* on *Sesamum indicum* Linn. in Uganda. *Nature* **213**: 102–103.

Davis, A. S., D. A. Landis, V. Nuzzo, et al. 2006. Demographic models inform selection of biological control agents for garlic mustard (*Alliaria petiolata*). *Ecological Applications* **16**: 2399–2410.

Davis, M., M. K. Chew, R. J. Hobbs, et al. 2011. Don't judge species on their origins. *Nature* **474**: 153–154.

De Clercq, P., P. G. Mason, and D. Babendreier. 2011. Benefits and risks of exotic biological control agents. *Biological Control* **56**: 681–698.

de Groot, M., M. Drenthen, and W. T. de Groot. 2011. Public visions of the human/nature relationship and their implications for environmental ethics. *Environmental Ethics* **33**: 25–44.

de Lange, W. J. and B. W. van Wilgen. 2010. An economic assessment of the contribution of biological control to the management of invasive alien plants and to the protection of ecosystem services in South Africa. *Biological Invasions* **12**: 4113–4124.

de Valpine, P., K. Scranton, J. Knape, et al. 2014. The importance of individual developmental variation in stage-structured population models. *Ecology Letters* **17**: 1026–1038.

DeBach, P. P. 1974. *Biological Control by Natural Enemies*. Cambridge University Press. London, UK.

Decker, D. J., A. B. Forstchen, J. F. Organ, et al. 2014. Impacts management: an approach to fulfilling public trust responsibilities of wildlife agencies. *Wildlife Society Bulletin* **38**: 2–8.

Delfosse, E. S. 2005. Risk and ethics in biological control. *Biological Control* **35**: 319–329.

Denoth, M. and J. H. Myers. 2005. Variable success of biological control of *Lythrum salicaria* in British Columbia. *Biological Control* **32**: 269–279.

Denslow, J. S. and C. M. D'Antonio. 2005. After biological control: assessing indirect effects of insect releases. *Biological Control* **35**: 307–318.

Desneux, N., R. Blahnik, C. J. Delebecque, and G. E. Heimpel. 2012. Host phylogeny and specialisation in parasitoids. *Ecology Letters* **15**: 453–460.

DeWalt, S. J. 2006. Population dynamics and potential for biological control of an exotic invasive shrub in Hawaiian rainforests. *Biological Invasions* **8**: 1145–1158.

Dickie, I. A., B. M. Bennett, L. E. Burrows, et al. 2014. Conflicting values: ecosystem services and invasive tree management. *Biological Invasions* **16**: 705–719.

Downey, P. O. 2011. Changing of the guard: moving from a war on weeds to an outcome-oriented weed management system. *Plant Protection Quarterly* **26**: 86–91.

Downey, P. O. 2014. Protecting biodiversity through strategic alien plant management: an approach for increasing conservation outcomes in protected areas, pp. 508–528. *In*: Foxcroft, L. C., P. Pysek, D. M. Richardson, and P. Genovesi (eds.). *Plant Invasions in Protected Areas: Patterns, Problems and Challenges*. Springer. Dordrecht, The Netherlands.

Duan, J. J., K. L. Abell, L. S. Bauer, et al. 2014. Natural enemies implicated in the regulation of an invasive pest: a life table analysis of the population dynamics of the emerald ash borer. *Agricultural and Forest Entomology* **16**: 406–416. doi: 10.1111/afe.12070.

Dudley, T. L. and D. W. Bean. 2012. Tamarisk biological control, endangered species risk and resolution of conflict through riparian restoration. *Biological Control* **57**: 331–347.

Eckberg, J. O., B. Tenhumberg, and S. M. Louda. 2014. Native insect herbivory limits population growth rate of a non-native thistle. *Oecologia* **175**: 129–138.

Ehrenfeld, J. G. 2003. Effects of exotic plant invasions on soil nutrient cycling processes. *Ecosystems* **6**: 503–523.

Elkinton, J. S., D. Parry, and G. H. Boettner. 2006. Implicating an introduced generalist parasitoid in the invasive browntail moth's enigmatic demise. *Ecology* **87**: 2664–2672.

Elkinton, J. S. and G. H. Boettner. 2012. Benefits and harm caused by the introduced generalist tachinid, *Compsilura concinnata*, in North America. *Biological Control* **57**: 277–288.

Ellner, S. P. and M. Rees. 2007. Stochastic stable population growth in integral projection models: theory and application. *Journal of Mathematical Biology* **54**: 227–256.

ERMA. 2010. *Investigating Biological Control and the HSNO Act*. Environmental Risk Managament Authority. Wellington, New Zealand.

Eschtruth, A. K. and J. J. Battles. 2008. Deer herbivory alters forest response to canopy decline caused by an exotic insect pest. *Ecological Applications* **18**: 360–376.

Folke, C., S. Carpenter, B. Walker, et al. 2004. Regime shifts, resilience, and biodiversity in ecosystem management. *Annual Review of Ecology Evolution and Systematics* **35**: 557–581.

Follett, P. A. and J. J. Duan. 2000. *Nontarget Effects of Biological Control*. Kluwer Academic Publishers. Dordrecht, The Netherlands.

Forister, M. L., L. A. Dyer, M. S. Singer, et al. 2012. Revisiting the evolution of ecological specialization, with emphasis on insect-plant interactions. *Ecology* **93**: 981–991.

Fowler, S. V., P. Syrett, and R. L. Hill. 2000. Success and safety in the biological control of environmental weeds in New Zealand. *Austral Ecology* **25**: 553–562.

Fowler, S. V., Q. Paynter, S. Dodd, and R. Groenteman. 2012. How can ecologists help practitioners minimize non-target effects in weed biological control? *Journal of Applied Ecology* **49**: 307–310.

Foxcroft, L. C., P. Pysek, D. M. Richardson, and P. Genovesi. 2014. *Plant Invasions in Protected Areas: Patterns, Problems and Challenges*. Springer. Dordrecht, The Netherlands.

Fujita, Y., H. O. Venterink, P. M. van Bodegom, et al. 2014. Low investment in sexual reproduction threatens plants adapted to phosphorus limitation. *Nature* **505**: 82–86.

Gill, J. L., J. W. Williams, S. T. Jackson, et al. 2009. Pleistocene megafaunal collapse, novel plant communities, and enhanced fire regimes in North America. *Science* **326**: 1100–1103.

Gill, J. L., J. W. Williams, S. T. Jackson, et al. 2012. Climatic and megaherbivory controls on late-glacial vegetation dynamics: a new, high-resolution, multi-proxy record from Silver Lake, Ohio. *Quaternary Science Reviews* **34**: 66–80.

Goldson, S. L., S. D. Wratten, C. M. Ferguson, et al. 2014. If and when successful classical biological control fails. *Biological Control* **72**: 76–79.

Gratton, C. and R. F. Denno. 2006. Arthropod food web restoration following removal of an invasive wetland plant. *Ecological Applications* **16**: 622–631.

Greathead, D. J. 1995. Benefits and risks of classical biological control, pp. 53–63. *In*: Hokkanen, H. M. T. and J. A. Lynch, (eds.). *Biological Control: Benefits and Risks*. Cambridge University Press. Cambridge, UK.

Greenwood, P. and N. J. Kuhn. 2014. Does the invasive plant, *Impatiens glandulifera*, promote soil erosion along the riparian zone? An investigation on a small watercourse in northwest Switzerland. *Journal of Soils and Sediments* **14**: 637–650.

Groenteman, R., S. V. Fowler, and J. J. Sullivan. 2011. St. John's wort beetles would not have been introduced to New Zealand now: a retrospective host range test of New Zealand's most successful weed biological control agents. *Biological Control* **57**: 50–58.

Gudynas, E. 2009. The political ecology of the biocentric turn in Ecuador's new constitution. *Revista De Estudios Sociales*: **32**: 34–46.

Hall, R. W., L. E. Ehler, and B. Bisabri-Ershadi. 1980. Rate of success in classical biological control of arthropods. *Etomological Society of America Bulletin* **26**: 111–114.

Hamann, A. and S. N. Aitken. 2013. Conservation planning under climate change: accounting for adaptive potential and migration capacity in species distribution models. *Diversity and Distributions* **19**: 268–280.

Hare, D. and B. Blossey. 2014. Principles of public trust thinking. *Human Dimension of Wildlife: An International Journal* **19**: 397–406.

Harris, P. 1988. Environmental impact of weed-control insects. *BioScience* **38**: 542–548.

Harris, P. 1991. Classical biological control of weeds: its definition, selection of effective agents, and administrative political problems. *The Canadian Entomologist* **123**: 827–849.

Harvey, J. A., M. G. X. De Embun, T. Bukovinszky, and R. Gols. 2012. The roles of ecological fitting, phylogeny and physiological equivalence in understanding realized and fundamental host ranges in endoparasitoid wasps. *Journal of Evolutionary Biology* **25**: 2139–2148.

Havill, N. P., G. Davis, D. L. Mausel, et al. 2012. Hybridization between a native and introduced predator of Adelgidae: An unintended result of classical biological control. *Biological Control* **63**: 359–369.

Haye, T., H. Goulet, P. G. Mason, and U. Kuhlmann. 2005. Does fundamental host range match ecological host range? A retrospective case study of a *Lygus* plant bug parasitoid. *Biological Control* **35**: 55–67.

Henneman, M. L. and J. Memmott. 2001. Infiltration of a Hawaiian community by introduced biological control agents. *Science* **293**: 1314–1316.

Henri, D. C. and F. J. F. Van Veen. 2011. Body size, life history and the structure of host-parasitoid networks. *Advances in Ecological Research* **45**: 135–180.

Higgs, E., D. A. Falk, A. Guerrin, et al. 2014. The changing role of history in restoration ecology. *Frontiers in Ecology and Environment* **12**: 499–506

Hoddle, M. S. 2004. Restoring the balance: uisng exotic species to control invasive exotic species. *Conservation Biology* **18**: 38–49.

Hoddle, M. S., C. C. Ramirez, C. D. Hoddle, et al. 2013. Post release evaluation of *Rodolia cardinalis* (Coleoptera: Coccinellidae) for control of *Icerya purchasi* (Hemiptera: Monophlebidae) in the Galapagos Islands. *Biological Control* **67**: 262–274.

Hoffmann, J. H. 1995. Biological control of weeds: the way forward, a South African perspective, pp.77–98. *In*: *British Crop Protection Council Symposium no. 64*. The British Crop Protection Council. Farnham, UK.

Hooper, D. U., E. C. Adair, B. J. Cardinale, et al. 2012. A global synthesis reveals biodiversity loss as a major driver of ecosystem change. *Nature* **486**: 105–109.

Howarth, F. G. 1991. Environmental impacts of classical biological control. *Annual Review of Entomology* **36**: 485–509.

Huffaker, C. B. and C. E. Kennett. 1959. A ten-year study of vegetational changes associated with biological control of Klamath weed. *Journal of Range Management* **12**: 69–82.

Hultine, K. R., J. Belnap, C. van Riper, et al. 2010. Tamarisk biological control in the western United States: ecological and societal implications. *Frontiers in Ecology and the Environment* **8**: 467–474.

Hunt, E. J., U. Kuhlmann, A. Sheppard, et al. 2008. Review of invertebrate biological control agent regulation in Australia, New Zealand, Canada and the USA: recommendations for a harmonized European system. *Journal of Applied Entomology* **132**: 89–123.

Impson, F. A. C., C. A. Kleinjan, J. H. Hoffmann, et al. 2011. Biological control of Australian *Acacia* species and *Paraserianthes lophantha* (Willd.) Nielsen (Mimosaceae) in South Africa. *African Entomology* **19**: 186–207.

IUCN. 2012. *IUCN Red List. Categories and Criteria: Version 3.1. Second edition*. Gland, Switzerland. Available from: http://www.iucn.org/publications [Accessed January 2016].

Jablonski, D. 2008. Extinction and the spatial dynamics of biodiversity. *Proceedings of the National Academy of Sciences of the United States of America* **105**: 11528–11535.

Jackson, S. T. 2007. Looking forward from the past: history, ecology, and conservation. *Frontiers in Ecology and the Environment* **5**: 455–455.

Jackson, S. T. 2013. Natural, potential and actual vegetation in North America. *Journal of Vegetation Science* **24**: 772–776.

Jackson, S. T. and R. J. Hobbs. 2009. Ecological restoration in the light of ecological history. *Science* **325**: 567–569.

Jackson, S. T. and D. F. Sax. 2010. Balancing biodiversity in a changing environment: extinction debt, immigration credit and species turnover. *Trends in Ecology & Evolution* **25**: 153–160.

Jackson, S. T. and J. W. Williams. 2004. Modern analogs in Quaternary paleoecology: here today, gone yesterday, gone tomorrow? *Annual Review of Earth and Planetary Sciences* **32**: 495–537.

Jacoby, K. 2001. *Crimes against Nature: Squatters, Poachers, Thieves and the Hidden History of American Conservation*. University of California Press. Berkeley, California, USA.

Janz, N. 2011. Ehrlich and Raven revisited: mechanisms underlying codiversification of plants and enemies. *Annual Review of Ecology, Evolution, and Systematics* **42**: 71–89.

Jepson, P. 2005. Governance and accountability of environmental NGOs. *Environmental Science and Policy* **8**: 515–524.

Johnson, B. B. and L. S. Horowitz. 2014. Beliefs about ecological impacts predict deer acceptance capacity and hunting support. *Society & Natural Resources* **27**: 915–930. doi:10.1080/08941920.2014.905887.

Johnson, M. T., P. A. Follett, A. D. Taylor, and V. P. Jones. 2005. Impacts of biological control and invasive species on a non-target native Hawaiian insect. *Oecologia* **142**: 529–540.

Jones, R., J. Leahy, M. Mahoney, et al. 2004. *Overview of the Ecological Risk Assessment Process in the Office of Pesticide Programs, U.S. Environmental Protection Agency: Edangered and Threatened Species Effects Determinations*. Office of Prevention, Pesticides and Toxic Substances Office of Pesticide Programs, Washington, DC.

Jongejans, E., O. Skarpaas, and K. Shea. 2008. Dispersal, demography and spatial population models for conservation and control management. *Perspectives in Plant Ecology, Evolution and Systematics* **9**: 153–170.

Kalisz, S., R. Spigler, and C. Horvitz. 2014. In a long-term experimental demography study, excluding ungulates reversed invader's explosive population growth rate and restored natives. *Proceedings of the National Academy of Sciences of the United States of America* **111**: 4501–4506.

Kardol, P., I. A. Dickie, M. G. St. John, et al. 2014. Soil-mediated effects of invasive ungulates on native tree seedlings. *Journal of Ecology* **102**: 622–631.

Keane, R. M. and M. J. Crawley. 2002. Exotic plant invasions and the enemy release hypothesis. *Trends in Ecology & Evolution* **17**: 164–170.

Keeley, J. E. 2006. Fire management impacts on invasive plants in the western United States. *Conservation Biology* **20**: 375–384.

Kettenring, K. M. and C. R. Adams. 2011. Lessons learned from invasive plant control experiments: a systematic review and meta-analysis. *Journal of Applied Ecology* **48**: 970–979.

King, C. B. A., W. P. Haines, and D. Rubinoff. 2010. Impacts of invasive parasitoids on declining endemic Hawaiian leafroller moths (Omiodes: Crambidae) vary among sites and species. *Journal of Applied Ecology* **47**: 299–308.

Knight, T. M., J. L. Dunn, L. A. Smith, et al. 2009. Deer facilitate invasive plant success in a Pennsylvania forest understory. *Natural Areas Journal* **29**: 110–116.

Kuhlmann, U., P. Mason, H. L. Hinz, et al. 2006. Avoiding conflicts between insect and weed biological control: selection of non-target species to assess host specificity of cabbage seedpod weevil parasitoids. *Journal of Applied Entomology* **130**: 129–141.

Lampert, A., A. Hastings, E. D. Grosholz, et al. 2014. Optimal approaches for balancing invasive species eradication and endangered species management. *Science* **344**: 1028–1031.

Langley, J. A. and J. P. Megonigal. 2010. Ecosystem response to elevated CO_2 levels limited by nitrogen-induced plant species shift. *Nature* **466**: 96–99.

Lankau, R. A. 2011. Rapid evolutionary change and the coexistence of species. *Annual Review of Ecology, Evolution, and Systematics* **42**: 335–354.

Lau, J. A. 2013. Trophic consequences of a biological invasion: do plant invasions increase predator abundance? *Oikos* **122**: 474–480.

LeMaitre, D. C., B. W. Van Wilgen, R. A. Chapman, and D. H. McKelly. 1996. Invasive plants and water resources in the Western Cape Province, South Africa: modelling the consequences of a lack of management. *Journal of Applied Ecology* **33**: 161–172.

Louda, S. M. and P. Stiling. 2004. The double-edged sword of biological control in conservation and restoration. *Conservation Biology* **18**: 50–53.

Louda, S. M., D. Kendall, J. Connor, and D. Simberloff. 1997. Ecological effects of an insect introduced for the biological control of weeds. *Science* **277**: 1088–1090.

Louda, S. M., R. W. Pemberton, M. T. Johnson, and P. A. Follett. 2003. Nontarget effects – the Achilles' heel of biological control: retrospective analyses to reduce risk associated with biological control introductions. *Annual Review of Entomology* **48**: 365–396.

Louhaichi, M., M. F. Carpinelli, L. M. Richman, and D. E. Johnson. 2012. Native forb response to sulfometuron methyl on medusahead-invaded rangeland in Eastern Oregon. *Rangeland Journal* **34**: 47–53.

Lynch, L. D. and M. B. Thomas. 2000. Nontarget effects in the biological control of insects with insects, nematodes and microbial agents: the evidence. *Biological Control News and Information* **21**: 117N–130N.

Lynch, L. D., H. M. T. Hokkanen, D. Babendreier, et al. 2001. Insect biological control and non-target effects: a European perspective, pp. 99–125. *In*: Wajnberg, E., J. K. Scott, and

P. C. Quimby (eds.). *Evaluating Indirect Ecological Effects of Biological Control*. CABI Publishing. Wallingford, UK.

MacDougall, A. S. and R. Turkington. 2005. Are invasive species the drivers or passengers of change in degraded ecosystems. *Ecology* **86**: 42–55.

Mack, R. N., D. Simberloff, W. M. Lonsdale, et al. 2000. Biotic invasions: causes, epidemiology, global consequences, and control. *Ecological Applications* **10**: 689–710.

Marchetto, K. M., K. Shea, D. Kelly, et al. 2014. Unrecognized impact of a biological control agent on the spread rate of an invasive thistle. *Ecological Applications* **24**: 1178–1187.

Martin, L. J. and B. Blossey. 2013a. Intraspecific variation overrides origin effects in impacts of litter-derived secondary compounds on larval amphibians. *Oecologia* **173**: 449–459.

Martin, L. J. and B. Blossey. 2013b. The runaway weed: costs and failures of *Phragmites australis* management in the USA. *Estuaries and Coasts* **36**: 626–632.

Martin, L. J., S.-K. Rainford, and B. Blossey. 2014. Effects of plant litter diversity, species, origin, and traits on larval toad performance. *Oikos* **124**: 871–879. doi: 10.1111/oik.01745

Mausel, D. L., S. M. Salom, L. T. Kok, and J. G. Fidgen. 2008. Propagation, synchrony, and impact of introduced and native *Laricobius* spp. (Coleoptera: Derodontidae) on hemlock woolly adelgid in Virginia. *Environmental Entomology* **37**: 1498–1507.

McEvoy, P. B. and E. M. Coombs. 1999a. Biological control of plant invaders: regional patterns, field experiments, and structured population models. *Ecological Applications* **9**: 387–401.

McEvoy, P. B. and E. M. Coombs. 1999b. Biological control of plant invaders: regional patterns, field experiments, and structured population models. *Ecological Applications* **9**: 387–401.

McEvoy, P. B. and E. M. Coombs. 2000. Why things bite back: unintended consequences of biological weed control, pp. 167–194. *In*: Follett, P. A. and J. J. Duan (eds.). *Non-target Effects of Biological Control*. Kluwer Academic Publishers. Boston, Massachusetts, USA.

McEvoy, P. B., K. M. Higgs, E. M. Coombs, et al. 2012. Evolving while invading: rapid adaptive evolution in juvenile development time for a biological control organism colonizing a high-elevation environment. *Evolutionary Applications* **5**: 524–536.

McFadyen, R. E. C. 1998. Biological control of weeds. *Annual Review of Entomology* **43**: 369–393.

Messing, R. H. and M. G. Wright. 2006. Biological control of invasive species: solution or pollution? *Frontiers in Ecology and the Environment* **4**: 132–140.

Meyer, J. Y. and M. Fourdrigniez. 2011. Conservation benefits of biological control: the recovery of a threatened plant subsequent to the introduction of a pathogen to contain an invasive tree species. *Biological Conservation* **144**: 106–113.

Miller, M. L. and G. H. Aplet. 2005. Applying legal sunshine to the hidden regulation of biological control. *Biological Control* **35**: 358–365.

Miller, T. E. X., S. M. Louda, K. A. Rose, and J. O. Eckberg. 2009. Impacts of insect herbivory on cactus population dynamics: experimental demography across an environmental gradient. *Ecological Monographs* **79**: 155–172.

Mitchell, C. E. and A. G. Powers. 2003. Release of invasive plants from fungal and viral pathogens. *Nature* **421**: 625–627.

Mitchell, C. E., A. A. Agrawal, J. D. Bever, et al. 2006. Biotic interactions and plant invasions. *Ecology Letters* **9**: 726–740.

Moran, V. C., J. H. Hoffmann, and H. G. Zimmermann. 2005. Biological control of invasive alien plants in South Africa: necessity, circumspection, and success. *Frontiers in Ecology and the Environment* **3**: 77–83.

Muir, D. C. G. and C. A. de Wit. 2010. Trends of legacy and new persistent organic pollutants in the circumpolar arctic: overview, conclusions, and recommendations. *Science of the Total Environment* **408**: 3044–2051.

Neuenschwander, P. 2001. Biological control of the cassava mealybug in Africa: a review. *Biological Control* **21**: 214–229.

Nichols, J. D. and B. K. Williams. 2006. Monitoring for conservation. *Trends in Ecology & Evolution* **21**: 668–673.

Nuzzo, V. A., J. C. Maerz, and B. Blossey. 2009. Earthworm invasion as the driving force behind plant invasion and community change in northeastern North American forests. *Conservation Biology* **23**: 966–974.

Obrycki, J. J. and T. J. Kring. 1998. Predaceous Coccinellidae in biological control. *Annual Review of Entomology* **43**: 295–321.

Obrycki, J. J., N. C. Elliott, and K. L. Giles. 2000. Coccinellid introductions: potential for an evaluation of non-target effects, pp. 127–145. *In*: Follett, P. A. and J. J. Duan (eds.). *Non-target Effects of Biological Control*. Kluwer Academic Publishers. Boston, Massachusetts, USA.

Paine, T. D., J. C. Millar, and K. M. Daane. 2010. Accumulation of pest insects on eucalyptus in California: random process or smoking gun. *Journal of Economic Entomology* **103**: 1943–1949.

Parry, D. 2009. Beyond Pandora's Box: quantitatively evaluating non-target effects of parasitoids in classical biological control. *Biological Invasions* **11**: 47–58.

Paynter, Q. E., S. V. Fowler, A. H. Gourlay, et al. 2004. Safety in New Zealand weed biological control: a nationwide survey for impacts on non-target plants. *New Zealand Plant Protection* **57**: 102–107.

Paynter, Q., A. H. Gourlay, P. T. Oboyski, et al. 2008. Why did specificity testing fail to predict the field host-range of the gorse pod moth in New Zealand? *Biological Control* **46**: 453–462.

Pearson, D. E. and R. M. Callaway. 2003. Indirect effects of host-specific biological control agents. *Trends in Ecology & Evolution* **18**: 456–461.

Pearson, D. E. and R. M. Callaway. 2008. Weed-biological control insects reduce native-plant recruitment through second-order apparent competition. *Ecological Applications* **18**: 1489–1500.

Pemberton, R. W. 2000. Predictable risk to native plants in weed biological control. *Oecologia* **125**: 489–494.

Perkins, L. B. and R. S. Nowak. 2013. Native and non-native grasses generate common types of plant-soil feedbacks by altering soil nutrients and microbial communities. *Oikos* **122**: 199–208.

Petit, J. N., M. S. Hoddle, J. Grandgirard, et al. 2009. Successful spread of a biological control agent reveals a biosecurity failure: elucidating long distance invasion pathways for *Gonatocerus ashmeadi* in French Polynesia. *Biological Control* **54**: 485–495.

Pratt, P. D. and T. D. Center. 2012. Biological control without borders: the unintended spread of introduced weed biological control agents. *Biological Control* **57**: 319–329.

Pungetti, G., G. Oviedo, and D. Hooke (eds.). 2012. *Sacred Species and Sites: Advances in Biocultural Conservation*. Cambridge University Press. Cambridge, UK.

Ramula, S. 2014. Linking vital rates to invasiveness of a perennial herb. *Oecologia* **174**: 1255–1264.

Rinella, M. J., B. D. Maxwell, P. K. Fay, et al. 2009. Control effort exacerbates invasive-species problem. *Ecological Applications* **19**: 155–162.

Ripple, W. J. and R. L. Beschta. 2007. Hardwood tree decline following large carnivore loss on the Great Plains, USA. *Frontiers in Ecology and the Environment* **5**: 241–246.

Rosenheim, J. A. 1998. Higher-order predators and the regulation of insect herbivore populations. *Annual Review of Entomology* **43**: 421–447.

Royama, T. 1996. A fundamental problem in key factor analysis. *Ecology* **77**: 87–93.

Sagarin, R. D. and M. Turnipseed. 2012. The Public Trust Doctrine: where ecology meets natural resources management. *Annual Review of Environment and Resources* **37**: 473–496.

Sax, J. L. 1970. The Public Trust Doctrine in natural resource law: effective judicial intervention. *Michigan Law Review* **68**: 471–566.

Schlaepfer, M. A., D. F. Sax, and J. D. Olden. 2011. The potential conservation value of non-native species. *Conservation Biology* **25**: 428–437.

Schooler, S. S., E. M. Coombs, and P. B. McEvoy. 2003. Nontarget effects on crepe myrtle by *Galerucella pusilla* and *G. calmariensis* (Chrysomelidae), used for biological control of purple loosestrife (*Lythrum salicaria*). *Weed Science* **51**: 449–455.

Schooler, S. S., B. Salau, M. H. Julien, and A. R. Ives. 2011. Alternative stable states explain unpredictable biological control of *Salvinia molesta* in Kakadu. *Nature* **470**: 86–89.

Schutzenhofer, M. R., T. J. Valone, and T. M. Knight. 2009. Herbivory and population dynamics of invasive and native *Lespedeza*. *Oecologia* **161**: 57–66.

Shea, K. and D. Kelly. 1998. Estimating biological control agent impact with matrix models: *Carduus nutans* in New Zealand. *Ecological Applications* **8**: 824–832.

Shea, K., D. Kelly, A. W. Sheppard, and T. L. Woodburn. 2005. Context-dependent biological control of an invasive thistle. *Ecology* **86**: 3174–3181.

Sheppard, A. W., R. L. Hill, R. A. DeClerck-Floate, et al. 2003. A global review of risk-benefit-cost analysis for the introduction of classical biological control agents against weeds: a crisis in the making? *Biological Control News and Information* **24**: 77N–94N.

Sheppard, A. W., R. D. van Klinken, and T. A. Heard. 2005. Scientific advances in the analysis of direct risks of weed biological control agents to nontarget plants. *Biological Control* **35**: 215–226.

Simberloff, D. 2012. Risks of biological control for conservation purposes. *Biological control* **57**: 263–276.

Simberloff, D. 2014. Eradication: pipe dream or real option? pp. 549–560. *In*: Foxcroft, L. C., P. Pysek, D. M. Richardson, and P. Genovesi (eds.). *Plant Invasions in Protected Areas: Patterns, Problems and Challenges*. Springer. Dordrecht, The Netherlands.

Simberloff, D. and P. Stiling. 1996a. Risks of species introduced for biological control. *Biological Conservation* **78**: 185–192.

Simberloff, D. and P. Stiling. 1996b. How risky is biological control? *Ecology* **77**: 1965–1974.

Simberloff, D., J. L. Martin, P. Genovesi, et al. 2013. Impacts of biological invasions: what's what and the way forward. *Trends in Ecology & Evolution* **28**: 58–66.

Skurski, T. C., B. D. Maxwell, and L. J. Rew. 2013. Ecological tradeoffs in non-native plant management. *Biological Conservation* **159**: 292–302.

Smith, C. A. 2011. The role of state wildlife professionals under the Public Trust Doctrine. *Journal of Wildlife Management* **75**: 1539–1543.

Smith, L. 2012. Host plant oviposition preference of *Ceratapion basicorne* (Coleoptera: Apionidae), a potential biological control agent of yellow starthistle. *Biological Control Science and Technology* **22**: 407–418.

Stehn, S. E., M. A. Jenkins, C. R. Webster, and S. Jose. 2013. Regeneration responses to exogenous disturbance gradients in southern Appalachian *Picea-Abies* forests. *Forest Ecology and Management* **289**: 98–105.

Stephens, A. E. A., D. S. Srivastava, and J. H. Myers. 2013. Strength in numbers? Effects of multiple natural enemy species on plant performance. *Proceedings of the Royal Society B-Biological Sciences* **280**. 20122756. http://dx.doi.org/10.1098/rspb.2012.2756.

Stiling, P. and T. Cornelissen. 2005. What makes a successful biological control agent? A meta-analysis of biological control agent performance. *Biological Control* **34**: 236–246.

Stireman, J. O. 2005. The evolution of generalization? Parasitoid flies and the perils of inferring host range evolution from phylogenies. *Journal of Evolutionary Biology* **18**: 325–336.

Strauss, G. 2009. Host range testing of the nearctic beneficial parasitoid *Neodryinus typhlocybae*. *Biological Control* **54**: 163–171.

Strauss, S. Y., J. A. Lau, and S. P. Carroll. 2006. Evolutionary responses of natives to introduced species: what do introductions tell us about natural communities? *Ecology Letters* **9**: 357–374.

Strauss, S. Y., J. A. Lau, T. W. Schoener, and P. Tiffin. 2008. Evolution in ecological field experiments: implications for effect size. *Ecology Letters* **11**: 199–207.

Strayer, D. L. 2012. Eight questions about invasions and eco-system functioning. *Ecology Letters* **15**: 1199–1210.

Stromberg, J. C., S. J. Lite, R. Marler, et al. 2007. Altered stream-flow regimes and invasive plant species: the Tamarix case. *Global Ecology and Biogeography* **16**: 381–393.

Strong, D. R. and J. M. Pemberton. 2000. Biological control of invading species – risk and reform. *Science* **288**: 1969–1970.

Strong, D. R. and R. W. Pemberton. 2001. Food webs, risks of alien enemies and reform of biological control, pp. 57–79. *In*: Wajnberg, E., J. K. Scott, and P. C. Quimby (eds.). *Evaluating Indirect Ecological Effects of Biological Control*. CABI Publishing. Oxford, UK.

Suckling, D. M. 2013. Benefits from biological control of weeds in New Zealand range from negligible to massive: a retrospective analysis. *Biological Control* **66**: 27–32.

Suckling, D. M. and R. F. H. Sforza. 2014. What magnitude are observed non-target impacts from weed biological control? *PLoS One* **9**(1): e84847. doi: 10.1371/journal.pone.0084847.

Takacs, D. 2008. The public trust doctrine, environmental human rights, and the future of private property. *NYU Environmental Law Journal* **16**: 711–765.

Tewksbury, L., R. Casagrande, B. Blossey, et al. 2002. Potential for biological control of *Phragmites australis* in North America. *Biological Control* **23**: 191–212.

Thomas, M. B. and A. J. Willis. 1998. Biological control - risky but necessary? *Trends in Ecology & Evolution* **13**: 325–329.

Thomas, M. B. and A. M. Reid. 2007. Are exotic natural enemies an effective way of controlling invasive plants? *Trends in Ecology & Evolution* **22**: 447–453.

Travis, J. M. J., M. Delgado, G. Bocedi, et al. 2013. Dispersal and species' responses to climate change. *Oikos* **122**: 1532–1540.

Tscharntke, T. 1992. Fragmentation of *Phragmites* habitats, minimum viable population size, habitat suitability, and local extinction of moths, midges, flies, aphids, and birds. *Conservation Biology* **6**: 530–536.

Turnipseed, M., L. B. Crowder, R. D. Sagarin, and S. E. Roady. 2009. Legal bedrock for rebuilding America's ocean ecosystems. *Science* **326**: 183–185.

Tylianakis, J. M., R. K. Didham, J. Bascompte, and D. A. Wardle. 2008. Global change and species interactions in terrestrial ecosystems. *Ecology Letters* **11**: 1351–1363.

van der Leeuw, S., R. Costanza, S. Aulenbach, et al. 2011. Toward an integrated history to guide the future. *Ecology and Society* **16**(4): 2.

Van Driesche, R. and T. Center. 2014. Biological control of invasive plants in protected areas, pp. 561–598. *In:* Foxcroft, L. C., P. Pysek, D. M. Richardson, and P. Genovesi (eds.). *Plant Invasions in Protected Areas: Patterns, Problems and Challenges*. Springer. Dordrecht, The Netherlands.

Van Driesche, R., M. Hoddle, and T. Center (eds.). 2008. *Control of Pests and Weeds by Natural Enemies*. Blackwell. Malden, Massachusetts, USA.

Van Driesche, R. G., R. I. Carruthers, T. Center, et al. 2010. Classical biological control for the protection of natural ecosystems. *Biological Control* **54**: S2–S33.

van Klinken, R. D. and O. R. Edwards. 2002. Is host specificity of weed biological control agents likely to evolve rapidly following establishment? *Ecology Letters* **5**: 590–595.

van Lenteren, J. C., D. Babendreier, F. Bigler, et al. 2003. Environmental risk assessment of exotic natural enemies used in inundative biological control. *Biological Control* **48**: 3–38.

van Lenteren, J. C., J. Bale, E. Bigler, et al. 2006. Assessing risks of releasing exotic biological control agents of arthropod pests. *Annual Review of Entomology* **51**: 609–634.

van Lenteren, J. C., A. J. M. Loomans, D. Babendreier, and F. Bigler. 2008. *Harmonia axyridis*: an environmental risk assessment for Northwest Europe. *Biological Control* **53**: 37–54.

van Wilgen, B. W., V. C. Moran, and J. H. Hoffmann. 2013. Some perspectives on the risks and benefits of biological control of invasive alien plants in the management of natural ecosystems. *Environmental Management* **52**: 531–540.

Vickery, W. L. 1991. An evaluation of bias in k-factor analysis. *Oecologia* **85**: 413–418.

Vila, M., J. L. Espinar, M. Hejda, et al. 2011. Ecological impacts of invasive alien plants: a meta-analysis of their effects on species, communities and ecosystems. *Ecology Letters* **14**: 702–708.

Wajnberg, E., J. K. Scott, and P. C. Quimby (eds.). 2001. *Evaluating Indirect Ecological Effects of Biological Control*. CABI Publishing. Wallingford, UK.

Wallington, T. J., R. J. Hobbs, and S. A. Moore. 2005. Implications of current ecological thinking for biodiversity conservation: A review of the salient issues. *Ecology and Society* **10**(1): 15.

Wardle, D. A., G. M. Barker, G. W. Yeates, et al. 2001. Introduced browsing mammals in New Zealand natural forests: above-ground and belowground consequences. *Ecological Monographs* **71**: 587–614.

Weed, A. S. and M. Schwarzländer. 2014. Density dependence, precipitation and biological control agent herbivory influence landscape-scale dynamics of the invasive Eurasian plant *Linaria dalmatica*. *Journal of Applied Ecology* **51**: 825–834.

Weiss, E. B. 1989. *In Fairness to Future Generations: International Law, Common Patrimony, and Intergenerational Equity*. The United Nations University. Tokyo.

Wilcove, D. S., D. Rothstein, J. Dubow, et al. 1998. Quantifying threats to imperiled species in the United States. *BioScience* **48**: 607–615.

Wilkinson, R. and G. Fitzgerald. 1997. Public perceptions of biological control of rabbits in New Zealand: some ethical and practical issues. *Agriculture and Human Values* **14**: 273–282.

Williams, B. K., J. D. Nichols, and M. J. Conroy. 2001. *Analyses and Management of Animal Populations*. Academic Press, New York.

Williams, J. W. and S. T. Jackson. 2007. Novel climates, no-analog communities, and ecological surprises. *Frontiers in Ecology and the Environment* **5**: 475–482.

Winston, R. L., M. Schwarzländer, H. L. Hinz, M. D. Day, M. J. W. Cock, and M. H. Julien, eds. 2014. *Biological Control of Weeds: A World Catalogue of Agents and Their Target Weeds*, 5th edition. USDA Forest Service, Forest Health Technology Enterprise Team, Morgantown, West Virginia, USA. FHTET-2014-04.

Yelenik, S. G. and C. M. D'Antonio. 2013. Self-reinforcing impacts of plant invasions change over time. *Nature* **503**: 517–520.

Yurkonis, K. A. and S. J. Meiners. 2004. Invasion impacts local species turnover in a successional system. *Ecology Letters* **7**: 764–769.

Zeddies, J., R. P. Schaab, P. Neuenschwander, and H. R. Herren. 2001. Economics of biological control of cassava mealybug in Africa. *Agricultural Economics* **24**: 209–219.

CHAPTER 9

Methods for evaluation of natural enemy impacts on invasive pests of wildlands

Roy G. Van Driesche

Department of Environmental Conservation, University of Massachusetts, USA

Biological control projects against wildland pests vary in size and duration owing to the combined effects of project goals, political decisions, and the availability of funding. Foundational population regulation theory has long been important to arthropod biological control and to the assessment of interactions between natural enemies and their hosts or prey. When reviewed in total, these theoretical models conclude that in successful biological control programs natural enemy populations increase at the expense of their declining host or prey populations until they reach a more or less stable equilibrium (see Hassell [1978] for review of these early arthropod predator-prey models). In practice, few projects have been large enough, or lasted long enough, to assess fully the impacts of the released biological control agents on the population growth rate of the target pest, to say nothing of quantifying resultant changes in the ecosystem. Whether population-level growth rate evaluations (as discussed in Chapter 8) are possible or not is also affected by the length of the life cycle of the target pest (generations per year or years in life span) and other aspects of the pest's biology that constrain what can be measured. A further factor affecting the types of impact evaluation outcomes used by past projects has been the goals set by the projects themselves. With these issues in mind, approaches are discussed here that have been used previously by researchers to measure the impacts of natural enemies in the field. Because of the differences in plant and insect life cycles, and the historical differences in the literature on evaluating effects of

biological control on these two groups, it is easier to discuss these taxa separately, pointing out similarities in evaluation approaches where they occur.

In the preceding chapter, evaluation of biological control programs was discussed conceptually, in the broadest of terms. Here, I present a more detailed view of the actual techniques available to measure the biological and ecological success of introduced natural enemies, stressing how the use of various techniques differs between insect and plant control efforts. I also discuss constraints on methods imposed by the biologies of the natural enemies and targeted pests and give examples of the use of the various techniques against pests of wildlands.

Types of outcomes used as measures of impact

Measures used to assess impacts of attack by biological control agents differ between insect and weed targets. For insect targets, biological control agents nearly always exert their effects by killing individuals. In a few cases, agents affecting adults also reduce their fecundity before killing them (e.g., Van Driesche and Gyrisco, 1979; Senger and Roitberg, 1992; Coombs and Khan, 1998). These changes to survival and fecundity interact with other sources of mortality and change the pest's population growth rate (see Chapter 8).

For weeds, natural enemies rarely directly kill individual plants (with the exception of seed predators)

Integrating Biological Control into Conservation Practice, First Edition. Edited by Roy G. Van Driesche, Daniel Simberloff, Bernd Blossey, Charlotte Causton, Mark S. Hoddle, David L. Wagner, Christian O. Marks, Kevin M. Heinz, and Keith D. Warner.
© 2016 John Wiley & Sons, Ltd. Published 2016 by John Wiley & Sons, Ltd.

but rather act instead through a wider set of effects such as reductions in plant growth and reproduction, increases in susceptibility to pathogens, lowered competitiveness with other plants, and decreased dispersal. These changes may be measured in various ways, such as through changes in levels of nutrient reserves in individual plants, decreases in population-level plant biomass, percent cover, or seed production, and reduced counts in seed banks, among others (Crawley, 1983; Tipping et al., 2009, 2012).

Evaluation methods for invasive insects

In insect biological control projects, the question of drivers vs. passengers (see Chapter 2) is usually less important than for invasive plants, although in some cases dense populations of introduced insects may be facilitated by drought (White, 1984), habitat disturbance (King and Tschinkel, 2008), or associations with ants (e.g., Gaigher and Samways, 2013). Similarly, human actions may stimulate outbreaks (acting as the true drivers of insect outbreaks) through the use of pesticides that destroy effective natural enemies. While this is common in crop fields (e.g., Kenmore et al., 1984; Buschman and DePew, 1990; Grafton-Cardwell et al., 2008), it rarely happens in natural areas because widespread, repeated applications of insecticides to conservation areas are uncommon. Examples of pesticide-driven insect outbreaks in natural areas from the past include the widespread application of pesticides against forest insect pests or pests of public health such as biting flies (e.g., Luck and Dahlsten, 1975).

Methods to measure the effects of natural enemies released for biological control of invasive insects have historically been of two general types: (1) the use of experimental comparisons of pest density with and without the natural enemies of interest (DeBach and Huffaker, 1971; Luck et al., 1988) and (2) analyses of pest life tables (Varley and Gradwell, 1971; Bellows et al., 1992) to determine the importance of particular sources of mortality. Experiments can establish cause and effect, but must be done at a local scale. Observational studies used to build life tables or models of pest populations help create a basis for extrapolation to more regional scales. These approaches can be combined and both applied to evaluate the same natural enemy for either insects (e.g., Van Driesche and Taub, 1983) or plants (McEvoy et al., 1991). Here I discuss the relative merits of several methods and discuss their use against arthropod pests of natural areas.

Natural enemy evaluation through experimental comparisons of pest density

The basic premise of this approach is that, for pest insects that directly attack native plants (and do not vector systemic pathogens), "density makes the pest" (or, alternatively, *impact* = distribution × *abundance* × per capita effect [Parker et al., 1999]). Therefore, if introduced natural enemies reduce the density of the targeted invasive insect, this action is *prima facie* evidence of a successful remedy of the problem. An exception to this rule of practice is the case of introduced insects whose actual feeding is not the issue, but rather their action as vectors of introduced systemic plant pathogens, as for example redbay ambrosia beetle (*Xyleborus glabratus* Eichoff) (Hanula and Mayfield, 2014). In such cases there can be poor linkage between the density of the vectoring insect and the level of disease transmission in the plant population. In general, such systems are difficult or impossible to remedy through biological control of the vector, although disease incidence may be reduced.

Given this use of population density as a valid measure of the impacts of invasive insects, the experimental approaches used for their evaluation then focus on assessing differences in density of the insect in the habitat to be restored under circumstances that have or lack the natural enemy to be assessed. These plots then become the treatments (with and without the natural enemy) in the experiment, with target density as the measured response variable of most interest. When parasitoids are being evaluated for impact, it is possible and customary also to measure levels of parasitism in each treatment to demonstrate that it was high in the "with parasitoid" plot and low or absent in the "without parasitoid" plot, using field observations, rearing, or dissection to detect parasitism (e.g., Neuenschwander et al., 1986) or, when parasitoids are not separable to species as larvae, this can be done with molecular markers (Greenstone, 2006; Soper, 2012). These types of measures provide the formal confirmation that the treatments were successfully achieved in the field. When the natural enemy of interest is a predator, rates of prey mortality caused by the predator are estimated by observing predator–prey encounters directly (or with video equipment), counting prey cadavers, or estimating rates of prey consumption per field-collected predator using a variety of molecular methods that analyze gut contents (Pfannenstiel and Yeargen, 2002; Frank et al., 2007; King et al., 2008).

DeBach et al. (1976) described four approaches to creating the desired plots (with and without the natural enemy to be evaluated): (1) the addition method (comparisons made before and after the release of natural enemy, or in control and release plots); (2) the exclusion method (through use of cages); (3) the interference method, based on use of selective pesticides; and (4) suppression of honeydew-tending ants that exclude natural enemies where present thereby allowing natural enemies to exert their full effect in the absence of ants.

Before/after or release/control methods depend on being able to compare pest densities in plots either (a) before and after natural enemies have established and had sufficient time to increase in numbers ("before/after"), or (b) in plots where the natural enemy of interest is released and control plots where it is not ("release/control"). These evaluation designs can be used when the project's start precedes or, for the second option, is at the same time as the release of the natural enemy to be evaluated. It is often difficult to apply these methods when one wishes to measure the effect of long-established introduced or native natural enemies that have wide distributions. Also, it is important to note that the release/ control method requires that the dispersal rate of the newly released natural enemies is low enough relative to the distance between the control and release plots that several pest generations worth of data can be collected before control plots are lost to the natural enemy's invasion. This depends on estimating the parasitoid's dispersal rate, which may be influenced by both the density of the pest and increasing natural enemy numbers. In such instances, new control plots will be needed, farther from the release plots, as occurred with the fire ant parasitoid *Pseudacteon tricuspis* Borgmeier (Diptera: Phoridae), which overran control plots spaced 20 km from release sites within one year (Morrison and Porter, 2005).

Before/after

Densities of many species of Hemiptera (particularly whiteflies, scales, and mealybugs) are usually limited by parasitoids, and invasive species in these groups commonly reach high densities when separated from their specialized parasitoids. Whiteflies, scales, and mealybugs are often nearly sessile in several important life stages and many species undergo multiple generations per year. These attributes make the assessment of impacts of releases of specialized parasitoids relatively easy by means of before/after density counts (e.g., Dowell et al., 1979; Gould et al., 1992a; Bento et al., 2000; Grandgirard et al., 2008).

In this method, it is typical to collect data on the pest insect (usually of the sessile stages) per leaf or other unit of habitat over several generations or years and to compare changes in density to increases in rates of parasitism by the released natural enemy (Figure 9.1). This before/after design is often combined

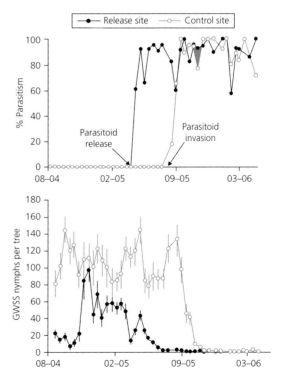

Figure 9.1 Evaluation methods "before/after" combined with "release/control" to measure impacts of the egg parasitoid *Gonatatocerus ashmeadi* on glassy-wing sharpshooter, *Homalodisca vitripennis*. Grandgirard et al. 2008. Reproduced with permission from Springer Science + Business Media.

with the release/control design discussed below and densities of the target pest may be measured in both types of plots, before and after natural enemy introduction (e.g., Gould et al., 1992a; Grandgirard et al., 2008), although control plots will very likely be invaded eventually by a successful natural enemy. This same combination of the before/after design, together with the release/control design (also called "near and far"), was used effectively by Huffaker and Kennett (1959) to evaluate the impact of *Chrysolina* species on the weed *Hypericum perforatum* L. in California.

For pests of natural areas, including forests, examples of use of the before/after method to evaluate insect biological control include studies of the larch case bearer (*Coleophora laricella* [Hübner]) (Lepidoptera: Coleophoridae) in Oregon (Ryan, 1997) (Figure 9.2), control of the glassy-winged sharpshooter, *Homalodisca vitripennis* (Germar) (Hemiptera: Cicadellidae), in Tahiti (Grandgirard et al., 2008), and re-evaluation in 2009–11 of the impact of the vedalia beetle (*Rodolia cardinalis* [Mulsant]) (Coleoptera: Coccinellidae) on cottony cushion scale (*Icerya purchasi* Maskell) (Hemiptera: Monophlebidae) in the Galápagos (Hoddle et al., 2013). In the case of the larch casebearer, the growth of a dominant forest tree was protected, while in those of the glassy-winged sharpshooter and cottony cushion scale, the health or survival of various foundational or rare native plants were restored.

Release/control

This design is useful when new natural enemies are being released for the first time in an area. Plots are designed where the density or other measures of the invasive insect's population dynamics are observed, and natural enemies are added in release, but not nearby control, plots. Control plots should be far enough away from release plots to slow the arrival of the natural enemy, but not so far as to risk being ecologically dissimilar to the release areas. The invasive insect's density and mortality are then monitored for a sufficient time for the natural enemy's population size to increase in response to the pest and to affect the pest's density. This will require something on the order of six to ten host generations, which, while for some Hemiptera may be a single year, in slowly developing pest species, such as many beetles or moths, may be up to 10 years or more. This approach may prove inadequate if natural enemies spread into control plots before sufficient time has passed to measure impacts in release plots (e.g., Morrison and Porter, 2005; Duan et al., 2015). In such cases, new control plots further away may be established (e.g., Morrison

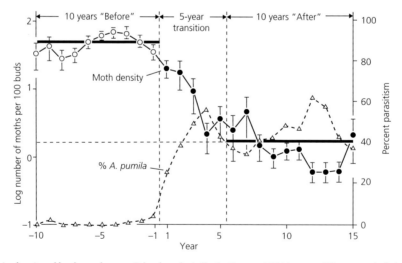

Figure 9.2 Changes in density of larch casebearer, *Coleophora laricella*, in Oregon (USA) over a 25-year period, in relation to the introduction of the parasitoid *Agathis pumila*, showing a drop in density of approximately 97% in the "after" period as compared to the "before" interval, demonstrating effective and sustained biological control of this pest. Ryan 1997. Reproduced with permission from Oxford University Press.

and Porter, 2005) or other methods of assessment, such as life table analyses, may be applied (e.g., for emerald ash borer, *Agrilus planipennis* Fairmaire [Duan et al., 2015]).

The exclusion method (caging)

In this approach, cages are used to isolate artificially created cohorts of insects from the attack of an already widely established introduced natural enemy, creating the "without natural enemy" treatment inside the cage for comparison to uncaged cohorts of the pest that naturally experience the presence of the natural enemy. This technique works well when the invasive insect involved is relatively sedentary (scales, mealybugs, whiteflies, aphids, psyllids, leafminers, gallmakers, etc.) and is capable of completing its immature development on a small section of the plant (a branch, a leaf, etc.) in a relatively short time period (3–6 weeks). In contrast, insects that move about appreciably during feeding (most caterpillars) or that take years to develop (wood-boring beetles) would be poorly suited for evaluation by this method, as it is difficult to exclude natural enemies in such circumstances or to maintain the integrity of exclusion cages over a number of years.

Cage effects, such as altered temperature or humidity, may or may not occur, depending on factors such as exposure of cages to direct sunlight. Small electronic measuring devices placed inside cages can be used to determine if such effects exist and their magnitude. If such effects occur, controls for cage effects (such as open bag treatments) may be used, a different enclosure type adopted, or the method may prove infeasible. When naturally occurring patches of pests are caged, care must be taken (based on timing of caging, selection of pest life stage or other approach) to ensure that no adult natural enemies or parasitized hosts are included in the cage at time of caging. Cages also, obviously, restrict movement. They are thus best used for species that are largely sessile (like scales) or which at least are in sessile stages during the period when the cages are deployed.

This approach has been used frequently to assess the impacts of natural enemies in crop fields (e.g., Van Driesche and Gyrisco, 1979; Prasad, 1989; Neuenschwander, 1996; Qureshi and Stansly, 2009) as, for example, to assess the impact of *Epidinocarsis lopezi* (De Santis) on cassava mealybug (*Phenacoccus manihoti* Matile-Ferrero) in Africa. In this case, caging of mealybugs to exclude the parasitoid caused parasitism rates to decrease from 23% to 1% and mealybug densities to increase 20-fold (from 7 to 140 per growing tip) (Neuenschwander et al., 1986). Similarly, Prasad (1989), for example, found strong effects of the predatory beetle *R. cardinalis* on the scale *I. purchasi* in South Australia (Figure 9.3).

Examples of use of this technique for evaluation of effects of natural enemies on invasive insects affecting natural areas include evaluation of the coccinellid beetle *R. cardinalis'* impact on the cottony cushion scale in the Galápagos (Calderón Alvarez et al., 2012). Cages were placed on scale-infested white mangrove (*Laguncularia racemosa* [L.] C.F. Gaertn.) early in 2002 to exclude the predatory beetle, which was released in the area immediately after cages were installed. During the first generation of the scale, densities dropped ca. 70% compared to densities inside cages. Unusual weather events (high rainfall), however, created physical conditions for scale inside cages such that the numbers in cages also declined in subsequent generations. However, scale densities remained extremely low on white mangrove through the reassessment done seven years later in 2009 (Hoddle et al., 2013).

Abell and Van Driesche (2011) used sleeve cages on hemlock branches to assess the importance of the aphelinid parasitoid *Encarsia citrina* (Crawford) as a source of mortality in the field to the invasive elongate hemlock scale (*Fiorinia externa* Ferris), an important invasive pest of eastern hemlock (*Tsuga canadensis* [L.] Carrière), and found that in the presence of this parasitoid only 8–11% of female scales reached sexual maturity – while inside exclusion cages 18–29% survived to reproduce. This result suggested that this parasitoid had a modest effect on the scale's population.

The interference method ("insecticide check method")

This approach protects cohorts of the target insect from the natural enemy to be evaluated by applying selective pesticides (that kill the natural enemy, but not the pest) to field plots (e.g., Braun et al., 1989). While selective pesticides may or may not exist for the study of a particular invasive insect, for the study of biological control of invasive plants this method is extremely easy to apply because nearly all insecticides are selective between insects and plants and can thus be used to remove the herbivorous insects used as

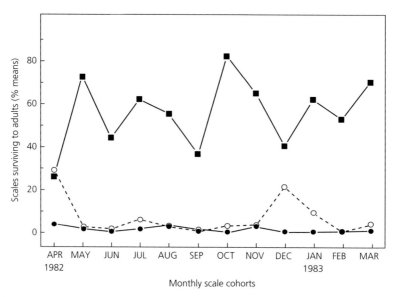

Figure 9.3 Use of exclusion cages in the native range (Australia) of cottony cushion scale (*Icerya purchasi*) to measure impact of the ladybird beetle *Rodolia cardinalis*. Prasad 1989. Reproduced with permission from Springer Science + Business Media.

biological control agents from plots (e.g., Tipping at al., 2008). Caution must be exercised to avoid materials causing phytotoxic effects and it needs to be recognized that all herbivores, not just the introduced natural enemies, are likely to be suppressed. This application is discussed in a following section on evaluation of weed biological control agents.

When selective pesticides are applied in an area where the natural enemy to be evaluated is abundant, the "with" and "without" plots required by the experimental method are readily obtained. This method was developed by Paul DeBach and applied to sucking pests affecting California crops, such as armored scales (e.g., DeBach et al., 1949; Huffaker et al., 1962). The insecticidal check method was used to evaluate the role of introduced parasitoids in regulating densities of olive scale (*Parlatoria oleae* [Colvée]) on olive trees in California; some trees were treated with DDT in a grove where the scale was under biological control and later compared to untreated trees in the grove. DDT applications resulted in scale densities on treated trees that were 75- to 1000-fold higher than on untreated trees, clearly showing the influence of natural enemies on the scale's density.

Examples of the use of selective pesticides for pests of natural areas are rare because of reluctance to apply insecticides in such places. However, in some semi-conserved areas such as forest reserves, unplanned experiments of this sort have occurred owing to pest control activities. In the forests surrounding South Lake Tahoe in northern California, for example, five years of malathion fogging for control of adult mosquitoes (weekly, June through mid September) led to an area-wide outbreak of pine needle scale (*Chionaspis pinifoliae* [Fitch]) on local pines. This outbreak subsided after fogging was discontinued (Luck and Dahlsten, 1975). While unintentional, these events demonstrated the regulation of this scale's density by natural enemies that were selectively suppressed by malathion, which had little effect on the scale.

Removal of ants

For some species, the presence or absence of ants can be manipulated to evaluate the effects of natural enemies. Some ant species defend scales, mealybugs, and other Hemiptera that produce honeydew from their natural enemy attacks. Removal of ants, with poison baits, can create plots where these natural enemies re-establish. Pest densities in these plots can then be compared to those in plots where ants are still present, allowing the effect of natural enemies to be measured. For example, bigheaded ant (*Pheidole megacephala* Fabricius) control

Figure 9.4 Recovery of *Pisonia grandis* on Cousine Island in the Seychelles after use of ant-baiting to break up mutualism supporting high density of invasive scale: left before baiting; right after baiting. Photos courtesy of René Gaigher.

was key to the restoration to a healthy condition of the native threatened tree *Pisonia grandis* R. Br. on Cousine Island in the Seychelles through suppression of the invasive (Gaigher and Samways, 2013). This ant tended and protected an invasive scale insect, *Pulvinaria urbicola* Cockerell, which in the absence of ants was suppressed to non-damaging levels by the parasitoid *Coccophagus ceroplastae* (Howard) (Smith et al., 2004). In the presence of ants (and thus the absence of the parasitoid), the scale reaches levels that debilitate and kill *P. grandis* trees. Data showed there was a 93% reduction in ant foraging in areas treated with ant bait, followed by a more than a 99% reduction in scale density, which allowed considerable overall improvement in *Pisonia* shoot condition and an increase in foliage density (Gaigher and Samways, 2013) (Figure 9.4).

Natural enemy evaluation using life tables

Life tables for analysis of insect populations were derived in concept from tables developed by the insurance industry for predicting human risk of death from particular factors. Because the age of insects is rarely known, insect life tables are lifestage-based instead. Southwood and Henderson (2000) provide a description of terminology used in insect life tables and their construction. Use of life tables to evaluate natural enemies is reviewed by Bellows et al. (1992). Early uses of life tables included both efforts to understand the effect of natural enemies on insect populations (Harcourt, 1969; Varley and Gradwell, 1970; Price, 1990) and to study the regulation of insect

densities in general (e.g., Varley and Gradwell, 1970; Royama, 1984; Carey, 1993).

At their simplest, life tables show how the numbers in an insect's population decrease across the species' life stages (numbers of live insects entering each stage) and how they died (numbers dying from particular factors). They therefore provide an opportunity to consider mortality caused by an introduced biological control agent in the context of other measured sources of mortality affecting the target insect. Life tables may be a snapshot of a single generation or may be collected for a series of generations. In the latter case, the search for regulating factors (ones whose mortality is positively dependent on the density of the pest in the susceptible stage) is possible through examination of how intensity of mortality caused by a factor varies across insect generations of differing density (see review by Stiling [1988]). Also, insect life tables allow better analysis of separate effects when two or more mortality factors act together on the same life stage, through the calculation of marginal rates of attack (Elkinton et al., 1992).

Life tables may be built either by sampling densities of naturally occurring populations of insects with discrete generations (e.g., Ryan, 1986) or, in the case of insects with overlapping generations, by following the fates of artificially created cohorts of same-age insects (e.g., Van Driesche and Taub, 1983; Abell and Van Driesche, 2011). The information that can be extracted from life tables increases as longer series of generations are studied, multiple sites are studied contemporaneously, or if life tables are constructed in paired plots having

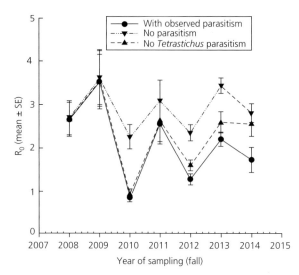

Figure 9.5 Net population growth rates (R_0) of emerald ash borer populations (*Agrilus planipennis*) estimated from life tables of each observed generation across different study sites in Michigan during the seven-year study (2008–14): solid black line represents R^0 values when all the observed parasitism (by both introduced parasitoid *Tetrastichus planipennisi* and all North American parasitoids); long-dashed line represents R_0 values when parasitism by *T. planipennisi* is absent, and short-dashed line represents R_0 values when all the parasitoid species were absent from the life tables. Duan et al. 2015. Reproduced with permission from John Wiley & Sons.

and lacking a natural enemy of special interest (e.g., Dowell et al., 1979; Van Driesche and Taub, 1983).

Life tables have been used to evaluate the influence of introduced or native natural enemies on densities and survival of pests of natural areas, especially pests of forests (e.g., Ryan, 1986; Olofsson, 1987; Duan et al., 2015) and amenity trees (Gould et al., 1992b; Girardoz et al., 2007). Life tables of invasive emerald ash borer populations in the north central United States were instrumental in tracking the decline in population growth rates of the pest and the correlation of the beetle's decline to increases in the attack rates on eggs and larvae by introduced and native parasitoids or predators (Duan et al., 2015) (Figure 9.5).

Evaluation methods for invasive plants

Overview

Assessing impacts of herbivores or pathogens (natural enemies) on invasive plants is more complex than is assessment of impacts of introduced parasitoids and predators on invasive insects for several reasons as addressed below. These comparisons are based on general life history patterns of insects and plants and they are offered as guidelines – obviously many exceptions exist.

1 *Death vs. injury*. For insects, most natural enemy attacks produce death of the attacked individual, while natural enemy attacks only sometimes kill plants and more typically just degrade plant performance, which may or may not translate into changes at the population level (Crawley, 1983). This reduction in performance may take a variety of forms, including reduced growth, reduced reproduction, lowered competitiveness with other plants, and increased susceptibility to stress or pathogens.

2 *Short vs. long generations*. For insects, new generations may arise in mere weeks, and even the longest-lived pest insects may require no more than one or two years per generation (with such obvious exceptions as cicadas). In contrast, plant life spans typically range from a single year (annuals) to decades or centuries (trees), although a few plants (species of *Azolla*, *Salvinia*, *Lemna*) approach insects in the brevity of their generations and thus are easy to study at the population level. The shorter generation time of insects allows densities of pest populations to adjust much faster to new natural enemies in comparison to weeds, especially long-lived perennials.

3 *Determinate vs. indeterminate body form.* The biomass of individual insects of a given species and life stage varies only narrowly around mean values because insects have relatively fixed body shapes and sizes. In contrast, plants of many species continue to grow over their whole life time. Consequently, individual plants may vary to a greater extent, with plants being large or small, depending on their past growth, which in turn affects the transition from effects on individuals to effects on populations. In addition, plant quality can vary greatly among individuals within a species, such that two plants of similar size may differ in key aspects, such as nutrient reserves or photosynthetic capability, and thus have different prospects for growth, defense from herbivory, and reproduction.

4 *Single stage vs. multiple stage reproduction.* Insects, with few exceptions, reproduce only as adults, while plants may reproduce over a range of body sizes, requiring a more complex accounting of when death or reductions in the performance of a plant from an agent occur relative to reproduction.

5 *Intraspecific competition.* In plants, unlike most insects, intraspecific competition between and within life stages is intense. This can strongly influence how plant populations respond to natural enemies, as recruitment compensation may counter the effects of natural enemies.

6 *Interspecific competition.* While interspecific competition between species of herbivorous insects clearly exists, insects are not as resource limited as plants, for which interspecific competition for light, water, and minerals is intense and constant. As a consequence, in insect biological control, natural enemies typically act alone, but in weed biological control, the competitive effects of other plants typically join with pressures from natural enemies to restrict growth and reproduction of targeted plants.

7 *Effects of disturbance and abiotic stress.* Plant germination from seed banks may be triggered by disturbance (e.g., McEvoy and Coombs, 1999), strongly affecting recruitment. While some insects respond to disturbance, most do not.

How to evaluate the effects of natural enemies on invasive plants has been reviewed by Syrett et al. (2000) and Morin et al. (2009), who discuss evaluation in terms of points in the life cycle of a biological control program, including (1) pre-release agent

selection (evaluating potential agents to choose ones most likely to affect plant populations given weaknesses in the targeted plant's population dynamics, e.g., McEvoy and Coombs [1999]); (2) post-release agent establishment; (3) agents' effects on plant performance (growth, seed set, etc.); (4) changes in plant population measures (reduction in cover or density) and their competition with other plants; (5) positive changes in the invaded plant community or the ecological processes formerly influenced by the invader (fire cycle, soil features, etc.); and (6) social or economic benefits of the project.

Below I expand on three of these: agents' effects on plant performance, changes in plant populations, and changes in the invaded community.

Measuring agents' effects on plant performance

At this level, it is the impacts on individual plants that are being assessed. While such assessments can be carried out before agent release (in quarantine greenhouses or in the native range) to estimate possible agent efficacy, pre-release experiments are not highly predictive of field impacts (because of environmental factors and population properties, such as resident natural enemy attacks on introduced agents, that cannot be included in laboratory tests); rather, these pre-release tests are mostly used to exclude agents that seem the least likely (of a group of candidates under study) to strongly debilitate plants.

Once an agent has been released and is established, the actual effects of populations of the agent on *individual* target plants can be measured by comparing plant attributes at sites with agent populations to nearby sites without the agent (*with/without design*), or within single sites by comparing plants with varying densities of the agent per plant, or by recording plant attributes at sites (ideally for several years) before agent release and then repeating these assessments after the agent is well established (*before/after design*), or some combination of these methods (e.g., Huffaker and Kennett, 1959). While confounding differences between sites, plants, or years might occur that would bias such comparisons and have to be considered, the purpose of such tests is to determine if agent populations, at naturally occurring densities, have measurable effects on the performance attributes of individuals of the target plant species.

Such effects are then either correlated to agent density, or compared between treatments plots (or years) with and without the agent. As a side note, in some cases, plant density may be unaffected by biological control agents, but if the agents reduce reproduction (seeding or other forms) drastically they may prevent further range expansion of the invasive plant, even if not reducing density of existing plants (e.g., Hoffmann and Moran, 1991; Tipping et al., 2008, 2009).

Such "with/without agent" comparisons can also be created by use of exclusion cages (*exclusion design*) or insecticides (*interference design*). The exclusion or interference design is needed if the agents to be evaluated are already established and widespread. Tipping et al. (2008), for example, used an insecticide to protect plots of young melaleuca trees (*Melaleuca quinquenervia* [Cav.] S. T. Blake) (a wetlands invasive tree) from locally established populations of two melaleuca biological control insects and compared tree performance between protected trees and ones exposed to the agents, which occurred naturally in the plots (interference design also called *insecticide check method*). Over a two-year period, large effects of the insecticide treatment were observed on tree height, woody biomass, leaf biomasss, and seed production. All measured parameters were enhanced by insecticide application, and by inference these changes were attributed to the biological control agents, as they were the only biotic agents present that both affected melaleuca and were susceptible to insecticides. Similar studies based on the insecticide check method include ones on puncturevine (*Tribulus terrestris* L.) (Kirkland and Goeden, 1978), prickly pear cacti (*Opuntia stricta* [Haw.] Haw.) (Hoffmann et al., 1998), and gorse (*Ulex europaeus* L.) (Davies et al., 2007), in various natural areas.

Assessing agent impacts on plant populations: attendant issues

While impacts on individual plants can translate into suppression of weed populations (as evidenced by successful biological control of some 40 species of invasive plants [McFadyen, 1998]), in some instances control of the plant population does not occur despite damage. This may be due to larger environmental factors, and the circumstances under which this failure to suppress the weed's population occurs need to be considered when implementing studies to look for impacts (Morin et al., 2009). While the list of such complications is potentially long, several are well known, including compensatory growth, interactions affecting compensation, and release from intraspecific competition among a plant's life stages. Below, these factors are briefly considered.

Compensation

One reason for failure of impacts on individual plants to result in impacts on the plant's population size, is that the attacked plants may simply tolerate the kind and degree of damage inflicted, resulting in no real harm to the plant. Or, in cases where significant damage (such as defoliation) does occur, the plant may be able to compensate for the injury through regrowth, potentially using stored reserves to reach even larger biomasses (Crawley, 1983; Rosenheim et al., 1997; Lu et al., 2014).

Factor interactions

Context can modify the efficacy of compensation responses of individual plants. In some instances, abiotic stresses may reduce the ability of plants to compensate for damage from weed biological control agents. Such is the case with alligator weed, *Alternanthera philoxeroides* Griseb., which can compensate for herbivory from *Agasicles hygrophila* Selman & Vogt (Coleoptera: Chrysomelidae) in terrestrial sites, but not under conditions of flooding (Lu and Ding, 2010), which determines the habitats in which this biological control has respectively succeeded (aquatic sites) and failed (terrestrial sites). Similarly, summer drought limits the ability of tansy ragweed *Senecio jacobaea* L. to compensate for herbivory caused by the cinnabar moth (*Tyria jacobaeae* [L.]) (Cox and McEvoy, 1983). Also, effects such as defoliation (which reduces root starch levels) can interact with other stresses, such as fire, to cause plant mortality and produce population-level effects that are more modest in the absence of the additional source of stress (e.g., Drus et al., 2014).

Release from intraspecific competition

Impacts at the plant level may not have population-level consequences because of release from intraspecific competition among plant stages, affecting rates of transition between stages. For example, reduction in seed production of annuals or biennials may or may not affect plant density in the subsequent generation if there are large seed banks or if the number of sites for

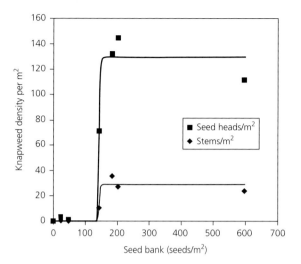

Figure 9.6 Relationship between seed bank densities and spotted knapweed (*Centaurea stoebe* L.) density, demonstrating the concept of processes acting with thresholds. Story et al. 2008. Reproduced with permission from Oxford University Press.

seedling establishment is more limiting than seed availability (for more discussion on this phenomenon, see review by Turnbull et al. [2000] and follow-up by McEvoy [2002]). Such effects may also be subject to thresholds, such that only once an effect like seed density reduction has reached a critical level will there be changes in subsequent adult plant densities (St\ory et al., 2008) (Figure 9.6). In a similar way, in some plant populations large cohorts of juvenile plants may exist that are suppressed by competition with older plants. In such cases, the death of adult plants from biological control agents may release rapid transition of juvenile plants to the adult stage, nullifying any population-level control from the agent (e.g., Ortega et al., 2012).

Assessing agent effects on plant populations: methods

As with assessment of impacts of introduced biological control agents on invasive insects, the broad goal of assessment of population-level effects of weed biological control agents is to measure differences in weed populations with and without the agents being evaluated. Using the same framework discussed above for insect targets, we can ask how well such methods work for plants and what exactly should be measured.

Release/control plots

For many weed biological control projects, new agents may be released in plots in several (or many) locations and compared to other sites where releases are not made. Paynter (2005), for example, evaluated the impact of the stem-mining moth *Carmenta mimosa* Eichlin and Passoa (Lepidoptera: Sesiidae) on mimosa (*Mimosa pigra* L.) by comparing nine stands where the moth was present to eight sites where it was absent. In this case, the variables chosen for evaluation included those measuring abundance of adult plants (% cover), reproductive output (seed rain and seed bank sizes), and establishment of new plants (rates of seedling establishment). While the authors found reductions in seed rain and seedling establishment, percent cover was not affected. A general limitation of this design is that many agents spread quickly and overrun control sites. Another limitation is that sites assigned to release or control may differ in ways that affect the comparison, emphasizing the need for random assignment of treatments to plots and adequate replication.

Before/after assessments

This is perhaps the most common approach to assessing outcomes of weed biological control. Plots where agents are released and establish are followed for a series of years and changes in the vegetation (declines in the target weed, increases in other plants) are tracked and correlated with abundance of the control agent. Plant population sizes are measured either as estimates of percent cover – in plots or over larger or remote areas with aerial photography (Nagler et al., 2012) – or for large species, as counts of numbers of plants per unit area. For example, declines in the invasive shrub/tree *Sesbania punicea* (Cav.) Benth. in South Africa were tracked for ten years; researchers counted plants and compared sites with one, two, or three natural enemy species, which attacked the plant in different ways. Large declines (nearly two orders of magnitude) were found in plots with all three agents and lesser effects in plots with two or one species (Figure 9.7) (Hoffman and Moran, 1998). In many early projects, photographs from fixed vantage points were used to measure declines in weed cover. This method may work very well for some species, such as waterhyacinth (*Eichhornia crassipes* [Mart.] Solms) or salvinia (*Salvinia molesta* D. Mitch.), whose mats are

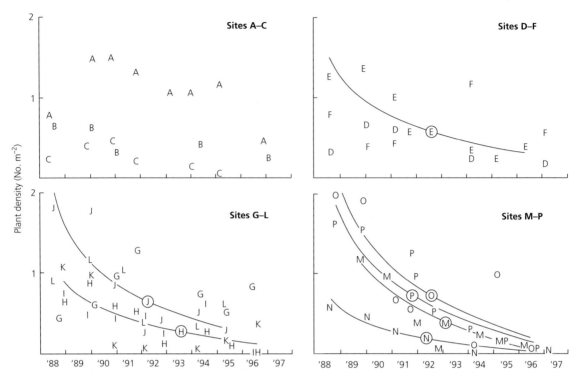

Figure 9.7 Relationship over time between density of *Sesbania punicea* at plots with one (A–C), two (D–F or G–L, each species), or three (M–P) biological control agents, in South Africa. Hoffman & Moran 1998. Reproduced with permission from Springer Science + Business Media.

replaced largely by clear water following successful biological control (Figure 9.8). However, while such photographs may be able to show declines in the target plant, they may not be effective at documenting the nature of the replacement vegetation. Another issue that arises in tracking plant density changes over time is that there is typically no way to separate natural enemy effects from other progressive changes (year effects) that might affect the region in which the work is done, such as a long period of drought, or changes in resource quality (e.g., improved water quality can contribute significantly to declines in waterhyacinth densities).

Cage or pesticide-exclusion plots

Exclusion methods are most easily used on a small scale for short periods and thus are best for assessing effects of established natural enemy populations on attributes of individual plants (e.g., Tipping et al., 2008). However, it is usually difficult or impossible to cage a population

of plants due to their large size, although it can be a viable option for species of modest size and simple architecture, such as annuals (forbs or grasses) or floating aquatic plants.

Life stage transition models

Models of plant populations can be constructed for several purposes. One is to understand the plant life cycle so as to identify transitions that are both potentially influential and, second, variable or amenable to manipulation, with the purpose of choosing agents that attack the plant at such stages in ways that have the greatest impact on the plant's success (e.g., Davis et al., 2006; Raghu and van Klinken, 2006; Dauer et al., 2012). Second, plant life stage models (Figure 9.9) help researchers understand more fully how the plant population copes with the environment in the broadest sense. Such models then provide a mechanistic structure of life stage transitions on top of which simulation modeling (e.g., Shea and Kelly, 1998) or field experimentation

Figure 9.8 Before (left) (1997) and after (right) (1999) control of waterhyacinth (Eichhornia crassipes) at Port Bell on Lake Victoria, Uganda. Photographs (left) by Ken L. S. Harley, Commonwealth Scientific and Industrial Research Organization, Bugwood.org; (right) by Mic Julien, Commonwealth Scientific and Industrial Research Organization, Bugwood.org. (*See Plate 6 for the color representation of this figure.*)

(McEvoy and Combs, 1999) can be used to overlay effects of specific sets of natural enemies and judge their importance against each other or against other factors (disturbance, competition, water regimes, etc).

Conceptually, the goal of matrix modeling (see Chapter 8) is to construct a dynamic model of how plants move through their life stages, driven by rates of transition (which are the equivalent of stage-specific survival rates in insect life tables) (Table 9.1). These transition rates, in principle, can be measured for real field populations of the plant. Such matrix models differ from insect life tables in that transitions are possible among many stages and the probabilities of these transitions can be expressed in the matrix (Figure 9.9). Of particular importance is that matrix models capture reproduction of all plant life stages able to produce progeny. This is unnecessary for insect populations, as insects reproduce only as adults. Also, for many insects, reproduction cannot be measured in the field and is estimated under laboratory conditions. In contrast, measurement of seed or other forms of reproduction in the field is a critical part of understanding plant population dynamics.

To use matrix modeling of plant populations to evaluate the effects of introduced natural enemies, model construction followed by simulated additions of natural enemies is one approach, as discussed above (e.g., Shea and Kelly, 1998). Such an approach was also used by Maines et al. (2013) to separate effects of intraspecific competition among plant life stages, precipitation, and effects of biological control agents on spotted knapweed (*Centaurea stoebe* L.) in North America. Their simulations suggested that only the inclusion of biological control agents (with other factors) led to plant population declines, supporting empirical evidence of the importance of these agents (Story et al., 2006; Knochel et al., 2010; Gayton and Miller, 2012).

Alternatively, models may be constructed for both populations that are subject to attack by the natural enemy of interest and populations that are not. This allows one to estimate the strength of the natural enemy's attack and its importance in the context of the plant's entire life cycle. One way to obtain this information is to measure outcomes experimentally when the natural enemy of interest is manipulated, either excluding it or raising its density. Models with

Table 9.1 Projection matrices, dominant eigenvalues, and elasticity matricies for *Carduus nutans* at two sites in New Zealand. Reproduced with permission from Shea and Kelly (1998).

	Midland					Argyll			
Projection matrix		SB	S	M	L		SB	S	M	L
	SB	0.0382	8.2499	179.4128	503.1428	SB	0.0382	5.3382	116.0907	0.0000
	S	0.1847	1.0906	22.1805	62.1848	S	0.1847	0.6725	14.3450	0.0000
	M	0.0000	0.0091	0.0000	0.0000	M	0.0000	0.0031	0.0000	0.0000
	L	0.0000	0.0056	0.0220	0.0000	L	0.0000	0.0005	0.0000	0.0000
λ (95% CI)		2.2142 (0.1256)					1.4398 (0.0190)			
Elasticity		SB	S	M	L		SB	S	M	L
	SB	0.0043	0.1984	0.0177	0.0311	SB	0.0089	0.3113	0.0146	0.0000
	S	0.2472	0.3090	0.0258	0.0453	S	0.3258	0.2976	0.0137	0.0000
	M	0.0000	0.0448	0.0000	0.0000	M	0.0000	0.0282	0.0000	0.0000
	L	0.0000	0.0752	0.0012	0.0000	L	0.0000	0.0000	0.0000	0.0000

Note: The matrix classes are "seed bank" (SB), and "small" (S), "medium" (M), and "large" (L) plants. See also Fig. 1.

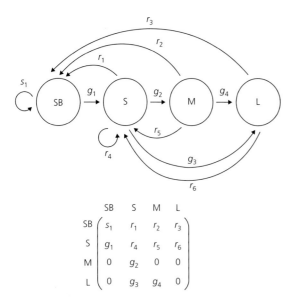

Figure 9.9 Sample life cycle diagram of *Carduus nutans* in New Zealand. Reproduced with permission from Shea and Kelly (1998).

and without the agent being evaluated can also be built for the plant population before the natural enemy's introduction and again after it is well established and has increased in number (*before/after, combined with matrix modeling*) or by comparing plots where the natural enemy was released to ones where it was not (*release/control, combined with matrix modeling*).

Other potential influences on the plant's success, such as the level of water availability or the level of soil fertility, can also be varied experimentally and those treatments can be crossed with the "with/without" natural enemy treatments to look for interactions. The differences between treatments that incorporate or omit the natural enemy of interest then reflect its impact at the population level, for a particular location and set of environmental conditions. Ideally, these impacts can be expressed as the enemies' influence on the plant population growth rates (lambda, as discussed more fully in Chapter 8). Models can also suggest why a given natural enemy might control a pest weed in one area but fail to do so in another (e.g., Shea et al., 2005). Application of matrix models to insect populations, rather than plants, is beginning (e.g., Lončarić and Hackenberger, 2013), but has not yet influenced evaluation of natural enemies of insects at a level comparable to that in weed biological control.

Changes in the invaded community

Biodiversity/dominance
Biodiversity or richness *per se* (i.e., number of species at a site within a group such as plants or insects) is not a useful measure of the benefits of biological weed control because increases in species number may be driven by colonization of unwanted invasive species, or be driven by other factors such as disturbance or succession.

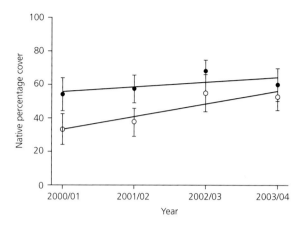

Figure 9.10 Increase in native plant richness in plots following biological control of mist flower (*Ageratina riparia*) in New Zealand, where the solid dots are for an invader plot and the hollow dots are for an invaded plot where the weed came under biological control. Barton et al. 2007. Reproduced with permission from Elsevier.

However, increases in the number of native species (found at the site pre-invasion), or their increase in abundance, reflects a measure of ecological recovery, particularly when the recovering native taxa are the high-value species identified as desirable. For example, native plant recovery was directly measured in plots where biological control agents suppressed invasive plants in the cases of biological control of melaleuca in Florida wetlands (Rayamajhi et al., 2009), mist flower (*Ageratina riparia* [Regel] R. M. King & H. Rob.) in New Zealand forests (Barton et al., 2007) (Figure 9.10), and forests in Tahiti invaded by the tree *Miconia calvescens* DC (Meyer et al., 2012). By contrast, in other cases suppression of an invasive weed may not lead to recovery of native vegetation, owing to factors such as site disturbance, changes in soil properties, or lack of native plant propagules. For example, following biological control of bridal creeper (*Asparagus asparagoides* [L.] Druce) in Australia, ground ceded by bridal creeper was occupied by a mixture of native and invasive plants in various proportions depending on the degree of disturbance at a site (Reid et al., 2008).

Ecological processes

Few biological control projects have attempted to measure restoration of ecological services following impacts on targeted environmental weeds. One exception is the biological control of saltcedar (especialy *Tamarix ramosissima* Ledeb.), which is believed to consume more water than the native vegetation it replaces through higher rates of evapotranspiration. Lowered rates of evapotranspiration, therefore, would be a predicted desirable consequence of the project. Field measurements in years and sites with high levels of saltcedar defoliation (90%) found that defoliation reduced stand evapotranspiration by 75% (Pattison et al., 2011). See also Chapter 14 and its discussion of loss of water flow from rivers in South Africa because of invasive woody plants and their reduction through biological control. Similarly, in South Africa, recovery of water flows in rivers occurred following an integrated program of biological control of several invasive trees or shrubs, together with manual clearance efforts.

Conclusions

Critical analyses of outcomes of biological control and restoration activities through experimentation and observation drive improvements to these sciences and are ignored at our own peril.

Acknowledgments

I thank Mark Hoddle, Kevin Heinz, David Wagner, Daniel Simberloff, and Peter McEvoy for reviews of Chapter 9.

References

Abell, K. J. and R. G. Van Driesche. 2011. The use of cohorts to evaluate the impact of *Encarsia citrina* (Hymenoptera: Aphelinidae) on *Fiorinia externa* (Hemiptera: Diaspididae) in the eastern United States. *Florida Entomologist* **94**: 902–908.

Barton, J., S. V. Fowler, A. F. Gianotti, et al. 2007. Successful biological control of mist flower (*Ageratina riparia*) in New Zealand: agent establishment, impact, and benefits to the native flora. *Biological Control* **40**: 370–385.

Bellows, T. S., Jr., R. G. Van Driesche, and J. S. Elkinton. 1992. Life-table construction and analysis in the evaluation of natural enemies. *Annual Review of Entomology* **37**: 587–614.

Bento, J. M. S., C. J. De Moraes, A. P. De Matos, and A. C. Bellotti. 2000. Classical biological control of the mealybug *Phenacoccus herreni* (Hemiptera: Pseudococcidae) in northeastern Brazil. *Environmental Entomology* **29**: 355–359.

Braun, A. R., A. C. Bellotti, J. M. Guerrero, and L. T. Wilson. 1989. Effect of predator exclusion on cassava infested with tetranychid mites (Acari: Tetranychidae). *Environmental Entomology* **18**: 711–714.

Buschman, L. L. and L. J. DePew. 1990. Outbreaks of Banks grass mite (Acari: Tetranychidae) in grain sorghum following insecticide applications. *Journal of Economic Entomology* **83**: 1570–1574.

Calderón Alvarez, C., C. E. Causton, M. S. Hoddle, et al. 2012. Monitoring the effects of *Rodolia cardinalis* on *Icerya purchasi* populations on the Galapagos Islands. *Biological Control* **57**: 167–179.

Carey, J. R. 1993. *Applied Demography for Biologists with Special Emphasis on Insects*. Oxford University Press. New York.

Coombs, M. and S. Khan. 1998. Fecundity and longevity of green vegetable bug, *Nezara viridula*, following parasitism by *Trichopoda giacomellii*. *Biological Control* **12**: 215–222.

Cox, C. S. and P. B. McEvoy. 1983. Effect of summer moisture stress on the capacity of tansy ragwort (*Senecio jacobaea*) to compensate for defoliation by cinnabar moth (*Tyria jacobaeae*). *Journal of Applied Ecology* **20**: 225–234.

Crawley, M. J. 1983. *Herbivory. The Dynamics of Animal–Plant Interactions*. Blackwell Scientific Publications. Oxford.

Dauer, J. T., P. B. McEvoy, and J. van Sickle. 2012. Controlling a plant invader by targeted disruption of its life cycle. *Journal of Applied Ecology* **49**: 322–330.

Davies, J. T., J. E. Ireson, and G. R. Allen. 2007. The impact of the gorse spider mites, *Tetranychus lintearius*, on the growth and development of gorse, *Ulex europaeus*. *Biological Control* **41**: 86–93.

Davis, A. S., D. A. Landis, V. Nuzzo, et al. 2006. Demographic models inform selection of biological control agents for garlic mustard (*Alliaria petiolata*). *Ecological Applications* **16**: 2399–2410.

DeBach, P. and C. B. Huffaker. 1971. Experimental techniques for evaluation of the effectiveness of natural enemies, pp. 113–140. *In*: Huffaker, C. B (ed.). *Biological Control*. Plenum. New York.

Debach, P., E. J. Dietrick, and C. A. Fleschner. 1949. A new technique for evaluating the efficiency of entomophagous insects in the field. *Journal of Economic Entomology* **42**: 546–547.

DeBach, P., C. B. Huffaker, and A. W. MacPhee. 1976. Evaluation of the impact of natural enemies, pp. 255–285. *In*: Huffaker, C. B. and P. S. Messenger (eds.). *Theory and Practice of Biological Control*. Academic Press. New York.

Dowell, R. V., G. E. Fitzpatrick, and J. A. Reinert. 1979. Biological control of citrus blackfly in southern Florida. *Environmental Entomology* **8**: 595–597.

Drus, G. M., T. L. Dudley, C. M. D'Antonio, et al. 2014. Synergistic interactions between leaf beetle herbivory and fire enhance tamarisk (*Tamarix* spp.) mortality. *Biological Control* **77**: 29–40.

Duan, J. J., L. S. Bauer, K. J. Abell, et al. 2015. Population dynamics of an invasive forest insect and associated natural enemies in the aftermath of invasion: implications for biological control. *Journal of Applied Ecology* in press.

Elkinton, J. S., J. P. Buonaccorsi, T. S. Bellows, and R. G. Van Driesche. 1992. Marginal attack rate, k-values and density dependence in the analysis of contemporaneous mortality factors. *Researches in Population Ecology* **34**: 29–44.

Frank, S. D., S. D. Wratten, H. S. Sandhu, and P. M. Shrewsbury. 2007. Video analysis to determine how habitat strata affects [sic] predator diversity and predation of *Epiphyas postvittana* (Lepidoptera: Tortricidae) in a vineyard. *Biological Control* **41**: 230–236.

Gayton, D. and V. Miller. 2012. Impact of biological control on two knapweed species in British Columbia. *Journal of Ecosystems and Management* **13**: 16–29.

Gaigher, R. and M. J. Samways. 2013. Strategic management of an invasive ant-scale mutualism enables recovery of a threatened tropical tree species. *Biotropica* **45**: 128–134.

Girardoz, S., D. L. J. Quicke, and M. Kenis. 2007. Factors favouring the development and maintenance of outbreaks in an invasive leaf miner, *Cameraria ohridella* (Lepidoptera: Gracillariidae): a life table study. *Agricultural and Forest Entomology* **9**: 141–158.

Gould, J. R., T. S. Bellows, Jr., and T. D. Paine. 1992a. Population dynamics *of Siphoninus phyllyreae* in California in the presence and absence of a parasitoid, *Encarsia partenopea*. *Ecological Entomology* **17**: 127–134.

Gould, J. R., T. S. Bellows, Jr., and T. D. Paine. 1992b. Evaluation of biological control of *Siphoninus phillyreae* (Haliday) by the parasitoid *Encarsia partenopea* (Walker), using life-table analysis. *Biological Control* **2**: 257–265.

Grafton-Cardwell, E. E., J. E. Lee, S. M. Robillard, and J. M. Gorden. 2008. Role of imidacloprid in integrated pest management of California citrus. *Journal of Economic Entomology* **101**: 451–460.

Grandgirard, J., M. S. Hoddle, J. N. Petit, et al. 2008. Engineering an invasion: classical biological control of the glassy-winged sharpshooter, *Homalodisca vitripennis*, by the egg parasitoid *Gonatocerus ashmeadi* in Tahiti and Moorea, French Polynesia. *Biological Invasions* **10**: 135–148.

Greenstone M. H. 2006. Molecular methods for assessing insect parasitism. *Bulletin of Entomological Research* **96**: 1–13.

Hanula, J. L. and A. E. Mayfield III. 2014. Redbay ambrosia beetle, pp. 199–311. *In*: Van Driesche and R. Reardons (eds.). *The Use of Classical Biological Control to Preserve Forests in North America*. FHTET- 2013-02. USDA Forest Service, Forest Health Technology Enterprise Team. Morgantown, West Virginia, USA. Available from: http://www.fs.fed.us/foresthealth/technology/pub_titles.shtml [Accessed January 2016].

Harcourt, D. G. 1969. The development and use of life tables in the study of natural insect populations. *Annual Review of Entomology* **14**: 175–196.

Hassell, M. P. 1978. The dynamics of arthropod predator-prey systems. *Monographs in Population Biology*. Volume **13**. Princeton University Press. Princeton, New Jersey, USA.

Hoddle, M. S., C. Crespo Ramírez, C. D. Hoddle, et al. 2013. Post release evaluation of *Rodolia cardinalis* (Coleoptera: Coccinellidae) for control of *Icerya purchasi* (Hemiptera: Monophlebidae) in the Galápagos Islands. *Biological Control* **67**: 262–-274.

Hoffmann, J. H. and W. C. Moran. 1991. Biological control of a perennial legume, *Sesbania punicea*, using a florivorous weevil, *Trichapion lativentre*: weed population dynamics with a scarcity of seeds. *Oecologia* **88**: 574–576.

Hoffmann, J. H. and W. C. Moran. 1998. The population dynamics of an introduced tree, *Sesbania punicea*, in South Africa, in response to long-term damage caused by different combinations of three species of biological control agents. *Oecologia* **114**: 343–348.

Hoffmann, J. H., W. C. Moran, and D. A. Zeller. 1998. Evaluation of *Cactoblastis cactorum* (Lepidoptera: Phycitidae) as a biological control agent of *Opuntia stricta* (Cactaceae) in the Kruger National Park, South Africa. *Biological Control* **12**: 20–24.

Huffaker, C. B. and C. E. Kennett. 1959. A ten-year study of vegetational changes associated with biological control of Klamath weed. *Journal of Range Management* **12**: 69–82.

Huffaker, C. B., C. E. Kennett, and G. L. Finney. 1962. Biological control of olive scale, *Parlatoria oleae* (Colvée) in California by imported *Aphytis maculicornis* (Masi) (Hymenoptera: Aphelinidae). *Hilgardia* **32**: 541–636.

Kenmore, P. E., F. O. Carino, C. A. Perez, et al. 1984. Population regulation of the rice brown planthopper (*Nilaparvata lugens* Stål) within rice fields in the Philippines. *Journal of Plant Protection in the Tropics* **1**: 19–37.

King, J. R. and W. R. Tschinkel. 2008. Experimental evidence that human impacts drive fire ant invasions and ecological change. *Proceedings of the National Academy of Sciences* **105**: 20339–20343.

King, R. A., D. S. Read, M. Traugott, and W. O. C. Symondson. 2008. Molecular analysis of predation: a review of best practices for DNA-based approaches. *Molecular Ecology* **17**: 947–963.

Kirkland, R. L. and R. D. Goeden. 1978. An insecticidal-check study of the biological control of puncturevien (*Tribulus terrestris*) by imported weevils, *Microlarinus lareynii* and *M. lypriformis* (Col.: Curculionidae). *Environmental Entomology* **7**: 349–354.

Knochel, D. G., N. D. Monson, and T. R. Seastedt. 2010. Additive effects of aboveground and belowground herbivores on the dominance of spotted knapweed (*Centaurea stoebe*). *Oecologia* **164**: 701–712.

Lončarić, Z. and B. K. Hackenberger. 2013. Stage and age structured *Aedes vexans* and *Culex pipiens* (Diptera: Culicidae) climate-dependent matrix population model. *Theoretical Population Biology* **83**: 82–94.

Lu, X-M. and J-Q. Ding. 2010. Flooding compromises compensatory capacity of an invasive plant: implications for biological control. *Biological Invasions* **12**: 179–189.

Lu, X-M., X. Shao, and J-Q. Ding. 2014. No impact of a native beetle on exotic plant performance and competitive ability due to plant compensation. *Plant Ecology* **215**: 275–284.

Luck, R. F. and D. L. Dahlsten. 1975. Natural decline of a pine needle scale (*Chionaspis pinifoliae* [Fitch]) outbreak at South Lake Tahoe, California following cessation of adult mosquito control with malathion. *Ecology* **56**: 893–904.

Luck, R. F., B. M. Shepard, and P. E. Kenmore. 1988. Experimental methods for evaluating arthropod natural enemies. *Annual Review of Entomology* **33**: 367–391.

Maines, A., D. Knochel, and T. Seastedt. 2013. Biological control and precipitation effects on spotted knapweed (*Centaurea stoebe*): empirical and modeling results. *Ecosphere* **4**: art80.

McEvoy, P. B. and E. M. Coombs. 1999. Biological control of plant invaders: regional patterns, field experiments, and structured population models. *Ecological Applications* **9**: 387–401.

McEvoy, P. B., C. Cox, and E. Coombs. 1991. Successful biological control of ragwort, *Senecio jacobaea*, by introduced insects in Oregon. *Ecological Applications* **1**: 430–442.

McEvoy, P. B. 2002. Insect–plant interactions on a planet of weeds. *Entomologia Experimentalis et Applicata* **104**: 165–179.

McFadyen, R. E. C. 1998. Biological control of weeds. *Annual Review of Entomology* **43**: 369–393.

Meyer, J.-Y., M. Fourdrigniez, and R. Taputuarai, R. 2012. Restoring habitat for native and endemic plants through the introduction of a fungal pathogen to control the alien invasive tree *Miconia calvescens* in the island of Tahiti. *Biological Control* **57**: 191–198.

Morin, L., A. M. Reid, N. M. Sims-Chilton, et al. 2009. Review of approaches to evaluate the effectiveness of weed biological control agents. *Biological Control* **51**: 1–15.

Morrison, L. W. and S. D. Porter. 2005. Testing for population-level impacts of introduced *Pseudacteon tricuspis* flies, phorid parasitoids of *Solenopsis invicta* fire ants. *Biological Control* **33**: 9–19.

Nagler, P. L., T. Brown, K. R. Hultine, et al. 2012. Regional scale impacts of *Tamarix* leaf beetles (*Diorhabda carinulata*) on the water availability of western U.S. rivers as determined by multi-scale remote sensing methods. *Remote Sensing of Environment* **118**: 227–240.

Neuenschwander, P. 1996. Evaluating the efficacy of biological control of three exotic Homopteran pests in tropical Africa. *Entomophaga* **41**: 405–424.

Neuenschwander, P., F. Schulthess, and E. Madojemu. 1986. Experimental evaluation of the efficiency of *Epidinocarsis lopezi*, a parasitoid introduced into Africa against the cassava mealybug, *Phenacoccus manihoti*. *Entomologia Experimentalis et Applicata* **42**: 133–138.

Olofsson, E. 1987. Mortality factors in a population of *Neodiprion sertifer* (Hymenoptera: Diprionidae). *Oikos* **48**: 297–303.

Ortega, Y. K., D. E. Pearson, L. P. Waller, et al. 2012. Population-level compensation impedes biological control of an invasive forb and indirect release of a native grass. *Ecology* **93**: 783–792.

Parker, I. M., D. Simberloff, W. M. Lonsdale, et al. 1999. Impact: toward a framework for understanding the ecological effects of invaders. *Biological Invasions* **1**: 3–19.

Pattison, R. R., C. M. D'Antonio, T. L. Dudley, et al. 2011. Early impacts of biological control on canopy cover and water use of the invasive saltcedar tree (*Tamarix* spp.) in western Nevada, USA. *Oecologia* **165**: 605–616.

Paynter, Q. 2005. Evaluating the impact of a biological control agent *Carmenta mimosa* on the woody wetland weed *Mimosa pigra* in Australia. *Journal of Applied Ecology* **42**: 1054–1062.

Pfannenstiel, R. S. and K. V. Yeargan. 2002. Identification and diel activity patterns of predators attacking *Helicoverpa zea* (Lepidoptera: Noctuidae) eggs in soybean and sweet corn. *Environmental Entomology* **31**: 232–241.

Prasad, Y. K. 1989. The role of natural enemies in controlling *Icerya purchasi* in South Australia. *Entomophaga* **34**: 391–395.

Price, P. W. 1990. Evaluating the role of natural enemies in latent and eruptive species: New approaches in life table construction, pp. 221–232. *In*: Watt, A. D., S. R. Leather, M. D. Hunter, and N. A. C. Kidd (eds.). *Population Dynamics of Forest Insects*. Intercept. Andover, Hampshire, UK.

Qureshi, J. A. and P. A. Stansly. 2009. Exclusion techniques reveal significant biotic mortality suffered by Asian citrus psyllid, *Diaphorina citri* (Hemiptera: Psyllidae), populations in Florida citrus. *Biological Control* **50**: 129–136.

Raghu, S. and R. D. van Klinken. 2006. Refining the process of agent selection through understanding plant demography and plant response to herbivory. *Australian Journal of Entomology* **45**: 308–316.

Rayamajhi, M., P. Pratt, T. Center, et al. 2009. Decline in exotic tree density facilitates increased plant diversity: the experience from *Melaleuca quinquenervia* invaded wetlands. *Wetlands Ecology and Management* **17**: 455–467.

Reid, A., L. Morin, and M. Neave. 2008. Impact evaluation of bridal creeper biological control in southern NSW, p. 267. *In*: van Klinken, R. D., V. A. Osten, F. D. Panetta, and J. C. Scanlan (eds.). *Proceedings of the 16th Australian Weeds Conference, Cairns Convention Centre, North Queensland, Australia, 18–22 May, 2008*. Queensland Weed Society. Brisbane, Queensland, Australia.

Rosenheim, J. A., L. R. Wilhoit, P. B. Goodell, et al. 1997. Plant compensation, natural biological control, and herbivory by *Aphis gossypii* on pre-reproductive cotton: the anatomy of a non-pest. *Entomologia Experimentalis et Applicata* **85**: 45–63.

Royama, T. 1984. Population dynamics of the spruce budworm, *Choristoneura fumiferana*. *Ecological Monographs* **54**: 429–462.

Ryan, R. B. 1986. Analysis of life tables for the larch casebearer (Lepidoptera: Coleophoridae) in Oregon. *The Canadian Entomologist* **118**: 1255–1263.

Ryan, R. B. 1997. Before and after evaluation of biological control of the larch casebearer (Lepidoptera: Coleophoridae) in the Blue Mountains of Oregon and Washington, 1972–1995. *Environmental Entomology* **26**: 703–715.

Senger, S. E. and B. D. Roitberg. 1992. Effects of parasitism by *Tomicobia tibialis* Ashmead (Hymenoptera: Pteromalidae) on reproductive parameters of female pine engravers, *Ips pini* (Say). *The Canadian Entomologist* **124**: 509–513.

Shea, K. and D. Kelly. 1998. Estimating biological control agent impact with matrix models: *Carduus nutans* in New Zealand. *Ecological Applications* **8**: 824–832.

Shea, K., D. Kelly, A. W. Sheppard, and T. L. Woodburn. 2005. Context-dependent biological control of an invasive thistle. *Ecology* **86**: 3174–3181.

Smith, D., D. Papacek, M. Hallam, and J. Smith. 2004. Biological control of *Pulvinaria urbicola* (Cockerell) (Homoptera: Coccidae) in a *Pisonia grandis* forest on North East Herald Cay in the Coral Sea. *General and Applied Entomology* **33**: 61–68.

Soper, A. 2012. *Biological Control of the Ambermarked Birch Leafminer (Profenusa thomsoni) in Alaska*, pp. 1–128. Ph.D. Dissertation. University of Massachusetts, Amherst.

Southwood, T. R. E. and P. A. Henderson. 2000. *Ecological Methods*, 3rd edn. Wiley. Oxford, UK.

Stiling, P. 1988. Density-dependent processes and key factors in insect populations. *Journal of Animal Ecology* **57**: 581–593.

Story, J. M., N. W. Callan, J. G. Corn, and L. J. White. 2006. Decline of spotted knapweed density at two sites in western Montana with large populations of the introduced root weevil, *Cyphocleonus achates* (Fahraeus). *Biological Control* **38**: 227–232.

Story, J. M., L. Smith, J. G. Corn, L. J. White. 2008. Influence of seed head-attacking biological control agents on spotted knapweed reproductive potential in western Montana over a 30-year period. *Environmental Entomology* **37**: 510–519.

Syrett, P., D. T. Briese, and J. H. Hoffmann. 2000. Success in biological control of terrestrial weeds by arthropods, pp. 189–230. *In*: Gurr, G. and S. Wratten (eds.). *Biological Control: Measures of Success*. Kluwer Academic Publishers. Dordrecht, The Netherlands.

Tipping, P. W., M. R. Martin, T. D. Center, et al. 2008. Suppression of growth and reproduction of an exotic invasive tree by two introduced insects. *Biological Control* **44**: 235–241.

Tipping, P. W., M. R. Martin, K. R. Nimmo, et al. 2009. Invasion of a West Everglades wetland by *Melaleuca quinquenervia* countered by classical biological control. *Biological Control* **48**: 73–78.

Tipping, P. W., M. R. Martin, R. Pierce, et al. 2012. Post-biological control invasion trajectory for *Melaleuca quinquenervia* in a seasonally inundated wetland. *Biological Control* **60**: 163–168.

Turnbull, L. A., M. J. Crawley, and M. Rees 2000. Are plant populations seed-limited? A review of seed sowing experiments. *Oikos* **88**: 225–238.

Van Driesche, R. G. and G. G. Gyrisco. 1979. Field studies of *Microctonus aethiopoides*, a parasite of the adult alfalfa weevil, *Hypera postica*, in New York. *Environmental Entomology* **8**: 238–244.

Van Driesche, R. G. and G. Taub. 1983. Field evaluation of the impact of parasites on *Phyllonorycter* leafminers on apple in Massachusetts, U.S.A. *Protection Ecology* **5**: 303–317.

Varley, G. C. and G. R. Gradwell. 1970. Recent advances in insect population dynamics. *Annual Review of Entomology* **15**: 1–24.

Varley, G. C. and G. R. Gradwell. 1971. The use of models and lifetables in assessing the role of natural enemies, pp. 93–110. *In*: Huffaker, C. B (ed.). *Biological Control*. Plenum. New York.

White, T. C. R. 1984. The abundance of invertebrate herbivores in relation to the availability of nitrogen in stressed food plants. *Oecologia* **63**: 90–105.

CHAPTER 10

Cases of biological control restoring natural systems

Roy G. Van Driesche[1], Paul D. Pratt[2], Ted D. Center[2], Min B. Rayamajhi[2], Phil W. Tipping[2], Mary Purcell[2], Simon Fowler[3], Charlotte Causton[4], Mark S. Hoddle[5], Leyla Kaufman[6], Russell H. Messing[6], Michael E. Montgomery[7], Rieks van Klinken[8], Jian J. Duan[9], and Jean-Yves Meyer[10]

[1] Department of Environmental Conservation, University of Massachusetts, USA

[2] USDA ARS Invasive Species Laboratory, Ft. Lauderdale, USA

[3] Landcare Research, Manaaki Whenua, New Zealand

[4] Charles Darwin Foundation, Galápagos Islands, Ecuador

[5] Department of Entomology, University of California, USA

[6] Department of Entomology, University of Hawaii, Manoa, USA

[7] Northern Research Station, USDA Forest Service (retired), USA

[8] Commonwealth Scientific and Industrial Research Organisation (CSIRO), Brisbane, Queensland, Australia

[9] USDA ARS Beneficial Insects Introduction Research Unit, Newark, USA

[10] Délégation à la Recherche, Tahiti, French Polynesia

The value of case histories

Here we present case histories of selected invasive insects and plants, discussing the damage they caused to native species, ecosystems, or biodiversity and why biological control was the approach chosen to attempt to reverse their effects. Cases have been chosen to include both examples in which biological control was successful and examples in which it failed. We also discuss several projects still in progress against pests of natural areas. Cases chosen focused on those in which the invasive species' impact on biodiversity or ecosystem function was high. Case histories provide insight into how natural enemies work, what problems might arise and preclude success, and how outcomes can be measured. Assessing the likelihood of non-target impacts of the biological control agents is not the focus of this chapter (rather, see Chapters 5 and 7). Similarly, while control of the invader *per se* is sometimes not sufficient for ecosystem restoration, which may require replanting of missing native plants or re-introduction of missing native animals, it is a critical first step that must precede such further efforts and is the focus here. For most chapters, one or more authors were among the scientists who carried out the biological control project and thus have personal familiarity with the events discussed.

Successes

These projects clearly met their ecological goals. In each case we discuss the ecological values at risk, the evidence for successful protection of these values, and any known evidence of unintended damage to the native ecosystem from the agents released.

Everglades preservation through biological control of *Melaleuca quinquenervia*

Paul D. Pratt, Ted D. Center, Min B. Rayamajhi, Phil W. Tipping, and Mary Purcell

Located in a transition zone between temperate and tropical ecosystems, the Florida Everglades is a 500,000 ha subtropical freshwater wetland (Craft et al., 1995). The topography is flat with a slight elevation change from the north to the south of only about 3–5 cm/km, creating a slowly southward-flowing system (ca. 0.8 km/d) emanating from the

Integrating Biological Control into Conservation Practice, First Edition. Edited by Roy G. Van Driesche, Daniel Simberloff, Bernd Blossey, Charlotte Causton, Mark S. Hoddle, David L. Wagner, Christian O. Marks, Kevin M. Heinz, and Keith D. Warner.
© 2016 John Wiley & Sons, Ltd. Published 2016 by John Wiley & Sons, Ltd.

southern end of Lake Okeechobee and terminating in the mangrove estuaries of Florida Bay (Kushlan, 1990). It is one of the largest freshwater marshes in North America and the largest single body of organic soils in the world (Loveless, 1959), encompassing marshes, sloughs, wet prairies, and tree islands. The global importance of the Everglades is reflected in its designations as an International Biosphere Reserve, a World Heritage Reserve, and a Wetland of International Importance (Maltby and Dugan, 1994).

Everglades plant communities contain species from both tropical (primarily Caribbean) and temperate floras along with many endemic species (Gunderson, 1994). These communities are largely defined by their hydrology, that is, the depth and duration of inundation (hydroperiod), which is governed by slight differences in elevation. Sawgrass (*Cladium jamaicense* Crantz), the quintessential Everglades plant community, covers about 70% of the area, either as a monoculture or intermixed with other emergent species (Loveless, 1959). Shallow-water sloughs, which traverse sawgrass marshes, are flooded year round and are dominated by floating and emergent aquatic species. Tree islands (small areas of higher elevation with diverse trees) are interspersed within a matrix of shorter vegetation, primarily sawgrass prairie (Rader and Richardson, 1992). Upland, drier habitats include tropical hardwood hammocks and pinelands (Gunderson, 1994).

Historically, Everglade habitats were drier in winter and wetter in summer. Drainage and water conservation programs, however, have largely reversed this pattern by retaining water during dry periods and discharging water through drainage canals during high rainfall events to meet urban and agricultural needs (Rader and Richardson, 1992). This reversal significantly harmed the flora and fauna and increased community susceptibility to invasion by non-indigenous species (Doren et al., 2009). Adjacent urban neighborhoods provide staging areas for the invasion of many non-native plant and animal species into Everglade ecosystems (Bodle et al., 1994; Gordon, 1998). While many of these invaders are seemingly benign, some are transformer species capable of altering the structure and function of the systems they invaded (Williamson and Fitter, 1996). The Australian tree *Melaleuca quinquenervia* (Cav.) S. T. Blake (Myrtales, Myrtaceae) is one such example owing to its ability to alter ecosystem structure and function (Gordon, 1998).

Melaleuca quinquenervia is native to northeastern Australia, New Caledonia, and parts of New Guinea. It has been present in south Florida (Figure 10.1) since the late

Figure 10.1 Melaleuca island in an invaded wetland. Photo courtesy of Francois Laroche, South Florida Water Management District. (*See Plate 7 for the color representation of this figure.*)

nineteenth century (Dray et al., 2006), and dispersal was assisted by nurserymen who are believed to have spread seeds deliberately into natural areas as a means of propagation (Austin, 1978). The US Army Corps of Engineers planted trees in the marshes of Lake Okeechobee during 1938–41 to create offshore tree islands to protect the southern levee from erosion (Dray et al., 2006). Altered hydrology from flood control and drainage projects during the 1950s undoubtedly contributed to the tree's invasion success. Stand coverage expands exponentially after initial colonization of suitable habitat (Laroche and Ferriter, 1992) so, by the late 1990s, *Melaleuca* infested about 200,000 ha and the Everglades was at risk of being totally overwhelmed (Laroche, 1998).

Although data are scant and some of the putative effects are dubious (e.g., increased transpiration [Allen et al., 1997]), *M. quinquenervia* clearly alters the vertical structure of plant communities, recruitment of native species, light availability, soil biogeochemistry, and nutrient availability (Gordon, 1998; Turner et al., 1998; Martin et al., 2009). Another negative ecological consequence of the plant's introduction is its effect on Everglade fire regimes. Sawgrass marshes are shallow-water communities that are well adapted to fire. They recover quickly after burning so long as water levels in the soil are not too deep and the organic soils do not burn (Kushlan, 1990; Lodge, 2004). However, *M. quinquenervia*, by virtue of its thick corky bark, also resists fire. Fires fueled by stands of this tree are very different in character from those fueled by sawgrass. In dense *M. quinqueneriva* stands, flames are quickly and explosively carried into the canopy as volatile essential oils in the foliage ignite (Flowers, 1991). These crown fires are extremely hot, kill other mature trees (pines and cypress), and ignite underlying muck soils, which can burn for weeks. The intense heat kills sawgrass and other native plants that normally survive the cooler ground fires that often occur in sawgrass-dominated areas. Intense fires induce massive seed release from *M. quinquenervia*, which retains seeds in persistent serotinous capsules on branches with individual trees storing millions of viable seeds (Rayamajhi et al., 2002). Lacking competition and surface litter, the dense carpets of *M. quinquenervia* seedlings that emerge prevent establishment of other plant species (Wade, 1981) (Figure 10.2). These recruitment events often give rise to nearly pure stands of mature trees with densities of up to 10,000 mature trees/ha (Rayamajhi et al., 2006, 2009).

Figure 10.2 Few to no plants can survive in the sterile habitat under melaleuca stands. Photo credit, Paul Pratt, USDA/ARS. (*See Plate 8 for the color representation of this figure.*)

Community transformation by *M. quinquenervia* in long-hydroperiod areas is driven by melaleuca's ability to accelerate soil accretion. As mentioned above, slight elevation differences determine hydroperiod durations and lead to large differences in plant communities. Individual *M. quinquenervia* trees growing in flooded environments produce adventitious "water" roots surrounding the base of the trunks up to the water line (Gomes and Kozlowski, 1980; Myers, 1983; McJannet, 2008). These roots directly add to the organic accumulation at the base of the tree, while also binding soil and trapping sediments (McJannet, 2008). In addition, litterfall adds as much as 12–25 MT/ha of undecomposed organic matter, which accumulates on the forest floor (Rayamajhi et al., 2010), leading to increased soil elevation. Soil accretion inevitably produces shorter hydroperiods over extensive areas, thus creating conditions conducive to further invasion. This "legacy effect" persists long after the trees are removed, so that while site rehabilitation may be possible, restored sites continue to differ, in varying degrees, from pre-invasion conditions.

Alarm over the deterioration of the Everglades led to a widespread desire to preserve and restore the system. Re-establishment of hydrological regimes was widely recognized by engineers as an essential foundation of restoration. However, biologists argued that correcting water flow patterns alone would not restore ecosystem function (Weaver, 2000) without addressing the invasive species problem (Doren et al., 2009). Chief among these was the need to reduce the effects of *M. quinquenervia*. Accordingly, a task force was assembled during the late 1980s to formulate a plan to reduce infestations of *M. quinquenervia*. This plan included biological control as one component of an overall management strategy. The plan called for traditional weed control measures (e.g., herbicide applications and mechanical harvesting) to remove the massive standing biomass and thus eliminate the tree from infested areas. However, anything done to kill the trees caused seed capsules to desiccate, resulting in mass seed releases, thus exacerbating the problem. To impede the re-invasion of cleared areas and to slow the rate of spread to new areas, a biological control program was designed with a primary goal of inhibiting stand regeneration. The biological control program, therefore, required herbivores that could prevent seed production or increase mortality of seedlings and saplings.

Preliminary surveys to find suitable candidate species for introduction against *M. quinquenervia* began in

Figure 10.3 The melaleuca weevil, *Oxyops vitiosa*, destroys seed production by more than 95%. Photo courtesy of Steve Ausmus, USDA/ARS. (*See Plate 9 for the color representation of this figure.*)

1986, but the most intensive faunal studies were done from 1989 to 1995, when more than 400 species of plant-feeding insects were recorded in field surveys (Center et al., 2012). Based on study of these insects and selection of species deemed both safe and potentially effective (for the goals described above), three biological control agents were released and established against the exotic tree in Florida: the weevil *Oxyops vitiosa* Pascoe (Figure 10.3), the psyllid *Boreioglycaspis melaleucae* Moore, and the gall midge *Lophodiplosis trifida* Gagné.

Oxyops vitiosa established readily at dry and seasonally wet sites but establishment failed at permanently flooded sites (Center et al., 2000). Initially, the weevil dispersed relatively slowly (Pratt et al., 2003), but it is now widely distributed (Balentine et al., 2009). Populations increased at rates comparable to other effective weed biological control agents but were influenced by availability of young shoots (Pratt et al., 2002, 2004). Damage to the stem tips dramatically reduced flowering and seed production (Pratt et al., 2005; Tipping et al., 2008). Coppicing from stumps was severely curtailed (Center et al., 2000), growth of saplings was markedly reduced, and termination of apical growth produced a bushier shape (Tipping et al., 2008). However, larvae became abundant mainly during winter and spring, coincident with seasonal production of young foliage (Center et al., 2000; Pratt et al., 2004). This allowed some "escape" at other times of the year. After attainment of large populations, weevils

(a)   (b)

Figure 10.4 (a) The melaleuca psyllid, *Boreioglycaspis melaleucae* Moore (adult) was the second agent released against melaleuca. Photo courtesy of Steve Ausmus, USDA/ARS. (b) Psyllid flocculence, showing a colony of the insect. Photo credit, Paul Pratt, USDA/ARS. (*See Plate 10 for the color representation of this figure.*)

regularly moved into flooded sites, causing significant damage. Although many fully grown larvae drowned, some managed to find pupation sites, allowing small populations to persist in permanently wet habitats (Center, pers. obs.).

The psyllid *B. melaleucae* (Figure 10.4) established quickly (Center et al., 2006) and dispersed rapidly throughout the range of *M. quinquenervia* in Florida. Populations spread at a rate of approximately 7 km/year and are now widely distributed (Balentine et al., 2009). Enormous populations developed during the spring dry season in all habitat types, but populations declined during the summer rainy season. This was probably more of an effect of high temperatures rather than of precipitation (Chiarelli et al., 2011). Psyllids caused high mortality of seedlings and premature leaf drop from mature trees (Franks et al., 2006; Morath et al., 2006). Mortality of coppicing stumps also increased in conjunction with infestations of *O. vitiosa* (Center et al., 2007).

The stem-galling midge *L. trifida* was initially released at 24 sites distributed throughout southern Florida in *M. quinqueneriva* stands of varying sizes and hydrology. Both small and large numbers of individuals were used in an attempt to determine an optimal release strategy. Establishment was universally successful (Pratt et al., 2013). Areas where *M. quinquenervia* stands were regenerating from seed or coppicing stumps were heavily

galled with a high percentage of the plants being killed, possibly because of interactions with fire or frost. While galls occurred most abundantly in the lower strata, they were also found as high as 13 m in the upper canopy (Pratt et al., 2013). Dispersal occurred at 20 km/year.

Various metrics can be used to quantify the cumulative efficacy of the introduced *Melaleuca* biological control agents. The program's primary objective was to curtail seed production, and the introduced herbivores met this goal by reducing the tree's reproductive capacity by as much as 99% (Pratt et al., 2005; Tipping et al., 2008). Seed suppression improved cost efficiency of conventional control tactics because post-herbicide follow-up treatments to remove recruited seedlings were less costly or unnecessary. *Melaleuca*-infested lands adjacent to cleared sites previously provided seed sources for re-invasion and were often inaccessible to land managers, but attack by introduced herbivores markedly limited seed movement into treated areas (Tipping et al., 2012).

Equally as important, no significant feeding on non-target species occurred (Center et al., 2007; Pratt et al., 2009, 2013). Reductions in reproduction and recruitment led to change in the population growth rate of *M. quinquenervia* in Florida. A demographic analysis of seven different *M. quinquenervia* stands showed negative regional population growth rate (Sevillano, 2010), indicating that mortality rates now outpace recruitment. These findings were further confirmed by a five-year study that found ca. 50% reduction in seedling and sapling density as a result of chronic feeding by the introduced herbivores (Tipping et al., 2009). In a third study, it was found that feeding by natural enemies accelerated mortality rates of *M. quinquenervia* growing in dense monocultures (Rayamajhi et al., 2007). Most importantly, this decline in melaleuca density was associated with a concurrent fourfold increase in plant species diversity within the study area (Rayamajhi et al., 2009) (Figures 10.5, 10.6).

Overall, this biological control program provided a sustainable pathway to the restoration and conservation of a unique ecosystem. Because *M. quinquenervia* has been suppressed at both the individual and population levels, habitat disturbances (fire intensity/frequency and soil accretion) will be mitigated as tracts of lands are revegetated by native species. Although full implementation of the restoration plan will take decades, it is clear that biological control has been a critical and necessary component in the conservation and restoration of the Florida Everglades.

Recovery of native vegetation in New Zealand forests following control of mist flower

Simon Fowler

Mist flower, *Ageratina riparia* (Regel) R. King and H. Robinson, was a weed in the northern part of North Island (New Zealand) in forest margins, open places, poorly managed pasture, wetlands, and river systems (Barton et al., 2007). Along tracks and riparian areas in native forests it was capable of forming large, dense mats of semi-woody stems that smothered native plants and prevented forest regeneration (Barton et al., 2007). In two specific examples, mist flower was a threat to the survival of two endemic plants, *Hebe bishopiana* (Petrie) Hatch and *H. acutiflora* Cockayne (Figure 10.7) (Barton et al., 2007). Mist flower was controlled biologically in Hawaii in the 1980s, and research indicated the program could be transferred to New Zealand (Morin et al., 1997).

Figure 10.5 Recovery of native vegetation in a melaleuca-dominated plot, following successful biological control, of melaleuca; note the dead trees and open canopy. Photo courtesy of Min Rayamajhi, USDA/ARS. (*See Plate 11 for the color representation of this figure.*)

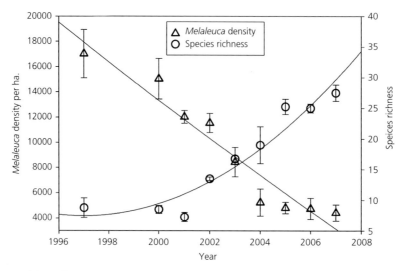

Figure 10.6 Decline in melaleuca density and increase in species diversity as biological control agents exerted their effect at a Florida research site. Unpublished data of Paul Pratt and Min Rayamajhi.

(a)   (b)   (c)

Figure 10.7 Control of mist flower (*Ageratina riparia*) by the introduced white smut fungus, *Entyloma ageratinae*, showing the same location (a) about a year after fungus release when plants were infected but not yet killed (January 27, 2000–summer) and (b) two years later when mist flower shrubs had died and been replaced by mostly native vegetation, particularly ferns and grasses (November 8, 2001–spring); site is Brookby on the edge of the Hunua Ranges (an important water catchment and biodiversity area, just southeast of Auckland, New Zealand), and (c) endangered New Zealand endemic plant, *Hebe acutiflora* Cockayne, no longer threated by mist flower. Photos courtesy of Jane Barton/Landcare Research (a, b) and Gillian Crowcroft (c). (*See Plate 12 for the color representation of this figure.*)

Releases of the white smut fungus *Entyloma ageratinae* Barreto and Evans began in New Zealand in 1998 and were followed by release of the gall fly *Procecidochares alani* Steyskal in 2001 (Barton et al., 2007). The monitoring results from 1998 to 2003 presented here are summarized from Barton et al. (2007) with a 10-year update in 2008 (Landcare Research, unpublished data). Infection of mist flower by the fungus increased dramatically in the first few years after its release, peaking at a mean of 67% of live leaves with lesions. Ten years after release the mean percentage of live leaves infected by the fungus was still 55% despite very large reductions in the percentage cover of mist flower. The gall fly, released three years after the fungus, spread more slowly than the fungus and had a much less dramatic impact on the target weed. However, two years after release it had reached a mean level of 1.96 galls/stem. In 2008, mean attack levels of 0.57 and 2.29 were measured in two areas. All these levels of attack for the gall fly were above the mean level of 0.46 galls/stem reported in Hawaii as contributing to mist flower suppression (Hapai, 1977; Barton et al., 2007). By 2008, mean plant height was reduced by 33% (compared to 1998) and the percentage of nodes showing regrowth to <2%, showing that the weed did not have the capacity through compensatory growth to overcome the agents' attack. In plots selected for high levels of mist flower infestation, mean percentage cover declined from >80% in 1998 to 1.5% in 2003. In randomly selected plots, mist flower cover was 6.1% in 1999, 0.86% in 2003, and 0.1% in 2008.

Mean species richness of native plants was initially 9.3 per plot where cover of mist flower was ≥70%, which was lower (by 44%) than the mean of 16.5 in paired plots that had not been invaded by mist flower. Between 1998 and 2003, the mean percentage cover of mist flower in the plots invaded by the weed decreased to 1.5%. Over the same period, both the species richness and percentage cover of native plants increased, and by 2003 these measures did not differ significantly from those in the paired plots that had never been invaded by mist flower. In contrast, there were no significant differences in the species richness or percentage cover of exotic plants species (excluding mist flower) between plots that were originally invaded by the weed and plots that had never been invaded. Many plant species colonizing the plots were important native mid- or late-successional shrubs or trees. The exotic plant species common in the plots were mostly not weeds that threatened native forest habitats. An exception was African club moss, *Selaginella kraussiana* (Kunze) A. Braun, which was invading the plots, and appeared to reduce the species richness of native plants. However, *S. kraussiana* was also invading the plots that had never had any mist flower present, and where mist flower had been suppressed by the fungus there was only a weak, non-statistically significant, "replacement weed effect."

Both of the endemic New Zealand plant species threatened by mist flower, *H. bishopiana* and *H. acuiflora*, are now considered under reduced threat because of the successful biological control of mist flower (P. de Lange, pers. comm.).

Protecting Galápagos plants from cottony cushion scale

Charlotte Causton, Mark S. Hoddle, and Roy G. Van Driesche

The Galápagos archipelago, a UNESCO World Heritage Site and Biosphere Reserve, is renowned for its unique fauna and flora. Geographic isolation and ongoing seismic and volcanic activity have led to the evolution of many species found nowhere else in the world, including 180 flowering plants and ferns. Galápagos plants, like other endemic species in the archipelago, however, are facing an increasingly uncertain future because of threats from invasive species and diseases (Gardener and Grenier, 2011). Three plant species are already extinct, and according to a 2006 evaluation using the threat categories of the International Union for Conservation of Nature (IUCN), more than half of the endemic plant species are in serious decline with at least 20 species facing possible extinction (Tye, 2008). Sixteen of these critically endangered species are restricted to the inhabited islands, which are undergoing severe ecological change as a result of invasive species (Tye, 2008).

Endemic Galápagos plants and other indigenous plants of conservation value are at high risk from invasive herbivorous insects, particularly the cottony cushion scale, *Icerya purchasi* Maskell (Hemiptera: Monophlebidae) (Figure 10.8), a cosmopolitan plant pest native to Australia and possibly New Zealand. Most widely known as a pest of citrus, this scale insect feeds on over 200 different plant species (Caltagirone and Doutt, 1989; Causton, 2001). Known locally as "el pulgon," it was first

Figure 10.8 White mangrove infested with cottony cushion scale. Photo credit, Mark Hoddle. (*See Plate 13 for the color representation of this figure.*)

discovered in the Galápagos Islands in 1982, where it was probably accidentally imported on plant material from mainland South America. Within 14 years of its discovery, this scale insect had spread by wind currents and the inter-island movement of plants to 15 of the 18 larger islands and had affected the growth and survival of many plant species of importance (Causton, 2001; Causton et al., 2004). High density *I. purchasi* populations can be especially damaging to native plants with restricted ranges, such as those found in the Galápagos Islands. Immature and adult *I. purchasi* cause damage directly to host plants by sucking sap and removing nutrients, and indirectly through the excretion of a sticky honeydew, which provides a substrate for black sooty molds that grow on leaves and stems disfiguring plants and preventing photosynthesis. Honeydew, a rich carbohydrate source, is also highly attractive to invasive ants (including the little fire ant, *Wasmannia auropunctata* [Roger], and the tropical fire ant, *Solenopsis geminata* [Fabricius]), which harvest this waste product for consumption and in turn help scale insect populations thrive by defending scale colonies from predators and transporting them between plants (Causton, 2001; Hoddle et al., 2013).

To prevent further damage to plants of conservation value, in 1996 conservation scientists at the Charles Darwin Foundation and Galápagos National Park Service managers formed a Technical Advisory Committee to evaluate the threats of the *I. purchasi* invasion and to develop a management strategy for the scale's control. Surveys across islands indicated that high-density populations of *I. purchasi* were restricting plant growth and possibly killing individuals of at least 62 native and endemic plant species, 16 of which were listed as threatened in the IUCN Red List of Threatened Species, including six Endangered or Critically Endangered species (Causton, 2001; Causton et al., 2004). Of particular concern was the high mortality reported on the native white mangrove (*Laguncularia racemosa* [L.] Gaertn. F.) (Figure 10.8), an important refuge for native freshwater and terrestrial invertebrates, and some birds, including the critically endangered mangrove finch, *Camarhynchus heliobates* Gould, which is restricted to small pockets of mangroves on the western coast of Isabela Island (Fessl et al., 2010). Experimental trials, in which potted plants were artificially infested with *I. purchasi*, confirmed that this scale insect was having a measurable impact on the leaf, shoot, and root growth of white mangrove and at least two other native plant species in the Galápagos Islands (*Acacia macracantha* Humb. & Bonpl. ex Willd. and *Phaseolus mollis* Hook f.) and was likely damaging others (Causton, 2001).

Several endemic daisy genera unique to Galápagos (*Darwiniothamnus* and *Scalesia*) (Figure 10.9) also appeared

Figure 10.9 Endemic Galápagan plants whose populations were threatened by cottony cushion scale: left *Darwiniothamnus tenuifolius*, and right, *Scalesia* sp. Photo credit, Mark Hoddle. (*See Plate 14 for the color representation of this figure.*)

to be heavily damaged by *I. purchasi*, including nine threatened species (Causton, 2001; Calderón Alvarez et al., 2012). Moreover, studies suggested that the damage caused by *I. purchasi* was affecting endemic plant-feeding Lepidoptera that are specialist feeders on some of these endemic plant species. For example, the local extirpation of three species of endemic Lepidoptera (*Platyptilia vilema* B. Landry, *Semiothisa cerussata* Herbulot, *Tebenna galapagoensis* Heppner & B. Landry) associated with *Darwiniothamnus tenuifolius* (Hook. f.) Harling (Darwin's Aster) on Alcedo Volcano may have been caused by *I. purchasi*-induced mortality of plants (Roque-Albelo, 2003).

The Technical Advisory Committee concluded that there was enough evidence to show that *I. purchasi* was having a significant impact on Galápagos ecosystems. Following an evaluation of management options, the committee decided that classical biological control using the coccinellid beetle *Rodolia cardinalis* Mulsant (Figure 10.10), a natural enemy of *Icerya* in its native range and a biological control agent used globally for the control of this pest in citrus (Bartlett, 1978), offered the best prospect for reducing the impact of *I. purchasi* on Galápagos plants. Biological control was considered to be the best option as it was the only management tool that was likely to cause a permanent and widespread suppression of this pest throughout the archipelago on both inhabited and uninhabited islands with minimal risk to non-target species. Studies in agricultural systems where *R. cardinalis* had been deliberately introduced (e.g., California) (Quezada and DeBach, 1973) and in its native range (Prasad, 1989) suggested that the beetle has a high degree of host specificity and can be

very effective at suppressing *I. purchasi* populations. Nevertheless, the deliberate importation and release of an exotic species that had the capacity to spread unassisted throughout the Galápagos Archipelago, a World Heritage Protected Area with a wealth of endemic species, was controversial (Causton, 2009). Rigorous host-specificity trials were considered necessary to determine whether the introduction of this beetle would pose risks to native and endemic insects.

In 1999, *R. cardinalis* was imported from Australia into a quarantine facility at the Charles Darwin Research Station for feeding trials. In quarantine, 16 insect species from three different orders (Hemiptera, Coleoptera, and Neuroptera) and nine families (Ortheziidae, Monophlebidae, Pseudococcidae, Eriococcidae, Coccidae, Diaspididae, Aphididae, Coccinellidae, and Chrysopidae) were tested for their suitability as prey for larvae and adults of *R. cardinalis*. The native species tested included those presumed to be at high risk of being preyed upon by *R. cardinalis* (Causton et al., 2004). Studies were also carried out to eliminate the possibility of toxic effects of the beetles on birds that might feed on them. This risk exists because some species of coccinellids produce a defensive fluid that contains an alkaloid that can be toxic to some vertebrates if ingested (e.g., Pasteels et al., 1973). Ornithologists were concerned that Darwin's finches and other passerines could be at risk should they consume *R. cardinalis* (Lincango et al., 2011). The results of these studies and a risk analysis suggested that *R. cardinalis* would not present a significant threat to either non-target insects or insectivorous birds (Lincango et al., 2011).

Following rigorous review, the Galápagos National Park Service approved release from quarantine, and between 2002 and 2005, 2206 *R. cardinalis* were released onto 10 different islands in the archipelago. This was the first time that Galápagos authorities had approved the introduction of a biological control agent. Follow-up studies soon after its introduction were considered important to document results for local stakeholders and community members who had participated in the release program. Directly after the release of *R. cardinalis*, a predator exclusion study and field observations were carried out on scale insect populations on white mangrove (*L. racemosa*) on Santa Cruz Island to document impact (Calderón Alvarez et al., 2012). In addition to this study, local residents and high school students participated in surveys to detect *R. cardinalis* on host plants in different habitat zones on Santa Cruz.

Figure 10.10 Adult *Rodolia cardinalis* preying on cottony cushion scale. Photo credit, Mark Hoddle. (*See Plate 15 for the color representation of this figure.*)

In less than three months after *R. cardinalis* was released, populations of *I. purchasi* on white mangrove that were exposed to the predator in the exclusion experiment, or were monitored in the field, had declined by 99–100% (Calderón Alvarez et al., 2012). Results suggested that *R. cardinalis* played a key role in this decline, possibly in combination with high rainfall that occurred around the same time. The coccinellid was also found on 10 other host plants infested by *I. purchasi* and by 22 weeks it had spread to the other side of Santa Cruz Island (45 km) and to the neighboring island of Baltra (separated from the northern coast of Santa Cruz by an ocean channel approximately 200 m wide) without human assistance (Calderón Alvarez et al., 2012).

In October 2009, seven years after the release of *R. cardinalis*, the biological control program was evaluated more fully to determine if there was successful continued suppression of the target pest, *I. purchasi*, and whether there had been any non-target impacts (Hoddle et al., 2013). The evaluation was carried out over two years and had three major objectives: (1) to survey islands for the presence of *I. purchasi* and *R. cardinalis* in urban, agricultural, and natural areas; (2) to measure the degree of suppression of *I. purchasi* by *R. cardinalis*; and (3) to investigate under field-like conditions the hypothesis that *R. cardinalis* has a limited prey range as predicted by quarantine laboratory studies. Results from this evaluation indicated that *R. cardinalis* survived and spread after its introduction in 2002. It is now widely present in many areas and habitats on at least six islands where it was released, and without purposeful human assistance it has colonized two other islands infested with *I. purchasi*. Furthermore it is found on a wide range of the plants infested by *I. purchasi* in the archipelago (at least 48 of the 112 known host plants) (Calderón Alvarez et al., 2012).

Monthly monitoring over a two-year period on Santa Cruz and San Cristóbal Islands in relatively undisturbed habitats (five study sites) and in a disturbed habitat (one study site) suggested that overall most *I. purchasi* populations were low and that *I. purchasi* has been suppressed to non-damaging levels on several important native host plants, including white mangrove, several *Acacia* species, *Waltheria ovata* Cav., *Prosopis juliflora* (Sw.) DC., and *Parkinsonia aculeata* L. (~60–98% reduction in *I. purchasi* densities depending on host plant and habitat when compared to comparably assessed pre-*Rodolia* pest densities) (Hoddle et al., 2013). For example, before the

release of *R. cardinalis*, white mangroves were heavily infested with *I. purchasi*. Following the release of *R. cardinalis* in 2002, *I. purchasi* populations declined on white mangroves (Calderón Alvarez et al., 2012) and continued to be very low on white mangroves over the two-year survey period between 2009 and 2011 (Hoddle et al., 2013) (Figure 10.11).

In some areas and on certain host plants, *I. purchasi* populations persisted at higher densities. The most noteworthy was the *I. purchasi* infestation on sea grape, *Scaevola plumieri* (L.) Vahl, at Tortuga Beach, Santa Cruz Island. At that study site, *I. purchasi* was always present, but the percentage of infested plants increased and decreased over the two-year survey period. As pest densities increased, *R. cardinalis* populations rebounded, and approximately two to six months later *I. purchasi* populations would decline again as predator populations peaked. Other notable exceptions existed when honeydew-collecting ants were present and guarded *I. purchasi* from *R. cardinalis* (Hoddle et al., 2013).

The feeding preferences and predation behavior of *R. cardinalis* were evaluated in a large walk-in field cage (Figure 10.12) at the Charles Darwin Research Station, October–December 2009. The cage held potted native plants (e.g., *P. aculeata*, *Gossypium* spp., *Acacia macracantha* Humb. & Bonpl. ex Willd., and *W. ovata*) that were infested with *I. purchasi* (the target) and non-target prey species (e.g., *Coccus viridis* [Green], *Ceroplastes* spp., mealybugs, aphids, and spider mites). Thirty starved

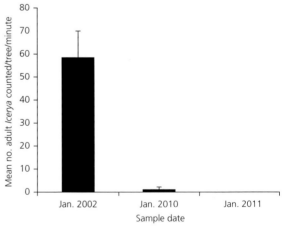

Figure 10.11 Reduction of cottony cushion scale on white mangrove, on Santa Cruz island, Galápagos, before (2002) and after (2010 and 2011) release of *Rodolia cardinalis*. Hoddle et al. 2013. Reproduced with permission from Elsevier.

Figure 10.12 Walk-in field cage (right) stocked with native plants (left) infested with non-target insects to assess feeding preferences of adult *Rodolia cardinalis*. Photo credit, Mark Hoddle. (*See Plate 16 for the color representation of this figure.*)

field-collected or laboratory-reared adult *R. cardinalis* were released individually into the cage and their foraging behaviors and prey choices were recorded visually for up to 60 min. There was no evidence of attack by free-ranging *R. cardinalis* adult beetles on non-target insects during more than 22 h of behavioral observations, and all recorded predation was solely on *I. purchasi*, the intended prey species of the biological control project. Furthermore, during separate field surveys, no evidence was found of *R. cardinalis* attacking non-*Icerya* prey species, even when non-target insect species were in close proximity to *I. purchasi* (Figure 10.13) (Hoddle et al., 2013). These results suggest that this program is unlikely to have had any significant adverse impact on non-target invertebrate fauna in the Galápagos Islands.

From the results of two years of population monitoring of *Icerya* in several distinct habitats of the Galápagos Islands, it appears that *I. purchasi* populations are being maintained at relatively low densities and are lower than those observed before the release of *R. cardinalis*. The level of control, however, depends on the plant species, habitat, season, and possibly on tending by invasive ants. The biological control program is successfully suppressing pest populations below non-damaging levels on many plant species in the Galápagos Islands, which in turn should facilitate restoration of the affected native ecosystems. Most noteworthy is the complete recovery of stands

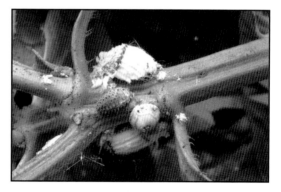

Figure 10.13 *Rodolia cardinalis* larva faced with a choice: cottony cushion scale or *Ceroplastes* sp. scale (top), attacks the cottony cushion scale (bottom). Photo credit, Mark Hoddle. (*See Plate 17 for the color representation of this figure.*)

of white mangroves on Santa Cruz Island that were severely damaged by *I. purchasi* and were declining before the introduction of *R. cardinalis* (Van Driesche et al., 2010). Populations of Darwin's aster, *D. tenuifolius*, also appear to be recuperating on the flanks of Alcedo Volcano according to observations by park rangers. Continued monitoring will allow us to understand better the benefits of this biological control program, but in the meantime, the results from rigorous studies before release, as well as considerable post-release assessment, show that biological control was an effective and safe tool for managing invasive species in natural ecosystems of high conservation value.

Preserving Hawaiian dry forests by protecting wiliwili trees from erythrina gall wasps

Leyla Kaufman and Russell H. Messing

Hawaii's dryland forests were once a source of great biotic diversity, hosting over 20% of all native plants in the Hawaiian Islands (World Wildlife Fund, 2013). However, impacts such as fire, grazing, agricultural and pasture development, and invasive weeds and ungulates have decimated these ecosystems. It is estimated that over 90% (and perhaps as much as 99%) of Hawaii's dry forest habitat has been lost (Cabin et al., 2000), and more than 25% of the species in this ecosystem are on the Federal Endangered Species List. Nevertheless, there is still hope for the preservation and recovery of some of these native communities, and many dryland restoration projects are in progress (Allen, 2000).

In the remaining stands of native dry forest, the endemic coral tree, or wiliwili (*Erythrina sandwicensis* O. Deg.) is a major overstory tree (Figure 10.14). It was once a dominant component of ancient endemic Hawaiian dryland forests. It is one of the few native trees in Hawaii that are deciduous, regularly losing its leaves during the summer in order to conserve water, and putting out new leaves in the fall. It grows up to 15 m tall, on leeward slopes of all the main islands, at elevations from sea level to just over 600 m. The trees form nitrogen-fixing symbioses with *Bradyrhizobium* species, thus naturally enriching forest soils.

Figure 10.14 Remnant patch of lowland dry forest, with wiliwili tree, *Erythrina sandwicensis*, in foreground at Puuwaawaa, Big Island, Hawaii. Photo credit, Juilana Yalemar. (*See Plate 18 for the color representation of this figure.*)

The wiliwili tree played an important role in native Hawaiian culture, and it is mentioned in many Hawaiian legends and proverbs. Its wood was used for surfboards, canoe outriggers, and fish net floats, and the bright red seeds were used for making leis (Wagner et al., 1990). Currently, however, the wiliwili is subject to sustained competitive pressure from non-native plants (such as *Pennisetum setaceum* [Forssk.] Chiov., *Melia azedarach* L. and *Grevillea robusta* A. Cunn. ex R. Br.), seed predation by an exotic bruchid weevil (*Specularius impressithorax* [Pic]), and herbivory by introduced ungulates.

Recently, the invasive Erythrina gall wasp, *Quadrastichus erythrinae* Kim (Hymenoptera: Eulophidae), (Figure 10.15) emerged as new, serious threat to the endemic wiliwili

and, by extension, the web of endangered native plants, birds, and insects that inhabit remaining stands of dryland forest (Heu et al., 2005). This species, which invaded Hawaii in 2005, was first described based on specimens from Mauritus, Reunion, and Singapore. The gall wasp occurs in East Africa, although its exact native range remains unknown (Rubinoff et al., 2010). Within a few months of its arrival in Hawaii, the gall wasp colonized all major islands, causing widespread damage to endemic *E. sandwicensis*, as well as exotic species of *Erythrina* occurring on the islands (Messing et al., 2009). The exotic ornamental *Erythrina variegata* L. was an important component of Hawaii's urban landscape; but since the gall wasp invasion, the City and County of Honolulu have had to remove over 1000 dead *Erythrina* trees.

Figure 10.15 Adult of erythrina gall wasp, *Quadrastichus erythrinae* (left, photo courtesy of M. Tremblay), and galled leaves from Koko Crater (right, photo credit, Leyla Kaufman). (*See Plate 19 for the color representation of this figure.*)

Female gall wasps deposit eggs in leaves, petioles, stems, inflorescences, and young seed pods, where subsequent larval feeding induces gall formation, which in heavy and or chronic infestations leads to defoliation and tree mortality (Kim et al., 2004; Heu et al., 2008). Given the extent of damage, dramatic reduction in flowering owing to high stress, and the fact that tree mortality was observed within a few months of the invasion, University of Hawaii researchers and staff at the Hawaii Department of Agriculture initiated efforts to search for biological control agents in Africa (Figure 10.16). In 2006, two agents were successfully reared and maintained under quarantine conditions in Hawaii: *Eurytoma erythrinae* Gates and Delvare (Hymenoptera: Eurytomidae) (Figure 10.16) and *Aprostocetus nitens* Prinsloo and Kelly (Hymenoptera: Eulophidae). These species were also previously unknown and were subsequently described by Gates and Delvare (2008) and Prinsloo and Kelly (2009), respectively.

After host-specificity testing and risk assessment were completed, *E. erythrinae* was released in November 2008. Female *E. erythrinae* lay their eggs in galled tissue, where the larvae feed externally on gall wasp larvae and pupae. Each *E. erythrinae* consumes 1 to 5 hosts to complete its larval development and then pupates inside a gall chamber. Monitoring sites were selected on the islands of Kauai, Oahu, Maui, and Hawaii. Pre-release monitoring data were collected for six months, and post-release evaluation continued for five years (through 2013). Several different monitoring techniques were used to assess the effectiveness of *E. erythrinae* in controlling gall wasp populations. These included ratings of infestation level in young shoots

Figure 10.16 Exploration in Madagascar (left, photo credit, Russell Messing) did not yield parasitoids; but additional collections in South Africa, Kenya, and Tanzania were successful, yielding several parasitoid species – including *Eurytoma erythinae* (right, photo courtesy of Walter Nagamine), which is helping to preserve rare endemic trees in Hawaii. (*See Plate 20 for the color representation of this figure.*)

using a 4-point scale (Messing et al., 2009), rearing parasitoids or gall wasps from galls, dissection of galled tissues, tagging and monitoring development of young shoots and inflorescences, and running germination tests on collected seeds.

The biological control agent became well established in the field within three to six months after release. Establishment of the biological control agent took longer at sites that had previously been treated with systemic insecticides; therefore, additional releases were made at those sites. Infestation levels of the gall maker in young shoots decreased significantly at all sites after release of the biological control agent. During the pre-release monitoring period, on average <10% of young shoots were rated as uninfested by the pest gall wasp (i.e., healthy). This figure changed dramatically post-release, with >80% of shoots rated as uninfested by the pest in the fourth year after release. Dissections of galled tissues revealed significant numbers of the biological control agent, confirming that they were indeed responsible for the reduction in foliage infestation levels.

Pest infestation levels in inflorescences also decreased post release, though this was at a slower rate compared to ratings in young shoots and varied significantly among sites. In order to evaluate the level of control, inflorescences were tagged early in the season and followed until mature seed pods were formed. Drier sites exhibited poor control by *E. erythrinae* in inflorescences. At some of these dry sites, 100% of the inflorescences tagged early in the season were rated as infested, with the highest infestation levels occurring at the end of the season. Tagged inflorescences that produced seeds were used in germination tests. Results from these tests showed that levels of infestation in mature seed pods were negatively correlated with germination rate (i.e., the higher the infestation level in seed pods, the lower the germination rate). There was no significant correlation between the level of infestation in seed pods and the number of seeds produced per seed pod; highly infested seed pods sometimes produce similar numbers of seeds as lightly infested or uninfested seed pods.

A census of several wiliwili tree populations conducted across the islands in 2012, four years after the release of the biological control agent, showed that 30–35% of the trees studied died because of the gall wasp invasion. This percentage could have been much higher if the biological control agent had not been released so promptly. Results from the census also showed that younger trees, or trees with smaller diameters, experienced higher mortality rates than trees with greater trunk diameters. Surviving trees exhibited improved health and more trees were able to flower. During our surveys, little-to-no tree recruitment was observed at any site, which may be caused by factors such as poor control by *E. erythinae* in inflorescences at many sites that renders seeds inviable. Other factors that may also play a role include widespread seed damage by the bruchid *S. impressithorax*, competition with invasive weeds such as fountain grass (*P. setaceum*), or seedling predation by feral ungulates (such as goats, pigs, and deer). The importance of these other factors has yet to be assessed, but if they prove important, it may take more than biological control alone to restore healthy populations of *E. erythrinae* in the Hawaiian Islands. The second biological control agent, *A. nitens*, is still (2014) under quarantine in Hawaii, and tests are being conducted to determine compatibility of the two agents.

In summary, while the biological control project has improved the health status of tree foliage dramatically, seed suppression caused by flower galling remains high. Further study of the system is required to promote recovery of the wiliwili tree at the population level.

Failures

Destruction of Fraser fir by balsam woolly adelgid in Appalachian "sky-island" forests

Michael E. Montgomery

Fraser fir, *Abies fraseri* (Pursh) Poir., is endemic to the Appalachian Mountains of the southeastern United States. It has a disjunct distribution of only 26,610 ha that is limited to elevations above 1600 m on six massifs in southwestern Virginia, western North Carolina, and eastern Tennessee. It is associated with red spruce, *Picea rubens* Sargent, and these two conifers along with an assortment of northern hardwoods, shrubs, and herbaceous species form the southern Appalachian spruce-fir "sky-island" forests. This unique ecosystem was listed as the second most endangered in the United States (Noss and Peters, 1995). The main threats to the ecosystem are air pollution and the balsam woolly adelgid, *Adelges piceae* (Ratzburg) (Hemiptera: Adelgidae) (Figure 10.17) (Dull et al., 1988).

Three decades after the discovery of *A. piceae* in the southern Appalachians, mortality of Fraser fir ranged from 44% to 91% among the six isolated "sky-island" populations of Fraser fir (Figure 10.18) (Dull et al., 1988).

Figure 10.17 Balsam woolly adelgid on trunk (left) and gouty twigs resulting from twig infestation (right). Photos by William M. Ciesla, Forest Health Management International, Bugwood.org. (*See Plate 21 for the color representation of this figure.*)

Figure 10.18 Fraser fir killed by balsam woolly adelgid, together with regenerating young trees not yet old enough to be susceptible. Photo credit, Michael Montgomery, USDA Forest Service. (*See Plate 22 for the color representation of this figure.*)

Red spruce is also affected because it cannot survive the high winds that blast higher elevations without the protection of the much sturdier Fraser fir. Without the thick canopy of spruce and fir, the understory dries out, input of coarse woody debris is changed, and the vegetative state is permanently altered (Stehn et al., 2011). Thus, the loss of Fraser fir has a cascading effect on the entire ecosystem.

While it seems unlikely that the Fraser fir will be extirpated, the forest type remains imperiled with an uncertain future. The young, understory fir trees grew rapidly following the initial wave of mortality. Currently, they are producing seed, but are also becoming susceptible to the adelgid (Lusk et al., 2010). Mortality of regrowth trees, however, is patchy, unlike the widespread mortality that occurred earlier. In the first wave of mortality, there was a complete loss of the overstory (McManamay et al., 2011), which caused the previously mesic to wet, much-shaded understory to be exposed to greater solar insolation and desiccating winds, covered only by dense stands of young trees. Air pollution and climate change also place additional stress on this ecosystem; thus, the future of this forest type hinges on the frequency and intensity of these three exogenous disturbances (Stehn et al., 2013).

Although the southern Appalachian spruce-fir forest resembles the northern boreal forests of North America, it differs in the species of spruce and fir, a denser and more diverse understory, and a warmer and wetter climate. Further, it is an ice age relic, a remnant of the vast spruce-fir forest that covered the southern United States during the last ice age, whereas the boreal forest is younger, forming after the retreat of the glaciers. The southern Appalachian spruce-fir ecosystem supports more than 30 endemic plant and animal species, as well as many disjunct species more common in boreal forests (DENR, 2010). Four of these are listed as federally endangered: the Carolina northern flying squirrel, *Glaucomys sabrinus coloratus* Handley; the spruce-fir moss spider, *Microhexura montivaga* Crosby & Bishop; the spreading avens, *Geum radiatum* Michx.; and the rock gnome lichen, *Gymnoderma lineare* (A. Evans) Yosham & Sharp. The ghost moth, *Gazoryctra sciophanes* (Ferguson) (Lepdioptera: Hepialidae), another globally rare taxon, is restricted to these forests and just a few sky islands in West Virginia (USA). Likely there are many additional unknown species of insects that are endemic to this ecosystem, since systematic surveys have been made only

for some Lepidoptera (Scholtens and Wagner, 2007) and carabid beetles in the genus *Trechus* (Barr, 1962). The latter is a group of flightless ground beetles restricted to cool, moist microhabitats such as caves, sink holes, moss carpets, and deep humus. Of the several species that Barr (1962) described, sixteen are each restricted to only one or two of the six "sky islands" above 1500 m in the southern Appalachians. The *Trechus* ground beetles are an example of the speciation and diversity that can occur in isolated habitats by species with low dispersal capability.

The balsam woolly adelgid is considered to be native to Europe and was first discovered in North America in 1908 in Maine. It was first detected in the southern Appalachians in 1957, and by 1968 it was in all naturally occurring Fraser fir stands. The adelgid attacks all native North American fir species causing a hypersensitive response at its feeding sites that blocks translocation vessels (Balch, 1952). Feeding on the twigs causes gout (Figure 10.18), whereas feeding on the bole produces "rotholz" deformities. The thin bark of Fraser fir makes it especially vulnerable. In the southern Appalachians, the adelgid has two or three parthenogenic generations each year (Arthur and Hain, 1984). It overwinters as a first instar nymph under the bark or bud scales and becomes active in April, going through two more nymphal instars before becoming an adult, with peak egg laying in mid June. The crawlers that hatch from these eggs find a feeding site, settle, and aestivate until mid July, then resume development and, in mid August, peak production of the second batch of eggs occurs. In developing a biological control strategy for *A. piceae*, it is important to recognize that it has two well-separated periods of egg production and that both periods should be targeted.

Biological control programs targeting the balsam woolly adelgid in Canada and the United States began in 1933 and more than 700,000 individuals representing 33 predator species were released before the project ended in 1966 (Montgomery and Havill, 2014). Worldwide, the family Adelgidae has no known parasitoids and no significant diseases. In the southern Appalachians, releases of 2914 individuals of 22 predator species were made from 1959 through 1966 (Amman, 1961; Amman and Speers, 1964, 1971). At the end of the program, the species considered to be established in both Canada and the United States were

Aphidecta obliterata (L.), *Aphidoletes thompsoni* (Moehn), *Cremifania nigrocellulata* Czerney, *Laricobius erichsonii* Rosenhauer, *Neoleucopis obscura* Haliday, and *Scymnus impexus* (Mulsant) (Clausen, 1978; Schooley et al., 1984). Except for *C. nigrocellulata*, these species were also released in the southern Appalachians, and *A. obliterata*, *A. thompsoni*, and *L. erichsonii* were reported as established (Amman and Speers, 1965), but a survey in 1968 found only a few *L. erichsonii* and that native mites were the most important predators (Fedde, 1972).

Many of the imported species released to control *A. piceae* may have been confused with native species. *Aphidoletes thompsoni* is likely a junior synonym of *Aphidoletes abietis* (Kieffer), a common, widespread species considered native to North America (Gagné, 2010). *Aphidoletes abietis* was reported in New York State (Felt, 1917) and has been collected recently from *A. piceae* in Canada and the United States (Gagné, 2010) and from *A. tsugae* in the eastern United States (Wallace and Hain, 2000). *Laricobius erichsonii* may have been confused with the native *Laricobius rubidus* (LeConte) since the adults are very similar in appearance and no morphological features distinguish larvae in the genus. Clark et al. (1971) considered the *Leucopis*/*Neoleucopis* released in North America to be a mixture of five species, two of which may be confused with native North American species. After *N. obscura* was released in New Brunswick, it was reported to have spread rapidly, including to neighboring Maine. Chamaemyiid flies native to eastern North America include *Neoleucopis pinicola* Malloch, *Leucopis americana* Malloch, and *Leucopis piniperda* Malloch. Later it was determined from vouchers that *Leucopis hennigrata* McAlpine and *Leucopis atratula* Ratzburg were also released in the Maritimes and that the former may be native to North America (McAlpine, 1971). The species most represented in Canadian museums from field collections made during the program are *L. atratula* and *L. piniperda* (McAlpine, 1971, 1978; Tanasijshuk, 2002). The native North American predator *L. americana* was reported to have been displaced by *N. obscura* in New Brunswick and Newfoundland (Balch, 1952; Bryant, 1963). However, both of these entities are dubious (Tanasijshuk, 2002). The "*N. obscura*" released in the southern Appalachians were field collected in Maine (Amman, 1961; Amman and Speers, 1964) and the identity of these remains uncertain. Until a revision of the taxonomy of the Chamaemyiidae is completed, voucher specimens studied, and a survey of existing predators made, it will remain unclear which imported species in this family were released and established, and which native species have made a host shift to *A. piceae*.

Despite a comprehensive 30-year biological control program, the balsam woolly adelgid remains a serious pest not only of Fraser fir but also other fir species in North America. Although it is generally accepted that biological control of *A. piceae* in North America has not been successful (Clark et al., 1971; Schooley et al., 1984), there is no conclusive explanation for the lack of success. The failure of several species can be attributed to poor climate matching and not vetting specificity to the target species. The first imported predators were collected in Great Britain, from adelgids on pine, and only one of the six species was specific to adelgids. None of the 15 species from India and Pakistan that were released in the southern Appalachians established (Amman and Spears, 1971). These Asian predators were mainly generalist species and included many species of uncertain identity. On the other hand, the importations made through the biological control laboratory of CABI in Switzerland exemplify a well-run classical biological control program where the natural enemy complex on the target host was studied and the most promising species were exported in large numbers. Although several of these species established, at least temporarily, none have provided effective control. That the European climate is warmer and drier than the release sites in North America has been suggested as a factor (Clark et al., 1971; Schooley et al., 1984). It should not be surprising, however, that the predators did not bring the adelgid below damaging thresholds, since in Europe the population dynamic of the entire predator complex is largely inversely density dependent and not regulative (Eichhorn, 1969a), and tree resistance and weather both have strong influences on the pest's population dynamics (Franz, 1956; Pschorn-Walcher and Zwölfer, 1956).

The ineffectiveness of the predators exported from Europe has also been attributed to poor synchronization with one or both adelgid generations, and predator preferences for adelgids on the stem versus on the twigs (Pschorn-Walcher and Zwölfer, 1956; Eichhorn, 1968). A main weakness of the predaceous beetles *L. erichsonii*, *A. obliterata*, and *S. impexus* is that they are univoltine, whereas the adelgid has two to three generations per year. These predators also appear after the adelgid has started to lay eggs in the spring (Pschorn-Walcher and Zwölfer,

1956). The chamaemyiid flies, however, are multivoltine, appear earlier in the spring, and attack both generations. One of these, *C. nigrocellulata* was not released in the southern Appalachians, and it is not certain what fly species was actually released because of problems with identification. Thus, there does seem to be an opportunity for additional introductions to establish biological controls that could reduce the damage to Fraser fir.

After the end of the balsam woolly adelgid biological control program, the Caucasus Mountains were explored for natural enemies, as it was felt that this may be the ancestral home of *A. piceae* (Eichhorn, 1969a, b). In this region of Turkey, Georgia, and Russia, *Adelges nordmannianae* (Eckstein), which is closely related to *A. piceae*, occurs on *Abies nordmanniana* (Steven) Spach (Ravn et al., 2013). These authors confirmed Eichhorn's (1969b, 2000) observation that the chamaemyiid *L. hennigrata* is the most important predator of the fir adelgid in Turkey. This is in contrast to North America and Europe, where the predator is also native, but seems to be uncommon. It is unknown if this different dynamic reflects differences in habitat, prey species, or genetic differences among populations of the fly.

Another source of potential biological control agents for the balsam woolly adelgid may be the hemlock woolly adelgid (*Adelges tsugae* [Annand]), currently the target of an active biological control program in the eastern United States (Onken and Reardon, 2011). *Laricobius nigrinus* Fender (Coeloptera: Derodontidae) and *Sasajiscymnus tsugae* Sasaji & McClure (Coleoptera: Coccinellidae) have been successfully established and are widespread. In laboratory tests, the lady beetle *S. tsugae* fed and oviposited equally on the balsam and hemlock adelgids, although survival of beetle larvae was higher on the hemlock woolly adelgid (Jetton et al., 2011). The lady beetle's life history may be better synchronized with *A. piceae* since both are inactive during the winter and active during the summer, whereas *A. tsugae* is active from fall until late spring and is in diapause during the summer.

A prerequisite to new biological control efforts is to resurvey the natural enemies now present and to make positive identifications of both established and candidate species. The confusion over identity of the chamaemyiid species that were released and recovered points to the importance of taxonomy in biological control (see also Chapter 6). There are several species of Chamaemyiidae in North America recovered from adelgids feeding on fir and hemlock (McAlpine, 1987;

Humble, 1994; Ross et al., 2011). Finding biological control candidates in Oregon, Washington, or Maine (rather than in foreign countries) would greatly simplify legal issues related to transporting insects and obtaining permits for their release. Since the Fraser fir seedlings that survived the first wave of *A. piceae* attack are now maturing and becoming susceptible to the pest, it seems timely to reassess the possibilities for biological control of balsam woolly adelgid.

A century of unsuccessful *Lantana camara* biological control effort in Australia

Rieks van Kinken

The neotropical shrub lantana, *Lantana camara* L. *sensu lato*, is listed among the world's worst 100 invasive species because of its impacts on biological diversity and human activities (Lowe et al., 2000). It has been distributed pantropically, largely as an ornamental, and is now considered invasive in approximately 60 countries or island groups (Day et al., 2003). Invasive lantana grows as individual clumps or dense monospecific stands, typically 1–4 m tall. As an environmental weed it invades disturbed natural forests and woodlands (Figure 10.19), including rainforest margins, where it disrupts succession, decreases biodiversity, and hinders regeneration efforts (Day and Zalucki, 2009, and references therein). This in turn can facilitate higher fire frequency and intensity, further favoring lantana (Fensham et al., 1994). Harm to native species richness occurs above a 75% cover threshold in wet sclerophyll forest in southeastern Australia (Gooden et al., 2009). One analysis suggests that lantana threatens more than 1400 native species in Australia, including 279 plants and 93 animals listed under state or national threatened species legislation, and 100 threatened ecosystems (Turner et al., 2007). Its primary impact as a weed of pastures is displacement of desirable species, but it also interferes with stock management, increases maintenance expenditure, fatally poisons stock under some circumstances, and reduces land values (Day et al., 2003; AEC Group, 2007; Day and Zalucki, 2009). It is estimated to cost the pastoral industry Aus $104,000,000 per year (2005–06 value) in lost productivity and management costs (AEC Group, 2007). It also affects Australian silviculture, accounting for an estimated 30% of establishment costs and up to 50% of the harvesting costs (Wells, 1984).

Figure 10.19 Lantana invading pastures and native forest understory in southeast Queensland, Australia. Photos courtesy of Michael Day, QDAFF. (*See Plate 23 for the color representation of this figure.*)

Lantana was first introduced into Europe from the Americas in 1636 (Day and Zalucki, 2009). Subsequent centuries of hybridization and selection among several American taxa have resulted in over 650 varieties that vary in morphology and ecology (Zalucki et al., 2007). As a result, the taxonomy is complicated and remains disputed. Morphological studies suggest it could be a single highly variable hybrid swarm (Sanders, 2012). Alternatively, recent genetic analyses suggest invasive lantana in Australia could be from a single, morphologically variable species with primary genetic influences from the Caribbean and Venezuela (R. Watts, unpublished data), which would be a subset of the native distribution of the plant in the larger sense, which extends from Florida and Texas through to northern Argentina and Uruguay (Zalucki et al., 2007).

In Australia, lantana has a relatively broad ecological niche. It is able to become dominant in habitats from temperate regions to wet–dry tropics and across a broad rainfall gradient (750–3000 mm) (Day et al., 2003). It does best in high-light environments. Although it can become dense in open woodlands, it performs poorly in forests. Furthermore, lantana requires disturbance to

invade and become dominant, such as through feral and domestic animal activity, fire, or clearing (Fensham et al., 1994; Day et al., 2003). Seeds have fleshy endocarps and are dispersed by birds and animals (Day et al., 2003). In Australia lantana probably occurs across most of its potential range, although infilling continues.

Many management tools are available, including herbicides, mechanical control, and fire (Stock et al., 2009). However, they are often not practical or economical, especially on steep hillsides, in extensive natural forests, pastures with low productivity, or along waterways (Haseler, 1963). It was even harder to manage in the early 1900s when management was largely restricted to manual control. Lantana, therefore, became one of the first targets for biological control in 1902, starting in Hawaii (Davis et al., 1992). It is also the longest running biological control program, with the search for new agents continuing (Day and Zalucki, 2009), and one of the most active, with 42 agents released in at least 41 countries (Zalucki et al., 2007).

In Australia, the biological control program began in 1914, and it has so far resulted in the release of 32 agents, with further species still being considered (Day,

2012). Economic analysis suggests that even a 5% reduction in lantana densities would result in substantial benefits (AEC Group, 2007). Several agents can be highly damaging seasonally and locally; however, few if any effects on lantana populations have been noted in Australia, and it continues to be recognized as one of the most serious environmental and pastoral weeds (Day, 2012). The case of lantana contrasts with many successful biological control programs where the pool of potential agents has been limited, including against weeds with similar life histories (Van Driesche et al., 2010). Is this failure caused by lantana being an inherently impossible target, or is it because of the way the biological control program was carried out? If the latter, success might still be possible.

Defining clear goals and criteria for success in a biological control program is important as a means to focus activities (van Klinken and Raghu, 2006; see also Chapters 8 and 15), but they have never been clearly articulated for lantana. For much of its history, the Australian lantana biological control program was almost certainly focused on reducing economic losses caused by lantana, largely to the pastoral industry (Haseler, 1963). However, reducing environmental damage is now also being considered (Stock et al., 2009). Although not defined, successful biological control was probably perceived as a significant reduction in lantana densities resulting in more pasture, and reduced negative effects on forest understories.

There have been no long-term demographic studies or recent experimental work in Australia to test for population-level benefits from lantana biological control. These studies are critical, not only to quantify changes in distribution and abundance, but also to test whether any observed reductions have resulted from biological control or other potential drivers such as succession, disturbance regimes, or land-management practices, all of which have also changed substantially in Australia over the past 100 years. Nonetheless, there is little evidence that biological control has significantly reduced damage from the weed in pastures or natural areas. Biological control agents do cause intense damage seasonally, but plants recover when agent densities decrease during winter or drought periods (Broughton, 2000; Day, 2012). Damage is also variable spatially and across years. Anecdotal reports suggest that late-season defoliation in Australia reduces lantana growth, flowering, and seeding, but the net results on annual

growth and annual seed production have not been quantified. Plant deaths have only occasionally been reported in Australia, but even then other contributing stressors such as drought are generally also implicated (Day, 2012).

Lack of evidence of population-level effects from biological control is not unique to Australia. For example, biological control has been demonstrated to cause high levels of leaf damage and loss of viable seed in Guam (Muniappan et al., 1996), but demographic consequences have not been quantified. An exception is the Big Island of Hawaii where lantana is no longer considered to be a problem in arid areas, apparently because of the combined effect of several biological control agents (Davis et al., 1992). All of those agents were also released in Australia, but to minimal effect (Zalucki et al., 2007).

Of the 32 agents released in Australia, 18 (58%) established, which is similar to weed biological control in general (60–64%) (Zalucki et al., 2007), and most of these (15 species) reached high densities at least somewhere and sometimes. Of those, nine caused at least seasonal and local damage (typically reported as significant defoliation or seed damage), but only one species, the sap-sucker *Aconophora compressa* Walker (Hemiptera: Membracidae) (Figure 10.20), resulted in sustained damage through the season in at least some years and places. Established agents are typically only patchily abundant (Day et al., 2003), so negative interactions between agents are considered unlikely (Day and Zalucki, 2009), and additive effects, for example on defoliation, are localized (Day, 2012).

So why has lantana biological control been apparently ineffective in Australia despite a century of effort, the release of 32 agents, and so many agents (48%) reaching high densities? There are two parts to this question. First, what damage to lantana plants is required to sustain the required demographic effects? Second, why are agents not reaching and sustaining the densities required to achieve that level of damage? Recent demographic modelling suggests that large effects on mortality and reproduction are required to regulate lantana populations (Osunkoya et al., 2013). This is supported by general observations that lantana plants are resilient to severe disturbances including hot fires, mechanical damage, and considerable and frequent defoliation (Day, 2012; Osunkoya et al., 2013) and that natural mortality is rare (Osunkoya et al., 2013; Yeates, 2013). For example, plants recovered

Figure 10.20 Lantana defoliated by the tree hopper *Aconophora compressa* in southeast Queensland, Australia and close up of insect. Photos courtesy of Michael Day, QDAFF. (*See Plate 24 for the color representation of this figure.*)

even after 100% of lantana leaves were removed manually every month for a one to two year period (Broughton, 2000). Demographic impacts in the absence of other stressors such as drought will therefore require prolonged, systemic damage such as can only be achieved through galling or pathogen attack. Alternatively, it will require prolonged loss of leaves throughout the growing season, although even this will not necessarily result in mortality (Broughton, 2000). Seed predation is unlikely to affect demography significantly unless most seeds are killed (van Klinken and Flack, 2008). These requirements contrast with the types of agents released in Australia: only 29% of agents attack the plant structurally (1 pathogen, 5 sap-suckers, 2 stem borers and 1 galler). In contrast, 48% (15 species) are leaf feeders (including leafminers) and 23% (7 species) are flower or fruit feeders. Most agents would therefore not be expected to cause the required demographic impacts, unless populations maintained exceptionally high densities for prolonged periods.

There is precedence for agents reaching and maintaining high densities, resulting in demographic impact on even long-lived resilient species (Van Driesche et al., 2010). Why then have agents not achieved this on lantana? A primary reason given is a poor climate match (Day and Neser, 2000; Day et al., 2003). This seems an unlikely reason for biological control failing throughout Australia given the diverse climates and environments in which lantana grows. Certainly most agents have been able to reach high densities in particular climate regions and habitats (under canopy or in the open). Nonetheless, climate does appear to be the primary factor preventing most agents from sustaining high densities, either directly or through their effects on plant quality. Most agents have multiple generations in a year and apparently emerge in low numbers at the start of the season, taking much of the season to reach high densities. Furthermore, for most species, high densities are typically reached in late summer and early autumn as the temperature drops or plant quality declines, allowing plants to then recover. Densities also drop following extreme weather events, such as drought, heavy rainfall, or heat waves. The sap-sucker *A. compressa* is the least coupled to season, but populations crash after heat waves (Day et al., 2003).

Another commonly cited reason for poor or variable agent performance is differences in host varietal preferences. Certainly differences have been observed in the laboratory in at least 10 biological control agents (Day et al., 2003), but observations in the field remain largely anecdotal (e.g., Day and Neser, 2000) and are easily misinterpreted. Predation and parasitism of agents is also a possible explanation. Although not considered important in Australia (Day et al., 2003), some biological control agents have reached high densities soon after release, only to crash subsequently, so it is nonetheless worth further investigation. For example, the leafminer *Calycomyza lantanae* (Frick) (Diptera: Agromyzidae) caused significant damage within four years of its release in 1974, but populations subsequently declined (Julien and Griffiths, 1999). Finally, host associations in the native range are likely to contribute to some failures or poor performance, with agents collected off *L. camara* and *Lantana urticifolia* Mill. establishing better than those from other *Lantana* species (Day and Zalucki, 2009, and refs therein). The centuries of artificial selection of lantana means that exact genetic matches may not exist in the field, although further searches in the putative source of genetic lineages (the Caribbean and Venezuela) may well yield further agents.

Lantana is clearly a difficult target for biological control, because of its demographic resilience and, perhaps, its genetic complexity. Nonetheless it does occur in diverse climates even within Australia. Is failure then the consequence of a lack of suitable agents or the difficult nature of the target? Or have the wrong agents been introduced? Certainly further agents are being considered, and further surveys in the native range may well yield additional agents. With the benefit of hindsight, the lantana biological control program has suffered from poorly defined goals, for example regarding whether the objective was to reduce lantana densities in open fields or in forest understories, and in which climatic region. Furthermore, there was limited explicit consideration of the likely ability of potential agents to achieve these outcomes through demographic impacts (but see Day and Urban, 2004), and relatively few follow-up studies to understand why biological control

agents have been largely unsuccessful. The biological control program is still ongoing. These limitations therefore need to be considered to maximize its likelihood of success and should be guided by recent demographic studies (Osunkoya et al., 2013), an improved understanding of why previous agents have been ineffective, and quantitative, long-term evaluation of the biological control program. It is yet to be seen whether an effective agent for lantana biological control in Australia can be found, that is, one that will cause the type and duration of damage necessary to achieve stated objectives for lantana biological control.

Projects still unfolding

These projects are not yet finished and may or may not achieve their ecological goals fully, but they illustrate the kind of cases in which biological control has potential to benefit damaged ecosystems.

Preserving ash-dependent biodiversity through biological control of emerald ash borer

Jian J. Duan

Ash trees (*Fraxinus* spp.) are dominant or subdominant forest components in many forest stands in the northeastern and midwestern United States as well as in the Pacific Northwest and other states. There are sixteen species of native ash in the United States (USDA PLANTS Database, 2013). Each *Fraxinus* species is adapted to a slightly different habitat within forest ecosystems. Several species tolerate poorly drained sites and wet soils, protecting environmentally sensitive riparian areas (D'Orangeville et al., 2008; Rockermann, 2011); for instance, pure stands of black ash (*Fraxinus nigra* Marshall) grow in bogs and swamps in northern areas, where they provide browse, thermal cover, and protection for wildlife such as deer and moose (*Alces alces* [L.]). The ecological effects of emerald ash borer (EAB) (Figure 10.21), *Agrilus planipennis* Fairmaire, are likely to include both loss of habitat and food resources for vertebrates and habitat loss for ash-specialized herbivorous insects and their specialized predators and parasitoids (Wagner, 2007; Wagner and Todd, 2015).

First detected in North America in Michigan in 2002 (Haack et al., 2002), emerald ash borer is a highly

Figure 10.21 Adults of emerald ash borer, *Agrilus planipennis*. Photo courtesy of Stephen Ausmus, USDA ARS. (*See Plate 25 for the color representation of this figure.*)

destructive invasive beetle that attacks and kills North American ash (Figure 10.22) (Cappeart et al., 2005). While adults feed on mature ash leaves and rarely cause any significant damage to the host tree, EAB larvae feed on phloem, creating extensive galleries under the bark (Figure 10.23). When emerald ash borer populations are high, larval consumption of tree phloem is substantial, resulting in tree girdling and death in three to five years (Poland and McCullough, 2006; McCullough et al., 2009).

EAB appears well adapted to the climatic conditions in North America, is now present in 21 US states and several Canadian provinces, and has caused high levels of ash tree mortality (Emerald Ash Borer Information, 2013; USDA PLANT Database, 2013). Ash species native to North American forests that are known to be susceptible to EAB include white ash (*Fraxinus americana* L.), green ash (*F. pennsylvanica* Marshall), and black ash (*F. nigra*), which are major components of forests; and blue ash

(*F. quadrangulata* Michx.) and pumpkin ash (*F. profunda* [Bush] Bush), which are less common species. There is increasing evidence that EAB will attack all North American *Fraxinus* spp., although susceptibility varies by species and, for ornamental forms, variety (Wie et al. 2004, 2007; Liu et al., 2007; Rebek et al., 2008; Duan et al., 2012a).

Initial efforts to reduce or contain emerald ash borer damage focused on eradication. However, because the population in the originally infested area (Michigan) was already high and widespread when first recognized, eradication proved both ineffective and cost-prohibitive and by 2009 efforts were abandoned. Subsequently, efforts focused on slowing the invasion of new areas (Poland et al., 2010; Mercader et al., 2011) by restricting movement of EAB-infested wood or plant materials, insecticide treatment or physical destruction of infested trees (including use of artificially girdled EAB trap trees), and biological control via introduction and release of natural enemies collected from EAB's native range (e.g., Liu et al., 2007; USDA APHIS, 2007; McCullough et al., 2009; Poland et al., 2010; Mercader et al., 2011). Although none of these approaches alone appear to be sufficiently effective in containing the spread of EAB, biological control via self-propagating and dispersing natural enemies has the potential to reduce EAB populations in forests, nurseries, and urban ash plantings.

The EAB biological control program began in 2007 with the release of three Asian hymenopteran parasitoids: *Spathius agrili* Yang (Braconidae), *Tetrastichus planipennisi* Yang (Eulophidae) (Figure 10.24a, b), and *Oobius agrili* Zhang and Huang (Encyrtidae) (Figure 10.25) (Yang et al. 2005, 2006; Zhang et al., 2006; Liu et al., 2007; USDA APHIS, 2007; Bauer et al., 2008, 2009, 2010). While the former two parasitoid species attack EAB larvae, the latter (*O. agrili*) parasitizes EAB eggs. With the establishment of the USDA APHIS EAB biological control rearing facility in Brighton, Michigan, in 2009, large numbers of these parasitoids have been released in ash-dominated forests in 16 EAB-infested states. All three parasitoid species have been recovered following their release in the United States; however, only *T. planipennisi* and *O. agrili* have been recovered consistently more than one year after release. At two sites in Michigan, parasitism by *O. agrili* increased from 3–4% in the year of release to 20–28% two years later. Establishment and spread of the stronger flying *T. planipennisi* has been

Figure 10.22 Dead ash, killed by emerald ash borer, lining Michigan roadside. Photo credit, Jian J. Duan, USDA ARS. (*See Plate 26 for the color representation of this figure.*)

even more impressive. At six intensively studied sites in Michigan, 92% of the trees at the release site contained at least one brood of *T. planipennisi* four years after release, and parasitism levels increased steadily to an average of over 21% four years after the field releases (Figure 10.26) (Duan et al., 2013a). Parasitism levels at the six control sites (at least 1 km away) also increased yearly to an average level of 13%. It is hoped that populations of these exotic parasitoids will continue to establish themselves in more EAB-infested areas, increase over time, and exert significant control of EAB populations within the next few years in the United States.

However, the effectiveness of *T. planipennisi* in controlling EAB is affected by the thickness of tree bark (Figure 10.27), which is directly related to tree size. Data from the earlier EAB parasitoid surveys in northeast China (Liu et al., 2007) and the Russian Far East (Duan et al., 2012b), as well as from areas in the United States where the parasitoid has recently been released (Duan et al., 2012a, 2013a), show that *T. planipennisi* is indeed more prevalent in smaller diameter

ash trees. The reason for this apparent bias was identified in a field experiment that showed that *T. planipennisi* rarely parasitizes EAB larvae in larger, thick-barked trees (>3.2 mm thick bark, typical of trees with >12 cm dbh) owing to the parasitoid's relatively short ovipositor (average 2–2.5 mm) (Abell et al., 2012). The thick bark of large ash trees provides a refuge for EAB larvae from attack by *T. planipennisi* (Abell et al., 2012). Abell et al. (2012) further suggest that *T. planipennisi* will be more effective in stands with younger trees (<12 cm dbh) such as in the Michigan study sites, or in natural ash regeneration found in some EAB-affected stands.

To control EAB successfully on both small and larger ash trees, other EAB parasitoids from the pest's native range that have longer ovipositors should be released in EAB-invaded areas in North America. Currently, a new braconid larval parasitoid (*Spathius galinae* Belokobylskij & Strazanac), collected from the Russian Far East, has been evaluated for host specificity and a petition filed for its release against EAB in the United States (Duan et al., 2012b; Gould and Duan, 2013). This Russian parasitoid is also likely to be more cold tolerant than *S. agrili*, which was collected farther south in Tianjin, China (east of Beijing on the coast) where temperatures are moderated by the China Sea. Climate-matching studies

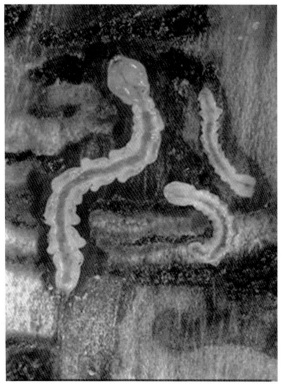

Figure 10.23 Emerald ash borer larvae feeding on ash cambium. Photo credit, Jian J. Duan, USDA ARS. (*See Plate 27 for the color representation of this figure.*)

indicated that the central part of the United States is potentially more suitable for *S. agrili* than the more northern areas where most releases have been made to date. Furthermore, climate-matching analysis indicates that the portion of the United States suitable for *S. galinae* extends farther north than is the case for *S. agrili* (Gould and Duan, 2013).

Besides parasitism by *T. planipennisi*, EAB larvae also suffer heavy losses in the United States from other biotic factors, including woodpeckers, mortality from plant defenses, diseases, and other parasitoids, mostly ones native to North America (e.g., Duan, 2010; 2012a, b; 2013a, b). For example, Duan et al. (2013a) report that among the other mortality factors detected across their study sites, woodpeckers were the most abundant factor in Michigan, removing up to 57% of immature and older EAB stages from feeding galleries and/or pupal chambers. Duan et al. (2013a) also show that putative tree resistance killed up to 15% of younger larvae and unknown diseases caused similar levels of mortality of larger larvae throughout the study, while parasitism by North American native parasitoids (*Atanycolus* spp. [Hymenoptera: Braconidae] and *Phasgonophora sulcata* Westwood [Hymenoptera: Chalcididae]) inflicted 18% larval mortality in Michigan. How these North American native parasitoids and other mortality factors will interact with the newly introduced *T. planipennisi* has yet to play out.

Finally, the success of EAB biological control programs in the United States may also hinge on the degree of ash tree resistance to EAB. Field studies in the pest's native range showed that EAB is rarely a

(a)

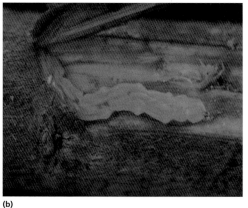

(b)

Figure 10.24 (a, b) *Tetrastichus planipennisi* adult and larvae inside host larva, an important parasitoid of emerald ash borer, introduced from China. Photo credits, Jian J. Duan, USDA ARS. (*See Plate 28 for the color representation of this figure.*)

serious pest on many Asian species of ash trees (e.g., *F. mandschurica* Rupr., *F. rhyncophylla* Hance), largely because of the combination of natural enemies (primarily parasitoids) and putative tree resistance that effectively killed immature stages of EAB (Liu et al., 2007; Duan et al., 2012b). In contrast, EAB frequently becomes a serious pest on introduced North American ash species in Asia (e.g., *F. velutina* Torr., *F. pennsylvanica*,

Figure 10.25 *Oobius agrili* ovipositing in an emerald ash borer egg; this is another important parasitoid introduced from China. Photo credits, Jian J. Duan, USDA ARS. (*See Plate 29 for the color representation of this figure.*)

and *F. americana*), and often kills infested trees grown in plantations or as landscape trees even with the presence of abundant parasitoids that may cause >70% parasitism of both eggs and larvae (Liu et al., 2007; Duan et al., 2012b). In North America, the common ash species such as *F. pennsylvanica*, *F. americana*, and *F. nigra* appear to have little resistance to EAB larvae and are readily killed by EAB infestations. However, recent field observations showed that some North American ash, such as blue ash, have not been heavily infested by EAB largely because of their high resistance to EAB larvae (Tanis and McCullough, 2012). In addition, some "lingering" healthy (often younger) trees of the susceptible ash species (*F. pennsyvanica* and *F. americana*) have been frequently observed surviving the wave of EAB invasion even in the epicenter of EAB invasions (e.g., in Michigan). Using experimentally established emerald ash borer cohorts, Duan et al. (2010) also showed that healthy young green and white ash trees in fact killed nearly 30% early instars of emerald ash borer larvae, indicating some level of host-tree resistance in those healthy young ash trees. Together, these observations provide hope that some level of host-tree resistance in North American ash may facilitate biological control in protecting North American ash and their dependent communities

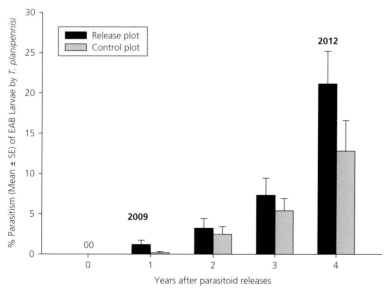

Figure 10.26 Increase in rates of parasitism of emerald ash borer larvae by the introduced parasitoid *Tetrastichus planipennisi* in Michigan, following its introduction in 2008. Reproduced with permission from Duan et al. (2013).

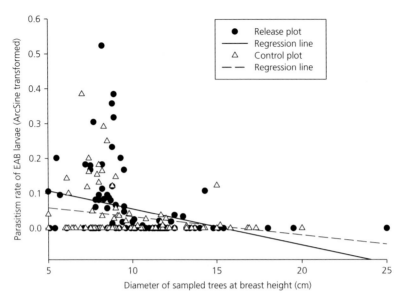

Figure 10.27 Parasitism by *Tetrastichus planipennisi* is limited by bark thickness relative to ovipositor length, such that only smaller trees are protected. Unpublished data of Jian J. Duan.

(Herms and McCullough, 2014). It is hoped that populations of introduced parasitoids will continue to establish themselves in more EAB-infested areas, increase over time, and exert significant control of EAB populations within next few years in the United States. If they do, the benefit of such a classical biological control program is also likely to extend to other native arthropods that depend on ash resources.

Restoring native vegetation in Tahitian forests dominated by *Miconia calvescens*

Jean-Yves Meyer

In Tahiti, a small tropical oceanic island located in the South Pacific, the massive invasion of native rainforests by *Miconia calvescens* DC (hereafter miconia) (Melastomataceae) was considered an intractable management situation in the 1980s. This small tree (6–12 m in height, up to 16 m) native to tropical Central and South America was introduced to Tahiti as a garden ornamental in 1937 for its large (80 × 30 cm) showy leaves with purple undersides (Meyer, 1996). After introduction, it spread to form dense forests, excluding native and endemic trees and preventing new seedling recruitment. A relatively shade-tolerant and fast-growing species (up to 1.5 m of height growth

per year), early reproducing (after four to five years of growth), and a prolific fruit producer (millions of seeds produced by a mature tree per reproductive season), miconia is a formidable competitor for the native flora. It is able to penetrate into relatively undisturbed native forests (Meyer, 1998); its rapid spread is aided by natural disturbances such as cyclones that open the forest canopy and promote faster recruitment and growth (Murphy et al., 2008). Frugivorous introduced and native birds facilitate the invasion by dispersing its small fleshy fruits over long distances, and birds consume more miconia where it is abundant, likely accelerating the rate of spread at sites where it is already invasive (Spotswood et al., 2012). Invading miconia density reaches two to four individuals per m², and its basal area attains 30 to 40 m²/ha. Light availability in miconia-invaded forest understories decreases to 0.4–0.6% compared to 1–3% in native forests (Meyer et al., 2007). By 1980, over 80,000 ha were considered invaded (two-thirds of the island's land surface), from near sea level to about 1400 m elevation. At high elevations, miconia encroaches on montane cloud forest and threatens 40–50 Tahitian endemic plant species (Meyer and Florence, 1996). The understory of miconia-invaded forests typically resembles a "biological desert" with only dead endemic tree fern trunks still standing.

Because of the extent of the infestation, the steep mountainous and often inaccessible slopes on Tahiti, and the failure of conventional manual and chemical control owing to the long duration (>15 years) of the plant's seed bank (Meyer et al., 2011), biological control was chosen as a potentially more effective management approach to reduce the impacts of miconia on native vegetation. However, many questions were asked when the biological control program was officially started in 1997, including whether effective, safe biological control agents existed in miconia's native range and whether native forests, once cleared of miconia, would be invaded by other weeds. The latter was viewed as a possible risk, based on outcomes of other biological control projects against pests of natural areas (e.g., Barton et al., 2007) and given that more than 420 naturalized exotic plant species are present on Tahiti, about 70 of which pose serious invasion threats (Fourdrigniez and Meyer, 2008).

Several trips to Central and South America to search for potential control agents yielded a fungal pathogen (*Colletotrichum gloeosporioides* [Penz] Sacc. f. sp. *miconiae* Killgore & L. Sugiyama) (Order Melanconiales, Class Coelomycetes, Subdivision Deuteromycetinae, hereafter *Cgm*) that was found to be highly specific to miconia during extensive testing in the quarantine laboratory of the Hawaii Department of Agriculture. The pathogen causes leaf anthracnosis and necrosis, leading to defoliation (Figure 10.28) and death of seedlings. This fungus appeared to be a promising agent (Killgore et al., 1999).

In 2000, a release was performed on Tahiti (Meyer and Killgore, 2000). *Cgm* established successfully and spread across the entire island in less than three years, infecting almost all miconia plants from sea level to high elevation. However, while mortality for very young seedlings was between 70–75% in the laboratory, the mortality rate was much lower under field conditions (30% for seedlings <50 cm in height at release sites [Meyer et al., 2008]). A relatively unexpected effect of the *Cgm* was partial defoliation of miconia canopy trees, with leaf damage of 6–36%, increasing with elevation (Meyer et al., 2008; Meyer and Fourdrigniez, 2011). The cooler climate found above 600 m in Tahiti proved to be the most favorable for *Cgm* reproduction, development, and efficiency.

Figure 10.28 Miconia leaf, showing injury owing to infestation by the fungal pathogen *Colletotrichum gloeosporioides* [Penz] Sacc. f. sp. *miconiae* at Nuku Hiva, Tahiti, 2007. Photo credit, J.-Y. Meyer. (*See Plate 30 for the color representation of this figure.*)

Having demonstrated that reproductive success and seedling recruitment of some critically endangered endemic trees and subshrubs increased with partial defoliation of miconia by *Cgm* (Meyer et al., 2007; Meyer and Fourdrigniez, 2011), changes were monitored in forest dynamics in permanent plots set up in 2005 at different elevations in rainforests and cloud forests, focusing on plant composition and abundance in the understory. Plant succession over a four-year period showed an increase of total native and endemic species richness and plant cover (mainly fern species) at all sites. The appearance of seedlings of endemic woody species that had been absent for the past decade in our permanent plots (Figure 10.29) located in miconia-invaded forests was noteworthy (Meyer et al., 2012). One of our study sites located at the lowest elevation (about 600 m) was re-invaded by other light-demanding pioneer herbs, shrubs, and trees, while at

Figure 10.29 Regrowth of native vegetation, including the rare endemic shrub *Psychotria* (Rubiaceae), in the understory of a miconia invaded forest in Tahiti, with 20% leaf damage on canopy miconia leaves caused by the introduced fungal pathogen *Colletotrichum gloeosporioides* [Penz] Sacc. f. sp. *miconiae*. Photo credit, J.-Y. Meyer. (*See Plate 31 for the color representation of this figure.*)

higher elevation (above 800 m elevation), most newly recruited species were native and endemic plants. These results demonstrate that biological control can be a useful tool in ecosystem management (Headrick and Goeden, 2001) and should be considered for restoration of highly invaded forests in situations where re-invasion by other weeds is unlikely to occur. Dense monotypic stands of other shade-tolerant invasive trees such as *Psidium cattleianum* Sabine and *Syzygium jambos* L. (Alston) (Myrtaceae), for example, are known to occur on many Indo-Pacific tropical islands (e.g., in the Mascarenes, the Seychelles, and the Hawaiian archipelago) with dramatic impacts on the native flora, and such species might potentially be controlled in a similar manner. However, it must be emphasized that the new plant assemblage after miconia control is different from the pre-invasion stage, and

long-term monitoring of forest dynamics will be still needed to study the trajectory of these newly created or "novel habitats."

Acknowledgments

We thank Don Strong and David Wagner for reviews of Chapter 10. Section acknowledgments are as follow: Mist flower/New Zealand: acknowledgments are made to the other authors of Barton et al. (2007) and to others credited in this paper; Cottony cushion scale/Galápagos: the success of this program was due to a collaborative effort (see participants listed in Calderón Alvarez et al., 2012 and Hoddle et al., 2013). We thank everybody who contributed to this project; Lantana-invaded communities/Australia: I thank Mike Day and Olusegun Osunkoya for comments on an earlier draft; Miconia-invaded Tahitian forests: Jean-Yves Meyer thanks Erica Spotswood (University of California at Berkeley) and Lloyd Loope (USGS, Haleakala National Park, Maui, Hawaii) for revising the English and for their comments that improved this short contribution based on 20 years of studies conducted in Tahiti with the financial support of the Government of French Polynesia.

References

Abell, K. J., J. J. Duan, L. Bauer, et al. 2012. The effect of bark thickness on host partitioning between *Tetrastichus planipennisi* (Hymen: Eulophidae) and *Atanycolus* spp. (Hymen: Braconidae), two parasitoids of emerald ash borer (Coleop: Buprestidae). *Biological Control* **63**: 320–325.

AEC Group. 2007. Economic impact of lantana on the Australian grazing industry. Department of Natural Resources and Water, Brisbane, Queensland, Australia. Available from: http://www.weeds.org.au/WoNS/lantana/docs/60_Lantana_Grazing_EIA_Final_Report_%28b%29.pdf [Accessed January 2016].

Allen, L. H., Jr., T. R. Sinclair, and J. M. Bennett. 1997. Evapotranspiration of vegetation of Florida: perpetuated misconceptions versus mechanistic processes. *Proceedings of the Soil and Crop Science Society of Florida* **56**: 1–10.

Allen, W. 2000. Restoring Hawaii's dry forests. *Bioscience* **50**: 1037–1041.

Amman, G. D. 1961. Predator introductions for control of the balsam woolly aphid on Mt. Mitchell, North Carolina. *Research Notes No. 153*. Southeastern Forest Experiment

Station Asheville, North Carolina, USDA, Forest Service. 2 pp.

Amman, G. D. and C. F. Speers. 1964. Release of predators of the balsam woolly aphid in North Carolina. *Research Note SE-32*, Southeastern Forest Experiment Station, Asheville, North Carolina, USDA, Forest Service. 4 pp.

Amman, G. D. and C. F. Speers. 1965. Progress in biological control of the balsam woolly aphid in North Carolina. *Southern Lumberman.* **211**(2632): 147–149.

Amman, G. D. and C. F. Speers. 1971. Introduction and evaluation of predators from India and Pakistan for control of the balsam woolly aphid (Homoptera: Adelgidae) in North Carolina. *The Canadian Entomologist* **103**: 528–533.

Arthur, F. H. and F. P. Hain. 1984. Seasonal history of the balsam woolly adelgid (Homoptera: Adelgidae) in natural stands and plantations of Fraser fir. *Journal of Economic Entomology* **77**: 1154–1158.

Austin, D. F. 1978. Exotic plants and their effects in southeastern Florida. *Environmental Conservation* **5**: 25–34.

Balch, R. E. 1952. Studies of the balsam woolly aphid, *Adelges piceae* (Ratz.) and its effects on balsam fir, *Abies balsamea* (L.) Mill. Canadian Department of Agriculture, Ottawa, Canada. 76 pp.

Balentine, K. M., P. D. Pratt, F. A. Dray, Jr., et al. 2009. Geographic distribution and regional impacts of *Oxyops vitiosa* (Coleoptera: Curculionidae) and *Boreioglycaspis melaleucae* (Hemiptera: Psyllidae), biological control agents of the invasive tree *Melaleuca quinquenervia*. *Environmental Entomology* **38**: 1145–1154.

Barr, T. C. 1962. The genus *Trechus* (Coleoptera: Carabidae: Trechini) in the southern Appalachians. *The Coleopterists Bulletin* **16**: 65–69.

Bartlett, B. R. 1978. Margarotidae, pp. 132–136. *In*: Clausen, C. P. (ed.). *Introduced Parasites and Predators of Arthropod Pests and Weeds: A World View.* Agricultural Handbook No. 480. USDA Agricultural Research Service. Washington, DC.

Barton, J., S. V. Fowler, A. F. Gianotti, et al. 2007. Successful biological control of mist flower (*Ageratina riparia*) in New Zealand: Agent establishment, impact and benefits to the native flora. *Biological Control* **40**: 370–385.

Bauer, L. S., H. P. Liu, D. L. Miller, and J. Gould. 2008. Developing a biological control program for emerald ash borer (*Agrilus planipennis*), an invasive ash pest in North America. *Michigan Entomological Society Newsletter* **53**: 38–39. Available from: http://www.nrs.fs.fed.us/pubs/jrnl/2008/nrs_2008_bauer_002.pdf [Accessed January 2016].

Bauer, L. S., H. P. Liu, and D. L. Miller, D. L., 2009. Emerald ash borer biological control: rearing, releasing, establishment, and efficacy of parasitoids, pp. 7–8. *In:* McManus, K. and K. Gottschalk (eds.). *Proceedings 20th U. S. Department of Agriculture Interagency Research Forum on Invasive Species, 2009 January 13–16, Annapolis, Maryland.* General Technical Report NRS-P-51. USDA Forest Service. Newtown Square, Pennsylvania, USA.

Bauer, L. S., J. Gould, and J. J. Duan. 2010. Can biological control of emerald ash borer save our ash? *Michigan Entomological Society Newsletter* **55**: 26–27.

Bodle, M. J., A. P. Ferriter, and D. D. Thayer. 1994. The biology, distribution, and ecological consequences of *Melaleuca quinquenervia* in the Everglades, pp. 341–355. *In*: Davis, S. M. and J. C. Ogden (eds.). *Everglades. The Ecosystem and Its Restoration.* St. Lucie Press. Delray Beach, Florida, USA.

Broughton, S. 2000. Review and evaluation of lantana biological control programs. *Biological Control* **17**: 272–286.

Bryant, D. G. 1963. *Adelges piceae* (Ratz.) studies in Newfoundland: Review of Associated predators, introduced and native from 1952 to 1962. Interim Report 1963-1, Forest Entomology and Pathology Laboratory, Corner Brook, Nfld. Available from: https://afc-fr.cfsnet.nfis.org/fias/pdfs/afc/nfld_interim_1963-1.pdf [Accessed 11 April 2011].

Cabin, R. J., S. G. Weller, D. H. Lorence, et al. 2000. Effects of long-term ungulate exclusion and recent alien species control on the preservation and restoration of a Hawaiian tropical dry forest. *Conservation Biology* **14**: 439–453.

Calderón Alvarez, C., C. E. Causton, M. S. Hoddle, et al. 2012. Monitoring the effects of *Rodolia cardinalis* on *Icerya purchasi* populations on the Galápagos Islands. *Biological Control* **57**: 167–179.

Caltagirone, L. E. and R. L. Doutt. 1989. The history of the vedalia beetle importation to California and its impact on the development of biological control. *Annual Review of Entomology* **34**: 1–16.

Cappeart, D., D. G. McCullough, T. M. Poland, and N. W. Siegert. 2005. Emerald ash borer in North America: a research and regulatory challenge. *American Entomologist* **51**: 152–165.

Causton, C. E. 2001. Dossier on *Rodolia cardinalis Mulsant* (Coccinellidae: Cocinellinae) a potential biological control agent for the cottony cushion scale *Icerya purchasi* Maskell (Margarodidae). Charles Darwin Foundation. Puerta Ayora, Galápagos, Ecuador. Unpublished report.

Causton, C. E. 2009. Success in biological control: the scale and the ladybird, pp. 184–190. *In*: De Roy, T. (ed.). *Galápagos: Preserving Darwin's Legacy.* Firefly Books. Auckland, New Zealand.

Causton C. E., M. P. Lincango, and T. G. A. Poulsom. 2004. Feeding range studies of *Rodolia cardinalis* (Mulsant), candidate biological control agent of *Icerya purchasi* Maskell in the Galápagos Islands. *Biological Control* **29**: 315–325.

Center, T. D., T. K. Van, M. Rayachhetry, et al. 2000. Field colonization of the melaleuca snout beetle (*Oxyops vitiosa*) in south Florida. *Biological Control* **19**: 112–123.

Center, T. D., P. D. Pratt, P. W. Tipping, et al. 2006. Field colonization, population growth, and dispersal of *Boreioglycaspis melaleucae* Moore, a biological control agent of the invasive tree *Melaleuca quinquenervia* (Cav.) Blake. *Biological Control* **39**: 363–374.

Center, T. D., P. D. Pratt, P. W. Tipping, et al. 2007. Initial impacts and field validation of host range for *Boreioglycaspis melaleucae* Moore (Hemiptera: Psyllidae), a biological control agent of the invasivtree *Melaleuca quinquenervia* (Cav.) Blake (Myrtales: Myrtaceae: Leptospermoideae). *Environmental Entomology* **36**: 569–576.

Center, T., M. Purcell, P. Pratt, et al. 2012. Biological control of *Melaleuca quinquenervia*: an Everglades invader. *Biological Control* **57**: 151–165.

Chiarelli, R. N., P. D. Pratt, C. S. Silvers, et al. 2011. Influence of temperature, humidity, and plant terpenoid profile on life history characteristics of *Boreioglycaspis melaleucae* (Hemiptera: Psyllidae), a biological control agent of the invasive tree *Melaleuca quinquenervia*. *Annals of the Entomological Society of America* **104**: 488–497.

Clark, R. C., D. O. Greenbank, D. G. Bryant, and J. W. E. Harris. 1971. *Adelges piceae* (Ratz.) balsam woolly aphid (Homoptera: Adelgidae), pp. 113–117. *In*: Anon. *Biological Control Programmes against Insects and Weeds in Canada, 1959–1968*. Technical Communication No. 4. Commonwealth Institute of Biological Control. Farnham Royal, UK.

Clausen, C. P. 1978. Chermidae, pp. 49–55. *In*: Clausen, C. P. (ed.). *Introduced Parasites and Predators of Arthropod Pests and Weeds: A World View*. Agricultural Handbook No. 480. USDA Agricultural Research Service. Washington, DC.

Craft, C. B., J. Vymazal, and C. J. Richardson. 1995. Response of Everglades plant communities to nitrogen and phosphorus additions. *Wetlands* **15**: 258–271.

Davis, C. J., E. Yoshioka, and D. Kageler. 1992. Biological control of lantana, prickly pear, and hamakua pamakani inhawah: a review and update, pp. 411–431. *In*: Stone, P., C. W. Smith, and J. T. Tunison (eds.). *Alien Plant Invasions in Native Ecosystems of Hawaii—Management and Research*. University of Hawaii Press. Manoa, Honolulu, Hawaii, USA.

Day, M. 2012. *Lantana camara* L. – lantana, pp. 334–346. *In*: Julien, M., R. McFadyen, and J. Cullen (eds.). *Biological Control of Weeds in Australia*. CSIRO Publishing. Melbourne, Australia.

Day, M. D. and S. Neser. 2000. Factors influencing the biological control of *Lantana camara* in Australia and South Africa, pp. 897–908. *In*: Spencer, N. R. (ed.). *Proceedings of the X International Symposium on Biological Control of Weeds, 4–14 July 1999*. Montana State University. Bozeman, Montana, USA.

Day, M. D., C. J. Wiley, J. Playford, and M. P. Zalucki. 2003. *Lantana: Current Management Status and Future Prospects*. ACIAR Monograph Series 102. Canberra, A.C.T, Australia.

Day, M. D. and A. J. Urban. 2004. Ecological basis for selecting biological control agents for lantana, pp. 81. *In*: Cullen, J. M., D. T. Briese, D. J. Kriticos, et al. (eds.). *Proceedings of the XI International Symposium on Biological Control of Weeds, 27 April–2 May 2003*. CSIRO Entomology. Canberra, A.C.T., Australia.

Day, M. D. and M. P. Zalucki. 2009. *Lantana camara* Linn. (Verbenaceae), pp. 211–246. *In*: Muniappan, R., G. Reddy, and A. Raman (eds.). *Biological Control of Tropical Weeds Using Arthropods*. Cambridge University Press. Cambridge, UK.

DENR. 2010. Spruce fir forests, 9 pp. *In*: North Carolina Ecosystem Response to Climate Change: DENR Assessment of Effects and Adaptation Measures. Available from: http://portal.ncdenr.org/c/document_library/get_file?uuid=c180b2d8-8135-4b86-a9f0-e1cb344b3604&groupId=61587 [Accessed January 2016].

D'Orangeville, L., A. Bouchard, and A Cogliastro. 2008. Post-agricultural forests: landscape patterns add to stand-scale factors in causing insufficient hardwood regeneration. *Forest Ecology and Management* **255**: 1637–1646.

Doren, R. F., J. H. Richards, and J. C. Volin. 2009. A conceptual ecological model to facilitate understanding the role of invasive species in large-scale ecosystem restoration. *Ecological Indicators* **9**: S150–S160.

Dray, F. A., Jr., B. C. Bennett, and T. D. Center. 2006. Invasion history of *Melaleuca quinquenervia* (Cav.) S.T. Blake in Florida. *Castanea* **71**: 210–225.

Duan, J. J., M. D. Ulyshen, L. S. Bauer, et al. 2010. Measuring the impact of biotic factors on populations of immature emerald ash borers (Coleoptera: Buprestidae). *Environmental Entomology* **39**: 1513–1522.

Duan, J. J., L. S. Bauer, K. J. Abell, and R. G. Van Driesche. 2012a. Population responses of hymenopteran parasitoids to the emerald ash borer (Coleoptera: Buprestidae) in recently invaded areas in north central United States. *Biological Control* **57**: 199–209.

Duan, J. J., G. Yurchenko, and R. Fuester. 2012b. Occurence of emerald ash borer (Coleoptera: Buprestidae) and biotic factors affecting its immature stages in the Russian Far East. *Environmental Entomology* **41**: 245–254.

Duan, J. J., L. S. Bauer, K. J. Abell, et al. 2013a. Establishment and abundance of *Tetrastichus planipennisi* (Hymenoptera: Eulophidae) in Michigan: potential for success in classical biological control of the invasive emerald ash borer (Coleoptera: Buprestidae). *Journal of Economic Entomology* **106**: 1145–1154.

Duan, J. J., P. P. Taylor, R. W. Fuester, et al. 2013b. Hymenopteran parasitoids attacking the invasive emerald ash borer (Coleoptera: Buprestidae) in Western and Central Pennsylvania. *Florida Entomologist* **96**: 166–172.

Dull, C., J. Ward, D. Brown, G. Ryan, et al. 1988. Evaluation of spruce and fir mortality in the Southern Appalachian Mountains. Protection Report R8-PR 13 edition. USDA Forest Service, Southern Region. 92 pp.

Eichhorn, O. 1968. Problems of the population dynamics of silver fir woolly aphids, genus *Adelges* (= *Dreyfusia*), Adelgidae. *Zeitschrift für angewandt Entomologie* **61**: 157–214.

Eichhorn, O. 1969a. Natüurliche Verbreitungsareale und Einschleppungsgebiete der Weisfstannen-Wolläuse (Gattung *Dreyfusia*) und die Möoglichkeiten ihrer biologischen Bekäampfung. *Zeitschrift für angewvandte Entomologie* **62**: 113–131.

Eichhorn, O. 1969b. Investigations on woolly aphids of the genus *Adelges*. (Homopt.: Adelgidae) and their predators in Turkey. *Commonwealth Institute of Biological Control, Technical Bulletin* **12**: 83–103.

Eichhorn, O. 2000. Untersuchungen uber Fichtengallenlause, *Dreyfusia* spp. (Hom., Adelgidae) und deren Predatoren in der Nord-Turkei [Studies on *Drefusia* spp. and their predators in northern Turkey]. *Journal of Pest Science (Anzeiger Schädlingskunde)* **73**: 13–16.

Emerald Ash Borer Information, 2013. Emerald ash borer information. Available from: http://emeraldashborer.info/ [Accessed January 2015].

Fedde, G. F. 1972. Status of imported and native predators of the balsam woolly aphid on Mt. Mitchell, North Carolina. USDA Forest Service Research Note SE-175, Southeastern Forest Experiment Station, Asheville, North Carolina, USA. 4 pp.

Felt, E. P. 1917. 33rd report of the state entomologist on injurious and other insects of the State of New York. *New York State Museum Bulletin* No. 202. Albany, New York, USA. 133 pp.

Fensham, R., R. Fairfax, and R. Cannell. 1994. The invasion of *Lantana camara* L. in Forty Mile Scrub National Park, north Queensland. *Australian Journal of Ecology* **19**: 297–305.

Fessl, B., H. G. Young, R. P. Young, et al. 2010. How to save the rarest Darwin's finch from extinction: the mangrove finch on Isabela Island. *Philosophical Transactions Royal Society B* **365**: 1019–1030.

Flowers, J. D. 1991. Tropical fire suppression in *Melaleuca quinquenervia*, pp. 151–158. *In*: Center, T. D., R. F. Doren, R. L. Hofstetter, et al. (eds). *Proceedings of a Symposium on Exotic Pest Plants, November 2–4, 1988, Miami, Florida*. Technical Report NPS/NREVER/NRTR-91/06. U.S. Department of the Interior, National Park Service, Washington, DC.

Fourdrigniez, M. and J.-Y. Meyer. 2008. Liste et caractéristiques des plantes introduites naturalisées et envahissantes en Polynésie française. Contribution à la Biodiversité de Polynésie française N°17. Délégation à la Recherche, Papeete, Tahiti. 62 pp.

Franks, S. J., A. M. Kral, and P. D. Pratt. 2006. Herbivory by introduced insects reduces growth and survival of *Melaleuca quinquenervia* seedlings. *Environmental Entomology* **35**: 366–372.

Franz, J. M. 1956. The effectiveness of predators and food in limiting gradations of *Adelges (Dreyfusia) piceae* (Ratz.) in Europe, pp. 781–787. *In*: Anon. *Proceedings of the Tenth International Congress of Entomology. Montreal, Canada, August 17–25, 1956*. Mortimer, LTD (Printer). Ottawa, Canada.

Gagné, R. J. 2010. Update for a Catalog of the Cedidomyiidae (Diptera) of the World. Digital Version 1. pp. 111–112. Available from: http://www.ars.usda.gov/SP2UserFiles/ Place/12754100/Gagne_2010_World_Catalog_ Cecidomyiidae.pdf. [Accessed January 2016].

Gardener, M. R. and C. Grenier. 2011. Linking livelihoods and conservation: challenges facing the Galápagos Islands, pp. 73–85. *In*: Baldacchino, G. and D. Niles (eds.). *Island Futures – Conservation and Development across the Asia-Pacific Region*. Springer. Tokyo.

Gates, M. and G. Delvare. 2008. A new species of *Eurytoma* (Hymenoptera: Eurytomidae) attacking *Quadrastichus* spp. (Hymenoptera: Eulophidae) galling *Erythrina* spp. (Fabaceae), with a summary of African *Eurytoma* biology and species checklist. *Zootaxa* **175**: 1–24.

Gomes, A. R. S. and T. T. Kozlowski. 1980. Responses of *Melaleuca quinquenervia* seedlings to flooding. *Physiologia Plantarum* **49**: 373–377.

Gooden, B., K. French, P. J. Turner, and P. O. Downey. 2009. Impact threshold for an alien plant invader, *Lantana camara* L., on native plant communities. *Biological Conservation* **142**: 2631–2641.

Gordon, D. R. 1998. Effects of invasive, non-indigenous plant species on ecosystem processes: lessons from Florida. *Ecological Applications* **8**: 975–989.

Gould, J. and J. J. Duan. 2013. Petition for release of an exotic parasitoid, *Spathius galinae* Belokobylskij & Strazanac, for the biological control of the emerald ash borer, *Agrilus planipennis* Fairmaire. Submitted to USDA APHIS (see reformulated document available from: http://www.aphis.usda.gov/plant_ health/ea/downloads/2015/spathius-galinae-eab-biocontrol. pdf [Accessed January 2016].

Gunderson, L. H. 1994. Vegetation of the Everglades: determinants of community composition, pp. 323–340. *In*: Davis, S. M. and J. C. Ogden (eds.). *Everglades. The Ecosystem and its Restoration*. St. Lucie Press. Delray Beach, Florida, USA.

Haack, R.A., E. Jendek, H. P. Liu, et al. 2002. The emerald ash borer: a new exotic pest in North America. *Newsletter of the Michigan Entomological Society* **47**: 1–5.

Hapai, M. N. 1977. The Biology and Ecology of the Hamakua Pamakani Gall Fly, *Procecidochares alani* (Steyskal). M.S. Thesis, University of Hawai'i, Honolulu, Hawaii, USA.

Haseler, W. 1963. Progress in insect control of lantana. *Queensland Agricultural Journal* **89**: 65–68.

Headrick, D. H. and R. D. Goeden. 2001. Biological control as a tool for ecosystem management. *Biological Control* **21**: 249–257.

Herms, D. A. and D. G. McCullough, D. G. 2014. Emerald ash borer invasion of North America: history, biology, ecology, impacts, and management. *Annual Review of Entomology* **59**: 13–30.

Heu, R. A., D. M. Tsuda, W. T. Nagamine, and T. H. Suh. 2005. New pest advisory No. 05-03; Department of Agriculture, State of Hawaii, Honolulu, Hawaii, USA.

Heu, R. A., D. M. Tsuda, W.T. Nagamine, et al. 2008. Erythrina gall wasp, *Quadrastichus erythrinae* Kim. New pest advisory No. 05-03. Updated December 2008. Hawaii Department of Agriculture, Honolulu, Hawaii. 2 pp. Available from: http:// hdoa.hawaii.gov/pi/files/2013/01/npa05-03-EGW.pdf [Accessed January 2016].

Hoddle, M. S., C. Crespo Ramírez, C. D. Hoddle, et al. 2013. Post release evaluation of *Rodolia cardinalis* (Coleoptera: Coccinellidae) for control of *Icerya purchasi* (Hemiptera: Monophlebidae) in the Galápagos Islands. *Biological Control* **67**: 262–274.

Humble, L. M. 1994. Recovery of additional exotic predators of balsam woolly adelgid, *Adelges piceae* (Ratzenburg) (Homoptera: Adelgidae), in British Colombia. *The Canadian Entomologist* **126**: 1101–1103.

Jetton, R. M., J. F. Monahan, and F. P. Hain. 2011. Laboratory studies of feeding and oviposition preference, developmental performance, and survival of the predatory beetle, *Sasajiscymnus tsugae* on diets of the woolly adelgids, *Adelges tsugae* and *Adelges piceae*. *Journal of Insect Science* **11**: article 68.

Julien, M. H. and M. W. Griffiths. 1999. *Biological Control of Weeds: A World Catalogue of Agents and their Target Weeds*. CAB International. Wallingford, UK.

Killgore, E. M., L. S. Sugiyama, R. W.Baretto, and D. E. Gardner. 1999. Evaluation of *Colletotrichum gloeosporioides* for biological control of *Miconia calvescens* in Hawaii. *Plant Disease* **83**: 964.

Kim, I., G. Delvare, and J. LaSalle. 2004. A new species of *Quadrastichus* (Hymenoptera: Eulophidae): A gall-inducing pest on *Erythrina* spp. (Fabaceae). *Journal of Hymenoptera Research* **13**: 243–249.

Kushlan, J. A. 1990. Freshwater marshes, pp. 324–363. *In*: Myers, R. L. and J. J. Ewel (eds.). *Ecosystems of Florida*. University of Central Florida Press.Orlando, Florida, USA.

Laroche, F. B. 1998. Managing melaleuca (*Melaleuca quinquenervia*) in the Everglades. *Weed Technology* **12**: 726–762.

Laroche, F. B. and A. P. Ferriter. 1992. The rate of expansion of melaleuca in south Florida. *Journal of Aquatic Plant Management* **30**: 62–65.

Lincango, P. M., C. E. Causton, C. Alvarez Calderón, and G. Jiménez-Uzcátegui. 2011. Evaluating the safety of *Rodolia cardinalis* to two species of Galapagos finch, *Camarhynchus parvulus* and *Geospiza fuliginosa*. *Biological Control* **56**: 145–149.

Liu, H. P., L. S. Bauer, D. L. Miller, et al. 2007. Seasonal abundance of *Agrilus planipennis* (Coleoptera: Buprestidae) and its natural enemies *Oobius agrili* (Hymenoptera: Encyrtidae) and *Tetrastichus planipennisi* (Hymenoptera: Eulophidae) in China. *Biological Control* **42**: 61–71.

Lodge, T. E. 2004. *The Everglades Handbook. Understanding the Ecosystem*. CRC Press, Boca Raton. Florida, USA.

Loveless, C. M. 1959. A study of the vegetation in the Florida Everglades. *Ecology* **40**: 1–9.

Lowe, S., M. Browne, S. Boudjelas, and M. De Poorter. 2000. 100 of the world's worst invasive alien species: a selection from the global invasive species database: Invasive Species Specialist Group. Auckland, New Zealand. Published by The Invasive Species Specialist Group (ISSG) a specialist group of the Species Survival Commission (SSC) of the World Conservation Union (IUCN), University of Auckland, Auckland, New Zealand. 12 pp.

Lusk, L., M. Mutel, E. S. Walker, and F. Levy. 2010. Forest change in high-elevation forests of Mt. Mitchell, North Carolina: Re-census and analysis of data collected over 40 years, pp. 104–112. *In*: Rentch, J. S. and T. M. Schuler (eds.). *Proceedings from the Conference on the Ecology and Management of High Elevation Forests in the Central and Southern Appalachian Mountains, May 14–15, 2009, Slatyfork, West Virginia*. Northern Research Station General, Technical Report NRS-P-64. USDA, Forest Service. Newtown Square, Pennsylvania. USA.

Maltby, E. and P. J. Dugan. 1994. Wetland ecosystem protection, management, and restoration: an international perspective, pp. 29–46. *In*: Davis, S. M. and J. C. Ogden (eds.). *Everglades: the Ecosystem and its Restoration*. St. Lucie Press. Delray Beach, Florida, USA.

Martin, M. R., P.W.Tipping, and J. O. Sickman. 2009. Invasion by an exotic tree alters above and belowground ecosystem components. *Biological Invasions* **11**: 1883–1894.

McAlpine, J. F. 1971. A revision of the subgenus *Neoleucopis* (Diptera: Chamaemyiidae). *The Canadian Entomologist* **103**: 1851–1874.

McAlpine, J. F. 1978. A new dipterous predator of balsam wooly aphid from Europe and Canada (Diptera: Chamaemyiidae). *Entomologica Germanica* **4**: 349–355.

McAlpine, J. F. 1987. Chamaemyiidae, pp. 965–971. *In*: J. F. McAlpine (ed.). *Manual of Nearctic Diptera, Vol. 2*. Research Branch, Agriculture Canada. Ottawa, Canada.

McCullough, D. G, N. W. Siegert, and J. Bedford. 2009. Slowing ash mortality: a potential strategy to slam emerald ash borer in outlier sites, pp. 44–46. *In*: McManus, K. and K.Gottschalk (eds.). *Proceedings 20th U. S. Department of Agriculture Interagency Research Forum on Invasive Species, 2009 January 13–16, Annapolis, Maryland*. Northern Research Station, General Technical Report NRS-P-51. USDA Forest Service Morgantown. West Virginia, USA.

McJannet, D. 2008. Water table and transpiration dynamics in a seasonally inundated *Melaleuca quinquenervia* forest, north Queensland, Australia. *Hydrological Processes* **22**: 3079–3090.

McManamay, R. H., L. M. Resler, J. B. Campbell, and R. A. McManamay. 2011. Assessing the impacts of balsam woolly adelgid (*Adelges piceae* Ratz.) and anthropogenic disturbance on the stand structure and mortality of Fraser fir [*Abies fraseri* (Pursh) Poir.] in the Black Mountains, North Carolina. *Castanea* **76**: 1–19.

Mercader, R. J., A. M. Siegert, N. W. Liebhold, and D. G. McCullough. 2011. Simulating the effectiveness of three potential management options to slow the spread of emerald ash borer (*Agrilus planipennis*) populations in localized outlier sites. *Canadian Journal of Forest Research* **41**: 254–264.

Messing, R. H., S. Noser, and J. Hunkeler. 2009. Using host plant relationships to help determine origins of the invasive Erythrina gall wasp, *Quadrastichus erythrinae* Kim (Hymenoptera: Eulophidae). *Biological Invasions* **11**: 2233–2241.

Meyer, J.-Y. 1996. Status of *Miconia calvescens* (Melastomataceae), a dominant invasive tree in the Society Islands (French Polynesia). *Pacific Science* **50**: 66–76.

Meyer, J.-Y. 1998. Observations on the reproductive biology of *Miconia calvescens* DC (Melastomataceae), an alien invasive tree on the island of Tahiti (South Pacific Ocean). *Biotropica* **30**: 609–624.

Meyer, J.-Y. and J. Florence. 1996. Tahiti's native flora endangered by the invasion of *Miconia calvescens* DC. (Melastomataceae). *Journal of Biogeography* **23**: 775–783.

Meyer, J.-Y. and E. M. Killgore. 2000. First and successful release of a bio-control pathogen agent to combat the invasive alien tree *Miconia calvescens* (Melastomataceae) in Tahiti. *Aliens* **12**: 8.

Meyer, J.-Y. and M. Fourdrigniez. 2011. Conservation benefits of biological control: the recovery of a threatened endemic plant subsequent to the introduction of a pathogen agent to control a dominant invasive tree on the island of Tahiti (South Pacific). *Biological Conservation* **144**: 106–113.

Meyer, J.-Y., A. Duplouy, and R. Taputuarai. 2007. Dynamique des populations de l'arbre endémique *Myrsine longifolia* (Myrsinacées) dans les forêts de Tahiti (Polynésie française) envahies par *Miconia calvescens* (Mélastomatacées) après introduction d'un champignon pathogène de lutte biologique: premières investigations. *Revue d'Ecologie (Terre Vie)* **62**(1): 17–33.

Meyer, J.-Y., R. Taputuarai, and E. M. Killgore. 2008. Dissemination and impact of the fungal pathogen *Colletotrichum gloeosporioides* f. sp. *miconiae* on the invasive alien tree *Miconia calvescens* (Melastomataceae) in the rainforests of Tahiti (French Polynesia, South Pacific), pp. 594–599. *In:* Julien, M. H., R. Sforza, M. C. Bon, et al. (eds.). *Proceedings of the XII International Symposium on Biological Control of Weeds, La Grande Motte, France, 22–27 April 2007.* CAB International. Wallingford, UK.

Meyer, J.-Y., L. L. Loope, and A-C. Goarant. 2011. Strategy to control the invasive alien tree *Miconia calvescens* in Pacific islands: eradication, containment or something else? pp. 91–96. *In:* Veitch, C. R., M. N. Clout, and D. R. Towns (eds.). *Island Invasives: Eradication and Management.* IUCN. Gland, Switzerland.

Meyer, J.-Y., M. Fourdrigniez, and R. Taputuarai, R. 2012. Restoring habitat for native and endemic plants through the introduction of a fungal pathogen to control the alien invasive tree *Miconia calvescens* in the island of Tahiti. *Biological Control* **57**: 191–198.

Montgomery, M. E. and N. P. Havill. 2014. Balsam woolly adelgid (*Adelges piceae* [Ratzeburg]) (Hemiptera: Adelgidae), pp. 9–19. *In:* Van Driesche, R. G. and R. Reardon (eds.). *The Use of Classical Biological Control to Preserve Forests in North America.* FHTET- 2013-02. USDA Forest Service, Morgantown, West Virginia, USA. Available from: http://www.fs.fed.us/forest health/technology/pub_titles.shtml [Accessed January 2016].

Morath, S. U., P. D. Pratt, C. S. Silvers, and T. D. Center. 2006. Herbivory by *Boreioglycaspis melaleucae* (Hemiptera: Psyllidae) accelerates foliar senescence and abscission in the invasive tree *Melaleuca quinquenervia*. *Environmental Entomology* **35**: 1372–1378.

Morin, L., R. L. Hill, and S. Matayoshi. 1997. Hawaii's successful biological control strategy for mist flower (*Ageratina riparia*) – can it be transferred to New Zealand? *Biological control News and Information* **18**: 77N–88N.

Muniappan, R., G. Denton, J. Brown, et al. 1996. Effectiveness of the natural enemies of *Lantana camara* on Guam: a site and seasonal evaluation. *Entomophaga* **41**: 167–182.

Murphy, H. T., B. D. Hardesty, C. S. Fletcher, et al. 2008. Predicting dispersal and recruitment of *Miconia calvescens* (Melastomataceae) in Australian tropical rainforests. *Biological Invasions* **10**: 925–936.

Myers, R. L. 1983. Site susceptibility to invasion by the exotic tree *Melaleuca quinquenervia* in southern Florida. *Journal of Applied Ecology* **20**: 645–658.

Noss, R. F. and R. L. Peters. 1995. *Endangered Ecosystems: A Status Report on America's Vanishing Habitat And Wildlife.* Defenders of Wildlife. Washington, DC. 133 pp.

Onken, B. and R. Reardon (eds.). 2011. *Implementation and Status of Biological Control of the Hemlock Woolly Adelgid.* U.S. Forest Service, Forest Health Protection. Morgantown, West Virginia, USA. 230 pp.

Osunkoya, O. O., C. Perrett, C. Fernando, et al. 2013. Modeling population growth and site specific control of the invasive *Lantana camara* L. (Verbenaceae) under differing fire regimes. *Population Ecology* **55**: 291–303.

Pasteels, J. M., C. Deroe, B. Tursch, et al. 1973. Distribution and activities of the defensive alkaloids of the Coccinellidae. *Journal of Insect Physiology* **19**: 1771–1784.

Poland, T. M. and D. G. McCullough. 2006. Emerald ash borer: invasion of the urban forest and the threat to North America's ash resource. *Journal of Forestry* **104**: 118–124.

Poland, T. M., D. G. McCullough, D. A. Herms, et al. 2010. Management tactics for emerald ash borer: chemical and biological control, pp. 46–49. *In:* McManus, K. and K. Gottschalk (eds.). *Proceedings 21th U. S. Department of Agriculture Interagency Research Forum on Invasive Species, 2010 January 12–15, Annapolis, Maryland.* USDA Forest Service Northern Research Station. General Technical Report # NRS-P-75. Newtown Square, Pennsylvania, USA.

Prasad, Y. K. 1989. The role of natural enemies in controlling *Icerya purchasi* in South Australia. *Entomophaga* **34**: 391–395.

Pratt, P. D., M. B. Rayachhetry, T. K. Van, and T. D. Center. 2002. Field-based rates of population increase for *Oxyops vitiosa* (Coleoptera: Curculionidae), a biological control agent of the invasive tree *Melaleuca quinquenervia*. *Florida Entomologist* **85**: 286–287.

Pratt, P. D., D. H. Slone, M. B. Rayamajhi, et al. 2003. Geographic distribution and dispersal rate of *Oxyops vitiosa* (Coleoptera:

Curculionidae), a biological control agent of the invasive tree *Melaleuca quinquenervia* in south Florida. *Environmental Entomology* **32**: 397–406.

Pratt, P. D., M. B. Rayamajhi, T. K. Van, and T. D. Center. 2004. Modeling the influence of resource availability on population densities of *Oxyops vitiosa* (Coleoptera: Curculionidae), a biological control agent of the invasve tree *Melaleuca quinquenervia*. *Biological Control Science and Technology* **14**: 51–61.

Pratt, P. D., M. B. Rayamajhi, T. K. Van, et al. 2005. Herbivory alters resource allocation and compensation in the invasive tree *Melaleuca quinquenervia*. *Ecological Entomology* **30**: 316–326.

Pratt, P. D., M. B. Rayamajhi, T. D. Center, et al. 2009. The ecological host range of an intentionally introduced herbivore: A comparison of predicted versus actual host use. *Biological Control* **49**: 146–153.

Pratt, P. D., M. B. Rayamajhi, P. W. Tipping, et al. 2013. Establishment, population increase, spread, and ecological host range of *Lophodiplosis trifida* (Diptera: Cecidomyiidae), a biological control agent of the invasive tree *Melaleuca quinquenervia* (Myrtales: Myrtaceae. *Environmental Entomology* **42**: 925–935.

Prinsloo, G. L. and J. A. Kelly. 2009. The tetrastichine wasps (Hymenoptera: Chalcidoidea: Eulophidae) associated with galls on *Erythrina* species (Fabaceae) in South Africa, with the description of five new species. *Zootaxa* **2083**: 27–45.

Pschorn-Walcher, H. and H. Zwölfer. 1956. The predator complex of white-fir woolly aphids (*Dreyfusia*, Adelgidae). *Zeitschrift für Angewandte Entomologie* **39**: 63–75.

Quezada, J. R. and P. DeBach. 1973. Bioecological and populations studies of the cottony cushion scale, *Icerya purchasi* Mask., and its natural enemies, *Rodolia cardinalis* Mul. and *Cryptochaetum iceryae* Will. in southern California. *Hilgardia* **41**: 631–688.

Rader, R. and C. Richardson. 1992. The effects of nutrient enrichment on algae and macroinvertebrates in the Everglades: A review. *Wetlands* **12**: 121–135.

Ravn, H. P., N. P. Havill, S. Akbulut, et al. 2013. *Dreyfusia nordmannianae* in northern and central Europe: Potential for biological control and comments on its taxonomy. *Journal of Applied Entomology* **137**: 401–417.

Rayamajhi, M. B., T. K. Van, T. D. Center, et al. 2002. Biological attributes of the canopy-held melaleuca seeds in Australia and Florida, U.S. *Journal of Aquatic Plant Management* **40**: 87–91.

Rayamajhi, M., T. Van, P. Pratt, and T. Center. 2006. Temporal and structural on litter production in *Melaleuca quinquenervia* dominated wetlands of south Florida. *Wetlands Ecology and Management* **14**: 303–316.

Rayamajhi, M., T. Van, P. Pratt, et al. 2007. *Melaleuca quinquenervia* dominated forests in Florida: analyses of natural-enemy impacts on stand dynamics. *Plant Ecology* **192**: 119–132.

Rayamajhi, M., P. Pratt, T. Center, et al. 2009. Decline in exotic tree density facilitates increased plant diversity: the experience from *Melaleuca quinquenervia* invaded wetlands. *Wetlands Ecology and Management* **17**: 455–467.

Rayamajhi, M., P. Pratt, T. Center, and T. Van. 2010. Exotic tree leaf litter accumulation and mass loss dynamics compared with two sympatric native species in south Florida, USA. *European Journal of Forest Research* **129**: 1155–1168.

Rebek, E. J., D. A. Herms, and D. R. Smitley. 2008. Interspecific variation in resistance to emerald ash borer (Coleoptera: Buprestidae) among North American and Asian ash (*Fraxinus* spp.). *Environmental Entomology* **37**: 242–246.

Rockermann, P. 2011. *Implications for Invasion by Emerald Ash Borer in New York: Ash Abundance in Riparian Areas and Moth Assemblages in Upland and Wetland Forests with High and Low Ash Densities*. M.S. Thesis, State University of New York, Environmental Science and Forestry, Syracuse, New York, USA.

Roque-Albelo, L. 2003. Population decline of Galápagos endemic Lepidoptera on Volcan Alcedo (Isabela Island, Galapagos Islands, Ecuador): an effect of the introduction of the cottony cushion scale? *Bulletin de L'institut Royal des Sciences Naturelles de Belgique* **73**: 1–4.

Ross, D. W., S. D. Gaimari, G. R. Kohler, et al. 2011. Chamaemyiid predators of the hemlock woolly adelgid from the Pacific Northwest, pp. 97–106. *In*: Reardon, R. and B. Onken (eds.). *Implementation and Status of Biological Control of the Hemlock Woolly Adelgid*. FHTET-2011-04, USDA, Forest Service. Morgantown, West Virginia, USA.

Rubinoff, D., B. S. Holland, A. Shibata, et al. 2010. Rapid invasion despite lack of genetic variation in the erythrina gall wasp (*Quadrastichus erythrinae* Kim). *Pacific Science* **64**(1): 23–31.

Sanders, R.W. 2012. Taxonomy of Lantana Sect. Lantana (Verbenaceae): II. Taxonomic revision. *Journal of Botanical Research of the Institute of Texas* **6**: 403–441.

Scholtens, B. and D. L. Wagner. 2007. Lepidopteran fauna of the Great Smoky Mountains National Park. *Southeastern Naturalist* **6** (Special Issue 1): 193–206.

Schooley, H. O., J. W. E. Harris, and B. Pendrel. 1984. *Adelges piceae* (Ratz.), balsam woolly adelgid (Homoptera: Adelgidae), pp. 229–234. *In*: Kelleher, J. S. and M. A. Hulme (eds.). *Biological Control Programmes against Insects and Weeds in Canada 1969–1980*. Commonwealth Agricultural Bureax. Farnham Royal, Slough, UK.

Sevillano, L. G. M. 2010. The Effects of Biological Control Agents on Population Growth and Spread of *Melaleuca quinquenervia*. Ph.D. Dissertation. Department of Biology, University of Miami, Florida, USA.

Spotswood, E., J. Y. Meyer, and J. Bartolome. 2012. An invasive tree alters the structure of seed dispersal networks between birds and plants on islands. *Journal of Biogeography* **39**: 2007–2010.

Stehn, S. E., C. R. Webster, M. A. Jenkins, and S. Jose. 2011. High-elevation ground-layer plant community composition across environmental gradients in spruce-fir forests. *Ecological Research* **26**: 1089–1101.

Stehn, S. E., M. A. Jenkins, C. R. Webster, and S. Jose. 2013. Regeneration responses to exogenous disturbance gradients in southern Appalachian *Picea-Abies* forests. *Forest Ecology and Management* **289**: 98–105.

Stock, D, K. Johnson, A. Clark, and E. van Oosterhout. 2009. Lantana: Best practice manual and decision support tool. Department of Employment, Economic Development, and Innovation, Yeerongpilly, Queensland, Australia. http://www.weeds.org.au/WoNS/lantana/docs/bpm-intro.pdf [Accessed January 2016].

Tanasijshuk, V. N. 2002. Studies on Nearctic species of *Leucopis* (Diptera: Chamaemyidae). I. The redescription of Nearctic *Leucopis* published before 1965. *Zoosystematica Rossica* **11**: 193–207.

Tanis S. R. and D. G. McCullough. 2012. Differential persistence of blue ash and white ash following emerald ash borer invasion. *Canadian Journal of Forest Research* **42**: 1542–1550.

Tipping, P. W., M. R. Martin, P. D. Pratt, et al. 2008. Suppression of growth and reproduction of an exotic invasive tree by two introduced insects. *Biological Control* **44**: 235–241.

Tipping, P. W., M. R. Martin, K. R. Nimmo, et al. 2009. Invasion of a west Everglades wetland by *Melaleuca quinquenervia* countered by classical biological control. *Biological Control* **48**: 73–78.

Tipping, P.W., M. R. Martin, R. Pierce, et al. 2012. Post-biological control invasion trajectory for *Melaleuca quinquenervia* in a seasonally inundated wetland. *Biological Control* **60**: 163–168.

Turner, C. E., T. D. Center, D. W. Burrows, and G. R. Buckingham. 1998. Ecology and management of *Melaleuca quinquenervia*, an invader of wetlands in Florida, U.S.A. *Wetlands Ecology and Management* **5**: 165–178.

Turner, P., M. Winkler, and P. Downey. 2007. Establishing conservation priorities for lantana; 2007. The 14th Biennial NSW Weeds Conference Proceedings, University of Wollongong, New South Wales, Australia.

Tye, A. 2008. The status of the endemic flora of Galapagos: the number of threatened species is increasing. Galápagos report, for 2006–2007, pp. 96–103. Charles Darwin Foundation, Galápagos National Park and INGALA, Puerto Ayora, Galápagos, Ecuador.

USDA APHIS. 2007. The proposed release of three parasitoids for the biological control of the emerald ash borer (*Agrilus planipennis*) in the continental United States: environmental assessment. *Federal Register* **72**: 28947 – 28948, Docket No. APHIS-2007-0060. Available from: http://www.emeraldashborer.info/files/07-060-1%20ea_leah_environassess.pdf [Accessed January 2016].

USDA PLANTS Database 2013. *United States Department of Agriculture, Nature Resource and Conservation Services,* Washington, DC. Available from: http://plants.usda.gov/java/ [Accessed January 2016].

Van Driesche, R. G., R. I. Carruthers, T. Center, et al. 2010. Classical biological control for the protection of natural ecosystems. *Biological Control* **54**: S2–S33.

van Klinken, R.D. and L. K. Flack. 2008. What limits predation rates by the specialist seed-feeder *Penthobruchus germaini* on an invasive shrub? *Journal of Applied Ecology* **45**: 1600–1611.

van Klinken, R. D. and S. Raghu. 2006. A scientific approach to agent selection. *Australian Journal of Entomology* **45**: 253–258.

Wade, D. D. 1981. Some melaleuca-fire relationships including recommendations for homesite protection, pp. 29–35. *In*: Geiger, R. K. (ed.). *Proceedings of Melaleuca Symposium*. Florida Division of Forestry. Tallahassee, Florida.

Wagner, D. L. 2007. Emerald ash borer threatens ash-feeding Lepidoptera. *Newsletter of the Lepidopterists Society* **49**:10–11.

Wagner, D. L. and K. J. Todd. 2015. Conservation implications and ecological impacts of the emerald ash borer in North America, pp. 15–62. *In*: Van Driesche, R. G. and R. Reardon (eds.). *The Biology and Control of Emerald Ash Borer*. FHTET 2014-09. USDA Forest Service. Morgantown, West Virginia, USA.

Wagner, W. L., D. R. Herbst, and S. H. Sohmer. 1990. *Erythrina*, pp. 671–672. *In*: Wagner, W. L. (ed.). *Manual of the Flowering Plants of Hawaii. Vol. 1.* University of Hawaii Press and Bishop Museum Press. Honolulu, Hawaii, USA.

Wallace, M. S. and F. P. Hain. 2000. Field surveys and evaluation of native predators of the hemlock woolly adelgid (Homoptera: Adelgidae) in southeastern United States. *Environmental Entomology* **29**: 638–644.

Weaver, E. 2000. Limitations by *Melaleuca quinquenervia* to Everglades restoration. *Restoration and Reclamation Review* **6**: 1–9.

Wells, C. 1984. Management of lantana in forest plantations, pp. 138–141. *In*: Anon. *Proceedings of Woody Weed Control Workshop, Gympie, Queensland.* The Weed Society of Queensland. Brisbane, Queensland, Australia.

Wie, X., R. Reardon, Y. Wu, and J.-H. Sun. 2004. Emerald ash borer, *Agrilus planipennis*, in China: a review and distribution survey. *Acta Entomologica Sinica* **47**: 679–685.

Wie, X., W. Yun, R. Reardon, et al. 2007. Biology and damage traits of emerald ash borer (*Agrilus planipennis* Fairmaire) in China. *Insect Science* **14**: 367–373.

Williamson, M. and A. Fitter. 1996. The varying success of invaders. *Ecology* **77**: 1661–1666.

World Wildlife Fund. 2013. Hawaii's Dry Forests. http://wwf.panda.org/about_our_earth/ecoregions/hawaiis_dry_forests.cfm [Accessed January 2016].

Yang, Z-Q., J. S. Strazanac, P. M. Marsh, et al. 2005. First recorded parasitoid from China of *Agrilus planipennis*: a new species of *Spathius* (Hymenoptera: Braconidae: Doryctinae). *Annals of the Entomological Society of America* **98**: 636–642.

Yang, Z-Q., J. S. Srazanac, Y-X. Yao, and X-Y.Wang. 2006. A new species of emerald ash borer parasitoid from China belonging to the genus *Tetrastichus* Haliday (Hymenoptera: Eulophidae). *Proceedings of the Entomological Society of Washington* **108**: 550–558.

Zalucki, M., M. Day, and J. Playford. 2007. Will biological control of *Lantana camara* ever succeed? Patterns, processes, and prospects. *Biological Control* **42**: 251–261.

Yeates, A. G. 2013. Gaining More from Invasive Plant Management than Just Weed Control. Ph.D. Dissertation, University of Queensland, Brisbane, Queensland, Australia.

Zhang, Y. Z., D. W. Huang, D. W. Uang, et al. 2006. Two new species of egg parasitoids (Hymenoptera: Encyrtidae) of wood-boring beetle pests from China. *Phytoparasitica* **33**: 253–260.

PLATE 1 (FIGURE 5.2) Biological control is a method of invasive species management that generally shows high levels of target specificity. Shown here are two brown clumps of purple loosestrife (*Lythrum salicaria*) selectively defoliated by *Galerucella* leaf beetles introduced to North America, while all remaining wetland vegetation remains untouched despite food limitation for developing larvae. Photo credit: Bernd Blossey.

Integrating Biological Control into Conservation Practice, First Edition. Edited by Roy G. Van Driesche, Daniel Simberloff, Bernd Blossey, Charlotte Causton, Mark S. Hoddle, David L. Wagner, Christian O. Marks, Kevin M. Heinz, and Keith D. Warner.
© 2016 John Wiley & Sons, Ltd. Published 2016 by John Wiley & Sons, Ltd.

PLATE 2 (FIGURE 5.3) Larvae of *Cactoblastis cactorum*, an herbivore that attacks various native and pest *Opuntia* cacti. Photo courtesy of Ignacio Baez, USDA Agricultural Research Service, Bugwood.org.

PLATE 3 (FIGURE 5.4) *Rhinocyllus conicus*, a polyphagous thistle-head feeding insect that is well known to attack a range of native thistles. Photo courtesy of Loke Kok, Bugwood.org.

PLATE 4 (FIGURE 5.5) Adult polyphemus moth, *Antheraea polyphemus* (Cramer), one of the giant silkmoths strongly attacked by the tachinid *Compsilura concinnata*. Photo courtesy of Lacy L. Hyche, Auburn University, Bugwood.org.

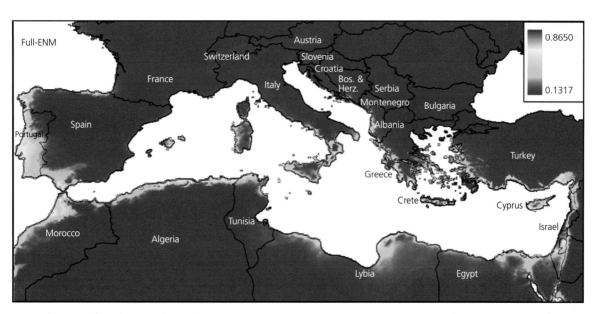

PLATE 5 (FIGURE 6.4) Ecological niche modeling (ENM) can be integrated with the results from molecular analyses to identify variation in environmental factors influencing the distributions of candidate biological control agents. Using ENM software, Lozier and Mills (2009) identified different distributions for evolutionary significant units (ESUs) of the parasitoid *Aphidius transcaspicus*. They then used these results to identify climatic variables that might restrict each of these ESUs and may influence their establishment as biological control agents. Lozier and Mills 2009. Used under CC-BY-4.0, http://creativecommons.org/licenses/by/4.0/.

PLATE 6 (FIGURE 9.8) Before (left) (1997) and after (right) (1999) control of waterhyacinth (Eichhornia crassipes) at Port Bell on Lake Victoria, Uganda. Photographs (left) by Ken L. S. Harley, Commonwealth Scientific and Industrial Research Organization, Bugwood.org; (right) by Mic Julien, Commonwealth Scientific and Industrial Research Organization, Bugwood.org.

PLATE 7 (FIGURE 10.1) Melaleuca island in an invaded wetland. Photo courtesy of Francois Laroche, South Florida Water Management District.

PLATE 8 (FIGURE 10.2) Few to no plants can survive in the sterile habitat under melaleuca stands. Photo credit, Paul Pratt, USDA/ARS.

PLATE 9 (FIGURE 10.3) The melaleuca weevil, *Oxyops vitiosa*, destroys seed production by more than 95%. Photo courtesy of Steve Ausmus, USDA/ARS.

(a)   (b)

PLATE 10 (FIGURE 10.4) (a) The melaleuca psyllid, *Boreioglycaspis melaleucae* Moore (adult) was the second agent released against melaleuca. Photo courtesy of Steve Ausmus, USDA/ARS. (b) Psyllid flocculence, showing a colony of the insect. Photo credit, Paul Pratt, USDA/ARS.

PLATE 11 (FIGURE 10.5) Recovery of native vegetation in a melaleuca-dominated plot, following successful biological control, of melaleuca; note the dead trees and open canopy. Photo courtesy of Min Rayamajhi, USDA/ARS.

(a)   (b)   (c)

PLATE 12 (FIGURE 10.7) Control of mist flower (*Ageratina riparia*) by the introduced white smut fungus, *Entyloma ageratinae*, showing the same location (a) about a year after fungus release when plants were infected but not yet killed (January 27, 2000–summer) and (b) two years later when mist flower shrubs had died and been replaced by mostly native vegetation, particularly ferns and grasses (November 8, 2001–spring); site is Brookby on the edge of the Hunua Ranges (an important water catchment and biodiversity area, just southeast of Auckland, New Zealand), and (c) endangered New Zealand endemic plant, *Hebe acutiflora* Cockayne, no longer threatened by mist flower. Photos courtesy of Jane Barton/Landcare Research (a, b) and Gillian Crowcroft (c).

PLATE 13 (FIGURE 10.8) White mangrove infested with cottony cushion scale. Photo credit, Mark Hoddle.

PLATE 14 (FIGURE 10.9) Endemic Galápagan plants whose populations were threatened by cottony cushion scale: left *Darwiniothamnus tenuifolius*, and right, *Scalesia* sp. Photo credit, Mark Hoddle.

PLATE 15 (FIGURE 10.10) Adult *Rodolia cardinalis* preying on cottony cushion scale. Photo credit, Mark Hoddle.

PLATE 22 (FIGURE 10.18) Fraser fir killed by balsam woolly adelgid, together with regenerating young trees not yet old enough to be susceptible. Photo credit, Michael Montgomery, USDA Forest Service.

PLATE 23 (FIGURE 10.19) Lantana invading pastures and native forest understory in southeast Queensland, Australia. Photos courtesy of Michael Day, QDAFF.

PLATE 20 (FIGURE 10.16) Exploration in Madagascar (left, photo credit, Russell Messing) did not yield parasitoids; but additional collections in South Africa, Kenya, and Tanzania were successful, yielding several parasitoid species – including *Eurytoma erythinae* (right, photo courtesy of Walter Nagamine), which is helping to preserve rare endemic trees in Hawaii.

PLATE 21 (FIGURE 10.17) Balsam woolly adelgid on trunk (left) and gouty twigs resulting from twig infestation (right). Photos by William M. Ciesla, Forest Health Management International, Bugwood.org.

PLATE 18 (FIGURE 10.14) Remnant patch of lowland dry forest, with wiliwili tree, *Erythrina sandwicensis*, in foreground at Puuwaawaa, Big Island, Hawaii. Photo credit, Juilana Yalemar.

PLATE 19 (FIGURE 10.15) Adult of erythrina gall wasp, *Quadrastichus erythrinae* (left, photo courtesy of M. Tremblay), and galled leaves from Koko Crater (right, photo credit, Leyla Kaufman).

PLATE 16 (FIGURE 10.12) Walk-in field cage (right) stocked with native plants (left) infested with non-target insects to assess feeding preferences of adult *Rodolia cardinalis*. Photo credit, Mark Hoddle.

PLATE 17 (FIGURE 10.13) *Rodolia cardinalis* larva faced with a choice: cottony cushion scale or *Ceroplastes* sp. scale (top), attacks the cottony cushion scale (bottom). Photo credit, Mark Hoddle.

PLATE 24 (FIGURE 10.20) Lantana defoliated by the tree hopper *Aconophora compressa* in southeast Queensland, Australia and close up of insect. Photos courtesy of Michael Day, QDAFF.

PLATE 25 (FIGURE 10.21) Adults of emerald ash borer, *Agrilus planipennis*. Photo courtesy of Stephen Ausmus, USDA ARS.

PLATE 26 (FIGURE 10.22) Dead ash, killed by emerald ash borer, lining Michigan roadside. Photo credit, Jian J. Duan, USDA ARS.

PLATE 27 (FIGURE 10.23) Emerald ash borer larvae feeding on ash cambium. Photo credit, Jian J. Duan, USDA ARS.

(a)   (b)

PLATE 28 (FIGURE 10.24) (a, b) *Tetrastichus planipennisi* adult and larvae inside host larva, an important parasitoid of emerald ash borer, introduced from China. Photo credits, Jian J. Duan, USDA ARS.

PLATE 29 (FIGURE 10.25) *Oobius agrili* ovipositing in an emerald ash borer egg; this is another important parasitoid introduced from China. Photo credits, Jian J. Duan, USDA ARS.

PLATE 30 (FIGURE 10.28) Miconia leaf, showing injury owing to infestation by the fungal pathogen *Colletotrichum gloeosporioides* [Penz] Sacc. f. sp. *miconiae* at Nuku Hiva, Tahiti, 2007. Photo credit, J.-Y. Meyer.

PLATE 31 (FIGURE 10.29) Regrowth of native vegetation, including the rare endemic shrub *Psychotria* (Rubiaceae), in the understory of a miconia invaded forest in Tahiti, with 20% leaf damage on canopy miconia leaves caused by the introduced fungal pathogen *Colletotrichum gloeosporioides* [Penz] Sacc. f. sp. *miconiae*. Photo credit, J.-Y. Meyer.

CHAPTER 11

Societal values expressed through policy and regulations concerning biological control releases

Andy W. Sheppard[1] and Keith D. Warner[2]

[1] Commonwealth Scientific and Industrial Research Organisation (CSIRO), ACT, Australia
[2] Center for Science, Technology, and Society, Santa Clara University, USA

Introduction

The evolution of policy and regulations for the importation of live biological organisms across national and other jurisdictional borders – under which biological control agent introductions are regulated – has mirrored changes in societal values and concerns about such deliberate introductions. In the 1800s, such introductions were unregulated and actively encouraged for global agriculture and for the acclimatization "benefits" of European colonists in those parts of the world recently colonized. In the early part of the twentieth century it was recognized that not all species introductions, be they deliberate or accidental, were benign in impact, at least in an agricultural context, as many agricultural plant pests were spreading to new regions. This was the driver of the first plant quarantine acts, particularly in countries with a strong, growing agricultural sector (1908 in Australia and 1912 in the USA). But these new laws did not regulate plant introductions, nor indeed animals without a clear, harmful reputation, and therefore they did not slow the arrival rate of non-agricultural pests. Nonetheless exotic invertebrate biological control agents were being used at this time and therefore required some scrutiny. This resulted in the first consideration of the risk of such introductions.

The USA was the first to recognize the need for protection of endangered vertebrates and associated habitat with the establishment of Yellowstone as the first National Park in the world in 1872, and Australia quickly followed suit in 1879, followed by a century of

creation of protected areas and, where necessary, reintroductions of extirpated native species. Concerns for the natural environment further intensified in the 1960s with the realization that agricultural production, increasingly built on chemical-based pest control, was detrimental to surrounding native communities and top predators, as popularized by the book *Silent Spring* by Rachel Carson (1962). As a consequence of such changing values, new environmental protection legislation was enacted around the developed world. In many cases, this included assessing the risks of a broad range of activities, including the introduction and release of some "beneficial" exotic organisms. This expanded review included consideration of the risks from invertebrate and pathogen weed biological control agents, which, for the first time, were to be comprehensively assessed against native plants as well as plants of economic importance. However, plants and plant products, aquaculture and livestock, and the pet trade were largely excluded from regulation to prevent impacts on trade and economic development. However, some countries, mainly Australia and New Zealand, had already recognized that plant introductions (largely for supposed pastoral and ornamental industry benefits) needed to be regulated.

Public awareness of environmental concerns continued to grow as human activities further degraded pristine natural habitats. Increasing global environmental concern led to the United Nations Convention on Biological Diversity (CBD) at the Rio environmental summit in 1992. Following the CBD Decision VI/23 on "alien species

Integrating Biological Control into Conservation Practice, First Edition. Edited by Roy G. Van Driesche, Daniel Simberloff, Bernd Blossey, Charlotte Causton, Mark S. Hoddle, David L. Wagner, Christian O. Marks, Kevin M. Heinz, and Keith D. Warner.
© 2016 John Wiley & Sons, Ltd. Published 2016 by John Wiley & Sons, Ltd.

that threaten ecosystems habitats and species," which recommended a "precautionary approach," the threats of invasive alien species to biodiversity through the continued deliberate and accidental movement of exotic organisms became increasingly recognized under national environmental protection legislation (see discussion of the precautionary approach in Chapter 13). Such legislation was intended to protect natural environments through the detection, exclusion, and management of potentially damaging invasive species. During this same period, attitudes toward classical biological control releases became increasingly risk averse.

Despite all this institutional activity, broader community attitudes toward novel organisms are best described as ambivalent (Perrings et al., 2010; Warner, 2011; Warner and Kinslow, 2013). More international trade agreements are now further increasing international movement of living organisms (agricultural commodities, live plants, and forest products) as the global economy seeks the associated benefits (Sheppard et al., 2011). Nonetheless, under the CBD and the International Plant Protection Convention (IPPC) such trade must increasingly adopt the "precautionary approach" goal of minimizing risks (of intentional and unintentional species introductions) to global agriculture and biodiversity (Perrings et al., 2010; Iverson and Perrings, 2012). As only a small proportion of introduced exotic organisms are known to have significant impacts on the natural environment, the challenge has been to identify which species could cause these impacts and prevent their introduction without overly restricting international trade (Leung et al., 2002; Perrings, 2005).

Proposals to introduce and release novel biological control agents have to be assessed within this increasingly risk-averse regulatory environment. The "precautionary approach" in national legislation and associated regulatory processes has led to a strictly risk-assessment focus for all importations, which for simplicity tends to follow internationally recognized Import Risk Assessment (IRA) protocols developed through the IPPC and recognized by the World Trade Organization (WTO). Under this approach there is no explicit evaluation of potential benefits, even though society wants the environmental and economic benefits of invasive species control. As a result, the review processes for biological control agent importation have become more rigorous and risk focused (Sheppard et al., 2003). Releases of weed biological control agents have historically been subject to greater scrutiny because of

potential risks to economically important plants. Nonetheless, release permits for agents targeting arthropods are also now more rigorously evaluated than in the 1990s (Hunt et al., 2008).

In the next section, we present the different national legislative and regulatory approaches applied to the release of biological control agents across six countries/regions where classical biological control is most commonly practiced against pests and weeds of the environment. Following that, we explore how some of the differing societal values and histories of dialogue and decision-making across countries have resulted in the different legislative processes that have emerged. We also look at the different perspectives of scientists and regulators in the process of decision-making for biological control releases. Finally, we briefly discuss some areas where research could further improve the approval processes.

National regulatory processes and differences

Five countries (Australia, Canada, New Zealand, South Africa, and the United States) have conducted the vast majority of the world's classical biological control agent introductions. Here we compare and highlight the distinctive aspects of the approaches taken by each country. In some countries there are separate but parallel regulatory processes for environmental and agricultural pests and weeds, whereas in others these processes have been successfully combined. We also consider the European Union. Although it has a history of insect biological control activity, the EU is only just starting to conduct classical weed biological control programs.

Australia and New Zealand are relatively isolated island states with unique biotas and strong agricultural export sectors. Their export crops gain economic advantage from their relatively pest-free status, and their societies have high levels of environmental awareness. They have collectively adopted a concept of "biosecurity" to manage the risks from trade and human movement of introduced species and novel genotypes to agriculture, forestry, fisheries, biodiversity, and human well-being (Takahashi, 2005). Biosecurity is a risk-based systems approach to quarantine that assesses and manages risks posed by diverse organisms using multiple pathways (Falk et al., 2011). Both Australia and

New Zealand created novel policy tools to manage the biosecurity risks from international trade after becoming parties to the WTO and signing the CBD (Goldson et al., 2010; Trampusch, 2014). Policymakers in other countries are beginning to adopt this terminology (Magarey et al., 2009). These policy tools represent the countries' desire to facilitate international trade of biological commodities, while maintaining a precautionary approach for biological importations (Perrings et al., 2005). Internationally, both countries champion strong international biosecurity standards to reduce biosecurity risks from international trade.

Australia

In Australia, organizations wishing to undertake biological control of a particular weed for the first time must submit a "Nomination of Target Weeds for Biological Control" to the Invasive Plants and Animals Committee (IPAC – formally the Australian Weeds Committee), which reports to the National Biosecurity Committee (NBC) within the Department of Agriculture. Required information includes the weed's pest status across each state, possible implications of the weed's biological control for affected industry (e.g., the horticultural industry), and other control methods that are available. This allows possible conflicts of economic interest to be identified and is the only part of the process that considers the possible benefits from the proposed biological control program. Such nominations must normally be accepted by the IPAC where there is negligible conflict of economic interest and must be endorsed by the NBC before any applications to release biological control agents can be approved. Arthropod targets (i.e., pest insects) do not need prior nomination.

Permits to import and release biological control agents are required from both the Department of Agriculture (DA) under the Quarantine Act (1908) and the Department of Environment (DE) under the Environment Protection and Biodiversity Conservation Act (EPBC, 1999). The EPBC act does not regulate microorganisms, which only need to be considered under DA processes. In practical terms, most aspects required for both processes have been streamlined and both require a period of public consultation. There is no formal requirement for host-specificity testing to be part of the risk assessment. Nonetheless, importation risk assessment of biological control agents is still the approach used. Applications can be made to import

potential agents into quarantine to complete risk assessment, which must include an assessment of the level of quarantine required for each agent. Release applications based on a standard format are submitted to both departments. These applications address only direct risk of agents to non-target organisms. The applications are made available for public comment and sent to representatives of expert organizations on the sister committee to IPAC, the Plant Health Committee (for both weed and arthropod biological control agents) for peer review. Once peer review has been completed and any public comments or concerns have been adequately addressed, so that the agency is assured that the agent presents a low risk, the DA issues a release permit under the Quarantine Act (1908) in the form of a letter changing the conditions on the importation permit.

Under the EPBC Act, applications seek to add proposed macroorganism biological control agents to the permitted (white) list of species allowed by the DE. The importer outlines draft terms of reference for an assessment of potential impacts of the proposed amendment on the Australian environment. The main aim of the assessment is to provide a summary analysis of key information (including impacts and benefits) and a detailed evaluation of the risks of the proposed release. The assessment is made available on the web both in the form of a proposal and as a completed assessment, and any public comments are addressed by the applicant. Voucher specimens of all tested agent material are lodged at a recognized institution. Once the final assessment has been accepted, the Federal Minister for the Environment tables it in parliament for consultation by other federal, state, and territory ministers considered appropriate and may also consult with other organizations or individuals before making a decision to amend the list. If the minister approves the amendment, the list will be formally amended to include the agent species through publication in the Australian *Government Gazette*, a change that can be reviewed within five years. Once the agent appears on the live importation list, importers of biological control agents can apply for the release permit. Groups negatively affected by the decision may request a written explanation.

The separation of pest target selection from biological control release decisions and the consistent call for public comment during the process allow for transparency in public agency decision-making. Australia is also the only country with biological control legislation: the Biological Control Act (1984) (Cullen and Delfosse, 1985). The act was the direct consequence of a legal challenge to the CSIRO around a particular biological control program, but it is now only applied (i.e., targets and agents are declared under the act, which leads to a requirement to consider the need for a public enquiry around risks, costs, and benefits) for high-profile or controversial biological control programs (e.g., the field release of pathogens against rabbits), where a formal process of public consultation from the start is deemed essential. No biological control agent of an arthropod pest has ever needed consideration under the Biological Control Act. The DA biosecurity risk assessment section fulfills an analogous function to the "pest risk assessment" branch of USDA-APHIS-Plant Protection and Quarantine. Despite the need for consideration under two contrasting pieces of legislation, biological control introductions into Australia have continued without significant confusion or difficulty, partly because of strong Australian perceptions of biological control benefits. The combined process currently takes on average six months to a year where the risks are low and public comment is minimal.

New Zealand

As a party to the WTO and the CBD, New Zealand developed novel policy tools to protect its biodiversity from the risks of international trade. It is among the most geographically isolated of nations, yet it has one of the highest ratios of exotic to native species of any country (Clout and Lowe, 2000). It now has the most comprehensive biosecurity regulations in the world, and its policy stance is inherently risk averse. New Zealand passed the Hazardous Substances and New Organisms Act (HSNO, 1996) to provide a common regulatory framework for all novel chemicals and living organisms, thereby creating the Environmental Risk Management Authority, or ERMA (Barratt and Moeed, 2005). This act initiated a new environmental risk evaluation procedure to address proposed new chemicals and novel organisms under ERMA, now renamed the Environmental Protection Authority with a broader role (EPA [NZ]), which has the responsibility of assessing and making decisions on standardized importation and release applications. For a discussion of New Zealand's regulations, see Environmental Risk Management Authority (2010) or Hill et al. (2013).

The HSNO Act is "enabling" legislation, meaning that the introduction of any organism is possible as long as EPA (NZ) can be convinced of its low impact and risk. Suitability of targets for biological control is judged through the risks, costs, and benefits as determined through the EPA (NZ) process. Permits are then issued by the Ministry of Primary Industries under the Biosecurity Act (1993). Scientists are not usually applicants for permits, but can be. Generally the relevant land management organization makes the application and pays the costs. Applicants often fund research agencies to conduct the pre-release testing and specialists to conduct Maori consultation and public outreach. This helps remove any (perceived or real) professional conflict of interest from the research scientist. Full consultation between applicant and the EPA (NZ) is encouraged from the start as a pre-application for comment is allowed during the preparation of the application to expedite the process.

An assistance package for applying to release a biological control agent under this process is available online (Better Border Biosecurity, 2014). An application for importation into quarantine requires a full risk analysis including an eradication strategy should escape occur. EPA (NZ) must respond to the application in 14–80 business days and there is no public consultation. The applicant must identify interested stakeholders likely to be concerned with proposed releases and engage them in discussions and in the assessment process. The defined risk, cost, and benefit categories are ecosystem health and intrinsic value, species conservation, social well-being (now and in the future), spiritual and cultural values, and public health (Barratt et al., 2010). The applicant is required to develop an ecologically based risk/cost: benefit analysis by identifying and analysing all reasonable potential hazards, risks, benefits, and conflicts of interest. Applicants must demonstrate that they have consulted affected members of the public, including indigenous Maori communities. Host-specificity testing and consultation on a test list are not requirements, but such testing has been quickly adopted as a standard for agents proposed for release against weeds and insect pests. Should EPA (NZ) require further risk assessment following application submission, this can be done in association with an extension to the application deadline.

Once the application is submitted, EPA (NZ) notifies the broader public through advertisements in four national newspapers and places the submission on its website for 90 business days. Concerned citizens may indicate in their submission that they wish to be heard, which requires EPA (NZ) to hold a public hearing at the cost of the applicant. In the meantime EPA (NZ) prepares a "Staff Assessment Report" of the application with the help of expert advice and peer review and feedback from the applicant, as required. The Department of Conservation and the Ministry of Primary Industries have a statutory obligation to comment on the application. The HSNO Act requires EPA (NZ) to consider an environmental "bottom line" in the form of minimum standards. The standards relevant for biological control agent release applications are that new organisms must not cause any of the following:

- significant displacement of any native species within its natural habitat;
- significant deterioration of natural habitats;
- significant adverse effect to New Zealand's inherent genetic diversity;
- disease, be parasitic, or become a vector for human, animal, or plant disease, unless the purpose of that importation or release is to import or release an organism to cause disease, be a parasite, or a vector for disease.

If full agreement is not reached, EPA (NZ) applies a "precautionary approach" in its review of the risks and costs versus benefits (with respect to direct or indirect impacts on native species and communities, a legislative requirement). EPA (NZ)'s decision is final. There is provision for the applicant to apply for an approval with controls, and EPA can also impose controls aimed at reducing risk. There is an opportunity for ministerial intervention in certain cases, and a judicial review of the process (but not the decision) can be undertaken.

Since the passage of HSNO, New Zealand has developed the world's most sophisticated decision-making process for evaluating novel organisms, balancing multiple scientific and social criteria, with explicit reference to biological control agent introductions (Harrison et al., 2005; Hill et al., 2013). The process requires clear decision-making criteria based on transparent and repeatable ecologically based risk/cost: benefit analysis, fixed time periods for decisions, and participatory public engagement. Stakeholders and scientists report that the EPA (NZ) review process has prompted broader and more explicit consultation and review of invasive species targeted and discussions of potential risks before any application is submitted and therefore enjoys greater public support. The average number of releases has not declined since the HSNO Act, and therefore it is evident that EPA (NZ) risk-analysis procedures have not caused undue restrictions on New Zealand's biological control programs, although under this process applicants have to pay application costs (Hill et al., 2013). New Zealand offers a working example of how risks associated with biological control introductions can be better identified, managed, and regulated globally (Barratt et al., 2010).

South Africa

In South Africa, the importation and release of biological control agents against invasive alien plants (weeds) and pests are subject to regulation by The Department of Agriculture, Forestry and Fisheries (DAFF) and The Department of Environmental Affairs (DEA). On a voluntary level, peer review within the South African weed biological control community, as well as by overseas counterparts, has contributed to the integrity of the science and practice of weed and insect biological control. Klein et al. (2011) reviewed the development of this regulatory system from the importation of the first weed biological control agent in 1913 through to 2010.

DAFF, through its Directorate of Plant Health (DPH), regulates the importation and release from quarantine of biological control agents under the Agricultural Pests Act of 1983. It is carried out in accordance with the IPPC and the relevant International Standards for Phytosanitary Measures developed by the Food and Agriculture Organisation (FAO) (IPPC, 2006, 2007). Before 1989, as well as between 2010 and 2014, DAFF was the only regulatory body involved in the importation and release of biological control agents. Decisions by DAFF-DPH on whether to permit the release of a particular biological control agent are based on recommendations by a panel of independent expert reviewers who are selected from an existing list on a case-by-case basis.

Since 1998, the DEA has had regulatory power over the importation and release of biological control agents, which were treated as potential threats to biodiversity. Biological control agent importations and releases are currently governed by the National Environmental Management: Biodiversity Act (NEM:BA, 2004), which deals with species and organisms posing potential threats to biodiversity. NEM:BA stipulates that, like all other alien species, biological control agents may not be introduced, reared, transported, or released into the environment without an assessment of risks and potential impacts on biodiversity. In practice, DEA has not exercised its powers of regulating biological control agent importations or releases since August 2010, because the necessary regulations for the implementation of NEM:BA have not yet been approved. The Alien and Invasive Species Regulations have been published in the *Government Gazette* (http://www.invasives.org.za/legislation/item/667-alien-and-invasive-species-regulations-published).

Since 2010, in the absence of regulations to govern the NEM:BA, DAFF has maintained responsibility for the review and approval of releases of biological control agents against weeds as well as arthropod pests. Initially a number of administrative processes restricted DAFF's ability to acquire the services of suitably qualified service providers (independent reviewers) to evaluate biological control release applications and make recommendations for approval of their release into the environment. The number of pending release applications submitted to DAFF and the cost of retaining potential biological control agents in quarantine for extended periods had built up to untenable levels.

To correct this problem, DAFF worked with members of the South African weed biological control community to devise a method of voluntary review of release applications by biological control and related science experts, both local and international. The biological control community urged members (local and international) to offer their services as reviewers of applications. Between DAFF and the biological control community a simple review matrix was devised. Each application was sent to three reviewers (two local and one international).

In order to consider the applications from biological control researchers and reviewers' comments, DAFF, with the assistance and encouragement of DEA, established a National Biological Control Release Application Review Committee. DAFF administers the review process, as well as the committee, which is chaired by a director from the South African National Biodiversity Institute (SANBI), a public entity regulated by NEM:BA.

DAFF ensures that future release applications for biological control agents are reviewed and considered by relevant departments, such as the DEA and Department of Water Affairs, by representation of these departments on the committee. Future release applications will have

to comply with these departments' requirements for review and consideration.

SANBI's role has strengthened the process of review and enabled parties to represent their views in the committee. SANBI has been able to appoint independent biological control experts onto the review committee, which has revised the guidelines for biological control release applications so that applicants address the specific focal areas that need to be considered by the departments involved, including stakeholder approval, identification of agents, and consultation with other experts in various matters, such as the development of the test species list. The reviewer's template has been revised and improved so that the committee receives sufficient information on which to base its decisions. However, the committee has the right to veto any recommendations by reviewers that are deemed to be unreasonable or requirements that are impossible to implement. Both departments, DAFF and DEA, have agreed to accept the recommendations of the committee. The prospects are good that finally, after many years of uncertainty and delay, a system is in place that will ensure professional, fair, and speedy reviews for all biological control agents in South Africa.

United States

Risk controversies in the United States have been acute relative to other countries, making the application process for federal permits for the introduction of classical biological control agents difficult, costly, and time consuming. The permitting of biological control agent introductions is governed by a patchwork of five different federal laws plus, in some cases, state laws, and the process differs according to whether the target is a weed or pest arthropod (Hunt et al., 2008). Introductions for classical biological control in the United States are guided by the Plant Quarantine Act (PQA, 1912), the Federal Plant Pest Act (FPPA, 1957), the Federal Noxious Weed Act (FNWA, 1974), the Plant Protection Act (PPA, 2000), and Executive Order 13112 (1999) for invasive species. Biological control organisms are explicitly defined under the Plant Protection Act (2000). Threats to native species are covered under the Endangered Species Act (ESA, 1973), the National Environmental Policy Act (NEPA, 1969), and the Coastal Zone Management Act (1972).

The USDA Agricultural Plant Health Inspection Service – Plant Protection & Quarantine section (APHIS-PPQ) is the federal agency charged with controlling invasive plant pests, and its regulatory responsibilities include oversight and permitting of the importation and release of natural enemies. Currently, USDA-APHIS-PPQ issues permits to import live natural enemies into approved US quarantine facilities, their release from quarantine, and their interstate movement. Before 1993, permit approval was relatively straightforward; however, in that year, APHIS legal staff interpreted existing laws to conclude that the agency lacked authority to regulate entomophagous natural enemies beyond their importation into quarantine. In 2000, the Plant Protection Act (PPA), passed as part of the Agricultural Risk Protection Act (2000), came into effect, granting USDA-APHIS-PPQ authority to regulate all biological control organisms. The Plant Protection Act consolidated prior legislation and regulatory authority. Unlike the previous Federal Plant Pest Act (FPPA 1957), the PPA broadly defines biological control agents and recognizes their potential to control plant pests, and thus under this Act, not only did all biological control organisms become subject to regulation, but USDA-APHIS-PPQ was charged with facilitating their use.

Petitions for the release of phytophagous arthropods or obligate plant pathogens for classical biological control are assessed through a consultation process under guidelines maintained by APHIS-PPQ. A formal Technical Advisory Group (TAG) to evaluate the suitability of a weed targets and non-target testing protocols, and to make recommendations to APHIS on agent release was formed (Cofrancesco and Shearer, 2004). The TAG is made up of representatives of 13 federal agencies that either support, conduct research on, or use weed biological control as part of their activities: five USDA agencies (including the Natural Resources Conservation Service), six US Department of the Interior (USDI) agencies (including the Fish and Wildlife Service [FWS], the National Park Service [NPS], the Bureau of Land Management [BLM], and the Bureau of Indian Affairs [BIA]), the US Environmental Protection Agency (EPA), and the Department of Defense (DOD). Representatives from Canada and Mexico also participate in TAG activities and provide necessary input on the likelihood of US-released natural enemies crossing into these neighboring countries and their potential effects there (there have been suggestions that the TAG should be expanded to include representatives from Caribbean nations).

Through this consultation process, TAG makes recommendations on target suitability, recommends lists of test organisms for agent host-specificity assessment, and comments on the advisability of permit issuance. Development of lists of test species for weed biological control requires at least informal independent consultation with FWS and/or the National Marine Fisheries Service (NMFS) on threats to federally listed endangered species. TAG serves as an intermediary review committee, articulating the views of the research community to the regulatory agencies involved in weed biological control. Its advisory role to APHIS-PPQ includes: (1) reviewing petitions with the aim of making conservative recommendations to APHIS-PPQ; (2) assessing risks associated with releases and recommending specific actions for the petitioner before a release application is submitted to APHIS-PPQ; (3) incorporating member agencies' concerns and perspectives into planning of biological control programs; and (4) assisting in defining a course of action when there might be a conflict of interest. A reviewer's manual defines the requirement of petitions, petitioners, and reviewers (Cofrancesco and Shearer, 2004; USDA, 2007).

The elected chair of the TAG builds consensus amongst TAG members, although there is no strict need for unanimity for a positive TAG recommendation. TAG, therefore, only recommends a release petition to APHIS-PPQ if there is sufficient justification that the benefits of possible successful biological control outweigh the risks and, perhaps most importantly, that the activities undertaken will not directly or indirectly jeopardize the continued existence of endangered species. When a positive recommendation is received from TAG, APHIS-PPQ invites the petitioner to submit another permit application to release natural enemies from quarantine.

To some, TAG engages scientific experts in a form of peer review to scrutinize safety testing before submitting a weed biological control permit application to the regulatory agency, providing a template for reviewing permits for biological control of arthropods (Strong and Pemberton, 2000). Critics have charged that the process is closed to public review and biased through over representation of pro-biological control scientists who manifest a conflict of interest (Louda, 1999; Strong and Pemberton, 2000; see also discussion of conflict of interest in Chapters 13, 14, and 15). Historically, weed biological control scientists report that the TAG process is generally effective and even handed, albeit slow.

Recently, however, TAG has recommended some biological control agents for permits that APHIS-PPQ has chosen not to issue, suggesting that the regulatory agency is becoming more risk averse.

For importing and releasing exotic entomophagous invertebrate biological control agents, the United States has no governmental body equivalent to the TAG (Messing, 2005; Messing and Wright, 2006), despite calls for such (Strong and Pemberton, 2000). As a proxy for a TAG review, APHIS asks the applicant to prepare a separate report based on "Guidelines for Petition for First Release of Non-indigenous Entomophagous Biological Control Agents" (NAPPO, 2015) and then submits this to the NAPPO (North American Plant Protection Organization) Biological Control Review Committee, which has been coordinated by Agriculture & Agri-Food Canada since 2003. This committee solicits expert peer review on submitted applications. Following review, the chair of the NAPPO Biological Control Review Committee prepares an assessment of the application based on peer review feedback and recommends a course of action for APHIS-PPQ to consider. APHIS has repeatedly attempted to establish regulations to address the need for published rules to implement PPA; however, these various proposed regulations were poorly constructed and difficult to implement, and never formally adopted. Therefore, current regulations for movement and release of entomophagous biological control agents, especially, remain obscure, causing not only frustration and inefficiency and but also lengthy delays in making final decisions on release from quarantine.

Before APHIS can issue a permit it must comply with the requirements of the National Environmental Protection Act (NEPA). In most cases, this includes preparing an "Environmental Assessment" and soliciting public comment. If the initial assessment indicates negative impacts on endangered species or critical habitats are likely to occur, the review process becomes protracted. DeLoach et al. (2000) is an example of a substantive Biological Assessment (Environmental Impact Statement). Where such effects have not been identified, a draft Environmental Assessment is prepared, typically by the applicant, which includes the rationale for the project and outcomes of host-range and specificity tests performed in quarantine. This provides a base document from which APHIS-PPQ proceeds with its review and evaluation of the application. The

report prepared by APHIS-PPQ includes consultation with F&WS on endangered species and with native American tribes, and may consider health risks to specific human populations. It may also include information provided by TAG, NAPPO, and university experts. The draft Environmental Assessment is then published in the Federal Register with at least 30 days for public comment. APHIS-PPQ considers the comments and consults with the relevant state plant regulatory officials and then (1) makes a "finding of no significant impact" (FONSI) and issues a release permit for the agent(s) from quarantine, or (2) asks the applicant to produce an Environmental Impact Statement when significant negative impacts are expected, or (3) advises that the project be discontinued. Once an agent is released from quarantine, it is considered to be in the environment. Additional permits may be needed for interstate movement, but the agent is considered precedented and the permit process is relatively simple and fast, serving primarily to notify state authorities and to track releases of the agent.

Lack of clear legislation and legislative language, the terrorist attacks of September 11, 2001, and associated political responses, the dual role of APHIS-PPQ to both facilitate and review the processing of applications, plus public agencies' fear of lawsuits, have led to meandering and contradictory regulations and rules over subsequent years. Procedures and rules have been proposed and withdrawn, or put in place and then revised, or suspended. Scientists, commodity groups, and legislators have criticized the agency for the duration of permit reviews, and the opacity of the review process and criteria. Reports by the USDA Inspector General have been harshly critical of the review and permitting processes applied to novel organisms. A year-long internal review resulted in the primary finding that practitioner complaints about permit review processes were, in general, justified. Shortcomings were identified in the implementation of agency procedures due to staff changes, deficiencies in record keeping, and weaknesses of evaluative criteria and policy. APHIS-PPQ has conducted numerous internal studies and convened different groups of expert external advisors. These institutional dynamics continue as of this writing.

For all biological control releases, individual US states may also require their own release permits under state-specific laws and regulations. State regulations may be more stringent than federal ones and are so in some states, such as Hawaii (Messing and Wright, 2006). The geographic size of the United States, the risk-averse stance of the US Federal government, the unevenness of state-specific laws, and the absence of stakeholder input suggest that US biological control regulation is not likely to be streamlined in the foreseeable future.

Canada

Most introductions of agents into Canada for classical biological control purposes have historically been conducted by federal scientists, working for the Research Branch of Agriculture and Agri-Food Canada (AAFC). A Canadian Food Inspection Agency, Plant Biosecurity and Forestry Division (CFIA-PBFD) import permit is required for candidate biological control agents brought into containment for evaluation. New populations of approved biological control agents brought in from overseas for release also require an importation permit. Releases of weed biological control agents in Canada are also issued by CFIA-PBFD. Biological control agents are regulated in the same way as pests under the Plant Protection Act (PPA, 1990) and associated regulations.

Proposed lists of non-target test species for host-specificity testing and release applications include a basic risk–benefit assessment compiled by the applicant. Such petitions are submitted to the National Manager, CFIA Invasive Alien Species and Domestic Programs and must conform to the format and substance of the NAPPO Regional Standards for Phytosanitary Measures for the import and release of phytophagous (RSPM 7 [NAPPO, 2001]) and entomophagous (RSPM 12 [NAPPO, 2015]) biological control agents. The petition is reviewed by the AAFC Biological Control Review Committee (BCRC), which consists of regulators and scientists from federal and provincial government agencies, universities, and industry representatives. The Committee's membership varies depending on the application, but it typically includes experts in taxonomy, biological control, and ecology.

There is no formal process for public consultation during petition evaluation. Petitions received are also reviewed by representatives from Mexico and the USA. In the case of weed biological control agents to be released in Canada and the USA, petitions are also submitted to TAG in the USA for comment. Although BCRC and CFIA-PBFD pay particular attention to non-target issues, there is currently no formal requirement to design and assess petitions in the way that is required

by the USDA-APHIS TAG and FWS, with respect to listed endangered species. Reviewers do take into account any concerns expressed by the Committee on the Status of Endangered Wildlife in Canada. The Chair of BCRC makes a recommendation to the Director of CFIA-PBFD and provides all reviewers' comments. The petition is then reviewed by the CFIA regulatory entomologists and CFIA-PBFD makes the decision about the release of the agent. In Canada, the time from petition submission to final decision is usually less than six months. For approved agents, voucher specimens of populations released must be deposited in the national collections of Canada, Mexico, and the USA. For weed BCAs, CFIA-PBFD considers comments from the US TAG committee to be important for its decision-making, but is not obliged to follow US regulations (e.g., the NEPA process). In some cases agents may be approved for release in Canada before being approved for release in the USA (e.g., some agents against *Cynoglossum officinale* L. [Mason and Huber, 2002]. Release of biological control agents may also be regulated by some Provinces, such as Newfoundland and British Columbia (Mason et al., 2013).

European Union

The release of biological control agents for insect pests in Europe is relatively commonplace, with over 170 species of agents released. Regulatory review remains the responsibility of national quarantine authorities, usually administered by the respective national departments of agriculture. Protocols for risk assessment are largely at the discretion of these national agencies and are often not implemented thoroughly. International discussions on the proposed agent releases are optional but are usually undertaken. New EU regulations on invasive species management may change these arrangements as there is recognition of the limitations of the current system (Bigler et al., 2005).

Licensing (release permits) for classical weed biological control agents in Europe can be carried out effectively under plant health legislation (EC Directive 2000/29/EC), but because of a lack of recognized process for review of biological control agent petitions, extensive discussion and negotiation are required to attain a mutually acceptable process in each country. This can be achieved through an appropriate risk analysis, and in the case of *Aphalara itadori* Shinji, the psyllid released against Japanese knotweed (*Fallopia japonica* [Houtt.] Ronse Decr.), the UK government used a slightly modified version of the European and Mediterranean Plant Protection Organisation Pest Risk Assessment (PRA) template, as available in 2008. While this seemed an counter-intuitive approach, as the agent needed to be considered a "beneficial pest" and half the questions asked in the process were not directly relevant to the intentional introduction of a biological control agent, it proved relatively straight-forward and had the benefit of being recognized by members of the Standing Committee on Plant Health, who were kept informed of the UK application for release. The publication of the risk assessment in a peer-reviewed journal (Shaw et al., 2009) helped acceptance of the PRA and enhanced the subsequent credibility of the research in the wider scientific community. Each European Member State has incor-porated some regulation of releases of exotic organ-isms into its own national legislation. In the United Kingdom, this was done through the 1981 Wildlife and Countryside Act. Under this act, a revised application to release a non-native organism was made to the government Advisory Committee on Releases to the Environment (ACRE), which requested further information before passing a positive recommendation to government. ACRE requested information on effi-cacy and possible secondary, tertiary, and community-level non-target impacts, as well as data on the impact per plant and likely responses of commercially avail-able generalist predators used as biological control agents for glasshouse pests. The application was then sent by ACRE to three independent experts and the Government's Chief Scientific Advisor for review, before being subject to a three-month public consulta-tion period. At the end of the process, the final autho-rization was granted by the Secretary of State after appropriate ministerial consideration. In effect, the PRA allowed the agent to be released from the plant health quarantine license under which it was being held, and the Wildlife and Countryside Act application was used to permit release into the wild. These licenses had to be granted concurrently, as both were required to allow release. Due care and attention was paid to the process in the case of the Japanese knotweed psyllid because this would likely set a precedent for future weed biological control activities in Europe under EU legislation. An *ad hoc* cross-departmental Advisory Committee was set up at the beginning of the

licensing process and was helpful in determining the right review path. Similar advisory committees should be considered by other European Member States for future releases.

Unfortunately, fungi are not dealt with by the same legislation and the path to release this category of biological control agent is less clear. However, the UK government is currently considering the release of the specialist rust fungus *Puccinia komarovii* Tranzschel. var *glandulifera var nov.* for the control of *Impatiens glandulifera* Royle (Balsaminaceae), and this should clarify the pro-cess for the rest of Europe. A number of other weed biological control agents are being considered for release by other EU countries, but none have yet been approved, largely because processes still need to be defined and accepted by each country. The definition of such processes is in progress for another country. Portugal is considering the release of the gall wasp *Trichilogaster acaciaelongifoliae* (Froggatt) against *Acacia longifolia* (Andrews) Willdenow.

Scientists and regulators: different perspectives

Broader recognition of the impacts of invasive species on the natural environment led to a scientific conceptual framework for consideration of invasive species and their effects in the 1980s under the Scientific Committee on Problems of the Environment (SCOPE) (Mooney and Drake, 1986; Office of Technology Assessment, 1993). Since then, invasion biology has developed into a new ecological science discipline, built on a scientific understanding of invasion processes and the impacts of exotic species on native species and communities. From a management perspective, the science of biological invasions has expanded to address different stages along the invasion continuum: (1) before arrival when *preven-tion* can be used to reduce future threats by better screening of proposed species introductions, (2) imme-diately upon arrival when early detection can lead to *eradication* of incipient populations, and (3) after estab-lishment when *management* of widespread or spreading pests and weeds can be applied to contain species' spread or reduce impacts of widespread invaders through biological control. In turn, this has evolved into a systems-based approach to risk assessment for preven-tion and mitigation of biological invasions, including social factors and economic drivers, which at least in

Australia and New Zealand is referred to as biosecurity science. The use of the term biosecurity, however, still varies greatly between countries.

An objective criticism of invasion biology is, however, that it tends to adopt and promulgate subjective, even alarmist, value-based vocabulary: alien, invasion, harmful, combat, arsenal, ammunition, war, weapons, and so on (Larson, 2005; Gobster, 2005). Additionally, biological control scientists, in promoting their science, have in some cases become overly zealous advocates for biological control releases. Scientists should be grounding policy with scientific evidence concerning risk and potential impacts, but should also recognize that scientific evidence is only one criterion societies might use to decide whether to make releases.

Regulatory decision-makers have the professional responsibility to consider a broader spectrum of decision-making criteria than just the scientific evidence, including societal expectations, legal interpretations of legislative responsibility, and political will. Most applications to import novel living organisms are for commercial purposes. Biological control agent release decisions are generally considered to be more in the public interest. But as they are a tiny fraction of importation applications, biological control release applications are often caught up in the complexities and contradictions of dual agricultural and environmental legislation-based risk-assessment frameworks that are more trade focused.

As described above, some countries with a long history of biological control and an *a priori* acceptance of its benefits have processes or laws in place to facilitate biological control. Nonetheless most countries, including the EU, for example, do not have regulatory processes that facilitate biological control, but rather have risk-averse policies designed to consider the risks associated with trade-associated species importation and, in such cases, agencies typically do not consider any potential public benefits of biological control releases. Because biological control releases cannot generally be reversed, they, like all species introductions, are inherently of potentially high risk and require a rigorous risk assessment (Sheppard et al., 2003). Indeed many of the attributes that make biological control agents effective at controlling pests (e.g., effective search for prey; high reproductive rate) are associated with successful invasive species. Risk assessment therefore needs to show that such inherent risk is tempered by a biological

dependence of the agent on its target for sustenance. As very few countries have the regulatory environment, technical capacity, or the background to undertake such a risk assessment independently, such countries often are more risk averse and place a lower social value on biological control. Many potential biological control agents are also new to science or come from making subspecific entities into full species (see also Chapter 6), and this too challenges the capacity of regulatory staff to evaluate permit applications.

Biological control practitioners therefore often, at least in private, exhibit frustration regarding overly complex or inappropriate regulatory processes preventing the gaining of potentially high public benefits, while regulatory decision-makers find the practitioners inconsiderate of the non-scientific assessment criteria related to public values and concerns.

Differences in national approaches and associated cultural values

Each nation's cultural history shapes the values held by its citizens, including values around nature and biological diversity. As nations are now highly interconnected through technology, travel, and global trade, national cultural values are no longer independent and international practices are harmonized through international agreements. Despite this, national differences remain and need to be recognized with respect to variation in acceptance of the introduction of exotic organisms as biological control agents, and science, as we have discussed, only has a partial role to play in the formulation of such values.

Australia and New Zealand are both developed large-island countries that suffered massive deliberate exotic species introductions into their unique environments during European colonization. In South Africa, Australia, and New Zealand, relatively new institutions have been assigned to protect biodiversity. As Southern Hemisphere countries and former English colonies, they suffer negative environmental impacts owing to the historical importation of an astonishing array of species (from Europe and the Americas). South Africa, New Zealand, and Australia are all signatories to the UN Convention on Biodiversity (CDB), and regulatory agencies in these countries reference this as a national commitment. Biodiversity protection in these countries is described as a public interest value by regulatory scientists, and it serves to foster national identity protection. Most other countries, in contrast, are largely encompassed by land borders with multiple neighbors. This makes the restriction of movement of organisms across borders more difficult, both practically and politically (e.g., European countries).

Many countries are harmonizing their goals with respect to reducing invasive species' impacts on the environment through their joint commitments under the CBD. The United States is not a signatory member to this treaty, and biodiversity protection there is governed under existing legislation, such as the Endangered Species Act of 1973. Attempts to repeal or reform this act have been politically deadlocked since 1995. However, invasive species policies and management practices have been incrementally improved over the past two decades, for the most part through executive orders or within the context of the pre-existing legislation (Lodge et al., 2006). Australia, with a small population but a large land mass and the highest level of mammal extinction in the world, is recognizing through proposed revisions to its EPBC Act that biodiversity legislation built around the hub of threatened and endangered species protection is increasingly untenable and unworkable and there is greater academic discussion about the need for an explicit triage approach (Possingham et al., 2002). In contrast, in the United States, an unreformed ESA remains the benchmark for action, such that as an increasing number of threatened cryptic species are genetically defined, resources and actions have to be allocated quickly for their protection (Williams, 2006).

The most striking comparison of national approaches to importation risk regulation is that between the United States and New Zealand. The New Zealand biological control agent review processes are currently the most scientifically rigorous in the world and also the most open to public participation. EPA (NZ) has emerged from New Zealand's distinct cultural values and political history (Miller, 2001; Young, 2004). New Zealand's small, well-educated population and associated limited diversity of cultures and beliefs may also have contributed to the adoption of a rational and participatory approach. Nonetheless, elements of its public-engaging, ecologically based evaluation system can broadly inform risk management in the public interest elsewhere. The EPA (NZ) structure balances multiple scientific and

social criteria, while requiring robust, genuine public engagement. EPA (NZ) imposes the burden of public engagement (not just consultation) on the permit petitioner, allowing concerned citizens to question the proposed introduction publicly within a defined scientific and temporal framework.

New forms of public engagement will be necessary to address invasive species management in the United States (Warner, 2011; Warner and Kinslow, 2013). It is hard to imagine an EPA (NZ) type of process working in the United States, but many of its features could be creatively adapted there, some by agency practice, some by rule making, and some by legislation (see Chapters 13 and 15).

Where can process improvements be made?

Weed biological control risk-assessment processes

There are different complex processes regulating the release of weed biological control agents across countries. In South Africa, regulatory inexperience under cumbersome new environmental legislation has slowed the agent release frequency, but delays are decreasing as solutions are found. In the United States, release applications must address a multitude of existing state and federal laws and ensure a strict prohibition against harm to endangered species under the ESA. Simplification could be considered where appropriate to speed up considerations of approvals. Application assessment and approval processes increasingly include some form of public consultation, at least in Australia and New Zealand. A professionally moderated public consultation process is in the interests of all stakeholders. Some countries (e.g., New Zealand, Canada, and Australia) have developed administratively efficient and balanced approval processes that are also scientifically rigorous and publicly justifiable. IRA approaches developed under accepted IPPC protocols have made a difference in providing simplified processes for countries with no regulatory mechanisms for deciding whether to release biological control agents. Continued skepticism in many countries about releasing arthropod and particularly plant pathogen weed biological control agents is often based on ill-informed public perceptions of risk, and this problem could be addressed by increasing public awareness of risks and potential benefits and direct engagement in biological control programs from the start.

Insect biological control risk-assessment processes

Only Australia and New Zealand have parallel processes for assessing the risks of weed and arthropod biological control agents. Scientific rigor would be improved if such parallel risk-assessment processes were also adopted by other countries. In the United States, this could be done by subjecting arthropod-targeted biological control agents to review by the TAG and when undertaking environmental assessments and during NAPPO consultation. Furthermore, panels assessing risks of agents targeting invertebrates would do well to include invertebrate systematists and conservationists. Requirements for adequate vouchering of both invertebrate targets and agents should receive special attention.

International sharing of biodiversity for public benefit

The development of the CBD decisions and targets around biodiversity protection led to the Nagoya Protocol (2010) (*Access to genetic resources and the fair and equitable sharing of benefits arising from their utilization*) concerning benefit-sharing of biodiversity. While this protocol primarily targets commercial usage, it has direct relevance to classical biological control where species native to one country are to be exported for the benefit of another country.

The protocol aims to provide a transparent legal framework, rules, and procedures for access, creating incentives to conserve and sustainably use genetic resources, and therefore enhance the contribution of biodiversity to development and human well-being. Utilization includes research and development of the genetic resources, as well as subsequent applications and commercialization. Sharing benefits may be monetary or non-monetary, such as royalties and the sharing of research results and help with development of in-country research capability and institutions. However, public benefit activities like biological control are not explicitly mentioned even though submissions were made (Cock et al., 2009, 2010). While such activities in the past have generally been undertaken through formal or informal collaborations directly between research scientists and institutions, under this protocol an expectation has been created of some kind of direct payment, royalties, or compensation. For example, Argentina, one of the countries where access to native natural enemies for biological control elsewhere has been increasingly difficult because of strict permitting procedures, has recently started asking for some form of financial or service payment. Benefit sharing around biological control can only occur here if, through this process, such countries can directly share agents for their exchanged pest species, but this rarely happens in time frames relevant for the protocol (Cock, 2010). Further consideration of access to biological control agents under the Nagoya Protocol would be valuable in the context of CBA Aichi target 9 (CBD, 2011).

Conclusions

Contrasting positions concerning appropriate biodiversity protection and low public awareness about the benefits of biological control in invasive species management are factors that have constrained the use of classical biological control as an invasive species management tool. This appears to be especially true in the United States. As Strong and Pemberton (2000) pointed out, "in the absence of reform, opposition to biological control – rational as well as irrational – will grow." In Australia and Canada, declines in biological control activity seem to be less concerned with risks and more associated with reduced investment in long-term invasive species management solutions, like biological control, in favor of other short-term politically attractive opportunities. As activity declines, however, public memory, recognition, and understanding of the benefits these approaches offer are declining with it. At the same time, a few countries like New Zealand and South Africa are hosts to much biological control activity, and in Europe some institutions are starting new biological control programs. Despite these legislative problems and financial retrenchments that have recently affected biological control, future continued expansion of trade is likely to increase the introduction of pest plants and insects, expanding also the need for biological control services.

Legislation and regulatory processes remain diverse and complex around the world for biological control agent release decision-making, driven by diverse national legislation and associated priorities, but New Zealand demonstrates how biological control regulation can be done efficiently based on an ecologically informed model that considers risks, costs, and benefits, while being inclusive and transparent to the public (Barratt et al., 2010).

In this chapter, we have outlined the different perspectives of both scientists and regulatory decision-makers and some of the reasons different countries have different attitudes and approaches to biological control. We have explained why obtaining release permits in those countries that recognize biological control benefits has, nonetheless, become more complex with globally evolving environmental values. This is not a concern, more a natural process of societal maturation around wanting greater input into the management of risks of releasing exotic biological control agents in the context of such values. The scientist's role remains to provide the best available information and evidence of the risks and potential benefits of biological control introductions. It is not the scientists' choice whether countries decide or not to attempt to realize such benefits by releasing the agents they recommend. Improvements can be made to the scientific basis of regulatory processes for agent approvals to help simplify or improve such processes, where relevant, without loss of science rigor behind the assessments. We offer some suggestions. Greater public consultation and engagement in biological control programs is in everyone's interest, whether or not extreme opinions emerge. Adopted processes should ensure the majority view holds sway. We are confident that biological control will remain a valuable tool for reducing threats to biodiversity and the environment caused by invasive species.

Acknowledgments

We would like to thank the following people for help with the country sections in Chapter 11: Mark Hoddle (UCR), Michael Montgomery (USDA Forest Service, retired), Dick Shaw (CABI), Hildegard Klein (ARC-PPRI), Philip Ivey (SANBI), Barbara Barrett (Ag Research NZ), and Helmuth Zimmerman and Rorisang Mahlakoana (South African DAFF), and also Bernd Blossey and David Wagner for reviews of the chapter. The writing of this chapter and the research on which it is written was supported by the Australian Government, the California Department of Food and Agriculture, and the US National Science Foundation (award 0646658).

References

Agricultural Risk Protection Act. 2000. US Public Law 106-224. Available from: http://www.aphis.usda.gov/brs/pdf/PlantProtAct2000.pdf [Accessed October 2014].

Barratt, B. I. P. and A. Moeed. 2005. Environmental safety of biological control: policy and practice in New Zealand. *Biological Control* **35**: 247–252.

Barratt, B. I. P., F. G. Howarth, T. M. Withers, et al. 2010. Progress in risk assessment for classical biological control. *Biological Control* **52**: 245–254.

Better Border Biosecurity. 2014. Biological Control Information Resource for EPA Applicants. Available from: http://b3.net.nz/birea/index.php [Accessed January 2016].

Bigler, F., J. S. Bale, M. J. W. Cock, et al. 2005. Guidelines on information requirements for import and release of invertebrate biological control agents in European countries. *Biological Control News and Information* **26**(4): 115N–123N.

Biological Control Act. 1984. An Act to make provision for the biological control of pests in the Australian Capital Territory, and for related purposes. Commonwealth Government Printer, 25 pp. Available from: www.opbw.org/nat_imp/leg_reg/aus/bcar.pdf 03 [Accessed January 2016].

Carson, R. 1962. *Silent Spring*. Houghton Mifflin. Boston, Massachusetts, USA.

Clout, M. N. and S. J. Lowe. 2000. Invasive species and environmental changes in New Zealand, pp. 369–383. *In*: Mooney H. A. and R. J. Hobbs (eds.). *Invasive Species in a Changing World*. Island Press. Washington, DC.

Coastal Zone Management Act. 1972. Congressional action to help manage our nation's Coasts. Available from: http://www.fema.gov/coastal-zone-management-act-1972 [Accessed January 2015].

Cofrancesco, A. F., Jr. and J. F. Shearer. 2004. Technical advisory group for biological control agents of weeds, pp. 38–41. *In*: Coombs, E. M., J. K. Clark, G. L. Piper, and A. F. Cofrancesco A.F., Jr. (eds.). *Biological Control of Invasive Plants in the United States*. Oregon State University Press. Corvallis, Oregon, USA.

Cock, M. J. W. 2010. Biopiracy rules should not block biological control. *Nature* **467**(7314): 369–369.

Cock, M. J. W., J. C. van Lenteren, J. Brodeur, et al. 2009. The use and exchange of biological control agents for food and agriculture. Commission for Genetic Resources for Food and Agriculture. FAO. Rome. Available from: ftp://ftp.fao.org/docrep/fao/meeting/017/ak569e.pdf [Accessed January 2016].

Cock, M. J. W., J. C. van Lenteren, J. Brodeur, et al. 2010. Do new access and benefit sharing procedures under the Convention on Biological Diversity threaten the future of biological control? *Biological Control* **55**: 199–218.

CBD (Convention on Biological Diversity). 2011. Aichi target 9: *"By 2020, invasive alien species and pathways are identified and prioritized, priority species are controlled or eradicated and measures are in place to manage pathways to prevent their introduction and establishment."* Available from: http://www.cbd.int/sp/targets/rationale/target-9/ [Accessed January 2016].

Cullen, J. M. and E. S. Delfosse. 1985. *Echium plantagineum*: catalyst for conflict and change in Australia, pp. 249–292. *In*: Delfosse, E. S. (ed.). *Proceedings of the VIth International Symposium on Biological Control of Weeds, August 1984, Vancouver, Canada*. Agriculture Canada. Ottawa, Ontario, Canada.

DeLoach, C. J., R. I. Carruthers, J. Lovich, et al. 2000. Ecological interactions in the biological control of saltcedar (*Tamarix* spp.) in the U.S.: toward a new understanding, pp. 819–874. *In*: Spencer, N. R. (ed.). *Proceedings of Xth International Symposium on Biological Control, July 4–14, 1999, Bozeman, Montana*. Montana State University. Bozeman, Montana, USA.

EPBC (Environment Protection and Biodiversity Conservation Act). 1999. Act No. 91. An act relating to the protection of the environment and the conservation of biodiversity, and for related purposes. Commenced 16 July 2000, as amended by Act No. 63 2002. 528 sections. Australian Commonwealth Government Printer. Available from: http://secure.environment.gov.au/epbc/about/history.html [Accessed January 2016].

ESA (Endangered Species Act). 1973. An act to provide for the conservation of endangered and threatened species of fish, wildlife, and plants, and for other purposes. Public Law 93-205, 16 U.S.C. pp. 1531-1544. Available from: http://www.epw.senate.gov/esa73.pdf [Accessed January 2016].

Environmental Risk Management Authority (ERMA) - New Zealand. 2010. *Investigating Biological Control and the HSNO Act*. Environmental Protection Agency of New Zealand, Wellington, New Zealand. Available from: http://www.epa.govt.nz/Publications/Investigating-Biological-Control-and-the-HSNO%20Act-ERMA-Report-2010).pdf [Accessed October 2010].

Executive Order 13112. 1999. Invasive species. Available from: http://www.gpo.gov/fdsys/pkg/FR-1999-02-08/pdf/99-3184.pdf [Accessed October 2010].

Falk, I., R. Wallace, and M.L. Ndoen. 2011. *Managing Biosecurity across Borders*. Springer. New York.

FNWA (Federal Noxious Weed Act). 1974. An Act to manage and control the spread of noxious weeds. Pursuant to the Act, the U.S. Secretary of Agriculture was given the authority to declare plants "noxious weeds", and limit the interstate spread of such plants without a permit. Available from: http://definitions.uslegal.com/f/federal-noxious-weed-act/ [Accessed January 2016].

FPPA (Federal Plant Pest Act). 1957. This act regulates the importation and interstate movement of plant pests and authorizes the Secretary of Agriculture to take emergency measures to destroy infected plants or materials. 7 U.S.C. pp.150aa-150jj 23 May, as amended 1968, 1983, 1988 and 1994. (superseded by the Plant Protection Act of 2000) Available from: https://www.aphis.usda.gov/brs/pdf/PlantProtAct2000.pdf [Accessed January 2016].

Gobster, P. H. 2005. Invasive species as ecological threat: Is restoration an alternative to fear-based resource management? *Ecological Restoration* **23**: 261–271.

Goldson, S. L., E. R. Frampton, and G. S. Ridley. 2010. The effects of legislation and policy in New Zealand and Australia on biosecurity and arthropod biological control research and development. *Biological Control* **52**: 241–244.

Harrison, L., A. Moeed, and A. Sheppard. 2005. Regulation of the release of biological control agents of arthropods in New Zealand and Australia, pp. 715–725. *In*: Hoddle, M. S. (ed). *Second International Symposium for the Biological Control of Arthropods, September 12–16, 2005. Davos, Switzerland*. FHTET-2005-08. USDA Forest Service. Morgantown, West Viriginia, USA. Available from: http://www.bugwood.org/arthropod2005/vol2/14e.pdf [Accessed January 2016].

HSNO (Hazardous Substances and New Organisms Act). 1996. June No. 30, 252 pp. and amendment No. 35, 4 pp. New Zealand Government. Available from: www.mfe.govt.nz/laws/hsno.html [Accessed January 2016].

Hill, R., D. Campbell, L. Hayes, et al. 2013. Why the New Zealand regulatory system for introducing new biological control agents works, pp. 75–83. *In*: Wu, Y., T. Johnson, S. Sing, et al. (eds.). *Proceedings of the XIII International Symposium on Biological Control of Weeds, September 11–16, 2011. Waikoloa, Hawaii, USA*. FHTET- 2012-07 USDA Forest Service. Morgantown, West Virginia, USA.

Hunt, E. J., A. Kuhlmann, A. Sheppard, et al. 2008. Review of invertebrate biological control agent regulation: recommendations for a harmonised European system. *Journal of Applied Entomology* **132**: 89–123.

IPPC (International Plant Protection Convention). 2006. Guidelines for the export, shipment, import and release of biological control agents and other beneficial organisms (2005). International Standards for Phytosanitary Measures No. 3. Food and Agriculture Organisation of the United Nations, Rome, Italy.

IPPC (International Plant Protection Convention). 2007. Framework for pest risk analysis (2007). International Standards for Phytosanitary Measures No. 2. Food and Agriculture Organisation of the United Nations, Rome, Italy.

Iverson, T. and C. Perrings. 2012. Precaution and proportionality in the management of global environmental change. *Global Environmental Change* **22**: 161–177.

Klein, H., M. P. Hill, C. Zachariades, and H. G. Zimmermann. 2011. Regulation and risk assessment for importations and releases of biological control agents against invasive alien plants in South Africa. *African Entomology* **19**: 488–497.

Larson, B. M. H. 2005. The war of the roses: demilitarizing invasion biology. *Frontiers in Ecology and the Environment* **3**: 495–500.

Leung, B., D. M. Lodge, D. Finnoff, et al. 2002. An ounce of prevention or a pound of cure: bioeconomic risk analysis of invasive species. *Proceedings of the Royal Society, London B* **269**: 2407–2413.

Lodge, D. M., S. Williams, H. J. MacIsaac, et al. 2006. Biological invasions: recommendations for U.S. policy and management. *Ecological Applications* **16**: 2035–2054.

Louda, S. M. 1999. Ecology of interactions needed in biological control practice and policy. *Bulletin of the British Ecological Society* **29**: 8–11.

Magarey, R. D., M. Colunga-Garcia, and D. A. Fieselmann. 2009. Plant biosecurity in the United States: roles, responsibilities, and information needs. *BioScience* **59**: 875–884.

Mason, P. G. and J. T. Huber (eds). 2002. *Biological Control Programmes in Canada, 1981–2000*. CABI Publishing. Wallingford, UK.

Mason, P. G., J. T. Kabaluk, B. Spencer, and D. R. Gillespie. 2013. Regulation of biological control in Canada, pp. 1–5. *In:* Mason, P. G. and D. R. Gillespie (eds.). *Biological Control Programmes in Canada, 2001–2012*. CAB International Publishing. Wallingford, UK.

Messing, R. H. 2005. Hawaii as a role model for comprehensive U.S. biological control legislation: the best and the worst of it, pp. 686–691. *In:* Hoddle, M. S. (ed.). *Proceedings of the Second International Symposium on Biological Control of Arthropods, Davos, Switzerland, 12–16 September 2005*. FHTET-2005-08. USDA, Forest Service. Morgantown, West Virginia, USA.

Messing, R. H. and M. G. Wright. 2006. Biological control of invasive species: solution or pollution? *Frontiers in Ecology and the Environment* **4**: 132–140.

Miller, R. (ed.). 2001. *New Zealand: Government and Politics*. Oxford University Press. Auckland, New Zealand.

Mooney, H. A. and J. A. Drake (eds.). 1986. *Ecology of Biological Invasions of North America and Hawaii*. Springer-Verlag. New York.

Nagoya Protocol. 2010. On access to genetic resources and the fair and equitable sharing of benefits arising from the utilization to the Convention on Biological Diversity. Available from: http://www.cbd.int/abs/doc/protocol/nagoya-protocol-en.pdf [Accessed January 2016].

NAPPO. 2001. Guidelines for Petition for Release of Exotic Phytophagous Agents for the Biological Control. Regional Standards for Phytosanitary Measures #7. Available from: www.nappo.org/index.php/download_file/view/79/260 [Accessed January 2016].

NAPPO. 2015. Guidelines for Petition for Release of Exotic Entomophagous Agents for the Biological Control. Regional Standards for Phytosanitary Measures #12. Available from: www.nappo.org/files/1814/4065/2949/RSPM12_30-07-2015-e.pdf [Accessed January 2015].

NEM:BA (National Environmental Management: Biodiversity Act). 2004. No. 10. & Draft Alien and Invasive Species Regulations. Available from: http://cer.org.za/virtual-library/national-environmental-management-biodiversity-act-2004 [Accessed January 2016].

NEPA (National Environmental Policy Act). 1969. 83 Stat. 852. Available from: http://www.gsa.gov/portal/content/104676?utm_source=PBS&utm_medium=print-radio&utm_term=nepa&utm_campaign=shortcuts [Accessed January 2016].

Office of Technology Assessment. 1993. *Harmful Non-Indigenous Species in the United States*. Office of Technology Assessment, U.S. Congress, Washington, DC. Available from: http://govinfo.library.unt.edu/ota/Ota_1/DATA/1993/9325.PDF [Accessed January 2016].

Perrings, C. 2005. Mitigation and adaptation strategies for the control of biological invasions. *Ecological Economics* **52**: 315–325.

Perrings, C., K. Dehnen-Schmutz, J. Touza, and M. Williamson. 2005. How to manage biological invasions under globalization. *Trends in Ecology & Evolution* **20**: 212–215.

Perrings, C., H. Mooney, and M. Williamson. 2010. *Bioinvasions and Globalization*. Oxford University Press. Oxford, UK.

PPA (Plant Protection Act). 1990. c.22 (P-14.8) An act to prevent the importation, exploitation and spread of pests injurious to plants into Canada and to provide for their control and eradication and the certification of plants and other things.19 June 1990. Available from: http://laws-lois.justice.gc.ca/eng/acts/P-14.8/index.html [Accessed January 2016].

PPA (Plant Protection Act). 2000. Plant Protection Act provides that the Secretary of the Department of Agriculture may issue regulations "to prevent the introduction of plant pests into the United States or the dissemination of plant pests within the United States. 7 USC 7702. 114 STAT. 438 Public Law 106–224—June 20, 2000. Available from: http://www.aphis.usda.gov/plant_health/plant_pest_info/weeds/downloads/PPAText.pdf [Accessed January 2016].

PQA (Plant Quarantine Act). 1912. An act to regulate the importation of nursery stock and other plants and plant products; to enable the Secretary of Agriculture to establish and maintain quarantine districts for plant diseases and insect pests; to permit and regulate the movement of fruits, plants, and vegetables, therefrom, and for other purposes. 20 Aug 1912. 7 U.S.C. pp. 154–167. Available from: https://archive.org/details/plantquarantinea00unit [Accessed January 2016].

Possingham, H. P., S. J. Andelman, M. A. Burgman, et al. 2002. Limits to the use of threatened species lists. *Trends in Ecology & Evolution* **17**: 503–507.

Quarantine Act. 1908. Act No. 3 as amended by Act No. 17 of 2002. 87 sections 195 pp. Australian Commonwealth Government Printer. Available from: www.opbw.org/nat_imp/leg_reg/aus/qar.pdf [Accessed January 2016].

Shaw, R. H., S. Bryner, and R. Tanner. 2009. The life history and host range of the Japanese knotweed psyllid, *Aphalara itadori* Shinji: potentially the first classical biological weed control agent for the European Union. *Biological Control* **49**: 105–113.

Sheppard, A.W., R. Hill, R. A. DeClerck-Floate, et al. 2003. A global review of risk-benefit-cost analysis for the introduction of classical biological control agents against weeds: a crisis in the making? *Biological control News & Information* **24**: 91N–108N.

Sheppard, A. W., I. Gillespie, M. Hirsch, and C. Begley. 2011. Biosecurity and sustainability within the growing global bio-economy. *Current Opinion in Environmental Sustainability*, **3**(1): 4–10.

Strong D. R. and R. W. Pemberton. 2000. Biological control of invading species – risk and reform. *Science* **288**: 169–179.

Takahashi, M. A. 2005. Are the Kiwis taking a leap? Learning from the biosecurity policy of New Zealand. *Temple Journal of Science, Technology and Environmental Law* **24**: 461–478.

Trampusch, C. 2014. 'Protectionism, obviously, is not dead': A case study on New Zealand's biosecurity policy and the causes-of-effects of economic interests. *Australian Journal of Political Science*. doi:10.1080/10361146.2014.894961.

USDA, 2007. Reviewer's manual for the technical advisory group for biological control agents of weeds: guidelines for evaluating the safety of candidate biological control agents. Manuals Unit, Plant Protection and Quarantine, Animal and Plant Health Inspection Service, United States Department of Agriculture. Available from: https://www.hsdl.org/?view&did=28977 [Accessed January 2016].

Warner, K. D. 2011. Public engagement with biological control of invasive plants: the state of the question, pp. 340–345. *In*: Wu, Y., T. Johnson, S. Sing, et al. (eds.). *Proceedings of the XIII International Symposium on Biological Control of Weeds, September 11–16, 2011 - Waikoloa, Hawaii, USA*. FHTET- 2012-07. USDA Forest Service. Morgantown, West Virginia, USA.

Warner, K. D. and F. Kinslow. 2013. Manipulating risk communication: value predispositions shape public understandings of invasive species science in Hawaii. *Public Understanding of Science* **22**: 203–218.

Williams, R. N. 2006. *Return to the River: Restoring Salmon to the Columbia River*. Elsevier Academic Press. London, UK.

Young, D. 2004. *Our Islands, Our Selves*. University of Otago Press. Dunedin, New Zealand.

CHAPTER 12

Managing conflict over biological control: the case of strawberry guava in Hawaii

M. Tracy Johnson

USDA Forest Service, Pacific Southwest Research Station, Volcano, USA

Introduction

Conflict can be a part of virtually any decision about management of natural resources, not least decisions regarding use of biological control. The shared, public character of natural resources and the permanent, irrevocable nature of biological control releases ensure that many people will have opinions, sometimes strongly held, regarding use of a particular biological control agent in an area they care about. Conflicts associated with the use of biological control have been reviewed elsewhere, along with approaches for resolving conflict and the challenges these present (Turner, 1985; Stanley and Fowler, 2004; Hayes et al., 2008). A variety of conflict-resolution approaches exist; however, the time and effort involved in gathering the necessary data and engaging in the required processes to resolve conflicts tend to be seen as onerous, motivating weed biological control scientists to select only targets with low potential for conflict (Stanley and Fowler, 2004), essentially a strategy of conflict avoidance. In cases where target-based concerns are unavoidable, for example biological control of invasive *Acacia* trees that nevertheless have timber value in South Africa, some programs have focused on biological control agents that minimize conflict, such as seed-feeding agents that limit tree reproduction and spread without harming existing timber stands of the targeted species (e.g., Dennill et al., 1999).

Avoiding targets with potential for high conflict may remain common, but for certain highly damaging invaders with no viable management alternatives, it may be necessary to consider biological control even when it is likely to generate conflict. Discussed here is a case study, strawberry guava (*Psidium cattleianum* Sabine) in Hawaii, in which conflicting views of the targeted plant engendered substantial opposition to biological control. The processes followed and lessons learned may be useful for others considering biological control as a management option for widespread plant invasions in natural areas, especially where there are groups with conflicting interests.

Strawberry guava in Hawaii

Following its introduction into Hawaii in the early 1800s, strawberry guava (Figure 12.1) was cultivated widely in gardens and used occasionally in efforts to reforest degraded landscapes (Degener, 1939; Wagner et al., 1990). Today strawberry guava continues to be planted as an ornamental, and jam from the fruit is sold at farmers' markets. However, it is not cultivated as an agricultural commodity, in contrast with its congener, common guava (*Psidium guajava* L.). Dense stands of strawberry guava infest tens of thousands of hectares of wet and mesic forest on all the major Hawaiian islands, and it continues to spread, potentially affecting an estimated 475,000 hectares, 90% of Hawaii's forests (State of Hawaii, 2011). It is considered a serious threat to native ecosystems and to dozens of threatened and endangered species (Mitchell et al., 2005) owing to its

---

Integrating Biological Control into Conservation Practice, First Edition. Edited by Roy G. Van Driesche, Daniel Simberloff, Bernd Blossey, Charlotte Causton, Mark S. Hoddle, David L. Wagner, Christian O. Marks, Kevin M. Heinz, and Keith D. Warner.
© 2016 John Wiley & Sons, Ltd. Published 2016 by John Wiley & Sons, Ltd.

Figure 12.1 Strawberry guava, *Psidium cattleianum*, thickets crowd out native forest species in Hawaii. Photo courtesy of Jack Jeffrey.

ability to form dense thickets even in relatively undisturbed wet forests (Huenneke and Vitousek, 1990). Strawberry guava is among the worst invasive species globally (IUCN, 2013), posing a threat to native ecosystems in Florida, Puerto Rico, Reunion, Mauritius, Guam, Cook Islands, Fiji, French Polynesia, Palau, Samoa, and Norfolk Island (Lorence and Sussman, 1986; Tunison, 1991; Space, 2013).

Invasion by strawberry guava has devastating consequences for Hawaiian ecosystems. In addition to loss of habitat for native species, forests dominated by strawberry guava lose more water to evapotranspiration than native forests: 27% more in a recent measurement, equivalent to a loss of 33 cm (13 inches) in annual precipitation from watersheds (Takahashi et al., 2011). Millions of pounds of fallen strawberry guava fruit each year are a primary breeding resource for oriental fruit flies, *Bactrocera dorsalis* (Hendel), a pest of Hawaiian agriculture (Vargas et al., 1989, 1990, 1995). Fruit flies can cause direct yield loss, but even more importantly, their infestation limits possibilities for export of Hawaiian produce to major markets such as California and Japan. Also, sustainable wild harvest and management of Hawaii's high value native hardwood *Acacia koa* A. Gray is impeded by dense stands of strawberry guava, which

rapidly colonize areas disturbed by logging and suppress koa regrowth (Dobbyn, 2003; Baker et al., 2009).

Effective management of natural areas invaded by strawberry guava requires repeated mechanical and chemical treatments to reduce its copious resprouting, but such intensive work is not practical across the vast areas of affected forest, much of it in remote, rugged terrain. Population growth rates of strawberry guava can be rapid: studies in native forest at 900 m elevation in Hawaii showed nearly 10% annual increases in strawberry guava stem density and basal area (Denslow et al., unpublished data). Although strawberry guava seeds are short lived (Uowolo and Denslow, 2008), its fleshy fruits are abundant and rapidly dispersed by introduced birds and pigs (Diong, 1982). The invasive success of strawberry guava has been attributed to its broad environmental tolerances, prolific fruit production, frequent vegetative spread, and an absence of natural enemies (Huenneke and Vitousek, 1990; Tunison, 1991).

Strawberry guava was one of the earliest species selected by a multi-agency group established in the late 1970s to identify key pests of Hawaiian forests and develop biological control remedies (Hodges, 1988). Exploration for natural enemies in strawberry guava's

Figure 12.2 *Tectococcus ovatus* scales make galls on young leaves of strawberry guava. Photo courtesy of USDA Forest Service.

Table 12.1 Impacts expected from biological control of strawberry guava.

- *Tectococcus ovatus* scales make galls on young leaves; defoliation is uncommon; wood will not be affected
- Trees will not be killed, but will grow slower and compete less with native plants
- Fruit and seed production will be lower by as much as 90%, slowing invasion into native forests
- Fewer pest fruit flies will be produced in areas with abundant strawberry guava
- Yard trees far from forests may not be affected due to limited dispersal of the scale
- Trees grown for fruit can be protected with common horticultural oil sprays
- Biological control can enhance chemical/mechanical control by slowing regrowth and re-invasion

native range, southern Brazil, began in 1988, and in-depth studies of several potential agents were conducted in Brazil in the 1990s. A leaf-galling scale insect, *Tectococcus ovatus* Hempel (Hemiptera: Eriococcidae) (Figure 12.2), was given top priority for further development and was imported to Hawaii in 1999 for quarantine studies. Impacts of this gall insect on strawberry guava in Brazil indicated that the scale had high potential to reduce the plant's growth and reproduction (Table 12.1). Testing in Brazil, Hawaii, and Florida demonstrated that *T. ovatus* was restricted to a narrow subset of *Psidium* species (Vitorino et al., 2000;

Johnson, 2005; Wessels et al., 2007). Furthermore, *T. ovatus* had never been recorded as a pest of any agricultural or ornamental plants, and never attacked common guava, which grows throughout its native range. In 2005, a petition for release of the agent was submitted by the Forest Service to the Hawaii Department of Agriculture, the state agency charged with regulating biological control introductions (Johnson, 2005).

Regulatory review of strawberry guava biological control

The proposed biological control of strawberry guava underwent a state-mandated process entailing review by expert committees, public hearings in each of the four counties, and finally a decision by the Hawaii Board of Agriculture in 2007 to place *T. ovatus* on a list of approved biological control agents. In addition, USDA-APHIS conducted a federal environmental assessment, standard to their review process, which was completed in early 2008 (USDA, 2008). Following these reviews, state and federal permits for environmental release of *T. ovatus* were issued to the Forest Service in April 2008. Throughout these processes, release of the strawberry guava agent was supported by almost all reviewers, with few expressions of concern and very little comment from the public. Release of

T. ovatus was delayed, however, pending completion of a state environmental assessment.

The Hawaii state environmental assessment (EA) process was applied to biological control releases for the first time in 2007. Although federal EAs had been conducted for many years, Hawaii's biological control practitioners decided voluntarily that it was also appropriate to submit proposed releases to the state environmental review process, which is triggered by use of state or county lands or funds, including resources and staff of state agencies involved in releasing biological control agents (Office of Environmental Quality Control, 2004). Two agents to be released against pest insects, plus an agent for fireweed (*Senecio madagascariensis* Poiret) biological control and *T. ovatus* for strawberry guava, were the first biological control agents subjected to this process (Office of Environmental Quality Control, 2008). There was very little public comment on the draft EAs for three of these releases, all proposed by the Pest Control Branch of the Hawaii Department of Agriculture. However, the draft EA for biological control of strawberry guava, submitted at the same time by the USDA Forest Service, received dozens of comments, both in support of and in opposition to the proposed release. The developing controversy over strawberry guava biological control attracted media attention and generated calls to local political leaders. It became evident that a more thorough process of public engagement was needed, and the strawberry guava draft EA was withdrawn for revision (Tummons, 2008).

State of Hawaii EAs are structured much like federal United States EAs, with detailed descriptions of alternative actions, including the proposed action, along with analyses of the expected impacts and proposed mitigation measures associated with each alternative (Office of Environmental Quality Control, 2004). Analyses are required to consider direct, indirect, and cumulative impacts expected from any proposed action. The Hawaii EA differs from the federal EA in some of the specific criteria that must be considered (Table 12.2) before the state can issue a Finding of No Significant Impact (FONSI), which allows a project to proceed without an Environmental Impact Statement. These state criteria emphasize assessment of possible impacts on people – for example through social, economic, or health effects – as well as on the environment. Hawaii also requires

each EA to include a cultural impact assessment, analyzing a project's expected impacts on cultural practices and features important to native Hawaiians and other ethnic groups (Office of Environmental Quality Control, 2004).

Proposing release of a biological control agent within the Hawaii environmental review process for the first time presented some unusual challenges. The strong emphasis on considering socio-cultural as well as biological impacts in the state EA process required additional consultations and expertise that were not well developed in the earlier federal EA (USDA, 2008). Furthermore, the environmental review process in Hawaii previously had been applied mainly to commercial or governmental construction or similar projects, with the geographic focus restricted to a particular site. In contrast, biological control releases have nearly always been statewide in scope, with impacts that are intended to extend everywhere the target pest may be found. This geographic breadth also

Table 12.2 Significance criteria specified for evaluation in a State of Hawaii Environmental Assessment: an Environmental Impact Statement is required if, to a significant level of expected impact, a project …

- **involves** an irrevocable commitment to loss or destruction of any natural or cultural resource
- **curtails** the range of beneficial uses of the environment
- **conflicts** with the state's long-term environmental policies or goals and guidelines as expressed in Chapter 344, HRS, and any revisions thereof and amendments thereto, court decisions, or executive orders
- **substantially** affects the economic or social welfare of the community or state
- **substantially** affects public health
- **involves** substantial secondary impacts, such as population changes or effects on public facilities
- **involves** a substantial degradation of environmental quality
- **is** individually limited but cumulatively has considerable effect upon environment or involves a commitment for larger actions
- **substantially** affects a rare, threatened or endangered species, or its habitat
- **detrimentally** affects air or water quality or ambient noise levels
- **affects** or is likely to suffer damage by being located in an environmentally sensitive area such as a flood plain, tsunami zone, beach, erosion-prone area, geologically hazardous land, estuary, fresh water, or coastal waters
- **substantially** affects scenic vistas and view planes identified in county or state plans or studies
- **requires** substantial energy consumption

tended to increase the required consultations and made analysis more complex.

Although more detailed than the three contemporaneous biological control EAs for the other agents mentioned above, the 2008 strawberry guava EA still fell short as a comprehensive consideration of the many criteria for a Hawaii assessment (Table 12.2), and it lacked a cultural impact assessment. To improve the document, the Forest Service contracted with a professional EA writer with long experience in the Hawaii system to develop the EA. In 2009, consultations with broader segments of the public were begun, including public meetings scheduled in each of Hawaii's four counties and targeted contacts to gather input for a cultural assessment. Meanwhile, the Forest Service and its partners began work to describe to the public more effectively the rationale and goals of the strawberry guava project. Outreach specialists involved in invasive species control and prevention, and many others within the Hawaii conservation community, actively collaborated to convey messages on the importance of managing strawberry guava and the role of biological control. This work was spurred initially by the vigorous campaign of early opponents who noted the inadequacy of the EA and questioned the necessity and safety of attempting biological control (Warner and Kinslow, 2013). Much of the information needed by biological control supporters was already available from previous Forest Service documents, such as "Frequently Asked Questions" on strawberry guava and biological control (USDA Forest Service, 2013a, b), but outreach specialists were able to reformulate this information so that it could be communicated more effectively. Over time, a variety of novel and effective communication pieces were produced by partners outside the biological control community who saw a crucial need to manage strawberry guava (Gon, 2009; Conservation Council for Hawaii, 2010; Hawaii Ecosystems at Risk, 2011).

The controversy following the 2008 strawberry guava EA immediately generated calls to local political leaders. Some responded by scheduling public meetings to share information about the proposed biological control and to hear the opinions of local citizens wanting to offer testimony (Tummons, 2008). The first meeting on Hawaii Island in June 2008 attracted approximately 150 people, and later meetings before the Hawaii County Council were similarly well attended. Hawaii County Council members proposed

two resolutions opposing the use of the biological control based on concern over the specificity of the agent and concern that it might evolve to use other plants. One of these resolutions eventually was adopted in August 2009 (County of Hawaii, 2009). In contrast, in March 2009 the Maui County Council adopted a resolution supporting use of biological controls for forest pests, including mention of strawberry guava as a key target for management (County of Maui, 2009). Meanwhile, bills in opposition to strawberry guava biological control were introduced in the state legislature in early 2009, but never received hearings (Hawaii State Legislature, 2009). In the end, none of these legislative actions changed the final decision-making process: whether to permit release of strawberry guava biological control remained a matter before the state administrative branch, pending completion of a state environmental assessment.

The revised Draft EA was completed and published in June 2010. This document incorporated the many statewide consultations over the preceding two years, including inputs from public meetings. It also included recent analyses of remote sensing and other studies of the strawberry guava invasion that strongly documented negative ecological effects of the weed (Asner et al., 2008; Zimmerman et al., 2008; Takahashi et al., 2011), an economic analysis of conventional control methods, and a cultural impact assessment (State of Hawaii, 2010). The Draft EA increased from 33 pages in 2008 to 133 pages in 2010. During the 30-day period for public comment, over 200 respondents sent letters, approximately 2 to 1 in support of biological control (Figure 12.3). According to Hawaii rules for the process, individual responses were written for each comment letter, and the EA was revised a final time and all comments and responses incorporated. This final EA was published in November 2011 with a FONSI issued by the Hawaii Department of Agriculture (State of Hawaii, 2011). Following this decision, the Department then re-issued a permit for release of *T. ovatus*, and in December 2011 the biological control agent was released for the first time from quarantine into the environment on Hawaii Island. Two years later the agent was released at forest sites on Oahu. As of 2014, strawberry guava biological control was established and spreading slowly at multiple sites on these two islands, and efforts had begun to establish it in invaded forests on Maui and Lanai.

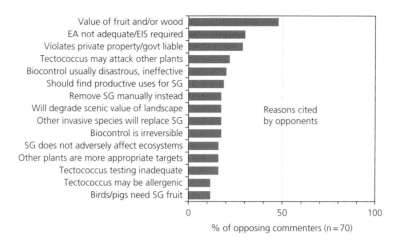

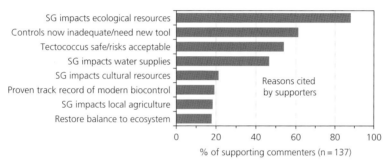

Figure 12.3 Relative frequency of public comments on 2010 Draft Environmental Assessment of proposed biological control of strawberry guava (State of Hawaii, 2011b).

Lessons learned about our public

People in Hawaii are probably unusual compared to most other Americans, in that a great number have at least a passing familiarity with the concept and practice of biological control. This might be expected from the state's very long and active history of biological control introductions: over 700 since 1902 (Follett et al., 2000). Unfortunately, it is failures that come to mind for most residents asked about biological control, with the great majority first citing the nineteenth-century introduction of mongoose to control rats in sugar cane fields. That this introduction was conducted by an individual cane farmer, with little scientific justification and no regulatory oversight, is not at all widely known (Baldwin et al., 1952; Reimer, 2002). Besides the regular visibility of mongoose in the islands and their popularity in local folklore and children's literature, the story of their introduction has great staying power as a cautionary tale about unforeseen consequences of introducing new

species to unique island ecosystems. This example has been memorized by generations of Hawaii residents, and it predates by a decade or two the modern era of targeted weed control using specialized natural enemies, inaugurated by Albert Koebele's 1902 introductions of agents into Hawaii from Mexico against lantana (*Lantana camara* L.). There are several recent biological control success stories that could be told in Hawaii (Trujillo, 2005; Hawaii Department of Agriculture, 2010), but none are yet so widely known as to effectively counter the mongoose story.

Negative views of species introductions also predominate in Hawaii because residents are highly attuned to the serious harm caused by invasive species. Many recent invasions – little fire ants (*Wasmannia auropunctata* [Roger]), coqui frogs (*Eleutherodactylus coqui* Thomas), coffee berry borer (*Hypothenemus hampei* [Ferrari]) (Coleoptera: Curculionidae), and axis deer (*Axis axis* [Erxleben]), among others – regularly earn headlines in local newspapers and affect many people's lives very tangibly. Negative attitudes toward invasive species are also well established because of longstanding educational efforts to highlight the problem and build support for enhanced biosecurity. Over the last 25 years multi-agency collaborations have alerted the public to the "silent invasion," and invasive species committees have worked steadily on each island toward control or eradication of the most damaging invasive species (Krauss and Duffy, 2010). So it is a great step forward that most people now understand well that introduced species can have serious ecological consequences. Unfortunately for biological control, most people do not clearly distinguish between the unplanned invasions of species arriving through commerce and the purposeful introductions of natural enemies for biological control of invasive species. Rather, all new introductions are viewed with suspicion and the expectation of bad outcomes. This is only compounded by widespread biases against arthropods and microorganisms, generally regarded as pests, which are the tools of biological control.

These existing biases against introduction of new species lead to the completely understandable default position among the great majority of our public: "Introducing new species is a bad idea, and introducing any new plant pest is crazy." However, almost everyone encountered is also open to asking "why would you ever do such a thing." Curiosity over the seemingly counterintuitive nature of biological control may even increase the likelihood that they will be receptive to an explanation. Usually this opportunity is all that is needed. Once one explains why biological control is necessary and how it is done safely, the typical result is that people move from some level of opposition to some level of acceptance.

Given that most people start from a position of skepticism, it is incumbent on scientists to convey the motives for managing invasive species with biological control. In the case of strawberry guava biological control, one flaw of the initial EA was that it did not adequately describe all the problems caused by this particular invasive plant. Instead of writing for an audience of managers and regulators who understood the issues surrounding strawberry guava, it would have been better to address the general public, who were largely unaware of the plant's impacts. In discussing with Hawaii residents the various motives for pursuing biological control of strawberry guava, it was interesting to see what reasons were most persuasive. For people concerned about native species and Hawaii's forests, the threat of strawberry guava has been generally well understood and was a strong motive for their approval of biological control. However most people did not share a deep appreciation of Hawaii's unique forest biota and instead tended to be moved by human needs such as economic well-being and health. For people generally indifferent to the plight of native forest species, the threats of strawberry guava to water supply and local agriculture – through increases in water loss (Takahashi et al., 2011) and higher fruit fly populations (Conservation Council for Hawaii, 2010) – were more persuasive reasons for management. In addition to making these connections clear, it was important to develop a detailed economic argument for the necessity of adding biological control as a tool for managing strawberry guava, instead of relying on increased efforts with other control methods. The revised EA addressed these points more explicitly and in greater detail and variety. For example, conventional methods of hand cutting strawberry guava and applying herbicide were described and efforts per acre mapped for specific areas, resulting in an overall estimated cost of more than $350 million for initial control of strawberry guava in conservation areas in the eastern half of Hawaii Island (State of Hawaii, 2010).

Another key reason why opinions tended to turn against biological control in Hawaii was because the public generally failed to grasp the concept of host specificity. This is a challenge familiar to all biological control scientists confronted with the question "won't the biological control agent move on to eat something else once its target food is exhausted." Building awareness of the fact that many organisms exist in extremely host-specific relationships is absolutely critical for gaining acceptance of biological control from a skeptical public. Once people recognized evidence of host specificity in everyday examples, most were readily convinced of the value and acceptability of managing invasive species with biological control.

Opponents

A certain segment of the public may remain opposed to using biological control in spite of our best efforts to articulate the economic and ecological justifications and to explain risks and benefits. Motivations of opponents varied in the case of strawberry guava, but seemed to center on valuing the tree, its wood, and especially its fruit for various uses (Figure 12.3). Such concerns over the value of an exotic plant are a common and straightforward source of conflict in weed biological control programs (Turner, 1985). Resolving this source of conflict is possible in some cases; for example, some opponents were persuaded by the explanation that yard trees valued for their fruit and attractive foliage might remain unaffected owing to the limited dispersal ability of *T. ovatus* and by the option to control the insect with horticultural sprays if it did reach their yard trees. The expectation that *T. ovatus* would not kill trees or reduce available wood from existing stands was an important mitigating factor for several common reasons for opposing management with biological control.

The value of strawberry guava fruit appeared to be the most common motivation for people who were actively opposed to biological control (Figure 12.3). People often expressed a fondness for the fruit as an occasional snack or in reminiscence of past experiences collecting and eating it, but less often mentioned it as a food source they depended on regularly. In either case, positive emotional connections with the fruit were the norm, with few people noticing a link between abundant fruit and pest insects. Those most vigorously opposed to biological control argued that, even if currently underutilized, the fruit was valuable as a free emergency food source – famine food, in the event of interrupted supply ships (State of Hawaii, 2011). Hawaii's dependence on imports for about 90% of its food has long been a major concern (Office of Planning, 2012), and arguments for maintaining strawberry guava as a wild food and potential crop appealed to the many proponents of small-scale subsistence agriculture, especially in rural Hawaii county. Strawberry guava fruits also were valued by many pig hunters as a food source for wild pigs, which are harvested as an important subsistence resource by some (Burrows et al., 2007).

Individuals actively opposed to strawberry guava biological control often expressed multiple rationales in addition to valuing the plant (Figure 12.3). Distrust of the science supporting biological control and of the government agencies conducting and reviewing the science were common underlying factors (Warner and Kinslow, 2013). Opponents typically expressed frustration at being left out of the decision-making process concerning biological control, starting with selection of

strawberry guava as a target for management. Many complained that they first heard of the proposed use of biological control and the problems caused by strawberry guava after biological control had already been in development for many years and approval of the release of an agent was imminent. Feeling powerless to influence events that seemed to be a foregone conclusion, some opponents became deeply inflamed, and their resentment over lack of engagement resonated with other anti-establishment sentiments (Warner and Kinslow, 2013). For these people, the fact that biological control is a permanent, irreversible, and relatively unbounded form of management became especially objectionable.

With the potential for passions to run high, the most active opponents of strawberry guava displayed remarkable savvy in tapping both emotional and rational arguments to garner support. Examples included a child holding a sign or writing a letter with a message such as "Don't take away my food." Often opponents' message points were poorly supported by facts, but had a strong emotional consistency, typically conveying fear, suspicion, and resentment (Warner and Kinslow, 2013). Such arguments tended to contrast sharply with the relatively dispassionate and factual language of scientists and agencies evaluating biological control. Emotional detachment, a usual characteristic of scientific analyses, in this case tended to widen the gulf between the two sides, adding to the distrust that many opponents felt toward the motives and wisdom of scientists and managers supporting biological control.

Supporters of biological control

Although a great majority of the public was silent and apparently uninterested in biological control of strawberry guava, there were substantial numbers of active supporters. Strong support for biological control came from Hawaii's conservation community, including land managers and biologists who were familiar with the ecological damage caused by strawberry guava and the extreme difficulty of managing it effectively over large areas of native forest. Advocates for release of the biological control agent included the Hawaii Conservation Council, Sierra Club, and, most significantly, the Hawaii Conservation Alliance (HCA), a consortium of public and private agencies working to preserve and restore native ecosystems (Conservation Council for Hawaii, 2010). Unanimity among the diverse members of HCA, ranging from The Nature Conservancy to the Office of Hawaiian Affairs, spoke to the depth of concern over strawberry guava and the support for use of biological control. Strong supporters also included ranchers who were familiar with successful use of biological control for other weeds and some agricultural specialists who were concerned about fruit flies and watershed protection (Conservation Council for Hawaii, 2010).

These various supporters were relatively quiet through the early development of biological control for strawberry guava, with only a few playing active roles in supporting or overseeing biological control research. However, when the proposed biological control came under vocal public criticism beginning in 2008, supporters rallied impressively. Their efforts resulted in a variety of communication products designed to reach the public in ways potentially more effective than the dryly scientific EA and release petition, including videos, blog discussions, online and printed testimonials, letters to the editor, and appearances at public meetings (Medeiros, 2008; Gon, 2009; Corrigan, 2009; Conservation Council for Hawaii, 2010). Through the coordinated efforts of specialists in invasive species control and public outreach, key messages were conveyed widely. By 2010, a broad coalition of active supporters were united behind the necessity of using biological control as a new tool for managing strawberry guava (Figure 12.3). Their many contributions strengthened the Final Environmental Assessment published in 2011, resulting in an evaluation of the proposed biological control release that was more comprehensive ecologically and socially (State of Hawaii, 2011).

Lessons learned about the process

No matter how thoroughly justified and well composed an Environmental Assessment, it will garner complaints from some segments of the population. In the face of opposition to biological control of strawberry guava, the best course of action appeared to be to adhere to a transparent, well established process for consulting public opinion and then provide the results to decision-makers. Designed for this purpose, the Hawaii EA process seemed to work well in the end. Improving the initial strawberry guava EA to meet Hawaii's regulatory laws better required substantial effort. The additional consultations

and documentation were much more comprehensive and inclusive than those for previous biological control EAs. It was critical to move beyond the central biological question of host specificity to address thoroughly the impacts of the invader and the expected results of biological control, considered as broadly as possible and from diverse viewpoints.

The work of professionals in EA writing and public outreach was tremendously important to the effort of improving the EA between 2008 and 2010, mainly by broadening and deepening consultations (State of Hawaii, 2010). Public meetings to share information and gather feedback involved teams of ten or more partners in each of four counties. Consultations with Hawaiian cultural practitioners by an experienced specialist were critical to completing the EA's cultural impact assessment. Dozens of scientists and managers from agencies across the state contributed expert opinions or analyses on economic, geographic, and ecological issues, for example interactions between strawberry guava and particular endangered species.

Although it can contribute importantly to sound decision-making, such a thorough EA process will add significantly to the time required to deliver biological control to managers. Our strawberry guava EA process lasted more than three years. In the absence of political and administrative complications that tended to delay research and writing, however, the Hawaii process might have taken only a few months to generate a comprehensive draft EA. The protracted course of the strawberry guava EA probably helped make it more inclusive and thorough, but many consultations could have taken place earlier, while the biological control agent was still in the final stages of testing. Accelerating a multi-disciplinary effort in this way, however, would require truly deft coordination of funds and expertise.

Conclusions

Given the divergent viewpoints on strawberry guava, conflict over biological control was probably unavoidable, and eruption of controversy in the media and local politics may have been inevitable. Release of *T. ovatus* was delayed at least three years as public concerns were addressed through outreach and the environmental assessment process. Surely these challenges could have been handled differently to lessen conflict; but how?

Topics of public concern around biological control of strawberry guava were not difficult to predict. Many of these concerns, such as questions over host specificity, would have applied to any biological control project, and none of the other objections to strawberry guava biological control were unanticipated (Figure 12.3). Proponents of biological control could understand and articulate most public concerns without conducting broad consultations with the public. However, it was the process of consultation itself that was a critical missing piece for many people skeptical of biological control (Warner and Kinslow, 2013). Just being able to hear about and discuss the issue of strawberry guava invasion and the use of biological control was important to many people.

Engagement with the public, especially when there is potential for significant conflict, needs to be broad and early. The case of strawberry guava underlines the desirability of a process to formally establish social consent for selection of a species as a target of biological control at the very beginning of a project (see Chapter 15). Two or three scientists making occasional presentations over a period of years to social organizations, horticultural fairs, and other public events clearly was too limited an outreach effort. Fortunately, the rise of internet-based media has opened a means to reach much larger audiences. Developing and deploying online communications early in a project is now essential. Actually engaging audiences in two-way discussions through social media presents an opportunity that deserves careful consideration. Even when time or institutional barriers prevent the scientists who are developing biological control from using such tools, the opportunity exists to share information with partners who do have time and expertise to discuss the goals and progress of biological control projects with online audiences. Widening the lines of communication gives everyone the opportunity to influence resource-management decisions, and deepening consultations in this way will ultimately strengthen the environmental assessment process.

In the United States and perhaps elsewhere, government-run and government-supported institutions, including those conducting biological control research and regulating its use, are widely viewed with mistrust, which seems unlikely to diminish in the foreseeable future. To overcome this challenge, biological control projects need strong support from independent groups that enjoy greater public trust. The swell of opposition to strawberry guava biological control in 2008 was most effectively countered when individuals and groups from outside the government, who could act as credible messengers, stepped forward in support of biological control (Gon, 2009; Corrigan, 2009; Conservation Council for Hawaii, 2010).

The potential to address issues of trust is an important reason for biological control programs to seek new avenues for public engagement (Warner and Kinslow, 2013). Hayes et al. (2008) described a specific process, based on facilitated face-to-face dialogues, for building trust among concerned groups with regard to biological control in New Zealand. They found that this form of public engagement involved substantial effort – more than initially expected – but that it might prove easier and cheaper in the long run than a less interactive approach. Intensive public engagement by biological control scientists and managers, to be manageable over the long term, certainly needs to be carefully targeted. In working toward conservation of native forests in Hawaii, it makes a great deal of sense to engage with selected small groups of pig hunters and Hawaiian cultural practitioners, citizens whose values are strongly tied to healthy forests but whose views sometimes diverge sharply from those of resource managers. Building relationships and understanding of different points of view can provide a means to get beyond disagreements and work collaboratively on shared interests.

Acknowledgments

Many individuals and agencies contributed to the development of strawberry guava biological control over the past 25 years, but particularly important was the work of Cliff Smith, J. H. Pedrosa Macedo, Marcelo Vitorino, Charles Wikler, and Robert Barreto in Brazil; Julie Denslow, Wendell Sato, Amanda Uowolo, Erin Raboin, and Nancy Chaney in Hawaii; and Jim Cuda and Frank Wessels in Florida. Working on outreach communications and public engagement were Pat Conant, Franny Kinslow, Anne Marie LaRosa, Evelyn Wight, Christy Martin, Jackie Kozak Thiel, Bob Masuda, and numerous other partners. Ron Terry expertly guided us through the environmental assessment process. My perspectives on dealing with conflict were influenced by the wisdom and aloha of Franny Kinslow,

Evelyn Wight, Keith Warner, Darcy Oishi, and others. I also thank Keith Warner, Bernd Blossey, David Wagner, and Dan Simberloff for reviews of Chapter 12.

References

Asner, G. P., R. F. Hughes, P. M. Vitousek, et al. 2008. Invasive plants alter 3-D structure of rainforests. *Proceedings of the National Academy of Sciences* **105**: 4519–4523.

Baker, P. J., P. G. Scowcroft, and J. J. Ewel. 2009. Koa (*Acacia koa*) ecology and silviculture. General Technical Report PSW-GTR-211. Pacific Southwest Research Station. USDA Forest Service. Albany, California.

Baldwin, P. H., C. W. Schwartz, and E. R. Schwartz. 1952. Life history and economic status of the mongoose in Hawaii. *Journal of Mammalogy* **33**: 335–356.

Burrows, C. P. M., C. L. Isaacs, and K. Maly. 2007. Pua'a (pigs) in Hawai'i, from traditional to modern. 7 pp. Fact sheet published by 'Ahahui Mālama I Ka Lōkahi, July 2007, Honolulu, Hawaii.

Conservation Council for Hawaii. 2010. Leveling the playing field in Hawaii's native forests: call to action. Available from: http://www.conservehi.org/documents/CCH_StrawberryGuava_ActionAlert.pdf [Accessed September 2014].

Corrigan, D. 2009. Video: Inside a strawberry guava thicket, Waiakea Forest Reserve, Hawaii. *Big Island Video News*. Available from: http://www.bigislandvideonews.com/2009/05/17/video-inside-a-strawberry-guava-thicket/ [Accessed September 2014].

County of Hawaii. 2009. Resolution No. 80 09, A resolution requesting a ban on the release of biological control agents on the island of Hawaii, including insects, fungi, bacteria, viruses, or other pathogens, for any tree species related to the ohi'a (*Metrosideros polymorpha*), including all species of the family Myrtaceae, such as the strawberry guava (*Spidium cattleianum*). Available from: http://records.co.hawaii.hi.us/Weblink8/0/doc/55424/Page1.aspx [Accessed September 2014].

County of Maui. 2009. Resolution No. 09-35, Supporting safe, effective biological control for Maui County's forest pests. Available from: http://www.co.maui.hi.us/documents/24/99/253/Reso%2009-035.PDF [Accessed September 2014].

Degener, O. 1939. *Flora Hawaiiensis: The New Illustrated Flora of the Hawaiian Islands*. Book 4, Family 273, Myrtaceae, *Psidium cattleianum* Sabine, 2 pp. (privately printed).

Dennill, G. B., D. Donnelly, K. Stewart, and F. A. C. Impson. 1999. Insect agents used for the biological control of Australian *Acacia* species and *Paraserianthes lophantha* (Fabaceae) in South Africa. *African Entomology Memoir* **1**: 45–54.

Diong, C. H. 1982. Population Biology and Management of the Feral Pig (*Sus scrofa*) in Kipahulu Valley, Maui. Ph.D. Dissertation, University of Hawaii, Honolulu, USA.

Dobbyn, P. 2003. Permit hurdles, litigation drag down plan to log koa in forest near Hilo. *Environment Hawai'i* **14**(6): 1–10.

Follett, P. A., J. J. Duan, R. H. Messing, and V. P. Jones. 2000. Parasitoid drift after biological control introductions: re-examining Pandora's box. *American Entomologist* **46**: 82–94.

Gon, S. O. 2009. Native Hawaiian forests vs. strawberry guava. Available from: http://www.webquest.hawaii.edu/kahihi/videos/NativeHawaiianForests.php?v=-pAh-At0HdM [Accessed September 2014].

Hawaii Department of Agriculture. 2010. Biological control in Hawaii: Working with nature to find a sustainable solution. 2 pp. Available from: http://hdoa.hawaii.gov/pi/files/2013/01/Biocontrol-Flier-4.27.10-cmps.pdf [Accessed September 2014].

Hawaii Ecosystems at Risk. 2011. Strawberry Guava Biological control: Restoring natural balance to Hawaii's forests and watersheds with the help of a bug. Available from: http://www.hear.org/strawberryguavabiocontrol/ [Accessed September 2014].

Hawaii State Legislature. 2009. Senate Resolution No. 108. Requesting a moratorium on the release of biological control agents for the environmental management of plant species that also serve as food resources. Available from: http://www.capitol.hawaii.gov/Archives/measure_indiv_Archives.aspx?billtype=SR&billnumber=108&year=2009 [Accessed September 2014].

Hayes, L. M., C. Horn, and P. O. B. Lyver. 2008. Avoiding tears before bedtime: how biological control researchers could undertake better dialogue with their communities, pp. 376–382. *In*: Julien, M. H., R. Sforza, M. C. Bon, et al. (eds.). *Proceedings of the XII International Symposium on Biological Control of Weeds*, La Grande Motte, France, 22–27 April 2007. CAB International. Wallingford, UK.

Hodges, C. S. 1988. Preliminary exploration for potential biological control agents for *Psidium cattleianum*. Technical Report 66. Cooperative National Park Resources Studies Unit. University of Hawai'i. Manoa, Hawaii, USA.

Huenneke, L. F. and P. M. Vitousek. 1990. Seedling and clonal recruitment of the invasive tree *Psidium cattleianum*: Implications for management of native Hawaiian forests. *Biological Conservation* **53**: 199–211.

IUCN. 2013. 100 of the world's worst invasive alien species, Global Invasive Species Database managed by the Invasive Species Specialist Group (ISSG). Available from: http://www.issg.org/database/species/search.asp?st=100ss&fr=1&str=&lang=EN [Accessed September 2014].

Johnson, M. T. 2005. Petition for field release of *Tectococcus ovatus* (Homoptera: Eriococcidae) for classical biological control of strawberry guava, *Psidium cattleianum* Sabine (Myrtaceae), in Hawaii, 31 pp. Submitted to Hawaii Department of Agriculture Pest Quarantine Branch, May 10 2005. Available from: http://www.fs.fed.us/psw/topics/ecosystem_processes/tropical/invasive/Petition%20for%20release%20of%20Tectococcus%20may05.pdf [Accessed September 2014].

Krauss, F. and D. C. Duffy. 2010. A successful model from Hawaii for rapid response to invasive species. *Journal for Nature Conservation* **18**: 135–141.

Lorence, D. H. and R. W. Sussman. 1986. Exotic species invasion into Mauritius wet forest remnants. *Journal of Tropical Ecology* **2**: 147–162.

Medeiros, A. C. 2008. Importing safe insects the only hope of saving Maui's native koa forests. *The Maui News,* November **8**, 2008.

Mitchell, C., C. Ogura, D. W. Meadows, A. Kane, L. Strommer, S. Fretz, D. Leonard, and A. McClung. 2005. Hawaii's Comprehensive Wildlife Conservation Strategy. Department of Land and Natural Resources. Honolulu, Hawai'i.

Office of Environmental Quality Control. 2004. *The Environmental Guidebook: A Guidebook for the Hawaii State Environmental Review Process.* State of Hawaii, Department of Health, Honolulu, Hawaii. 133 pp.

Office of Environmental Quality Control. 2008. *The Environmental Notice,* April 23, 2008, pp. 11–12. Available from: http://oeqc.doh.hawaii.gov/Shared%20Documents/Environmental_Notice/Archives/2000s/2008_Env_Notice/2008-04-23.pdf [Accessed September 2014].

Office of Planning. 2012. Increased Food Security and Food Self-Sufficiency Strategy: A State Strategic/Functional Plan Prepared in Accordance with HRS Chapter 226 Hawaii State Plan and the Hawaii Comprehensive Economic Development Strategy. Department of Business Economic Development & Tourism in cooperation with the Department of Agriculture. State of Hawaii.

Reimer, N. J. 2002. Review and permit process for biological control releases in Hawaii, pp. 86–90. *In*: Smith, C. W., J. Denslow, and S. Hight (eds.). *Proceedings of Workshop on Biological Control of Native Ecosystems in Hawaii.* Pacific Cooperative Studies Unit, University of Hawaii at Manoa, Department of Botany, Technical Report 129, Honolulu, Hawaii, USA. 122 pp. Available from: http://www.hear.org/pcsu/techreports/129/ [Accessed January 2016].

Space, J. C. 2013. USDA Forest Service, Pacific Island Ecosystems at Risk (PIER). Available from: http://www.hear.org/pier/ [Accessed January 2016].

Stanley, M. C and S. V. Fowler. 2004. Conflicts of interests associated with the biological control of weeds, pp. 322–340. *In*: Cullen, J. M., D. T. Briese, D. J. Kriticos, et al. (eds.) *Proceedings of the XI International Symposium on Biological Control of Weeds, Canberra, Australia, 27 April–2 May 2003.* CSIRO Entomology, Canberra, Australia.

State of Hawaii. 2010. Draft Environmental Assessment: biological control of strawberry guava by its natural control agent for preservation of native forests in the Hawaiian Islands. 131 pp. Available from: http://oeqc.doh.hawaii.gov/Shared%20Documents/EA_and_EIS_Online_Library/Statewide/2010s/2010-06-23-ST-DEA-Biocontrol-Strawberry-Guava.pdf [Accessed September 2014].

State of Hawaii. 2011. Final Environmental Assessment: Biological control of strawberry guava by its natural control agent for preservation of native forests in the Hawaiian Islands. 712 pp. Available from: http://oeqc.

doh.hawaii.gov/Shared%20Documents/EA_and_EIS_Online_Library/Statewide/2010s/2011-11-08-FEA-Biological control-Strawberry-Guava.pdf [Accessed September 2014].

Takahashi, M., T. W. Giambelluca, R. G. Mudd, et al. 2011. Rainfall partitioning and cloud water interception in native forest and invaded forest in Hawai'i Volcanoes National Park. *Hydrological Processes* **25**: 448–464.

Trujillo, E. E. 2005. History and success of plant pathogens for biological control of introduced weeds in Hawaii. *Biological Control* **33**: 113–122.

Tummons, P. 2008. Controversy flares over proposal to control waiawi with scale insect. *Environment Hawai'i* **19**(1): 1, 8–9, July 2008.

Tunison, J. T. 1991. Element stewardship abstract for *Psidium cattleianum*. The Nature Conservancy. Available from: http://wiki.bugwood.org/Psidium_cattleianum [Accessed September 2014].

Turner, C. E. 1985. Conflicting interests and biological control of weeds, pp. 203–225. *In*: Delfosse, E. S. (ed.). *Proceedings of the VI International Symposium on Biological Control of Weeds, Vancouver, Canada, August 1984.* Agriculture Canada. Ottawa, Canada.

Uowolo, A. A. and J. S. Denslow. 2008. Characteristics of the *Psidium cattleianum* Sabine (Myrtaceae) seed bank in Hawaiian lowland forests. *Pacific Science* **62**: 129–135.

USDA. 2008. Field release of *Tectococcus ovatus* (Homoptera: Eriococcidae) for biological control of strawberry guava, *Psidium cattleianum* Sabine (Myrtaceae), in Hawai'i. Environmental Assessment, March **7**, 2008. Available from: http://www.aphis.usda.gov/plant_health/ea/downloads/tectococcus_ovatus.pdf [Accessed September 2014].

USDA Forest Service. 2013a. Frequently asked questions, biological control. Available from: http://www.fs.fed.us/psw/topics/biological control/strawberryguava/biological control.shtml [Accessed September 2014].

USDA Forest Service. 2013b. Frequently asked questions, strawberry guava. Available from: http://www.fs.fed.us/psw/topics/biological control/strawberryguava/strawberry_guava.shtml [Accessed September 2014].

Vargas, R. I., J. D. Stark, and T. Nishida. 1989. Abundance, distribution, and dispersion indices of the oriental fruit fly and melon fly on Kauai, Hawaiian Islands. *Journal of Economic Entomology* **82**: 1609–1615.

Vargas, R. I., J. D. Stark, and T. Nishida. 1990. Population dynamics, habitat preference, and seasonal distribution patterns of oriental fruit fly and melon fly in an agricultural area. *Environmental Entomology* **19**: 1820–1828.

Vargas, R. I., L. Whitehand, W. A. Walsh, et al. 1995. Aerial releases of sterile Mediterranean fruit fly (Diptera: Tephritidae) by helicopter: dispersal, recovery and population suppression. *Journal of Economic Entomology* **88**: 1279–1287.

Vitorino, M. D., J. H. Pedrosa-Macedo, and C. W. Smith. 2000. The biology of *Tectococcus ovatus* Hempel (Heteroptera: Eriococcidae) and its potential as a biological control agent of *Psidium cattleianum* (Myrtaceae), pp. 651–657. *In*: Spencer,

N. R. (ed.). *Proceedings of the X International Symposium on Biological Control of Weeds, Bozeman, Montana, 4–14 July 1999.* Montana State University. Bozeman, Montana, USA.

Wagner, W. L., D. R. Herbst, and S. H. Sohmer. 1990. *Manual of the Flowering Plants of Hawai'i.* University of Hawaii Press. Honolulu, Hawaii, USA.

Warner, K. D. and F. Kinslow. 2013. Manipulating risk communication: value predispositions shape public understandings of invasive species science in Hawaii. *Public Understanding of Science* **22**: 203–218.

Wessels, F. W., J. P. Cuda, M. T. Johnson, and J. H. Pedrosa-Macedo. 2007. Host specificity of *Tectococcus ovatus* (Hemiptera: Eriococcidae), a potential biological control agent of the invasive strawberry guava, *Psidium cattleianum* (Myrtales: Myrtaceae), in Florida. *Biological Control* **52**: 439–449.

Zimmerman N., R. F. Hughes, S. Cordell, et al. 2008. Patterns of primary succession of native and introduced plants in lowland wet forests in eastern Hawai'i. *Biotropica* **40**: 277–284.

CHAPTER 13

An ethical framework for integrating biological control into conservation practice

Keith D. Warner

Center for Science, Technology, and Society, Santa Clara University, USA

Why integration of biological control into conservation practice is an ethical issue

The introduction of an organism to a new ecosystem has the potential to be simultaneously an ecologically and ethically consequential act. Deliberate introductions of exotic species have contributed greatly to human welfare, being the basis of virtually all agriculture and greatly expanding local options for forestry and fisheries, and useful in restoring ecosystem integrity (e.g., *Cactoblastis cactorum* [Bergroth]) in Australia [Dodd, 1940]). Accidental species introductions, in contrast, have generally been neutral or trivial in their consequences. However, in a few cases, novel organisms, either accidental or deliberate in origin, have caused great harm to native species diversity and ecosystem function (Perrings et al., 2010). The introduction of a novel organism can be consistent with conservation principles if it is for a conservation purpose, such as the protection of endangered species or ecosystem function. The release of a novel biological control agent into wildlands is a special category of introduction that should be justified by social values and conservation goals. Today, in most cases, a proposed agent is scientifically evaluated, reviewed by scientists working for regulatory agencies, and approved by public authorities (see Chapter 11). However, if the released organism successfully establishes, as intended, it might potentially have unanticipated effects on the novel ecosystem. This chapter will explain why the introduction of a biological control agent is necessarily ethical in character, describe how environmental ethics evolved to encompass biological control as an ethical issue, analyze decision-making processes that are based on consideration of risk, and propose an ethical framework for more effectively integrating ethics in biological control decision-making.

The field of environmental ethics emerged to provide guidance to people seeking to protect the Earth and its resources. In the 1980s, conservationists began to raise alarms about the extent of harm to biodiversity caused by certain introduced, non-native species (Drake et al., 1989). Some conservationists extended their concern to include classical biological control agents (Simberloff et al., 2005). Ensuing controversies touched on issues of evidence and statistical probabilities, but also regulation, oversight, ethics, and social values.

Criticisms of biological control practice that emerged around 1985 can be explained to a considerable degree by the progressive forced merging of two paradigms that had previously been somewhat distinct. Biological control science grew out of the need to provide practical and economic tools to control non-native organisms causing economic or social harm (Sawyer, 1996). Conservation biology emerged as a specialized subfield of ecology intended to promote environmental conservation and management (Takacs, 1996). Formerly, these two communities of scientific practice worked autonomously in different landscapes. With the rise of the invasive species as a prominent conservation biology concern, however, some branches

Integrating Biological Control into Conservation Practice, First Edition. Edited by Roy G. Van Driesche, Daniel Simberloff, Bernd Blossey, Charlotte Causton, Mark S. Hoddle, David L. Wagner, Christian O. Marks, Kevin M. Heinz, and Keith D. Warner.
© 2016 John Wiley & Sons, Ltd. Published 2016 by John Wiley & Sons, Ltd.

of these groups in some places came into conflict. These groups share many common elements, including use of the scientific method, reliance on data, and biological concepts and theories. The critics of biological control practice, generally people aligned with conservation biology, have argued against the release of biological control agents on the grounds that they pose unwarranted risks, are irrevocable, sometimes ill-considered, or hasty in design (Howarth, 1991; Lockwood, 1993, 1996, 1997; Lockwood et al., 2001; Simberloff and Stiling, 1996a, b; Wajnberg et al., 2001). Subsequently, biological control scientists (among others) have argued that the spread of invasive species and the resultant economic harm and ecological disruption they cause is so great that humans are obligated to take action, including through biological control when appropriate (Hoddle, 2002, 2004; Delfosse, 2005; Messing and Wright, 2006). Some very vocal conservation scientists have argued for greater restraint in the use of classical biological control agents, while some biological control scientists have argued in favor of human intervention using biological control agents. As will be discussed below, most research scientists, practitioners, land managers, and regulatory scientists, of course, hold a relatively pragmatic view of biological control introductions, especially those who have landscape management responsibilities. However, vehement disagreements between some representatives of these schools of thought have created a cloud of uncertainty, making the practice of biological control more difficult and costly. These disagreements have been most vigorous between conservation biologists and biological control scientists in the United States, but this cloud of uncertainty has touched most other countries considering such introductions. The ongoing integration of biological control into conservation practice will require careful scientific work but also the development of a negotiated consensus on the conditions under which it is justified (see Chapter 12 for a case study example of conflicts). These points suggest there is a need for ethical analysis of this issue.

Conservation biology and biological control are both value laden. All scientific practice that seeks a social or environmental goal is necessarily ethically laden, meaning that it is linked to a social goal (Shrader-Frechette and McCoy, 1993). The emergence of conservation biology as a field has been explicitly guided by environmental ethics toward the protection of biodiversity (Meffe and Carroll, 1997). Conservation biology in its origin and by its orientation is an ethically laden science, but so too is the practice of biological control, for both seek to protect environmental resources for society. Both claim to act in the public interest, on behalf of the public, and thus to be public interest science.

Public interest science is evaluated on the basis of good science but also by its practical accomplishments. Raffensperger et al. (1999) proposed three criteria to define public interest science: (1) research is conducted with input from or in collaboration with the public or an active citizenry, (2) research products are made freely available (not proprietary or patented), and (3) the primary, direct beneficiaries are society as a whole or specific groups unable to carry out research on their own behalf.

By this definition, public interest science necessarily relies upon some expression of public consent and results in a non-commodity product (Warner et al., 2011). Raffensperger et al.'s (1999) definition was originally created for agricultural, rather than environmental sciences, but its key points are nonetheless helpful for the purposes of this chapter. In the case of private science, individuals or companies identify the question to be investigated and pay for research to be done, and the benefits are accrued by the paying client. In contrast, public interest scientists have the challenge of determining a beneficiary, or client, who stands to benefit. This chapter presents the unusual case of two scientific communities who understand their work to be done in the public interest, but who have outspoken members who have clashed publicly about scientific practice, at times vehemently. Devising appropriate forms of public consultation and input is critical to the development of public interest science projects (Hayes et al., 2008; Warner et al., 2008; Warner, 2012), as is their ethical analysis.

The proposal to introduce a biological control agent is evaluated by government officials using potentially conflicting criteria: societies want the environmental and economic benefits of invasive species control, but wish to avoid risk to native biodiversity and beneficial species from introduced agents. Classical biological control agent permitting is caught up in this broader social ambivalence. Better data can address this, but cannot fully resolve it, because different social groups bring different values and assumptions to their understanding of the environment. For example, when invasive species control is not generally supported by public opinion, any management effort might conflict with democratic values.

Better science alone cannot resolve conflicts between values, and divergent social values are part of the cloud of uncertainty. Divergent assumptions have played an important role in these controversies. Here again, tools from ethics can be helpful. Ethics is one of three major branches of philosophy, along with metaphysics (the study of the fundamental nature of reality, beyond the empirical) and epistemology (the study of the nature of knowledge; the philosophy of science is in this branch) (Willot and Schmidtz, 2001). Ethics is a field of academic inquiry that investigates the nature of right and wrong, how to live a good life, and how to make good decisions. Unlike natural science, ethics does not rely primarily on empirical quantification (although it does use evidence). Instead, ethics describes the nature of goodness, what constitutes good action, and good human decision-making. Like the natural sciences, the field of ethics is not static. As societal values change, so too ethical perspectives evolve. Concepts from ethics can help analyze the composition and structure of human values, and content of moral arguments, and how these shape human behavior. The field of ethics relies heavily on language skills to analyze social values. To make an argument based chiefly on facts suggests one is operating in the sciences, but when one argues on the basis of social or environmental values, one has entered the realm of ethics. Scientists make empirical or evidentiary claims, and these are evaluated by peer review. However, if one makes an argument that a person or an organization "should" or "ought to" do something – whether a normative principle is explicit or implicit – one has made an ethical claim. Asserting that ethical principles should favor the conservation of species is not terribly difficult, but given the complex interplay of factors that threaten biodiversity and ecosystem integrity, ethical principles by themselves offer little guidance about specific management options. To navigate this complexity, dynamic decision-making processes are needed, informed by the threats posed by the invasive species, but then capable of weighing the risks of all management alternatives, including biological control. Environmental ethics can help design the structure and orientation of these processes, but does not offer recommendations for specific actions.

The halting yet progressive integration of two previously distinct paradigms – classical biological control and conservation biology – can be studied using the tools of descriptive and practical ethics. Paradigms can be subjected to ethical analysis because as mental models of how the world works, scientific paradigms orient people and society toward certain kinds of actions and decision-making. Descriptive ethics studies the opinions or beliefs that people have about the nature of right and wrong. The second part of this chapter will use descriptive ethics tools to explain how some conservation scientists mounted an ethical critique of biological control practice. This chapter will also reference the ethical and scientific views reported by biological control scientists and the critics of this practice, drawn from 183 personal interviews from 4 countries.

Criticisms and defenses of biological control practice have been made on the grounds of practical ethics. Practical ethics focuses on specific examples of how humans apply – or could apply – ethical values in their decision-making. Practical ethics can be subdivided into social ethics (how society makes decisions), professional ethics (e.g., how doctors, lawyers, scientists, and business leaders make decisions), and environmental ethics. Democratic participation is one example of a social ethic. Critics of biological control practice drew on professional and environmental ethics (Howarth, 1991; Miller and Aplet, 1993; Lockwood, 1996; Lockwood et al., 2001). Society has conferred significant decision-making authority on the professions because their members develop specialized, expert knowledge (which makes evaluations by nonmembers difficult). In exchange for significant decision-making autonomy, society expects professionals to internalize a code of ethics of service to society, above and beyond financial compensation (Larson, 1977). A medical analogy illustrates this tradeoff. Society accords medical doctors the authority to prescribe medications and assumes they will do so in patients' best interest. However, if a doctor were to own stock in a pharmaceutical company that produced medicines he or she prescribed, this would call into question whether the doctor could always hold a patient's best interest paramount. In this case, the doctor has to weigh his or her professional responsibilities to patients against personal (financial) gain. In the field of ethics, this situation is known as a "conflict of interest." This does not mean that a professional has acted improperly or unethically; it merely means that a professional is faced with a decision in which he or she might have to choose between personal gain and professional duties. Recognizing that some criticism of biological control practice has been made on the basis of an ethical conflict of interest (professional or scientific) is important for understanding these debates.

Part three of this chapter describes the ethical dimensions of the biological control permit and regulatory systems of several countries. It draws on the ethical principle of prudence to critique the limitations of the risk-assessment framework in the United States and other countries. Part four draws from the field of ethics to propose an idealized framework for making more integrated decisions about biological control introductions.

The fundamental question of environmental ethics focuses on how we should value nature. Environmental ethics extends moral significance beyond human beings to include other forms of life and nature (Willot and Schmidtz, 2001). Someone or something has moral significance when it has value, is considered to be worthy of moral consideration, or is included in ethical decision-making. For example, all societies have prohibitions against killing fellow human beings. However, a person who considers the killing of wild animals to be wrong has extended the principle of moral significance to wild animals, and thus asserts them worthy of consideration in ethical decision-making (DesJardins, 2006). While many members of the public support the conservation of charismatic megafauna, few members of the public accord moral significance to insect species, even those that are endangered (Samways, 2007). A primary task, therefore, of environmental ethics is to determine what duties human beings have to which aspects of the environment and why.

How environmental thought evolved to encompass classical biological control as an ethical issue

Over the past 50 years, environmental thought has evolved quickly and expansively. This is not to say that all human attitudes and behavior toward the environment have undergone transformation, but rather that the environmental implications of many human behaviors have entered the awareness of most well-educated people. Five decades ago, few people considered the environmental consequences of their food and fiber choices; now many do. Rachel Carson (1962)'s *Silent Spring* is among the most important books to shape

environmental thought. Her critiques of the devastating impact of pesticide pollution on wildlife, and more broadly of the science, technology, and public policy that facilitated this widespread use of toxins, had broad impacts on American thought and institutions for a generation (Perkins, 1982; Andrews, 1999; Warner, 2007). Carson's work prompted millions of people around the world to think about nature and wildlife in more explicitly ethical ways. She challenged people to consider their pest management choices in light of their responsibilities to future generations. Although it was written in a popular style, *Silent Spring* presented wildlife and nature as morally significant. Carson pointed to the "extraordinary array of alternatives," based on ecological science. "Much of the necessary knowledge [about alternatives] is now available but we do not use it" (Carson, 1962: p. 11). In the ultimate chapter of *Silent Spring*, Carson presented biological control and other forms of biologically based pest control as socially and ethically preferable. Many biological control scientists were inspired to join this field by Carson's work (Perkins, 1982; Palladino, 1996; Warner, 2007; Warner et al., 2011).

The most influential arguments for expanding ethical concerns to include the environment have not been made by formally trained ethicists, but rather deeply passionate individuals. In the United States, the contribution of Aldo Leopold (1887–1948) to environmental thought has profoundly shaped the ethical thought of conservation leaders through his articulation of a "land ethic," the principle that human beings have duties to other forms of life, to wilderness, the land, and ecosystems (Leopold, 1949). He wrote: "A thing is right when it tends to preserve the integrity, stability, and beauty of the biotic community. It is wrong when it tends otherwise" (p. 262). Thus, Leopold extended human duties not only to individual organisms, but to ecological communities and the functioning of their ecosystems. Quite simply, no one had expressed environmental ethical thought in these terms before Leopold. Carson's work focused public attention on pesticide alternatives and the need for environmental policy, while Leopold's land ethic profoundly shaped conservation thought.

The earliest expressions of concern about negative consequences of biological control organisms appeared in a letter from a US entomologist regarding a program in Hawaii in 1899, and the need to monitor its effects (cited in Howarth [1991]). The rise of DDT and synthetic insecticides in the post-World War II era led to marked decline in biological control research (Gurr et al., 2000). The publication of *Silent Spring* and the rise of the environmental movement prompted a renewed interest in biological control research and its application (Warner et al., 2011). In Australia, the problems caused by the introduction of the cane toad (*Rhinella marina* [L.]) raised the public profile of species introductions. Conflicts over biological control releases between different interest groups emerged in the 1980s, first over the potential release of phytophagous insects. In eastern Australia, pastoralists requested work be done to control the invasive plant *Echium plantagineum* L., which they named "Paterson's Curse" because it is toxic to cattle. Beekeepers had named the same plant "Salvation Jane" because it survived drought in some spots and provided a key source of nectar and pollen for bees. The conflict between these economic interests led Australia to pass the first (and still the only) dedicated Biological Control Act in 1984 (Cullen and Delfosse, 1985). In California, biological control researchers raised questions with other members of the biological control community about the risks that some phytophagous biological control agents posed to native flora, but these concerns were scarcely noted outside of this research community (Andres, 1981, 1999; Pemberton, 1985a, b; Stanley and Fowler, 2004). It is important here to distinguish between different meanings of "conflict of interest." The field of ethics uses the term "conflict of interest" in a narrow, technical sense to describe the situation in which a professional person has divergent personal incentives and professional duties. However, the biological control community generally uses it to describe conflicts between social interests or between economic interests and social values (Andres, 1981; Bennett, 1985; Sheppard et al., 2006).

Also during the 1980s, conservation scientists began to articulate the threat that non-native, introduced species posed to ecosystem function and threatened and endangered species. The Scientific Committee on Problems of the Environment (SCOPE) organized international conferences that facilitated shared understanding of the seriousness of this problem (Drake et al., 1989). Howarth's (1991) article was the first widely cited scientific publication to argue that the risks of a biological control agent introduction should be evaluated as any other species introduction. His paper caught the attention of conservation scientists, and subsequently several other papers expanded the scope of inquiry to other ecological systems.

Howarth advanced four categories of criticism against biological control based on environmental and professional ethics. These criticisms were that (1) native biodiversity should be protected from introduced control agents ("biological control introductions are part of the much larger problem of the invasion of new areas by alien species, which are recognized as a major factor in species extinctions," p. 486) (but see Chapter 7); (2) biological control practitioners should recognize their ethical responsibilities ("individuals and agencies who attempt to introduce an organism into a new land undertake a grave responsibility," p. 500); (3) scientific practice should be improved: pre- and post-release studies should be mandated, and "… the goal is to make applied ecology a predictive science" (p. 501); and (4) policy should be improved ("the environmental risks associated with the introduction of alien organisms for biological control are sufficient to justify the creation of legal safeguards. Releases should only be made in the public interest after the ecological consequences are considered," p. 500). Howarth's arguments did not attempt to provide a balanced analysis of all the different types of risks posed by species introductions (e.g., from horticultural plant and pet introductions, international agricultural and aquacultural commodity trade, etc.). Rather, he argued that the risks of biological control introductions should be considered in a common framework of species introductions.

Howarth thus argued that conservation should be an ethical norm guiding biological control scientists and their practice, and that greater scrutiny of proposed biological control agents should be conducted by agencies on behalf of the public interest. These four categories of critique persisted and framed the criticism of biological control introductions in four other key papers. Howarth (1991) was cited by 288 articles 1992–2010, with attention peaking in 2005. Two articles further developed elements of Howarth's critique of biological control practice and policy (Lockwood, 1993; Louda et al., 1997), and two articles extended the scope of concern with additional cases and policy recommendations (Simberloff and Stiling, 1996a; Strong and Pemberton, 2000). These five are the most cited articles in this subject area. A total of 609 unique references cite one or more of these five key critical articles, and these 609 articles constitute the core body of the scientific

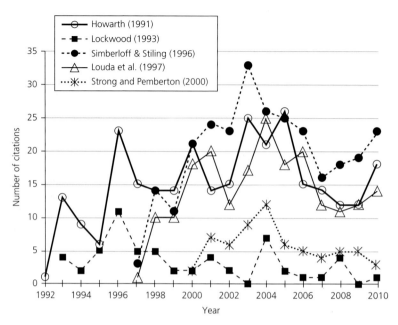

Figure 13.1 Distribution by year of citations of five major articles in the biological control controversy literature. Unpublished data of Keith Warner.

literature (peer-reviewed, scientific journal) on this controversy. These five principal articles arguing for the existence of risk from biological control introductions (Figure 13.1) showed their highest levels of citation between 2002 and 2006, with a subsequent decrease in citation. All the authors and co-authors of these five papers were based in the United States, and a majority of the 609 articles that cited these core articles were written by lead authors based at US institutions. This can be partially explained by relatively larger budgets for biological control research in the United States. Furthermore, disputes between conservation biologists and biological control scientists in the United States appear to have been particularly pointed.

Lockwood (1993) critiqued a USDA proposal to release exotic natural enemies to control native grasshoppers on the rangeland of the United States. He found that proposed action to be ethically unsupportable because it would establish a population of exotic natural enemies that had the potential to drive a native species to extinction. He went on to develop a substantive set of ethical concerns about biological control introductions, explicitly referencing environmental ethics (Lockwood, 1996, 1997). He was the first to discuss in the entomological literature the need to link environmental ethics

(concern for native species) with the issues posed by biological control. Lockwood ethically critiqued the establishment of a population of exotic natural enemies, the precise feature of classical biological control that makes it sustainable over time. This was followed by the discovery of an introduced biological control agent affecting the reproduction of native, non-target plants in a nature reserve and a national park (Louda et al., 1997) and a literature review of non-target impacts of biological control agents, which concluded with explicit recommendations for reform of research and regulatory agency practices (Simberloff and Stiling, 1996a). Strong and Pemberton (2000) outlined specific policy reforms to increase the safety of biological control introductions and to preserve the use of classical biological control as a management tool. These recommended policy reforms built on the risk assessment methods developed previously by weed biological control scientists.

Disputes over the interpretation of evidence, the definition of adverse effects, the characterization of the risks of biological control, and the responsibilities of scientists and regulators have taken place at meetings of the Entomological Society of America (Follett and Duan, 2000), at special meetings sponsored by the US Department of Agriculture (Carruthers and D'Antonio, 2005), at

International Congresses of Entomology (Lockwood et al., 2001), at special meetings of the International Organization of Biological Control (Wajnberg et al., 2001), and at quadrennial international scientific symposia dedicated to the biological control of weeds (12 between 1969 and 2013) and arthropod pests (4 between 2002 and 2013). In response to criticism, biological control scientists have argued that they are fulfilling their professional responsibility as scientists (Carruthers and Onsager, 1993; Carruthers, 2004). Some scientists have responded that critics are applying anachronistic ethical criteria to prior decisions (Follett and Duan, 2000), that modern testing protocols adequately predict risk (Messing and Wright, 2006), and that generalized claims of the risk of biological control introductions confuse the public (Delfosse, 2004). Biological control scientists and their critics agree that the review process for proposed agent introduction became more rigorous in most countries in the 1990s. Weed biological control permit applications have historically been subject to scrutiny because of the potential risk to economic plants (Sheppard et al., 2003), and thus the review process became incrementally more rigorous. Aware of the threat posed to their science by poor practice of any one of their members, weed biological control scientists have discussed, debated, and broadly agreed to an international code of best practices (Balciunas and Coombs, 2004). Historically, public agency review of agents targeting arthropods was less rigorous, but now proposed introductions are evaluated with much more rigor, at least in Australia, New Zealand, and South Africa (Hunt et al., 2008). Delfosse (2005) and Messing and Wright (2006) argued that an unintended consequence of misplaced concerns over risk will be regulatory gridlock, potentially resulting in greater spread of invasive species and their damage, as well as additional pesticide use. Biological control scientists in the United States report that controversies over perception of levels of risk have made federal permit applications for the introduction of classical biological control agents more difficult, time consuming, and costly since 2001.

Risk, precaution, prudence, and policy

Most deliberations over the proposed introduction of biological control agents have been in terms of risk assessment. The standard expert conceptualization of risk is the statistical probability of an adverse event that can be objectively quantified by risk-assessment process (National Research Council, 1996), usually expressed in probabilistic terms, such as "risk = hazard X exposure" (Delfosse, 2005). In the United States and most other countries, risk assessment is the primary environmental policy decision-making paradigm (National Research Council, 1983). Plant quarantine was an early, crude form of risk assessment. Similarly, antecedents of modern risk assessment were practiced by most weed biological control scientists throughout the twentieth century. Public agencies and government officials began to apply formalized risk assessment to biological control agent decision-making in the early 1990s (Coulson and Soper, 1989; Coulson, 1992, 1999; Coulson et al., 2000). It is important to recognize that regulatory systems in different countries conceptualize risk assessment differently (Jasanoff, 2005). These differences in regulatory systems provide meaningful national-scale cases for comparative analysis and also reveal how social priorities differ among nations.

The ethical justification for public oversight is rooted in the ethic of prudence, or the ability to anticipate the likely future outcomes of present decisions and make wise choices in light of this foresight. The concerns of conservation biologists and environmental ethicists about the potential harms caused by the introduction of any novel species, including a biological control agent, are consistent with the ethical value of prudence. When successful, the establishment of an agent is effectively irreversible; it cannot realistically be eradicated. Some of the critics of biological control introductions have raised concerns about the potential of introduced agents to shift hosts through evolution to non-target organisms (Louda et al., 1997). Although expansion of fundamental host ranges through evolution must occur in nature, it is rarely observed over historical periods of time, and in the literature this concept is extensively confused with the ecological process of adding new hosts as they are contacted, but which involves no change in the fundamental host range (Sheppard et al., 2005). The threat that a novel control agent introduction could pose to already threatened or endangered species requires careful consideration. An action that is irreversible and potentially harmful is a special ethical case, meaning that extra scrutiny and careful deliberation are required to make a good decision. Exposing threatened or endangered species to further harm that results in their population decline is ethically unjustifiable. Thus, a prudent approach to reviewing a proposed introduction is merited.

To better inform environmental policy decisions, Jonas (1985) proposed that environmental decisions should be guided by the German concept of *vorsogeprinzip*, which has no single equivalent term in English. It has been translated by some as "precautionary principle," and this term has been used by policymakers (Harremoës et al., 2002; Whiteside, 2006). It is better translated as "foresight principle," or "responsibility principle," but even these phrases do not convey the word's full meaning. *Vorsogeprizip* conveys an aspect of foresight, such as taking prudent action in anticipation of a winter storm. In a traditional sense, it conveys a sense of restraint (e.g., do not travel during a winter storm) but also a sense of obligation to act (e.g., stockpile food and fuel). When applied to environmental action and policy, it could be used to justify the reduction of pesticide use for better biological control. It could be used to justify no action, but it could also be used to assert an obligation to act, such as to introduce an effective biological control agent to control a highly damaging invasive species.

Jonas (1985) asserted that human society now produces novel technologies that cause irreversible harm to human health and the environment, and he argued that this demands more sustained ethical consideration from all stakeholders, including both those who benefit and those who are harmed by these technologies. He perceived a gap between our technological prowess and our capacity for exercising moral responsibility towards other forms of life and future generations. He critiqued conventional, inherited understandings of ethics as unable to provide society with the guidance necessary to make adequate decisions regarding the use of such technologies. Jonas (1985) argued that decision-making in relation to potentially catastrophic environmental risks carries with it a special moral responsibility, for which only an ethical principle, not a utilitarian balancing of the interests of the many versus the few, is appropriate for decision-making. Policymakers draw on the precautionary principle to address novel hazardous technological products and waste; it was not developed with management of invasive species as its goal. The dynamic nature of problems posed by invasive species mean that any policy or practice – including the application of the precautionary principle – that has embedded in it assumptions of a static natural world will likely be ineffective (National Research Council, 2002; Perrings et al., 2010). A dynamic decision-making structure that allows for consideration of the threat posed by the invasive species first and then considers the risks of all management alternatives, including biological control, would serve society better than merely looking at the risks of a biological control introduction in isolation from the larger context.

From an ethics perspective, "risk assessment" as applied in most regulatory systems has merit, but is problematic because in practice, the process generally frames issues too narrowly (Slovic, 2001; Slovic et al., 2004; Delfosse, 2005; Messing and Wright, 2006). In the case of a proposed biological control agent introduction, alternatives are not effectively considered, and analysis excludes consideration of benefits of any proposed introduction. In the United States, applicants for a permit to introduce an agent must, in effect, demonstrate that the introduction poses no meaningful risk of harm to the environment. Thus, the dynamics of invasive species progressively disrupting ecosystems or threatening endangered species are effectively excluded from decision-making processes, which should be a concern for conservation biology. The result is a partial, selective, or biased assessment of the proposed action. Without any means for considering the desired conservation benefits of a proposed agent introduction, the structure of decision-making is biased.

The USDA Agricultural Plant Health Inspection Service – Plant Protection & Quarantine (APHIS-PPQ) – is the federal agency charged with reviewing permit applications for the introduction of most novel organisms to the United States. Before 1993, permit issuance for biological control agents targeting plants was relatively straightforward and uncontested, and introducing agents targeting arthropods did not require formal review. To coordinate the requirements of the National Environmental Policy Act (NEPA), the Endangered Species Act, and the Plant Protection Act, APHIS-PPQ created the Technical Advisory Group (TAG) to evaluate the safety of weed biological control agents and make recommendations concerning the details of non-target testing protocols and issuance of permits for proposed introductions of new agents (Cofrancesco and Shearer, 2004). The TAG, however, is composed only of representatives of various federal agencies, and critics have charged that the process is closed to public review and biased favorably toward biological control as several of the agencies on the TAG employ biological control on their lands (Miller and Aplet, 2005). Weed biological control scientists reported that before 2001, the TAG process was effective and even-handed, even if slow. Since 2001, however, the TAG has recommended some biological control agents for release permits that APHIS-PPQ has declined to issue. Some scientists report that the opaque and shifting rules and criteria have in some cases caused years of research and hundreds of thousands of dollars of public funds to be wasted without scientific justification, and they feel that this is a counter-incentive for invasive plant control research in the United States. Arthropod biological control scientists in the United States have described similar challenges in obtaining federal permits.

When compared to permit review processes in other countries, the US permit decision-making process has several features that make it ethically problematic, above and beyond the bias resulting from excluding the consideration of potential benefits of the project from the decision-making. First, relative to other countries, US environmental regulatory processes are more consistently adversarial in character (Andrews, 1999). This makes negotiation and compromise much more difficult in environmental decisions in the United States.

Second, in the United States the research scientist applies for the permit to release an agent. This means that the individual research scientist functionally becomes, within this decision-making process, an advocate for the biological control release. This exposes the scientist to the perception of an ethical professional conflict of interest because he or she has the best knowledge of the organism proposed for release, but has requested a permit to release it, and therefore cannot be a neutral source of expertise. In other countries (such as New Zealand), land management organizations petition for releases (with the expert advice of scientists). The land management organization takes a public stance in favor of environmental management and invasive species control, and the request for biological control is in support of these goals.

Third, original data on the organism proposed for release may be held confidentially by APHIS-PPQ at the request of the research scientist, since the research scientist has not yet published the data in scientific literature (which is in the scientist's professional interest). However, this means that evidence used to make a

public decision of a public agency may not be available to the other scientists or the public. This is ethically problematic because it places the professional interests of the individual scientist above the democratic value of transparent decision-making by agencies that are supposed to be making decisions in the public's interest. In contrast, South Africa has long had a single, top-level institution responsible for coordinating biological control work (the Agricultural Research Council – Plant Protection Research Institute – Biological Control Division) but with active partnerships with virtually all university researchers in the country (Moran et al., 2005). This scientific network provides transparent yet critical peer review at every project stage, from problem definition through safety testing protocols and permit application. All South African biological control scientists participate in this network, and other scientists participate as well. The network has managed the risks of this collective scientific work because its participants recognize that in the absence of such a coordinated, critical review, there could be negative consequences for their entire community of practice and its work (Klein et al., 2011).

Fourth, there is no forum for public participation in biological control regulatory decisions in the United States, Canada, or Mexico (Mason et al., 2005). Public agency decisions are functionally inaccessible to the public, and have remained so despite calls for greater transparency, peer review, and public input (Strong and Pemberton, 2000). Thus, in the United States, the public has no meaningful mechanism to share its opinions about decisions made regarding a scientific practice that is supposedly in the public's own interest. In contrast, New Zealand has created participatory public processes for identifying targets and cultivating support for biological control projects, and has created a new agency to review proposed introductions of all novel organisms, including biological control agents.

New Zealand has created a responsive, transparent, democratic, and ecologically informed approach to biological control agent introductions since its legislature passed the Hazardous Substances and New Organisms (HSNO) Act in 1996 (see also Chapter 11). This act provided a common regulatory framework for all novel chemicals and living organisms (Barratt and Mooed, 2005). HSNO created the Environmental Risk Management Authority (ERMA) (now EPA-NZ) to regulate all novel organisms and hazardous substances, including the introduction of biological control agents (ERMA New Zealand, 2010). ERMA has two main components. "The Authority" is an autonomous quasi-judicial decision-making body that functions as the governing board for ERMA. "The Agency" consists of the staff providing operational support for the Authority. HSNO directed ERMA to impose public transparency on the decision-making criteria and process, to develop a common risk assessment for introduction of genetically modified organisms and biological control agents, and to require petitioners to provide evidence of risk, benefits, and public consultation (Barratt et al., 2010). ERMA has the world's most sophisticated decision-making process for evaluating novel organisms, with explicit reference to biological control agent introductions. It mandated clear decision-making criteria based on transparent and replicable ecologically based risk/cost : benefit analysis, fixed time periods for decisions, and participatory public engagement (Harrison et al., 2005). In contrast to APHIS-PPQ, all applications for and ERMA decisions regarding biological control introductions are posted on the internet.

The review process created by ERMA provides a more transparent and ecologically informed decision-making process for managing risk in the public interest. It balances multiple scientific and social criteria and requires applicants for a release permit to conduct public consultation before requesting a permit, the results of which are reported to the decision-making authorities (ERMA New Zealand, 2010). In the case of biological control introductions, the benefits must be considered and weighed against risks of harm (Harrison et al., 2005). In New Zealand, scientists are not applicants for permits. Instead, the land management agency or organization is the petitioner, and these organizations fund research agencies to conduct the pre-release testing and fund consultants to facilitate public outreach. This removes the research scientist from the situation in which he or she is vulnerable to charges of professional conflict of interest in advocating for a release. Petitioners must demonstrate that they have consulted affected members of the public, including indigenous Maori communities (Hayes et al, 2008). Concerned citizens may speak directly to the ERMA decision-makers, and if the petition merits a hearing, the applicant must pay for this. Permit applications average 18 public comments each, with as many as 59 recorded. For the two applications recording the highest number of public comments,

more than 90% of comments expressed support for releasing the agent (ERMA New Zealand, 2010). Stakeholders and scientists report that the ERMA review process has promoted broad and explicit consultation and review of the rationale for the selection of particular invasive species as targets and discussion of potential risks of particular natural enemies proposed for release, all of which occurs before any application is actually submitted.

The New Zealand model of biological control review has several ethical features that are superior to the US system. The risk/cost : benefit approval criteria are difficult to fulfill but quite clear: benefits must be scientifically demonstrated to outweigh risks and costs. Early in the application process, ERMA advises the applicant how to meet these criteria, but ERMA is required to follow a fixed schedule for making a decision by statute. By requiring consultation with affected communities before submitting an application, petitioners can receive early feedback from those who have concerns about the proposed introduction. Members of the public can provide input directly to the decision-makers. Thus, in contrast to the US model, deliberations about biodiversity protection are structured into several key points in the decision-making process.

A framework to improve ethics of biological control decisions

The effective integration of classical biological control into conservation practice could be advanced by enhanced, appropriate, democratic participation in decision-making. When a government agency creates a structured and transparent decision-making framework that supports appropriate democratic participation, conflicts can be avoided and better decisions can be made. The New Zealand system demonstrates that improving the structure of decision-making can result in improved biological control practice, as well as greater public support for invasive species control. The New Zealand case also suggests that the design of a regulatory system to accommodate democratic values can lead to a decision-making framework that integrates social and environmental ethics. Other countries are unlikely to adopt New Zealand's approach in its entirety; however, the negotiations among beneficiaries, scientists, critics, and public decision-makers illustrate how multiple ethical criteria can be balanced in decision-making. This in turn suggests several features for a model framework that could, ideally, enhance the integration of biological control for conservation practice more broadly (Warner, 2012).

A democratic political ethic proposes that increasing public participation in decisions that affect the public will result in better outcomes and more support for the substance of the decision itself. The effective articulation of science, democracy, and environmental ethics is not a simple problem. Balancing criteria from these three sources is critical for any public interest science. Expression of public support enhances the ability of scientists to claim they are acting in the public interest, just as an environmental advocacy group needs some expression of public support to claim legitimacy. All decisions by governments about environmental policy face a dilemma: science is not democratic. Scientific facts are not mutable to public opinion. The public might doubt that human actions are disrupting our planet's climate, but that skepticism does not change evidence thereof.

Scholars of science policy have articulated a new decision-making framework for connecting scientists and their institutions to society at large: participatory public engagement (McCallie et al., 2009). Participatory public engagement with science and technology (hereafter shortened to "public engagement") facilitates mutual learning among the public, scientists, and regulatory officials with respect to the development and application of science and technology in modern society (Mooney, 2010). Public engagement is a semi-structured, transparent, deliberative process that establishes consensus views on evidence, method, interpretation, and social value frameworks as the basis for making a scientifically informed decision (Rowe and Frewer, 2005). Public engagement is more costly in terms of time and resources. However, members of the public are challenging publicly funded researchers and regulatory agencies to be more transparent in their decision-making, and many existing models of public communication do not support effective responses (Warner, 2012). Any scientific community claiming to work on behalf of the public has a collective professional ethical responsibility to engage the public: to explain and listen to, and exchange information with, members of the public or organizations that represent the public.

As public interest sciences, conservation biology and biological control could benefit from cultivating more than passive public acceptance. Since the public funds most invasive species control programs through taxes, it is reasonable to expect scientists to inform, educate, and engage citizens on a continuing basis. Public engagement processes allow scientists to speak directly about the situation in the environment, but require scientists to communicate environmental conditions and a rationale for any conservation action (involving biological control or not) in terms that can be understood by nonscientists. Disputes surrounding the introduction of biological control agents can arise in high profile and costly restoration projects, and the application of classical biological control as an invasive species management practice is complicated by controversies (see Chapter 12). To be effective, public engagement must (1) foster greater, objective social understanding of the threat invasive species pose to conservation; (2) create greater social consensus on the need to control invasive species and the conditions under which biological control is an appropriate strategy to consider; and (3) incrementally increase the public's trust that government agencies are upholding the public's interest through appropriate regulatory review (see also Chapter 15).

Public engagement challenges scientists and their institutions to develop skills in public communication, and challenges public regulatory institutions to facilitate appropriate public review with biological control release decisions. Many scientists and public agencies are reluctant to engage the public and the media for very good reasons: they do not want to have scientific evidence put to a vote, engage in arguments with nonscientists, nor have their conclusions sensationalized. However, the cumulative effect of individual scientists and agencies not effectively communicating with nonscientists has the potential to perpetuate the problem of public misunderstanding of environmental problems and solutions (Mooney, 2010; Warner and Kinslow, 2013).

Quite reasonably, scientists and public agencies that use science in decision-making need forms of public engagement that do not (1) interfere with scientific research and practice, (2) impose significant additional burdens on their own time, (3) require them to take up a position of advocacy compromising their professional ethics, or (4) needlessly delay regulatory agency review. To avoid these problems, public engagement should distinguish invasive species *target selection* from *scientific research activities* from *public agency decision-making*. At a minimum, scientists working on behalf of the public interest should engage the public in the problem definition process. This requires greater initial public communication and consultation, and more transparency by agencies in the application of their decision-making criteria. However, these additional costs can be offset if they reduce subsequent controversy and lead to increased public support for invasive species control. Historically, virtually all biological control decisions were made within one government agency. Target selection, agent selection, testing criteria, and release permitting were all internal to one institution, usually a department of agriculture. In the United States and South Africa, review and decision points are now spread across multiple public agencies, making the process cumbersome, lengthy, and expensive (but see Chapter 11 for latest changes). In both these countries, there is no effective means for public input. In both these countries, biological control scientists and critics of introductions express varying degrees of frustration with public agency decision-making.

Public engagement differs from public hearings or consultation in that it requires bi-directional communication between scientists, public decision-makers, and lay publics (McCallie et al., 2009). It is a deliberative "dialogue" in which publics and scientists both benefit from listening to and learning from one another, which can be described as mutual learning (McCallie et al., 2009). Public engagement requires members of the public to do more than question experts. It requires scientists to do more than merely present their knowledge and perspectives. Public engagement requires lay publics to learn about science and policy, and scientists to learn what members of the lay public know and don't know about science, but also about social values. Thus, "engagement" in this sense includes both political engagement and educational engagement. Participants from a variety of perspectives participate over a sustained period of time, guided by shared goals and agreed upon guidelines for respectful communication.

The United States was a pioneer in early models of public participation; however, the legislation as it was developed in the 1970s required only public communication and gathering of public comments. This model is now unable to support social expectations of transparency and the need to cultivate active public participation

in these decisions (Yafee, 1997; Smith, 2004; Perrings et al., 2010). With ERMA, New Zealand has a participatory public process for identifying targets and cultivating support for biological control projects and has created a new agency to review proposed introductions of all novel organisms, including biological control agents. New Zealand has a national extension system for the biological control of weeds (Hayes, 1999; Hayes et al., 2008). This is much more than a technology transfer program, because it facilitates the exchange of information about invasive species threats to conservation at the landscape level among local land managers and regional public agencies with environmental management responsibilities. It also facilitates public education at the local level. This has fostered active public support among some members of the public for invasive species control on the basis of conservation values. Thus, the creation of ERMA and its decision-making processes, when combined with the open and participatory approaches to land management, demonstrate that the principles of public engagement and the ethical value of democratic participation can contribute to the integration of biological control into conservation practice.

The following are key elements of a practical ethical framework to improve decision-making, informed by principles of professional ethics, environmental ethics, and public engagement.

Public advocacy for invasive species control should be separate from selection of specific control strategies

Organizations, individuals, communities, or stakeholders should speak to the broader public about the need for controlling an invasive species. The problem of invasive species, first defined as a problem by scientists, needs to be explained to the public, and the public should express some form of consent before any specific management action is proposed. Without some expression of concern about the problem of invasive species, there is no reason for the public to support the introduction of a biological control agent or, indeed, any form of management of the invasive species.

Decision-making criteria and processes should be transparent

Public agencies, stakeholder organizations, and scientists should devise public engagement processes that enhance the capacity of stakeholders to understand science and agency decision-making processes. For public engagement to succeed, it must convene a structured co-learning process in which everyone, from critics to supporters, participates over time in establishing the same scientific information about the invasive species and possible control methods. Public engagement fails if parties have divergent information about the problem and possible remedies. Some public concerns about biological control are founded, at least loosely, on popular conservation concerns such as (1) is the invasive species really a problem; (2) why is it necessary to introduce another organism; (3) what other organisms will the agent attack; and (4) what will the agent do when it consumes most of its hosts. These questions touch on scientific, democratic, and ethical concerns. Many non-scientist stakeholders are not able to contribute to decision-making with the knowledge that they initially bring to such a process; therefore, education of stakeholders is essential to any kind of engagement.

Beneficiaries (stakeholders, not researchers) should explain why control of the invasive species is in the public's interest

When invasive species cause direct economic harm, those who wish to alleviate that economic harm are potential beneficiaries. When invasive species cause harm to ecosystem function or endangered species, human organizations must be able to speak on behalf of their conservation. Conservation groups and public conservation agencies should, ideally, speak on behalf of the public or society at large (for example, see Chapter 11). Creating greater consensus on the need to take such conservation actions is a critical first step that is fundamental to success. For example, Australia has a national weeds strategy that justifies action (Natural Resource Management Ministerial Council of Australia, 2007). In New Zealand, regional councils serve as critical intermediaries between tax payers (or rate payers) as stakeholders with research institutions (Hayes, 1999). This arrangement insulates researchers from public suspicions of conflict of interest, that is, that the researcher might lose objectivity by promoting a project that advances his or her career. In the United States, public/private coalitions of stakeholders, which often include land

managers, speak on behalf of the public's interest in invasive species control.

Beneficiaries should present an ecologically-based risk/cost : benefit analysis justifying biological control on conservation grounds

In New Zealand, permit applicants articulate an economic justification that makes clear to tax payers the advantages of biological control over other forms of control. In the New Zealand regulatory system, these regional councils are generally the body that petitions for invasive plant biological control release permits, and they are better positioned to articulate these advantages and to engage in discussions over conflicts of interest. The scientist serves as scientific expert advisor and is never the advocate for controlling a pest (ERMA New Zealand, 2010). Legislation imposes the burden of public consultation and engagement on the petitioner for a permit. Although this appears costly, in practice it appears that this is more than offset by decreased costs and conflicts associated with the actual regulatory decision (ERMA New Zealand, 2010). Other countries could benefit from this approach, although in the United States it would mean going beyond what is minimally required by law.

Public agencies should gather stakeholder input on how their criteria apply to a specific permit application

The New Zealand permitting system is efficient because any decision to release a biological control agent is made on a very narrow basis. It presumes that there has been prior public engagement about the desirability of targeting the invasive plant and the suitability of the biological control agent. Then, the question upon which the decision is made is straightforward: are the anticipated benefits greater than the costs and risks? In New Zealand, this has frontloaded costs and public engagement efforts, but has made release decisions less contested and more democratic.

The above list of key elements draws on more than one branch of ethics and links all participants in decision-making, which is appropriate for negotiating the multiple ethical considerations relevant to enhancing the integration of biological control into conservation practice.

Conclusion

Conservation biology and biological control are public interest sciences. They are chiefly funded by governments and are carried out on behalf of the public. To determine when it is appropriate to introduce a biological control agent requires negotiating conflicting social values. This depends upon excellent scientific data, prudence, and good judgment when data are lacking or limited, but also upon clear-eyed ethical analysis, including a shared understanding of the public's interest in conservation.

Some form of public consent is necessary for the application of public interest science. To foster sustained public engagement over time, the designation of particular invasive species as needing control should be separated from specific management actions, including the proposed solution of a biological control introduction. Fostering social consensus on the need to control the invasive species is a prerequisite for public action. Public engagement requires careful attention to devising appropriate roles for stakeholders, and opportunities for public input in decision-making processes. An integrated risk/cost : benefit analysis does a better job of modeling the implications of these management decisions on real world landscapes, and this approach should inform policy and regulatory reform. This is not the case in the United States, and this fragmentary model of public agency decision-making must be changed to realize the potential of integrating biological control into conservation practice.

The field of ethics can help fashion a more responsive and democratic decision-making model to integrate biological control into conservation practice. Careful attention to the professional responsibilities of scientists and regulatory agency personnel in such a decision-making process, combined with the practical application of prudence, can result in broader support for appropriate invasive species management and conservation goals.

Acknowledgments

I thank Cliff Moran, Bill Bruckart, Bernd Blossey, Andy Sheppard, David Wagner, and Bernd Blossey for reviews of Chapter 13. Research for this chapter was supported by the California Department of Food and Agriculture and the US National Science Foundation (award 0646658).

References

Andres, L. A. 1981. Conflicting interests and the biological control of weeds, pp. 11–20. *In*: Delfosse, E. S. (ed.). *Vth International Symposium on the Biological Control of Weeds, July 22–29, 1980, Brisbane, Australia*. CSIRO Entomology. Brisbane, Australia.

Andres, L. A. 1999. History of weed biological control in North America, pp. 8–9. *In*: Isaacson, D. and M. H. Brookes (eds.). *Weed Biological control: Extended Abstracts from the 1997 Interagency Noxious-Weed Symposium*. FHTET 98-12. USDA Forest Service. Morgantown, West Virginia, USA and Oregon Department of Agriculture. Salem Oregon. Available from: http://www.fs.fed.us/foresthealth/technology/pdfs/interagency1997.pdf [Accessed January 2016].

Andrews, R. 1999. *Managing the Environment, Managing Ourselves*. Yale University Press. New Haven, Connecticut, USA.

Balciunas, J. and E. M. Coombs. 2004. International code of best practices for classical biological control of weeds, pp. 130–136. *In*: Coombs, E. M., J. K. Clark, G. L. Piper, and A. F. Cofrancesco, Jr. (eds.). *Biological Control of Invasive Plants in the United States*. Oregon State University Press. Corvallis, Oregon, USA.

Barratt, B. I. P. and A. Mooed. 2005. Environmental safety of biological control: Policy and practice in New Zealand. *Biological Control* **35**: 247–252.

Barratt, B. I. P., F. G. Howarth, T. M. Withers, et al. 2010. Progress in risk assessment for classical biological control. *Biological Control* **52**: 245–254.

Bennett, F. D. 1985. Conflict of interest in CIBC Biological Control of Weeds Program, pp. 241–247. *In*: Delfosse, E. S. (ed.). *VI International Symposium for the Biological Control of Weeds, August 19–25, 1984, Vancouver, British Columbia, Canada*. Agriculture Canada. Vancouver, British Columbia, Canada.

Carruthers, R. I. 2004. Biological control of invasive species, a personal perspective. *Conservation Biology* **18**: 54–57.

Carruthers, R. I. and C. M. D'Antonio. 2005. Science and decision making in biological control of weeds: Benefits and risks of biological control. *Biological Control* **35**: 181–182.

Carruthers, R. I. and J. A. Onsager. 1993. A perspective on the use of exotic natural enemies for biological control of pest grasshoppers. *Environmental Entomology* **22**: 885–903.

Carson, R. 1962. *Silent Spring*. Houghton Mifflin. Boston, Massachusetts, USA.

Cofrancesco, A. F. and J. F. Shearer. 2004. Technical advisory group for biological control agents of weeds, pp. 38–41. *In*: Coombs, E. M., J. K. Clark, G. L. Piper, and A. F. Cofrancesco (eds.). *Biological Control of Invasive Plants in the United States*. Oregon State University Press. Corvallis, Oregon, USA.

Coulson, J. R. 1992. Documentation of classical biological control introductions. *Crop Protection* **11**: 195–206.

Coulson, J. R. 1999. Developing safe weed biological control in the United States, pp. 2–3. *In*: Isaacson, D. and M. H. Brookes (eds.). *Weed Biological control: Interagency Noxious Weed*

Sympoisum in 1997. FHTET 98-12. USDA Forest Service. Morgantown, West Virginia, USA and Oregon Department of Agriculture, Salem Oregon, USA. Available from: http://www.fs.fed.us/foresthealth/technology/pdfs/interagency 1997.pdf [Accessed January 2016].

Coulson, J. R. and R. S. Soper. 1989. Protocols for the introduction of biological control agents in the U.S., pp. 1–35. *In*: Kahn, R. P. (ed.). *Plant Protection and Quarantine Volume III. Special Topics.* CRC Press. Boca Raton, Florida, USA.

Coulson, J. R., M. E. Vail, W. C. Kauffman, and M. E. Dix (eds.). 2000. *110 Years of Biological Control Research and Development in the United States Department of Agriculture.* USDA, Washington, DC.

Cullen, J. M. and E. S. Delfosse. 1985. *Echium plantagineum*: catalyst for conflict and change in Australia, pp. 249–292. *In*: Delfosse, E. S. (ed.). *VI International Symposium for the Biological Control of Weeds, August 19–25, 1984, Vancouver, British Columbia, Canada.* Agriculture Canada. Vancouver, British Columbia, Canada.

Delfosse, E. S. 2004. Introduction, pp. 1–11. *In*: Coombs, E. M., J. K. Clark, G. L. Piper, and A. F. Cofrancesco (eds.). *Biological Control of Invasive Plants in the United States.* Oregon State University Press. Corvallis, Oregon, USA.

Delfosse, E. S. 2005. Risk and ethics in biological control. *Biological Control* **35**: 319–329.

DesJardins, J. R. 2006. *Environmental Ethics: An Introduction to Environmental Philosophy.* Wadsworth. Belmont, California, USA.

Dodd, A. P. 1940. *The Biological Campaign against Prickly Pear.* Commonwealth Prickly Pear Board Bulletin. Brisbane, Australia. 177 pp.

Drake, J. A., H. A. Mooney, F. Di Castri, et al. 1989. *Biological Invasions: A Global Perspective*, vol. **37**. John Wiley. Chichester, UK.

ERMA New Zealand. 2010. Investigating Biological Control and the HSNO Act. Environmental Protection Agency of New Zealand, Wellington, New Zealand. Available from: http://www.epa.govt.nz/Publications/Investigating-Biological-Control-and-the-HSNO%20Act-ERMA-Report-2010).pdf [Accessed October 2010].

Follett, P. A. and J. J. Duan. 2000. *Nontarget Effects of Biological Control.* Kluwer Academic Publications. Norwell, Massachusetts, USA.

Gurr, G. M., N. D. Barlow, J. Memmott, et al. 2000. A history of methodological, theoretical and empirical approaches to biological control, pp. 3–37. *In*: Gurr, M. and S. D. Wratten (eds.). *Biological Control: Measures of Success.* Kluwer Academic Publishers. Dordrecht, The Netherlands.

Harremoës, P., D. Gee, M. MacGarvin, et al. (eds.). 2002. *The Precautionary Principle in the 20th Century: Late Lessons from Early Warnings.* Earthscan. London.

Harrison, L., A. Mooed, and A. Sheppard. 2005. Regulation of the release of biological control agents of arthropods in New Zealand and Australia, pp. 715–725. *In*: Hoddle, M. S. (eds.). *Proceedings of the Second International Symposium on the Biological Control of Arthropods, September 12–16, 2005, Davos, Switzerland.* FHTET Publication 2005-08, USDA Forest Service. Morgantown, West Virginia, USA. Available from: http://www.bugwood.org/arthropod2005/vol2/14e.pdf. [Accessed June 2013].

Hayes, L. M. 1999. Technology transfer programmes for biological control of weeds – the New Zealand experience, pp. 719–727. *In*: Spencer, N. R. (ed.). *Proceedings of the X International Symposium on Biological Control of Weeds, July, 4–9, 1999, Bozeman, Montana. USA.* USDA. Sidney, Montana, USA. Available from: http://www.invasive.org/publications/xsymposium/Session9.html. [Accessed January 2016].

Hayes, L. M., C. Horn, and P. O. B Lyver. 2008. Avoiding tears before bedtime: how biological control researchers could undertake better dialogue with their communities, pp. 376–383. *In*: Julien, M. H., R. Sforza, M. C. Bon, et al. (eds.). *Proceedings of the XII International Symposium on Biological Control of Weeds, April 22–27, 2007, La Grande Motte, France.* CABI. Wallingford, UK.

Hoddle, M. S. 2002. Classical biological control of arthropods in the 21st century, pp. 3–16. *In*: Van Driesche, R. G. (ed.). *Proceedings of the First International Symposium on Biological Control of Arthropods, January 14–18, 2002, Honolulu, Hawaii, USA.* FHTET 2003-05, USDA Forest Service. Morgantown, West Virginia, USA. Available from: http://www.bugwood.org/arthropod/ [Accessed January 2016].

Hoddle, M. S. 2004. Restoring balance: Using exotic species to control invasive exotic species. *Conservation Biology* **18**: 38–49.

Howarth, F. G. 1991. Environmental impacts of classical biological control. *Annual Review of Entomology* **36**: 485–509.

Hunt, E. J., U. Kuhlmann, A. Sheppard, et al. 2008. Review of invertebrate biological control agent regulation in Australia, New Zealand, Canada and the USA: recommendations for a harmonized European system. *Journal of Applied Entomology* **132**: 89–123.

Jasanoff, S. 2005. *Designs on Nature: Science and Democracy in Europe and the United States.* Princeton University Press. Princeton, New Jersey, USA.

Jonas, H. 1985. *The Imperative of Responsibility: In Search of an Ethics for the Technological Age.* University of Chicago Press. Chicago, Illinois, USA.

Klein, H., M. P. Hill, C. Zachariades, and H. G. Zimmermann. 2011. Regulation and risk assessment for importations and releases of biological control agents against invasive alien plants in South Africa. *African Entomology* **19**: 488–497.

Larson, M. S. 1977. *The Rise of Professionalism: A Sociological Analysis.* University of California Press. Berkeley, California, USA.

Leopold, A. 1949. *A Sand County Almanac.* Oxford University Press. London.

Lockwood, J. A. 1993. Environmental issues involved in the biological control of rangeland grasshoppers (Orthoptera: Acrididae) with exotic agents. *Environmental Entomology* **22**: 508–518.

Lockwood, J. A. 1996. The ethics of biological control: understanding the moral implications of our most powerful ecological technology. *Agriculture and Human Values* **13**: 2–19.

Lockwood, J. A. 1997. Competing values and moral imperatives: an overview of ethical issues in biological control. *Agriculture and Human Values* **14**: 205–210.

Lockwood, J. A., F. G. Howarth, and M. F. Purcell. 2001. *Balancing Nature: Assessing the Impact of Non-native Biological Control Agents.* Entomological Society of America. Landham, Maryland, USA.

Louda, S. M., D. Kendall, J. Connor, and D. Simberloff. 1997. Ecological effects of an insect introduced for biological control of weeds. *Science* **277**(5329): 1088–1090.

Mason, P. G., R. G. Flanders, and H. A. Arrendondo-Bernal. 2005. How can legislation facilitate the use of biological control of arthropods in North America? pp. 701–713. *In:* Hoddle, M. S. (eds.). *Proceedings of the Second International Symposium on the Biological Control of Arthropods, September 12–16, 2005, Davos, Switzerland.* FHTET 2005-08, USDA Forest Service, Morgantown, West Virginia, USA. Available from: http://www.bugwood.org/arthropod2005/vol2/14e.pdf. [Accessed June 2013].

McCallie, E., L. Bell, T. Lohwater, et al. 2009. *Many Experts, Many Audiences: Public Engagement with Science and Informal Science Education.* Center for Advancement of Informal Science Education (CAISE), Washington, D.C. Available from: http://digitalcommons.calpoly.edu/cgi/viewcontent.cgi?article=1011&context=eth_fac [Accessed January 2016].

Meffe, G. and C. R. Carroll. 1997. *Principles of Conservation Biology.* Sinauer Associates. New York.

Messing, R. H. and M. G. Wright. 2006. Biological control of invasive species: solution or pollution? *Frontiers in Ecology and the Environment* **4**: 132–140.

Miller, M. and G. Aplet. 1993. Biological control: a little knowledge is a dangerous thing. *Rutgers Law Review* Winter issue: 285–334.

Miller, M. and G. Aplet. 2005. Applying legal sunshine to the hidden regulation of biological control. *Biological Control* **35**: 358–365.

Mooney, C. 2010. *Do Scientists Understand the Public?* American Academy of Arts & Sciences, Washington, DC. Available from: http://www.amacad.org/pdfs/scientistsUnderstand.pdf [Accessed July 2010].

Moran, V. C., J. H. Hoffmann, and H. G. Zimmermann. 2005. Biological control of invasive alien plants in South Africa: necessity, circumspection, and success. *Frontiers in Ecology and the Environment* **3**: 77–83.

National Research Council. 1983. *Risk Assessment in the Federal Government: Managing the Process.* National Academy Press. Washington, DC.

National Research Council. 1996. *Understanding Risk: Informing Decisions in a Democratic Society.* National Academy Press. Washington, DC.

National Research Council. 2002. *Predicting Invasions of Nonindigenous Plants and Plant Pests.* National Academies Press. Washington, DC.

Natural Resource Management Ministerial Council of Australia. 2007. *Australian Weeds Strategy - A National Strategy for Weed Management in Australia.* Commonwealth of Australia. Canberra, Australia. Available from: http://www.weeds.gov.au/publications/strategies/pubs/weed-strategy.pdf [Accessed January 2016].

Palladino, P. 1996. *Entomology, Ecology and Agriculture: The Making of Scientific Careers in North America, 1885–1985.* Harwood Academic Publishers. Amsterdam.

Pemberton, R. 1985a. Native plant considerations in the biological control of leafy spurge, pp. 365–390. *In:* Delfosse, E. S. (ed.). *VI International Symposium for the Biological Control of Weeds, August 19–25, 1984, Vancouver, British Columbia, Canada.* Agriculture Canada. Vancouver, British Columbia, Canada.

Pemberton, R. 1985b. Native weeds as candidates for biological control research, pp. 869–877. *In:* Delfosse, E. S. (ed.). *VI International Symposium for the Biological Control of Weeds, August 19–25, 1984, Vancouver, British Columbia, Canada.* Agriculture Canada. Vancouver, British Columbia, Canada.

Perkins, J. H. 1982. *Insects, Experts, and the Insecticide Crisis.* Plenum Press. New York.

Perrings, C., H. A. Mooney, and M. Williamson. 2010. *Bioinvasions and Globalization: Ecology, Economics, Management, and Policy.* Oxford University Press. London.

Raffensperger, C., S. Peters, F. Kirschenmann, et al. 1999. *Defining Public Interest Research.* White Paper written for the Science and Environmental Health Network; The Center for Rural Affairs; and the Consortium for Sustainable Agriculture, Research and Education. Available from: http://www.sehn.org/defpirpaper.html [Accessed January 2016]:

Rowe, G. and L. J. Frewer. 2005. A typology of public engagement mechanisms. *Science, Technology & Human Values* **30**: 251–290.

Samways, M. J. 2007. Insect conservation: a synthetic management approach. *Annual Review of Entomology* **52**: 465–487.

Sawyer, R. C. 1996. *To Make a Spotless Orange: Biological Control in California.* Iowa State University Press. Ames, Iowa, USA.

Sheppard, A.W., R. Hill, A. DeClerck-Floate, et al. 2003. A global review of risk-benefit-cost analysis for the introduction of classical biological control agents against weeds: a crisis in the making? *Biological Control News and Information* **24**: 91N–108N.

Sheppard, A.W., R. D. van Klinken, and T. Heard. 2005. Scientific advances in the analysis of direct risks of weed biological control agents to nontarget plants. *Biological Control* **35**: 215–226.

Sheppard, A. W., R. H. Shaw, and R. Sforza. 2006. Top 20 environmental weeds for classical biological control in Europe: a review of opportunities, regulations and other barriers to adoption. *Weed Research* **46**: 93–117.

Shrader-Frechette, K. and E. D. McCoy. 1993. *Method in Ecology: Strategies for Conservation*. Cambridge University Press. Cambridge, UK.

Simberloff, D. and P. Stiling, P. 1996a. How risky is biological control? *Ecology* **77**: 1965–1974.

Simberloff, D. and P. Stiling. 1996b. Risks of species introduced for biological control. *Biological Conservation* **78**: 185–192.

Simberloff, D., I. M. Parker, and P. N. Windle. 2005. Introduced species policy, management, and future research needs. *Frontiers in Ecology and the Environment* **3**: 12–20.

Slovic, P. 2001. *The Perception of Risk*. Earthscan. London.

Slovic, P., M. L. Finucane, E. Peters, and D. G. MacGregor. 2004. Risk as analysis and risk as feelings: some thoughts about affect, reason, risk, and rationality. *Risk Analysis* **24**: 311–322.

Smith, Z. A. 2004. *The Environmental Policy Paradox*. Prentice Hall. Upper Saddle River, New Jersey, USA.

Stanley, M. C. and S. V. Fowler. 2004. Conflicts of interest associated with the biological control of weeds, pp. 322–340. *In*: Cullen, J. M., D. T. Briese, D. J. Kriticos, et al. (eds.). *Proceedings of the XIth International Symposium on Biological Control of Weeds, April 27–May 2, 2003, Canberra, Australia*. CSIRO Entomology. Canberra, Australia.

Strong, D. R. and R. W. Pemberton. 2000. Biological control of invading species: risk and reform. *Science* **288**: 1969–1970.

Takacs, D. 1996. *The Idea of Biodiversity: Philosophies of Paradise*. Johns Hopkins Press. Baltimore, Maryland, USA.

Wajnberg, E., J. K. Scott, and P. C. Quimby. 2001. *Evaluating Indirect Ecological Effects of Biological Control*. CABI Publishing. Wallingford, UK.

Warner, K. D. 2007. *Agroecology in Action: Extending Alternative Agriculture through Social Networks*. MIT Press. Cambridge, Massachusetts, USA.

Warner, K. D. 2012. Fighting pathophobia: how to construct constructive public engagement with biological control for nature without augmenting public fears. *Biological Control* **57**: 307–317.

Warner, K. D and F. M. Kinslow. 2013. Manipulating risk communication: value predispositions shape public understandings of invasive species science in Hawaii. *Public Understanding of Science* **22**: 203–218.

Warner, K. D., J. N. McNeil, and C. Getz. 2008. What every biological control researcher should know about the public, pp. 398–402. *In*: Julien, M. H., R. Sforza, M. C. Bon, et al. (eds.). *XII International Symposium on the Biological Control of Weeds, April 22–27, 2007, La Grande Motte, France*. CAB International. Wallingford, UK.

Warner, K. D., K. M. Daane, C. M. Getz, et al. 2011. The decline of public interest agricultural science and the dubious future of crop biological control in California. *Agriculture and Human Values* **28**: 483–496.

Whiteside, K. 2006. *Precautionary Politics: Principle and Practice in Confronting Environmental Risk*. MIT Press. Cambridge, Massachusetts, USA.

Willot, E. and D. Schmidtz. 2001. *Environmental Ethics: What Really Matters, What Really Works*: Oxford University Press. Oxford, UK.

Yafee, S. L. 1997. Why environmental policy nightmares recur. *Conservation Biology* **11**: 328–337.

CHAPTER 14

Economics of biological control for species invading wildlands

Roy G. Van Driesche[1] and Kevin M. Heinz[2]

[1] Department of Environmental Conservation, University of Massachusetts, USA
[2] Department of Entomology, Texas A & M University, USA

Introduction

Economic issues, both as impacts and constraints, are entwined with species invasions and societal responses to them. Here we examine economic issues related to invasions of wildlands, as distinct from the more tangible economics of invasive species in settings such as health, industry, agriculture, and plantation forestry. Our goal is not to review comprehensively this larger topic, which has been addressed previously (e.g., Naylor, 2000; Pimentel, 2002; Williams et al., 2010; Murbuah et al., 2014), but rather to focus on the distinctive features of the economics of invasions of wildlands.

Invaders of wildlands and waters (hereafter "wildland invaders" for short) cause economic losses both of both tangible goods and intangible services from natural ecosystems (Section one, "Losses from wildland invaders"). Value of lost goods (wood, forage, water, electricity, fish harvests, etc.) is more easily quantified than the value of lost services (normal fire or flood cycles, or soil nutrient levels, reductions in levels of air pollution, retention of biodiversity of cultural or scientific importance), but both must be taken into account.

Control measures taken against such species also entail costs and such costs often become prohibitive for measures other than biological control and certain genetic control approaches once the opportunity for eradication is lost and the invasive covers large areas. Biological control programs are suitable for pests with large distributions, but programs entail considerable costs in the form of required public financing, review

and permitting, and professional expertise. Economic analyses of cost: benefit ratios in general suggest that use of biological control (Section two, "Benefit: cost ratios for biological control") is good public policy, although many such analyses have focused on projects carried out against economic pests other than those of wildlands.

Biological control alone, however, particularly in the case of weed biological control, may not be enough to meet projects' ecological goals. Additional costs may be required to restore adequately wildland sites affected by invasive species, such as for replanting of missing native species or physical modification or reconstruction of the affected environment, and these costs must also be taken into account (Section three, "Additional costs of restoration to desired conditions").

High costs, therefore, mean that not all invasives in wildlands can be addressed and target selection is important. Economics have in the past affected target selection, choices often being driven by economic interests injured by invasive species, and such an approach to target selection may be misaligned to address key pests of wildlands (Section four, "Effects of economic interests on target selection for biological control").

Finally, how biological control programs are paid for is a critical issue. Availability of funds to mount biological control programs against pests is often driven by pressures from groups facing economic losses from invaders and, as such, the importance of addressing pests causing losses in biodiversity, products, or services from wildlands is often undervalued. Policies concerning funding

Integrating Biological Control into Conservation Practice, First Edition. Edited by Roy G. Van Driesche, Daniel Simberloff, Bernd Blossey, Charlotte Causton, Mark S. Hoddle, David L. Wagner, Christian O. Marks, Kevin M. Heinz, and Keith D. Warner.
© 2016 John Wiley & Sons, Ltd. Published 2016 by John Wiley & Sons, Ltd.

sources for biological control (whether sourced from general revenues, specialized taxation, or NGO contributions) (Section five, "How should biological control in wildlands be financed?") currently do not link industries from which invaders arise with an obligation to provide funds to remediate unanticipated invasions; such a link, however, might provide a more stable means of addressing problems from wildlands invasions and that idea is explored here.

Losses from wildland invaders

Economic losses from invasive species in several areas or contexts have been estimated (e.g., Naylor, 2000; Pimentel, 2002; Pimentel et al., 2000, 2005; Lovell and Stone, 2005; Holmes et al., 2009; Williams et al., 2010; Murbuah et al., 2014). Estimates with larger scope (all species, whole countries, e.g., Pimentel et al., 2000) are more difficult to construct than estimates for single-species invasions (e.g., Leung et al., 2002), and accurate estimation demands use of advanced economic modeling (Cook et al., 2007; Holmes et al., 2009), especially if losses, not just of products but also of ecosystem services, are to be estimated (Charles and Dukes, 2007). While past estimates have been criticized over aspects of methods used (see discussion of Pimentel et al. [2000] in Holmes et al. [2009]), and exact numbers may be inaccurate (e.g., US$120 billion/yr for species in United States [Pimental et al., 2005]), it appears that such losses are quite large and are composed of both commodity losses and non-market losses (ecosystem services), although economic substitutions may reduce some direct commodity loss (Holmes et al., 2009).

Losses owing to invasions of wildlands are a subset of all invasive species losses and have not been measured separately, although estimates exist for invasive aquatic species (Lovell and Stone, 2005) and are discussed for forest pests (Holmes et al., 2009) in the United States. One part of these estimates is direct losses of commodities like wood, water, electricity, and so on, while another aspect is the less tangible value of lost ecosystem services such as maintenance of desired soil features or development of unfavorable fire regimes because of the invader. While comprehensively reviewing such losses is not possible here, we offer examples below of such losses from wildland invaders, bearing in mind that most such estimates lack the use of

sophisticated economic modeling. Finally, the values of services or products recovered, following biological control of the invader and use of other restoration methods, have been estimated in some cases and we offer examples.

Tangible wildland products lost to invaders

Products from uncultivated wildlands and waters lost because of insect or plant invasions include wood, forage, water, electricity, or fish and game, among others (Van Driesche et al., 2010). Some of these invasive species have been wholly or partly suppressed through biological control (Van Driesche et al. [2010] and references therein), permitting the recovery of all or part of the lost productivity. Some examples follow, along with notes on the agents responsible for the invasive species' control.

Wood from native forests

Some invasive insects can reduce the ability of native forests to yield wood products by lowering growth rates of native trees or increasing their mortality, especially in North America where extensive native forests are traditional sources of timber products, but have also experienced many insect invasions from Europe or Asia. Most other regions' native forests are either less extensive or less invaded, or the region's wood products come principally from forest plantations of either native or exotic tree species, or biological control has been infrequently employed. Consequently we focus this discussion on North America.

From the 1860s to the 1930s, a series of exotic moths and sawflies reached North America and damaged various native tree species in the region's forests, for example, the introduced pine sawfly (*Diprion similis* [Hartig]), European spruce sawfly (*Gilpinia hercyniae* [Hartig]), larch casebearer (*Coleophora laricella* [Hübner]), winter moth (*Operophtera brumata* [L.]), and gypsy moth (*Lymantria dispar* [L.]), all of which were eventually suppressed, at least in part, through natural enemy importations from the species' native ranges (Van Driesche and Reardon, 2014). The history of these invasions illustrates the severe threat that invasive herbivorous insects have posed to the region's forests.

In 1869, gypsy moth escaped from captivity and established in Massachusetts, eventually spreading widely in the oak-dominated forests of the eastern United States and Canada. Large areas were defoliated in 1980–82 and 1989–91 (10.7 and 5.8 million ha, respectively) (USDA Forest Service website: http://www.nrs.fs.fed.us/pubs/gtr/gtr_nrs6.pdf [see page 4]). Rates of mortality of defoliated trees varied with tree health. In 1911–31, 35% of oaks in poor condition died from a single defoliation, compared to 7% of trees in good condition (Campbell and Sloan, 1977). Since the 1990s, gypsy moth outbreaks have been small and localized in New England because of an exotic fungus (*Entomophaga maimaiga* Humber, Shimazu et Soper) and introduced parasitoids (Fuester et al., 2014); however, defoliation still occurs in western and southern parts of the infestation.

Around 1886, larch casebearer invaded New England and quickly spread to the Canadian Maritimes and British Columbia (Drooz, 1985). From the 1900s to the 1950s, outbreaks occurred frequently (Felt and Bromely, 1932; Webb and Quednau, 1971), until brought under biological control (Ryan and Van Driesche, 2014). Beginning in the 1950s, defoliation of western species of larch occurred in Idaho and Oregon until the invader was suppressed by natural enemies (Ryan, 1990). Repeated defoliation in northern Idaho reduced tree growth by 80%, but growth returned to normal after successful biological control (Long, 1988).

During 1914–22, both the introduced pine sawfly and European spruce sawfly invaded eastern North America (Coppel et al., 1974; Drooz, 1985). In 1935, European spruce sawfly defoliated 1.5 million ha spruce in New Brunswick, sometimes causing high spruce tree mortality. At its most damaging, this sawfly infested 363,000 km^2 (Van Driesche et al., 1996). European spruce sawfly eventually was suppressed to minor status through the deliberate introduction of parasitoids from its European range and the accidental introduction of an associated nuclear polyhedrosis virus (MacQuarrie, 2014). The introduced pine sawfly in 1981 defoliated 2.2 million ha of pines in Virginia and North Carolina (Drooz et al., 1979; Ghent et al., 1982). While many parasitoids were released into North America for control of this and other conifer-feeding sawflies, only two species (*Monodontomerus dentipes* [Dalman] and *Exenterus amictorius* [Panzer]) (Hymenoptera: Torymidae and Ichneumonidae, respectively) are now regarded as regulating populations of *D. similis*, and of these, *M. dentipes* arrived accidently into North America with its host and became established before its deliberate introduction. Therefore, the only additional impact of the biological

control program was the establishment of *E. amictorius* (Lyons, 2014).

In the 1930s, winter moth invaded the Canadian Maritimes and, later, British Columbia (1970s) and Massachusetts (2000s), defoliating many species of deciduous trees. In Nova Scotia an outbreak in the 1950s strongly affected red oak (*Quercus rubra* L.), with 40% of oaks dying in some stands over several years (Embree, 1967). In 1954–56, winter moth infestations were present in 60% of Nova Scotia, killing most oaks in areas of high defoliation (Cuming, 1961). In 1977, 120 km^2 of oaks were defoliated in British Columbia (Gillespie et al., 1978), and in 2003 60,325 ha were defoliated in Massachusetts (Anonymous, 2004). Following the introduction of the tachinid fly *Cyzenis albicans* Fallén, winter moth populations declined to minor levels in both Nova Scotia and British Columbia (Elkinton and Boettner, 2014), and a similar introduction in Massachusetts has resulted in establishment of the parasitoid although there has not yet been sufficient time to see if the pest will be suppressed.

While exact details of wood loss (as a result of some tree mortality, reduced growth after defoliation, and altered forest composition) are not calculated or presented here, and indeed would be hard to do so for the whole region because of variation in insect population densities and forest stand conditions and the need to apply economic models that consideravailability and cost of substitutes in the various time periods, it is clear that wood growth in the region was at times significantly affected by these invasive insects, reducing output of certain types of wood from native forests until pests were suppressed through biological control.

Since the 1970s, new introductions of highly damaging forest defoliators into North America declined significantly through better regulation of plant importations, but were replaced in part by invasions of wood-boring insects vectored by wooden packing materials used in trade. In terms of reductions of wood production, one of the most important of these was the emerald ash borer (*Agrilus planipennis* Fairmaire) (Coleoptera: Buprestidae), which invaded the north central United States in the 1990s. More than 30 million ash (*Fraxinus* spp.) trees were killed by 2010 (Poland and McCullough, 2006; EAB, 2009). The loss in one state alone (Michigan) was estimated at 42.5 million cubic meters of wood (18 million board feet) valued at US$4–9 million (Michigan Department of Natural

Resources website: http://www.michigan.gov/dnr/). Releases of parasitoids from Asia (the pest's native range) resulted in establishment of two species, and research on biological control of this pest is ongoing (Bauer et al., 2014).

Forage from native grasslands

Natural grasslands in many parts of the world have been extensively used for grazing. Their quality and area, however, have been reduced by land conversion, over-grazing, introduction of less useful grasses that alter fire cycles, and invasions of less palatable forbs or shrubs. While only one strand in the skein of influences, many of the most pernicious invasive plants in grasslands have been targets of biological control in North America, New Zealand, Australia, or South Africa. Such invasions can reduce the "carrying capacity" (stock supported per ha) by changing fire regimes, competitively reducing forage growth, being toxic to livestock, or fostering conversion to shrublands. Grasslands of the western United States and Canada have been invaded by a group of toxic or thorny forbs from Europe that cause an estimated US$2 billion losses each year (DiTomaso, 2000). At least 14 of these species have been targeted for biological control in one or more countries (Nechols et al., 1995; Julien and Griffiths, 1998; Van Driesche et al., 2002), including knapweeds (*Centaurea* spp.), yellow starthistle (*Centaurea solstitialis* L.), tansy ragwort (*Senecio jacobaea* Gaertn.), St. Johnswort (*Hypericum perforatum* L.), spurges (*Euphorbia* spp.), and thistles (*Carduus* and *Circium* spp.).

In California, moderate infestations of yellow starthistle (20–31% total vegetation) reduced carrying capacity for cattle by 10–15%, and heavier infestations reduced forage by up to 50% (Connor, 2003), while in western Montana (USA), spotted knapweed infestations reduced production of the dominant native forage (*Pseudoroegneria spicata* [Pursh] Á. Löve) by up to 88% (Watson and Renney, 1974), and diffuse knapweed (*Centaurea diffusa* Lam.) infestations in British Columbia reduced forage by up to 90% (Harris and Cranston, 1979; Strang et al., 1979). In Australia in the 1920s and 1930s, the invasive cactus *Opuntia stricta* (Haw.) Haw. covered 25 million ha, and infested land had no economic value (Walton, 2005) until cacti were suppressed by the pyralid moth *Cactoblastis cactorum* (Bergroth) (Dodd, 1940).

In addition, toxic or thorny invasive plants are often avoided by livestock. Spotted (*Centaurea maculosa* Lam.)

and diffuse knapweeds, yellow starthistle, leafy spurge (*Euphorbia esula* L.), Russian knapweed (*Acroptilon repens* [L.] DC.), houndstongue (*Cynoglossum officinale* L.), tansy ragwort, and St. Johnswort affect livestock in this way (Kingsbury, 1964; DiTomaso, 2000). Successful biological control of tansy ragwort in Oregon (USA) reduced animal poisonings from tansy (Turner and McEvoy, 1995; Coombs et al., 1999). In Australia, in the 1980s, Patterson's curse (*Echium plantagineum* L.), which is not grazed by cows or horses, infested 5.2 million ha in Australia. Its successful biological control provided benefits estimated as AUS$1.2 billion (from 1972 to 2005) (Paige and Lacey, 2006).

Finally, some invasive species convert grasslands to shrublands of lesser value. Prickly acacia (*Vachellia nilotica* subsp. *indica* [Benth.] Kyal. & Boatwr.) infested 6 million ha in western Australia (Mackey, 1997), and heavily infested areas often could not be reclaimed for grazing. While woody plants such as prickly acacia have been difficult targets for biological control, many of the forbs, as discussed above, have been greatly reduced in density by biological control efforts, allowing restoration of infested grasslands when coupled with better management of grazing and reseeding with native grasses.

Water

In drier areas of the world, invasive woody plants can reduce the flow of rivers in heavily infested watersheds. Plants such as Australian *Acacia*, northern hemisphere pines, Asian *Tamarix* (saltcedar) (Nagler et al., 2008; van Wilgen et al., 2008), and European reeds (*Arundo donax* L.) (Seawright et al., 2009) may reduce water flow by clogging channels or increasing evapotranspiration compared to native vegetation. Such effects may be thought of either as the loss of a product (water) or of an ecological service. In South Africa, plant invasion models predicted that up to 58% of the nation's water was at risk from invasive trees and shrubs (van Wilgen et al., 2008). In the mountains of the Western Cape in South Africa, invasive *Pinus*, *Acacia*, *Hakea*, and *Sesbania* species locally reduced river outflow by 30–80% (van Wilgen et al., 1992; Le Maitre et al., 1996). "Working for Water," a South African program to remove such invasive trees, employed cutting, burning, and application of herbicides for quick reductions in biomass, while biological control agents slowed spread and delayed re-infestation of cleared areas. While biological control agents did not usually kill the mature trees, they lowered seed production and/or seedling survival. When combined with mechanical removal, biological control agents suppressed several species, including *Acacia longifolia* (Andr.) Willd., *Acacia saligna* (Labill.) H. L. Wendl., *Acacia pycnantha* Benth., and *Sesbania punicea* (Cav.) Beth. (Hoffmann and Moran, 1998; Dennill et al., 1999; Moran et al., 2005). *Sesbania punicea*, which once formed dense bands 20–30 m wide along South African rivers, experienced >95% reduction in seeding because of introduced biological control agents (Hoffmann and Moran, 1991, 1998), and as stands died and were not replaced, stretches of rivers once clogged with *S. punicea* became open and free flowing (Hoffmann, pers. comm.).

Electricity generation

Electric power generation can be reduced by invasive plants that create blockages or flow shortages, as for example those caused by waterhyacinth (*Eichhornia crassipes* [Mart.] Solms-Laub.), water lettuce (*Pistia stratiotes* L.), and giant salvinia (*Salvinia molesta* D.S. Mitchell) (Holm et al., 1969; Liabunya, 2007). The economic life of reservoirs may also be shortened owing to increased siltation and build-up of organic detritus (Moorhead et al., 1988; Liabunya, 2007). For example, the Jebel Aulia Dam on the White Nile in Sudan became clogged by waterhyacinth within 10 years of its invasion (Holm et al., 1969; Beshir and Bennett, 1985). Biological control of waterhyacinth, water lettuce, and giant salvinia – the weeds most disruptive to electrical power generation in warm regions – has been widely achieved in tropical areas (Coetzee et al., 2009; Julien et al., 2009; Neuenschwander et al., 2009), relieving this difficulty.

Reductions in fishing or hunting

The harvest of fish and game animals from natural areas may be reduced by invasive plants. Floating invasive plants that form mats, such as waterhyacinth, make use of nets difficult and destroy fish breeding grounds. In Benin (Africa) this caused US$84 million/yr in lost harvest (de Groote et al., 2003). Once waterhyacinth came under biological control, fishing rebounded (Ajuonu et al., 2003), and harvest values increased US$30.5 million/yr (de Groote et al., 2003). Fishing in the Winam Gulf of Lake Victoria (Kenya), valued at US$83 billion and supporting 15 million people, was blocked by a waterhyacinth infestation (Opande et al., 2004) until that infestation also came under biological

control. In Papua New Guinea, salvinia infestations prevented the netting of fish (Mitchell, 1981), reducing salt fish production by 30% until biological control removed salvinia mats (Thomas and Room, 1986).

Effects of invasive plants on harvest levels of wild game are less well documented. In Idaho (USA), areas infested by yellow starthistle supported only two-thirds as many chukars (*Alectoris chukar*), a popular game bird, as uninfested areas (Lindbloom et al., 2004). In western Montana (USA), elk (*Cervus elaphus*) activity in areas infested by spotted knapweed was reduced by 98% compared to areas with bluebunch wheatgrass (*Pseudoroegneria spicata* [Pursh] Á. Löve) (Hakim, 1979).

Non-commodity ecosystem services lost to invaders

Ecosystem services can be reduced by invasive species. Which services are so affected and the value of the lost services are difficult to estimate, but the Millennium Ecosystem Assessment (2005) provides a framework for categorizing effects of invasive species on values obtained from ecosystems (here termed *wildlands* for convenience). The scheme uses four categories: *provisioning services*, *regulating services*, *cultural services*, and *supporting services* (Charles and Dukes, 2007). Of these, the first category corresponds to "tangible products" as discussed above. The remaining groups are non-commodity goods. *Regulating services* include regulation of air quality, climate, hydrological cycles (water discharge cycles, flooding, etc.), and fire regimes. *Cultural services* are relatively intangible values, such as the recreational, spiritual, educational, or scientific values of a place, while *supporting services* are overarching, long-term matters, such as soil and mineral recycling, fertility, and photosynthesis. Using these categories, Pejchar and Mooney (2009) discuss impacts of invasive species for South Africa (effects of woody plants on the fynbos), the Great Lakes of North America (effects of zebra mussel [*Dreissena polymorpha* {Phallas}] invasion), and Hawaii (effects of feral pigs on native forest) as examples. While the potential scope of impact of the whole suite of invasive species on ecosystem services is broad and complex (being net impacts of both positive and negative influences), here we focus on a subset of invasive species for which biological control has been attempted as a means of restoring such services (see Denslow and D'Antonio [2005] for examples). While the whole cycle of data collection, from demonstrating loss of a service because of an invader, to demonstrating biological control of the invader, to demonstrating recovery of the lost service, has only been achieved in a few cases, examples do exist. One such case is that of woody plants in the South African fynbos, whose effects on ecosystem services have been estimated (Le Maitre et al., 1996; de Lange and van Wilgen, 2010), whose suppression (including through biological control) has been achieved and documented (Moran et al., 2005), and the value of the recovered services estimated. For this and other systems in South Africa, de Lange and van Wilgen (2010) estimated that some US$19.7 billion in ecosystem services were saved annually through invasive species suppression and that, of those savings, 5–75% was owing to biological control, depending on the target plant species.

Here, while topic complexity does not allow a full accounting, we give some examples of the types of ecosystem services protected (if not fully quantified) through biological control of invasive species, focusing on protection of soils, water and fire conditions, and air quality.

Maintenance of soils

Conservation of soil attributes, upon which native plants depend, is classified as a *supporting service*. Many soil attributes can be affected by invasive plants, but impacts on erosion, moisture, salinity, and nitrogen are of particular interest. Change in vegetative cover, as follows from some plant invasions (sometimes coupled with other events like overgrazing), can increase erosion by reducing ground cover, as occurred, for example, with invasive knapweeds (Lacey et al., 1989) in Montana (USA) and rubber vine (*Cryptostegia grandiflora* [Roxb. ex R. Br.] R. Br.) in Australia (Vogler and Lindsay, 2002). Progress has been made on achieving biological control of both rubber vine (Tomley and Evans, 2004) and some knapweeds (Myers et al., 2009), but changes in erosion rates as a consequence of control have not been measured.

In the southwestern United States, dam construction and water removal have led to drier, saltier soil, which has provided a competitive advantage for invasive species of saltcedar (*Tamarix* spp. and hybrids) (Glenn and Nagler, 2005), which in turn promotes increased fire (Busch and Smith, 1993), further increasing dryness. Reversing these outcomes will require re-establishment of pulse flooding, substantial reductions in saltcedar

coverage, and efforts to re-establish native willows (*Salix* spp.) and cottonwoods (*Populus* spp.). Biological control of saltcedar is emerging, with widespread defoliation occurring at sites where agents have been released (DeLoach et al., 2008) and decreased regrowth and lowered carbohydrate stores following repeated defoliations (Hudgeons et al., 2007). Defoliation and fire synergistically increase *Tamarix* mortality (Drus et al., 2014). However, it is not yet clear that reductions in *Tamarix* alone will restore lost ecosystem services with regard to soil condition and water budgets.

While from an agricultural point of view, maintenance of high soil fertility is desirable, in some wildlands soil fertility may be naturally low, and increases caused by invasions of nitrogen-fixing plants can threaten important biodiversity. Examples of such interactions are found in young volcanic soils such as in parts of Hawaii and very old, highly leached soils, such as in the fynbos of South Africa, both of which have important endemic plant communities dependent on low-nitrogen soils. The continuation of such conditions would be classified as a supporting service (Millennium Ecosystem Assessment, 2005). In South Africa, several *Acacia* species, which as a group are nitrogen fixers, have been introduced from Australia for forestry or use as ornamentals. Several are now invasive in low-nitrogen landscapes such as the fynbos, where they are raising soil organic matter and nitrogen levels (Stock et al., 1995). In the Riverlands Nature Reserve in the Western Cape, for example, levels of nitrogen returned to the soil by *A. saligna* were an order of magnitude greater than by native fynbos plants (Yelenik et al., 2004), thereby reducing growth of some fynbos plants (Lamb and Klaussner, 1988). Control of *A. saligna* by the introduced fungus *Uromycladium tepperianum* (Saccado) reduced its density by 87–98% (Wood and Morris, 2007). Experimental mechanical clearance of *A. saligna* from fynbos areas created a pulse of available nitrogen, which was discharged in groundwater (Jovanovic et al., 2009). Since tree death from biological control is slower, the pulse of release nitrogen should presumably be less pronounced, but has not been studied.

Maintenance of hydrological conditions

Invasive plants can affect flood crests or diminish river outflows. To date, such effects of invasive plants have not been changed by biological control but may be in the future. For example, in its invaded range in the United States, dense stands of giant reed (*A. donax*) can increase flooding by increasing sedimentation, slowing flood discharges and clogging channels (Frandsen and Jackson, 1994; Graf, 1978). Biological control against giant reed in the United States has begun (Tracy and DeLoach, 1999), but no population-level impacts on the plant have yet been achieved.

In dry areas, invasive woody plants can reduce river outflows (Hope et al., 2009) if they use more water than the native vegetation. In South Africa's Western Cape, native *Protea* shrubs were displaced by invasive woody trees, increasing plant biomass 10-fold (Versfeld and van Wilgen, 1986) and reducing river discharge by 30–80% (van Wilgen et al., 1992; Le Maitre et al., 1996). Many of these invasive trees have been controlled by a mixture of biological control (Moran et al., 2005), and cutting and poisoning.

Maintenance of fire regimes
Conservation of fire regimes is classified as a regulating service. Some invasive plants can change fire regimes by altering fuel characteristics or timing (Brooks et al., 2004), and such changes can strongly affect plant communities. Several fire-altering invasive plants have been targets of biological control. *Chromolaena odorata* (L.) King & H. E. Robinson, for example, is a shrub from the Americas that is invasive in warm areas with a distinct dry season, including parts of Africa and Asia, where its pithy stems and oily leaves raise fuel loads and increase fire (McFadyen, 2004). Biological control of the biotype found in Asia has largely suppressed the plant in Papua New Guinea (Day and Bofeng, 2007) and East Timor (Zachariades et al., 2009), through the introduction of the gall fly *Cecidochares connexa* (Macquart) (Diptera: Tephritidae).

In southern Florida (USA), the tree melaleuca (*Melaleuca quinquenervia* [Cav.] S.T. Blake) has invaded several habitats, and in pine and cypress stands, oils in melaleuca litter and standing trees brought canopy fires into areas normally experiencing only ground fires, killing native trees (Wade, 1981). Biological control agents reduced seeding and spread of melaleuca, while tree cutting and application of herbicides in existing stands reduced the infested area by nearly half (Tipping et al., 2008, 2009; Center et al., 2012).

Also invasive in many south Florida habitats is Old World climbing fern, *Lygodium microphyllum* (Cav.) R.

Br., which forms thick skirts of foliage on trees that move ground fires into the canopy, increasing tree mortality in bald cypress (*Taxodium distichum* [L.] Rich.) stands (Pemberton and Ferriter, 1998). Biological control of this fern is developing (Smith et al., 2014) but has not yet been achieved.

With all of the above examples, what is missing is information determining whether or not fire regimes have reverted to conditions resembling historical norms following the biological control of these plants. While intuitively it seems that this should be the case, studies to document this outcome would be useful and become possible as the invasive plants are suppressed.

Reduction of air pollution
Preservation of clean air is classified as a regulating service and invasive plants may lower air quality by increasing soil erosion or emitting large amounts of allergenic pollen. For example, in the dry-land crops of eastern Washington (USA), tillage is used if fields are heavily infested with tumbleweeds (*Salsola* spp.) (Young, 2006). The resulting fine dust (<10 lm) is an air pollutant according to US-EPA definitions (Sharratt and Lauer, 2006). Biological control of tumbleweeds is in progress (Smith et al., 2009) but has not yet been achieved.

Benefit: cost ratios for biological control

There has long been a considerable interest in quantifying in monetary terms the value achieved through biological control in order to guide public policy. Methodologies for making benefit: cost estimates for biological control projects are given by Culliney (2005) and further discussed in terms of economic modeling by Holmes et al. (2009). For pests that cause monetary loss to agriculture, forestry, or livestock grazing, avoided losses have been measured many times. Estimates of benefit: cost ratios have been as high as 7405 : 1 for individual projects (Hill and Greathead, 2000); 4000 : 1 for individual weed projects (Culliney, 2005), 112 : 1 for Australian projects (Tisdell, 1990; see also Lubulwa and McMeniman [1998]), 23 : 1 for Australia weed projects (McFadyen, 2008), and 50 : 1 and 3726 : 1 for subtropical shrubs and invasive Australian trees, respectively, in South Africa (de Lange and van Wilgen, 2010). These ratios refer variously to pests of wildlands, pests of agriculture, or mixes of the two groups.

For biological control of weeds, financial benefits from loss of products (protection of *provisioning services*) may be measureable if the invasive species reduces yields of tangible assets, such as water or rangeland forage. Examples of such benefits (*provisioning services*) include the pasture pest tansy ragwort in Oregon (benefit: cost ratio 13 : 1) (Coombs et al., 1996), the grassland range pest leafy spurge in the northern Great Plains of the United States (Bangsund et al., 1999), waterhyacinth in Benin (Africa) (benefit: cost ratio 124 : 1) (de Groote et al., 2003), and red waterfern (*Azolla filiculoides* Lamarck) (benefit: cost ratio 13–15 : 1) (McConnachie et al., 2003).

van Wilgen et al. (2004) estimated benefits from six weed biological control projects in South Africa (*Opuntia aurantiaca* Lindl. [jointed cactus], *S. punicea* [red sesbania] *Lantana camara* L. [lantana], *A. longifolia* [long-leaved wattle], *A. pycnantha* [golden wattle], and *Hakea sericea* Schrad. & J.C.Wendl. [silky hakea]) in terms of three losses that were ameliorated: (1) loss of water from increased vegetative evapotranspiration of infested land; (2) reduced economic value of land (for commodity production) from weed infestation; and (3) loss of commodities harvested from nature because of reduced biodiversity. Note that all three losses are products (thus all benefits are from preservation of *provisioning services*); also, note that to estimate total value from change in land value, the spread of the invasive plants had to be estimated by the authors for the defined periods for which the estimate was made. Benefits were calculated both per ha and countrywide (assuming spread of weed to its ecological limits). On a per ha basis, the benefits of preventing weed infestation ranged from 300 to 3600 rand (ca US$39–468) per ha per year for jointed cactus and golden wattle, respectively. Impacts of jointed cacti were mainly on loss of land value, while impacts of wattles were on loss of stream outflow (water). Losses owing to biodiversity products not available for harvest were minor in comparison for all weeds, except jointed cactus. On a countrywide level, estimated from the beginning of biological control research on a particular weed to the year 2000, benefits of biological control of jointed cactus were estimated at 10,269,000 rand (ca US$1,334,000), while the benefits of control of long-leaved and golden wattles were 3,483,000 and

3,365,000 rand (US$452,000 and US$437,000), respectively. Note that no value was placed on biodiversity per se, nor were losses of *regulating*, *cultural*, or *supporting services* considered.

In contrast, de Lange and van Wilgen (2010), when estimating the value of weed biological control in South Africa, incorporated effects of invaders on both products (water, grazing) and local biodiversity, including both use values of extracted products and non-use values. The monetary value of biodiversity protected through biological control of invasive plants was calculated using the Biodiversity Intactness Index developed by Scholes and Biggs (2005). This index focuses on impacts of a set of factors (including invasive species) on plants and vertebrates. The index integrates richness estimates and area-weighted average impacts on native plants, mammals, birds, reptiles, and frogs in a given area to estimate the percent intactness of a given biome. For their purposes, de Lange and van Wilgen (2010) assumed a linear relation between intactness and services. Thus, while not explicit as to what ecosystems services are being protected, biodiversity acts as a surrogate for a potential package of services of either a provisioning, regulating, cultural, or supporting nature. To explain further: while biodiversity is not *per se* an ecological service (*sensu* Charles and Dukes, 2007), it may embody such services. For example, some aspects of biodiversity may be of scientific interest (as Darwin's finches in the Galápagos), thus being a *cultural service*, or other aspects of biodiversity (such as organisms that restrain pests) may be involved in pest control (as for example parasitoid diversity in Mexican tropical forests that likely contributes to control of pest fruit flies in orchards [see Aluja et al., 2014]), thus being a *regulating service*.

Ecological values of South African biomes were determined by van Wilgen et al. (2008), and those estimates were used by de Lange and van Wilgen (2010) to determine what portion of that ecological service's value was being protected by weed biological control. This assessment is unique in that it was made not species by species, but holistically, considering for each biome the whole suite of invasive plants that might act interchangeably. The study considered four groups of invasive plants: fire-adapted trees, perennial invasive Australian trees, cacti, and subtropical shrubs. The annual value of benefits from biological control was estimated (in millions of rand, in 2010) as follows: 67

(fire-adapted trees), 8330 (invasive Australian trees), 2977 (invasive succulents), and 205 (subtropical shrubs), equivalent to 8.7, 1082, 387, and 27 million US$, respectively. This approach also took note of likely species replacements (such as pines [not subject to biological control] being likely to replace *Hakea* [suppressed by biological control] in the fynbos) and reduced estimated values of certain biological control efforts accordingly. Using the biodiversity intactness index, van Wilgen et al. (2008) found that the five major biomes in South Africa were relatively intact (71–89%), but that this would decrease with time because of plant invasions, to about 30% for savanna, fynbos, and grassland biomes, and decline even further (13 and 4%) for the karoo biomes (succulent karoo and nama karoo). This estimate of impending loss then formed the basis for the estimates of ecosystem services preserved by weed biological control in South Africa (de Lange and van Wilgen, 2010). This study is the most comprehensive assessment of the value of biological control activities for protecting goods and services from ecosystems on wildlands.

Additional costs of restoration to desired conditions

In planning a biological control program in wildlands, it is essential to bear in mind from the beginning that the goal is restoration of the affected native species or ecosystem, not just suppression of the targeted invader. This distinction may be of less importance for invasive forest insects, which are typically drivers of ecological degradation (they kill their host trees) and often act in relatively undisturbed ecosystems – at least those not requiring other forms of restoration. However, for invasive plants, further restoration actions or pest control efforts are commonly necessary to bring the native ecosystem back to the desired condition. Therefore it is important that the full range of such potential costs be considered at the time of the initial project planning. That is, we should take a holistic approach to restoring the affected species or community and plan accordingly from the beginning, rather than react in an unplanned way subsequent to the biological control program.

One additional cost that may arise in some biological control projects is the need to supplement biological control with mechanical or chemical control measures taken against life stages of the target plant that are not affected by the biological control agents. The clearest example of this is in the biological control of invasive trees. Often the biological control agents used against such species, as for example those for melaleuca in the Everglades of Florida (USA) or used against the invasive trees in the fynbos of South Africa, do not affect adult trees. The biological control agents that ultimately were used against melaleuca affected primarily seed production and seedling or stump sprout survival. Defoliating agents affecting survival of adult trees that were considered sufficiently risk free were not found, although a gall-forming midge was released (Pratt et al., 2013), but has not yet affected survival of adult trees. Consequently, the biological control program lowered population reproduction to the point that spread of the tree to new areas was curtailed (Tipping et al., 2012a) and sprouts on stumps of cut trees were killed. Stands of mature trees, however, survived. Their removal on public conservation lands was achieved by publicly financed cutting, burning, or herbiciding (Center et al., 2012), removing mature melaleuca from about half of the total infested area in the state. These efforts had costs, but these costs were not recurring because the biological control program prevented re-infestation of cleared sites (Tipping et al., 2012b). The remaining stands left at the end of the cutting program were largely on private lands and were no longer a threat to public lands owing to the biological control program. In the case of biological control of invasive trees in South Africa (species of *Acacia*, *Sesbania*, *Hakea*), some of the targeted species were valued as ornamentals or for plantation forestry. Consequently, it was largely seed production that was targeted by the biological control agents, although a gall-making agent targeted against *A. longifolia* also significantly reduced vegetative growth (Dennill, 1988). Since the goal of the project was restoration of river outflow, removal of unaffected adult trees was needed and was achieved with mechanical, chemical, and manual clearance efforts (Hosking and du Preez, 1999).

Another consideration that has economic implications for use of biological control of weeds in wildlands is that there may be a suite of weed species that affect the habitat in a similar way (a group of vines, or a group of floating plants, or a group of nitrogen-fixing shrubs).

In tropical rivers, for example, control may be needed for alligatorweed (*Alternanthera philoxeroides* Griseb.), waterhyacyinth, and salvinia, and if only one is suppressed through biological control, one of the others, if present, may fill the gap created by biological control. Therefore, at the planning stage for system restoration, this potential has to be recognized and plans made accordingly. For example, in the fynbos region of South Africa affected by silky hakea, while that invasive tree was effectively suppressed by biological control, pines were not. Because economic interest groups were unwilling to allow seeds of *Pinus* species to be targeted by biological control (Hoffmann et al., 2011), it is expected that invasive pines will eventually dominate areas formerly infested by silky hakea (de Lange and van Wilgen, 2010).

It may also be the case that restoration of a system affected by invasive plants cannot be achieved by suppression of the targeted weed alone. The habitat may be in some way deficient and not favorable for the native plants whose return is a key part of the restoration goal. This may be because some key processes have been altered in ways that are unfavorable for native plants. For example, restoration of some western US riparian areas blanketed by monocultures of tamarisk (*Tamarix* spp.) may require re-institution of pulse flooding on dammed rivers because overbank flooding is needed for cottonwood (*Populus* sp.) and willow (*Salix* spp.) seed dispersal and seedling establishment.

Further efforts may also be needed for restoration, including the replanting of desired native plants either no longer found in the system or not in sufficient numbers. Many saltcedar stands in the southwestern United Stands, for example, may require such replantings as the biological control agents defoliate saltcedar plants, making them less competitive, as pure stands of saltcedar are likely to have few native plants. Similarly, re-introduction of missing native animal species may be required to meet restoration goals as, for example, introduction of engineering species such as the beaver (*Castor canadensis*), to slow water flow and create wetlands.

Finally, some invasive species, for which biological control is not feasible, may need to be suppressed or eradicated, such as larger mammals, rodents or invasive ants on islands. This may be unrelated to biological control of targeted plants, but is related if it is required to restore the affected ecosystem. One must plan restoration programs holistically and incorporate a comprehensive view of the likely costs of the whole suite of actions that may be necessary.

Effects of economic interests on target selection for biological control

Another economic issue connected to biological control of invasive pests of wildlands is how species are selected for control. The source of funds is an important influence on which pests become targets for biological control. While approaches vary by country, funding comes either from governments or, for agricultural pests only, from associations of producers (sugarcane growers, grazers, coffee producers, etc.), or a mixture of both.

In the United States, both sources have been involved. Sugarcane growers, for example, funded much of the early work in Hawaii on pests of sugarcane and other commodity associations funded work on pests affecting their industries. In contrast, much of the funding for biological control of minor crop and forestry pests has traditionally come from state or federal agencies. Some overlap may exist between pests of wildlands and crop pests. For example, glassy-wing sharpshooter, *Homalodisca vitripennis* (Germar), now invading various Pacific island nations, affects wild plants as well as crops (Petit et al., 2008), as does cottony cushion scale, *Icerya purchasi* Maskell, mainly known as a pest of citrus, but which became a critically important pest of wildlands after its invasion of the Galápagos Islands (Causton, 2005). In both cases, wildlands benefited from projects originally undertaken for protection of crops (Grandgirard et al., 2008; Calderón Alvarez et al., 2012; Hoddle et al., 2013). An important difference between targeting crop vs. wildland pests is that crop pests will have vocal advocates seeking access to government funds for biological control, but pests that affect only wildlands may not.

In Australia, weed biological control programs were traditionally funded publicly. In 1996 a new program to fund biological control of rangeland weeds, addressing weed impacts on both grazing and biodiversity, was launched. This initially made substantially greater funding available for weed biological control projects (Martin and van Klinken, 2006). However, later, government funding was reduced, and by 2014 weed biological control work was funded almost solely through trade associations, such as that of grazers, and was directed towards pests affecting the groups' economic interests. Projects funded in this manner targeted several invasive plants that had dual importance as both pests of grazing land and of natural areas, namely bellyache bush (*Jatropha gossyipifolia* L.), prickly acacia, and parkinsonia (*Parkinsonia aculeata* L.) (Rieks van Klinken, pers. comm.). With virtually no government funding (2014) currently available for weed biological control, it seems reasonable to surmise that pests chosen as targets will be those of greatest concern to the trade group's economic interests and that any benefits to wildlands will be secondary. Consequently, species of great importance to wildlands would often not be addressed if they did not also cause economic losses to citizens.

In New Zealand, resources for biological control programs against plants infesting wildlands come primarily from the National Biological Control Collective, which consists of local governmental bodies called Councils, which represent portions of the country, and the Department of Conservation. The Collective agrees on weed biological control priorities and provides funding for Landcare Research (a crown research institute) to undertake the selected projects. The National Biological Control Collective has provided a consistent level of funding for about two decades (1995–2015), enabling capability to be maintained at Landcare Research and providing the necessary evidence for stakeholder support that then enables Landcare Research to obtain central government research funding for scientific research that runs alongside the more operational projects. Choice of weed targets is based on a decision framework that measures the weed's impact (importance), the likelihood of successful biological control (feasibility), and the likely effort (cost) (Hayes et al., 2013). Most of the targets selected for funding by the Biological Control Collective since about 1995 have been weeds of wildlands.

In South Africa, projects are identified and carried out by the Plant Protection Research Institute (PPRI), the equivalent of the USDA in the United States, on behalf of the country with public funds.

While the above approaches work reasonably well for selection of economically damaging species as targets for biological control to protect farming, forestry, or grazing interests, they may neglect pests of natural areas whose impacts are not directly economically damaging. As an alternative, we suggest that a system to rank pests of natural areas be created, in cooperation with groups such as the Invasive Pest Plant Councils in the United States. Such groups, working regionally with conservation biologists familiar with regional ecosystems and invasive plants or insects, could identify which invasive species are most injurious to native ecosystems' function or biodiversity (see Chapter 2 for an example of how pests affecting the floodplain forest ecosystem of the Connecticut River watershed were ranked by Christian Marks of The Nature Conservatory and Chapter 15 for further ideas on how to choose and fund biological control projects). Such a system would prevent direct competition for funding between projects to protect crops and those needed to protect wildlands. This approach would also engage a source of expertise (conservation biologists) that while fully engaged in the process in South Africa, New Zealand, and Australia, is not currently involved in the United States. Such input would be particularly helpful when issues of being a driver vs. passenger of observed ecological change need clarification or ecosystems are complex or biological control alone is not sufficient for their restoration.

How should biological control in wildlands be financed?

Principles

Rather than just report here on how repair of ecosystems damaged by invasive species is funded, we speculate from principles on what might be a more optimal system. This situation is a clear example of the externalization of environmental costs flowing from economic activity: the benefits of growing interregional and international trade accrue predominantly to importers and exporters and their suppliers and customers, while the environmental and economic costs of the resulting invasive species introductions tend to be borne by society at large (Wallner, 1996; Chornesky et al., 2005; Burgiel et al., 2006; Humble, 2010; Lonsdale, 2010; Wingfield et al., 2010) (see also Chapter 15). Reversing this dynamic through prevention measures that require the beneficiaries of economic trade to internalize costs associated with invasive species would provide a direct economic incentive to those parties to reduce introductions of invasive alien species (see Simpson, 2010). Doing so is consistent with the "introducer- or polluter-pays" principle that is central to global environmental governance (Bailey et al., 2012). While it is possible to tax entities that legally sell, trade, or import non-native species (Jenkins, 2002; Knowler and Barbier, 2005; Mérel and Carter, 2008; Barbier et al., 2011), implementing such a tax requires information on how long it takes introduced species to become established, their rates of spread, and the damage they cause to the environment (Barbier et al., 2013).

A repeated conclusion in the invasive species literature is that funding is insufficient to manage or eradicate invasive species, regardless of the source. As a result, there has been rather substantial effort placed on developing protocols for prioritization of invasive species against which limited resources could be allocated in an effort to protect wildlands from their unwanted effects. Approaches vary from the use of decision-making algorithms (e.g., Pyke et al., 2010; Darin et al., 2011; Larson et al., 2011), bioeconomic modeling (e.g., Fleischer et al., 2013; Frid et al., 2013), case pattern analysis (e.g., Schmidt et al. 2012; Iacona et al. 2014), and pathways of introduction risk analysis (e.g., Leung et al. 2014). A limitation of all approaches is in their ability to capture all factors potentially necessary to make the most informed decision. However, each approach can simplify a complex array of factors while generating a reliable prediction.

Invasive species management and identification of the financial resources necessary to execute such programs will likely fail without a coordinated effort that incorporates substantial public participation (Perrings et al., 2002). Sharp et al. (2011) used an analytical approach for evaluating public knowledge of, attitudes toward, and support for invasive species management. With this approach, the researchers hope ecologically appropriate and socially acceptable funding and execution of invasive species management programs can minimize conflict by identifying stakeholder characteristics and synthesizing public preferences. Based on an extensive survey of visitors to Cumberland Island National Seashore, Georgia, USA. Sharp et al. (2011) identified two distinct groups of visitors. The first group consisted of younger individuals who were less educated and had less experience in parks and believed that all living things have a right to coexist without disruption. This group favored a hands-off management approach. The second group consisted of older individuals who were more educated and, with more experience in parks, believed that some degree of human interference is

necessary to maintain ecosystem integrity. They favored hands-on management. Yet when given the opportunity to select the most acceptable management options, both groups preferred adaptive on-site management of invasive species as the most acceptable and least divisive management option relative to leaving the invader alone or its complete eradication.

Self-regulation through a set of voluntary codes of conduct is unlikely to lead to predictable and effective management of invasive species. In California, a coalition of environmental groups, scientists, government agencies, and the horticultural industry are partnering with retail nurseries and growers on voluntary measures to reduce the sale of invasive plants and promote non-invasive alternatives (Brusati et al., 2014). While the success of the California coalition in reducing the flow of invasive plant species is unknown, evidence suggests that few nursery professionals have implemented a majority of the measures because of cost, lack of information, and limited personnel or time (Burt and Muir, 2007).

It has been proposed that biological invasions should be managed as natural disasters since both have similar characteristics (Ricciardi et al., 2011). Alexander (1993) defines a natural disaster as a rapid and damaging socioeconomic impact caused by the natural environment that arises when a hazard, concentrated in space and time, threatens a society "with major unwanted consequences as a result of failed precautions, which had hitherto been accepted as adequate." To the degree that biological invasions are fundamentally analogous to natural disasters, both require similar management strategies, unwavering political devotion, and sufficient financial obligations. Sapir and Lehat (1986) argue that biological invasions have more persistent impacts and a greater scope of ecological and economic damage than do natural disasters. As a result, there should be consideration for similar allocations of resources necessary for effective preventative and remedial management by nations and governments to rare natural events and rare invasion events that can potentially inflict enormous socio-economic costs (Ricciardi et al., 2011).

Some countries have adopted measures requiring exporters to conduct impact assessments that cover the full range of domestic ecosystem considerations when they seek to introduce new alien species (Burgiel et al., 2006). In addition to being consistent with long standing practice in the quarantine area, burden- and cost-shifting appear consistent with the World Trade Organization Agreement on the Application of Sanitary and Phytosanitary Measures (SPS Agreement of 1995). Although that agreement requires countries adopting prevention measures to ensure that measures are based on scientific evidence and risk assessments, nothing in the agreement specifies that the country undertaking the measure must bear all costs and burdens associated with that measure.

In the United States, federal and state authorities may aggressively prosecute individuals believed to be responsible for starting wildfires. A fire-starter's intent, past criminal history, insurance policy or wealth, willingness to report or fight the fire, and other case-specific circumstances are some of the considerations prosecutors weigh in deciding whether to pursue a criminal case. While this is not currently the case, in principle, the person or business responsible for an biological invasion, if known, could bear some financial responsibility for the consequences, depending on whether or not the pest species was (1) deliberately introduced (e.g., plants, pets) or (2) an unknown hitchhiker (wood-boring insects, insects and pathogens on live plants). In the latter case, the degree of responsibility could depend on whether or not proper preventative procedures to minimize risks from a given pathway were followed or not and how the business responded after the invader was detected. As a first step in this direction, exact sources of introductions (pathway and business source) should be identified if possible and the information made public on a website dedicated to such data. Such information may then be useful in inducing voluntary changes in behavior, either of the importing business or its customers, to lower the risk of the vectoring process. Second, governments would need to pass statutes outlining the degree of responsibility businesses should bear, define how responsibility should be legally determined (what standard of proof might be required), and enable conservation groups like the NRDC (Natural Resources Defense Council) to bring suits against invasive species importers who have acted willfully or without due care.

Species whose introduction is deliberate

Certain industries (trade in live plants, pet trade, and the aquaculture industry) and certain government agencies (fish and game departments, rangeland or forest organizations, and authorities in charge of biological control) import exotic species intentionally. In the case of the industries cited above, nations should pass statutes that define the extent of responsibility an importing business bears for "its" plant or animal species after individuals of the species have been sold to customers. For example, should the pet importation businesses that sold Burmese pythons (*Python bivittatus* Kuhl) to customers in Florida (a region with a climate suitable for the snake's establishment) be required to pay part of the mounting costs of python suppression efforts now that the snake has established breeding populations in the Florida Everglades? Currently no mechanism to enforce such a responsibility exists, but it could be established by law or civil suits. The problem of multiple sources (many businesses importing the same species over time) would certainly arise, but it could be dealt with in a manner similar to the way companies in the United States are held liable under Superfund legislation for cleanup for toxic sites polluted by several businesses.

In the case of government agencies that deliberately establish populations in the wild of species that later prove to undesirable invaders, a system of lawsuits and monetary fines seems less likely to be effective in changing behavior as the individuals involved change over time. A more effective approach may be to establish, by law, standards that aim to discourage introduction of exotic species, coupled with a reporting system that would require annual reporting by agencies of their introductions, the expected benefits vs. risks, and actual outcomes. Such reports should be subject to independent outside review and should be posted on government websites for public scrutiny. While past errors, such as the wide dissemination of rainbow trout (*Oncorhynchus mykiss*) into waters occupied by other native trout species and sowing of natural grasslands with exotic grass species to increase forage production, are regrettable, current agencies cannot be held accountable today for such past mistakes. However, this does not mean that introductions of new species or of previously moved species into yet more new areas should not be closely scrutinized and debated before being permitted or denied.

Species whose introduction is accidental

A separate question is the matter of who is responsible when importations of legitimate goods in trade lead to the establishment of damaging hitchhiking species. If, for example, a business rearing trout or other fish species imports diseased stock, which then leads to escape of the pathogen (or parasite) to the wild, who should pay the costs of remediation? In this case, the importer surely did not intend to import a pathogen, so the question becomes (1) was such a risk foreseen and mitigated in the design of the facility or not (or should it have been?), or were there design or maintenance flaws that permitted escape, and (2) once the pathogen or parasite's presence was known, did the business contact appropriate authorities and act responsibly to contain it? For example, abalone producers in California accidentally imported abalone infected with the parasitic sabellid worm, *Terebrasabella heterouncinata* (Fitzhugh and Rouse) (Culver and Kuris, 2000), which escaped in water discharged from the rearing facility directly into an ocean cove and became established in turban shell snails. The business' culpability lay not in the actual importation of the parasite, which was clearly not intentional, but in constructing its waste water disposal system so that it drained into water bodies instead of dead ending into a leach field and thus having no defense against such an event. The requirement for such preplanned defenses would be analogous to the currently required cement holding areas built around industrial oil depot tanks to contain tank leaks. Similarly, importers of nursery stock into California infected with what later came to be called the sudden oak death pathogen (*Phytophthora ramorum* [S. Werres, A. W. A. M. de Cock & W. A. Man in't Veld]), did so unwillingly and unaware, as the imported plant species were asymptomatic for the pathogen or nearly so. Therefore, the importers' culpability or innocence was determined by their later degree of cooperation with state officials to control runoff of water from contaminated nurseries that could bear the water-borne fungus away to natural habitats.

A further example of the need for courts to adjudicate degrees of culpability is found in a hypothetical case concerning importation of Siberian conifer logs for milling in underused Oregon sawmills that was proposed in the 1990s. Such logs, depending on their condition and degree of inspection, would potentially have posed a significant risk of introducing Asian pink moth (formerly Asian gypsy moth) (*Lymantria mathura* [Moore]), a forest defoliator. The policy concerning such log imports and the degree of APHIS inspection required was ultimately determined by Congress. But if during this legislative process, the sawmill owners had

lobbied for less stringent inspection, they would then have incurred some culpability in the event that the moth was introduced and established, since their actions would have led to the design of a less effective preventative system of inspection

While the concept of holding businesses financially responsible to some degree for invasive species invasions resulting from their business activities is not yet a notion established in US culture, parallels exist with product liability, clean-up of toxic waste sites (in the United States), and funding of environmental action programs that might be a model for managing risks associated with invasive species. The legal basis for imposing such responsibility should be explored by a law research article in one of the nation's law review journals. Taxes imposed on importing industries could provide funds for control activities, from eradication to biological control (see Chapter 15 for more on this idea).

Conclusions

From this consideration of how economics affects the use of biological control against invasive species of wildlands, we draw several general conclusions: (1) better documentation of economics of wildland invaders would be valuable in policy decisions. (2) A better process of selecting species as targets of biological control to reflect their true ecological impacts would be helpful. (3) Greater as well as new funding mechanisms are required if we are to tackle more of the important invasive species affecting native species and ecosystems.

Acknowledgments

We thank Keith Warner, David Wagner, and Mark Lonsdale for reviews of Chapter 14.

References

Ajuonu, O., V. Schade, B. Veltman, et al. 2003. Impact of the weevils *Neochetina eichhorniae* and *N. bruchi* (Coleoptera: Curculionidae) on waterhyacinth, *Eichhornia crassipes* (Pontederiaceae), in Benin, West Africa. *African Entomology* **11**: 153–161.

Alexander, D. E. 1993. *Natural Disasters*. Kluwer Academic Publishers. Norwell, Massachusetts, USA. 633 pp.

Aluja, M., J. Sivinski, R. G. Van Driesche, et al. 2014. Pest management through tropical tree conservation. *Biodiversity and Conservation* **24**: 831–853.

Anonymous. 2004. Forest Insect and Disease Conditions in the United States 2003. United States Department of Agriculture, Forest Service, Forest Health Protection. Available from: http://www.fs.fed.us/foresthealth/publications/Conditions Report_2003.pdf [Accessed January 2016].

Bailey, P., E. McCullough, and S. Suter. 2012. Can governments ensure adherence to the polluter pays principle in the long-term CCS liability context? *Sustainable Development Law & Policy* **12**(2): 46–51, 69–70.

Bangsund, D. A., F. L. Leistritz, and J. A. Leitch. 1999. Assessing economic impacts of biological control of weeds: the case of leafy spurge in the northern Great Plains of the United States. *Journal of Environmental Management* **56**: 35–43.

Barbier, E. B., J. Gwatipedza, D. Knowler, and S. H. Richard. 2011. The North American horticultural industry and the risk of plant invasion. *Agricultural Economics* **42**: 113–130.

Barbier, E. B., D. Knowler, J. Gwatipedza, et al. 2013. Implementing policies to control invasive plant species. *Bioscience* **63**: 132–138.

Bauer, L. S., J. J. Duan, and J. R. Gould. 2014. Emerald ash borer (*Agrilus planipennis* Fairmaire) (Coleoptera: buprestidae), pp. 189–209. *In*: Van Driesche, R. G. and R. Reardon (eds.). *The Use of Classical Biological Control to Preserve Forests in North America*. FHTET- 2013-02. USDA Forest Service, Morgantown, West Virginia, USA. Available from: http://www.fs.fed.us/foresthealth/technology/pub_titles.shtml [Accessed January 2016].

Beshir, M. O. and F. D. Bennett. 1985. Biological control of waterhyacinth on the White Nile, Sudan, pp. 491–496. *In*: Delfosse, E. S. (ed.). *Proceedings of the VI International Symposium on Biological Control of Weeds, August, 1984*. Agriculture Canada. Ottawa, Canada.

Brooks, M. L., C. M. D'Antonio, D. M. Richardson, et al. 2004. Effects of invasive alien plants on fire regimes. *BioScience* **54**: 677–688.

Brusati, E., D. W. Johnson, and J. DiTomaso. 2014. Predicting invasive plants in California. *California Agriculture* **68**(3): 89–95.

Burgiel, S., G. Foote, M. Orellana, and A. Perrault. 2006. *Invasive Alien Species and Trade: Integrating Prevention Measures and International Trade Rules*. Defenders of Wildlife and the Center for International Environmental Law. Washington, DC. 54 pp.

Burt, J. W. and A. A. Muir. 2007. Preventing horticultural introductions of invasive plants potential efficacy of voluntary inititatives. *Biological Invasions* **9**: 909–923.

Busch, D. E. and S. D. Smith. 1993. Effects of fires on water and salinity relations of riparian woody taxa. *Oecologia* **94**: 186–194.

Calderón Alvarez, C., C. E. Causton, M. S. Hoddle, et al. 2012. Monitoring the effects of *Rodolia cardinalis* on *Icerya purchasi* populations on the Galápagos Islands. *Biological Control* **57**: 167–179.

Campbell, R. W. and R. J. Sloan. 1977. Forest stand responses to defoliation by the gypsy moth. *Forest Science Monograph* **23**(2): monograph 19. 34 pp.

Causton, C. E. 2005. Evaluating risks of introducing a predator to an area of conservation value: *Rodolia cardinalis* in Galápagos, pp. 64–76. *In*: Hoddle, M. S. (ed.). *Second International Symposium on Biological Control of Arthropods, 12–16 September, 2005. Davos, Switzerland*. FHTET-2005-08. USDA, Forest Service. Morgantown, West Virginia, USA.

Center, T., M. Purcell, P. Pratt, et al. 2012. Biological control of *Melaleuca quinquenervia*: an Everglades invader. *Biological control* **57**: 151–165.

Charles, H. and J. S. Dukes. 2007. Impacts of invasive species on ecosystem services, pp. 217–233. *In*: Nentwig, W. (ed.). *Biological Invasions* (Ecological Studies, vol. 193). Springer-Verlag. Berlin, Germany.

Chornesky, E. A., A. M. Bartuska, G. H. Aplet, et al. 2005. Science priorities for reducing the threat of invasive species to sustainable forestry. *BioScience* **55**: 335–348.

Coetzee, J. A., M. P. Hill, M. H. Julien, et al. 2009. *Eichhornia crassipes* (Mart.) Solms-Laub. (Pontederiaceae), pp. 183–210. *In*: Muniappan, R., G. V. Reddy, and A. Raman (eds.). *Biological Control of Tropical Weeds Using Arthropods*. Cambridge University Press. New York.

Connor, J. M. 2003. Impacts of invasive species on rangelands. *Proceedings California Weed Science Society* **55**: 26–31.

Cook, D. C., M. B. Thomas, S. A. Cunningham, et al. 2007. Predicting the economic impact of an invasie species on an ecosystem service. *Ecological Applications* **17**: 1832–1840.

Coombs, E. M., H. Radtke, D. L. Isaacson, and S. P. Synder. 1996. Economic and regional benefits from the biological control of tansy ragweed, *Senecio jacobaea*, in Oregon, pp. 491–494. *In*: Moran, V. C. (ed.). *Proceedings of the IX International Symposium on the Biological Control of Weeds, 19–26 January 1996, Stellenbosch, South Africa*. University of Cape Town. Rondebosch, South Africa.

Coombs, E. M., P. B. McEvoy, and C. E. Turner. 1999. Tansy ragwort, pp. 389–400. *In*: Sheley, R. L. and J. K. Petroff (eds.). *Biology and Management of Noxious Rangeland Weeds*. Oregon State University Press. Corvallis, Oregon, USA.

Coppel, H. C., J. W. Mertins, and J. W. E. Harris. 1974. The introduced pine sawfly, *Diprion similis* (Hartig) (Hymenoptera: Diprionidae). A review with emphasis on studies in Wisconsin. Research Bulletin R2393. University of Wisconsin. Madison, Wisconsin, USDA. 74 pp.

Culliney, T. W. 2005 Benefits of classical biological control for managing invasive plants. *Critical Reviews in Plant Sciences* **24**: 131–150.

Culver, C. S. and A. M. Kuris. 2000. The apparent eradication of a locally established introduced marine pest. *Biological Invasions* **2**: 245–253.

Cuming, F. G. 1961. The distribution, life history, and economic importance of the winter moth, *Operophtera brumata* (L.) (Lepidoptera: Geometridae) in Nova Scotia. *The Canadian Entomologist* **93**: 135–142.

Darin, G. M. S., S. Schoenig, J. N. Barney, et al. 2011. WHIPPET: A novel tool for prioritizing invasive plant populations for regional eradication. *Journal of Environmental Management* **92**: 131-139.

Day, M. D. and I. Bofeng. 2007. Biological control of *Chromolaena odorata* in Papua New Guinea, pp. 53–67. *In*: Lai, P.-Y., G. V. P. Reddy, and R. Muniappan. (eds.). *Proceedings of the 7th International Workshop on the Biological Control and Management of Chromolaena odorata and Mikania micrantha, September 12–15, 2006, Taiwan*. National Pingtung University of Science and Technology. Taiwan.

DeLoach, C. J., P. J. Moran, A. E. Knutson, et al. 2008. Beginning success of biological control of saltcedars (*Tamarix* spp.) in the southwestern USA, pp. 535–539. *In*: Julien, M. H., R. Sforza, M. C. Bon, et al. (eds). *Proceedings of the XII International Symposium on Biological Control of Weeds, April 22–27, 2007, La Grande Motte, France*. CAB International. Wallingford, UK.

Dennill, G. B. 1988. Why a gall former can be a good biological control agent: the gall wasp *Trichilogaster acaciaelongifoliae* and the weed *Acacia longifolia*. *Ecological Entomology* **13**: 1–9.

Dennill, G. B., D. Donnelly, K. Stewart, and F. A. C. Impson. 1999. Insect agents used for the biological control of Australian *Acacia* species and *Paraserianthes lophantha* (Willd.) Nielsen (Fabaceae) in South Africa. *African Entomology Memoir* **1**: 45–54.

Denslow, J. S. and C. M. D'Antonio. 2005. After biological control: assessing indirect effects of insect releases. *Biological Control* **35**: 307–318.

DiTomaso, J. M. 2000. Invasive weeds in rangelands: species, impacts and management. *Weed Science* **48**: 255–265.

Dodd, A. P. 1940. *The Biological Campaign against Prickly Pear*. Commonwealth Prickly Pear Board Bulletin. Brisbane, Australia. 177 pp.

Drooz, A. T. (ed.). 1985. *Insects of Eastern Forests*. Miscellaneous Publication No. 1426. USDA Forest Service. Washington, DC.

Drooz, A. T., C. A. Doggett, and H. C. Coppel. 1979. The introduced pine sawfly, a defoliator of white pine new to North Carolina. USDA Forest Service Research Note SE-273, Southeast Forest Experiment Station. Ashville, North Carolina. 3 pp.

Drus, G. M., T. L. Dudley, C. M. D'Antonio, et al. 2014. Synergistic interactions between leaf beetle herbivory and fire enhance tamarisk (*Tamarix* spp.) mortality. *Biological Control* **77**: 29–40.

EAB, 2009. Emerald Ash Borer. Available from: http://www.emeraldashborer.info/ [Accessed January 2016].

Elkinton, J. S. and G. H. Boettner. 2014. Winter moth (*Operophtera brumata* L.) (Lepidoptera: Geometridae), pp. 221–230. *In*: Van Driesche, R. G. and R. Reardon (eds.). *The Use of Classical Biological Control to Preserve Forests in North America*. FHTET- 2013-02. USDA Forest Service. Morgantown, West Virginia, USA. Available from: http://www.fs.fed.us/foresthealth/technology/pub_titles.shtml [Accessed January 2016].

Embree, D. G. 1967. Effects of the winter moth on growth and mortality of red oak in Nova Scotia. *Forest Science* **13**: 295–299.

de Groote, H., O. Ajuonu, S. Attignon, et al. 2003. Economic impact of biological control of water hyacinth in southern Benin. *Ecological Economics* **45**: 105–117.

de Lange, W. J. and B. W. van Wilgen. 2010. An economic assessment of the contribution of biological control to the management of invasive alien plants and to the protection of ecosystem services in South Africa. *Biological Invasions* **12**: 4113–4124.

Felt, E. P. and S. W. Bromely. 1932. Observations on shade tree insects. *Journal of Economic Entomology* **28**: 390–393.

Fleischer, A., S. Shafir, and Y. Mandelik. 2013. A proactive approach for assessing alternative management programs for an invasive alien pollinator species. *Ecological Economics* **88**: 126–132.

Frandsen, P. and N. Jackson. 1994. The impact of *Arundo donax* on flood control and endangered species, pp. 13–16. *In*: Jackson, N. E., P. Frandsen, and S. Duthoit (eds.). Arundo donax *Workshop Proceedings, 19 November, 1993, Ontario, California*. Team Arundo and California Exotic Pest Plant Council. Pismo Beach, California.

Frid, L., D. Knowler, J. H. Myers, et al. 2013. A multi-scale framework for evaluating the benefits and costs of alternative management strategies against invasive plants. *Journal of Environmental Planning and Management* **56**: 412–434.

Fuester, R. W., A. E. Hajek, J. S. Elkinton, and P. W. Schaefer. 2014. Gypsy moth (*Lymantria dispar* L.) (Lepidoptera: Erebidae: Lymantriinae), pp. 49–82. *In*: Van Driesche, R. G. and R. Reardon (eds.). *The Use of Classical Biological Control to Preserve Forests in North America*. FHTET- 2013-02. USDA Forest Service, Morgantown. West Virginia, USA. Available from: http://www.fs.fed.us/foresthealth/technology/pub_titles.shtml [Accessed January 2016].

Ghent, J. H., C. M. Huber, and R. S. Williams. 1982. Status report of the introduced pine sawfly in the southern Appalachians. Forest Pest Management Report 83-1-1. USDA, Forest Service, State and Private Forestry, Southeast Area. Atlanta, Georgia, 18 pp.

Gillespie, D. R., T. Finlayson, N. V. Tonks, and D. A. Ross. 1978. Occurrence of the winter moth, *Operophtera brumata* (Lepidoptera: Geometridae), on southern Vancouver Island, British Columbia. *The Canadian Entomologist* **110**: 223–224.

Glenn, E. P. and P. L. Nagler. 2005. Comparative ecophysiology of *Tamarix ramosissima* and native trees in western U.S. riparian zones. *Journal of Arid Environments* **61**: 419–446.

Graf, W. L. 1978. Fluvial adjustments to the spread of tamarisk in the Colorado Plateau region. *Geological Society of America Bulletin* **89**: 1491–1501.

Grandgirard, J., M. S. Hoddle, J. N. Petit, et al. 2008 Engineering an invasion: classical biological control of the glassy-winged sharpshooter, *Homalodisca vitripennis*, by the egg parasitoid *Gonatocerus ashmeadi* in Tahiti and Moorea, French Polynesia. *Biological Invasions* **10**: 135–148.

Hakim, S. E. A. 1979. Range condition on the Threemile Game Range in Western Montana. M.S. Thesis. University of Montana, Missoula, Montana.

Harris, P. and R. Cranston. 1979. An economic evaluation of control methods for diffuse and spotted knapweed in western Canada. *Canadian Journal of Plant Science* **59**: 375–382.

Hayes, L., S. V. Fowler, Q. Paynter, et al. 2013. Biological control of weeds: achievements to date and future outlook, pp. 375–385. *In*: Dymond J. R. (ed.). *Ecosystem Services in New Zealand – Conditions and Trends*. Manaaki Whenua Press Lincoln, New Zealand.

Hill, G. and D. Greathead. 2000. Economic evaluation in classical biological control, pp. 208-223. *In*: Perrings, C., M. Williamson, and S. Dalmazzone (eds.). *The Economics of Biological Invasions*. Edward Elgar. Cheltenham, UK.

Hoddle, M. S., C. Crespo Ramírez, C. D. Hoddle, et al. 2013. Post release evaluation of *Rodolia cardinalis* (Coleoptera: Coccinellidae) for control of *Icerya purchasi* (Hemiptera: Monophlebidae) in the Galápagos Islands. *Biological Control* **67**: 262–274.

Hoffmann, J. H. and V. C. Moran. 1991. Biological control of *Sesbania punicea* (Fabaceae) in South Africa. *Agriculture, Ecosystems, and Environment* **37**: 157–173.

Hoffmann, J. H. and V. C. Moran. 1998. The population dynamics of an introduced tree, *Sesbania punicea*, in South Africa, in response to long-term damage caused by different combinations of three species of biological control agents. *Oecologia* **114**: 343–348.

Hoffmann, J. H., V. C. Moran, and B. W. van Wilgen. 2011. Prospects for the biological control of invasive *Pinus* species (Pinaceae) in South Africa. *African Entomology* **19**: 393–401.

Holm, L. G., L. W. Weldon, and R. D. Blackburn, R.D., 1969. Aquatic weeds. *Science* **166**: 699–709.

Holmes, T. P., J. E. Aukema, B. Von Holle, et al. 2009. Economic impacts of invasive species in forests. *Annals of the New York Academy of Science* **1162**: 18–38.

Hope, A., A. Burvall, T. Germishuyse, and T. Newby. 2009. River flow response to changes in vegetation cover in a South African fynbos catchment *Water SA* **35**(1). Available from: http://www.scielo.org.za/scielo.php?pid=S1816-79502009000100007&script=sci_arttext&tlng=pt [Accessed January 2016].

Hosking, S. G. and M. du Preez. 1999. A cost-benefit analysis of removing alien trees in the Tsitsikamma mountain catchment. *South African Journal of Science* **95**: 442–448.

Hudgeons, J. L., A. E. Knutson, K. M. Heinz, et al. 2007. Defoliation by introduced *Diorhabda elongata* leaf beetles (Coleoptera: Chrysomelidae) reduces carbohydrate reserves and regrowth of *Tamarix* (Tamaricaceae). *Biological Control* **43**: 213–221.

Humble, L. 2010. Pest risk analysis and invasion pathways – insects and wood packing revisited: What have we learned? *New Zealand Journal of Forestry Science* **4** (suppl.): S57–S72.

Iacona, G. D., F. D. Price, and P. R. Armsworth. 2014. Predicting the invadedness of protected areas. *Diversity and Distributions* **20**: 430–439.

Jenkins, P. T. 2002. Paying for protection from invasive species. *Issues in Science and Technology* **19**: 67–72.

Jovanovic, N. Z., S. Israel, G. Tredoux, et al. 2009. Nitrogen dynamics in land cleared of alien vegetation (*Acacia saligna*) and impacts on groundwater at Riverlands Nature Reserve (Western Cape, South Africa). *Water South Africa* **35**(1): 37–44.

Julien, M. H. and M. W. Griffiths. 1998. *Biological Control of Weeds: A World Catalogue of Agents and their Target Weeds*. CABI Publishing. Wallingford, UK.

Julien, M. H., M. P. Hill, and P. W. Tipping. 2009. *Salvinia molesta* D.S. Mitchell (Salviniaceae), pp. 378–407. *In*: Muniappan, R., G. V. P. Reddy, A. Raman (eds.). *Biological Control of Tropical Weeds Using Arthropods*. Cambridge University Press. New York.

Kingsbury, J. M. 1964. *Poisonous Plants of the United States and Canada*. Prentice Hall, Inc. Englewood Cliffs. New Jersey, USA.

Knowler, D. and E. B. Barbier. 2005. Importing exotic plants and the risk of invasion: Are market-based instruments adequate? *Ecological Economics* **52**: 341–354.

Lacey, J. R., C. B. Marlow, and J. R. Lane. 1989. Influence of spotted knapweed (*Centaurea maculosa*) on surface runoff and sediment yield. *Weed Technology* **3**: 627–631.

Lamb, A. J. and E. Klaussner. 1988. Response of the fynbos shrubs *Protea repens* and *Erica plukenetii* to low levels of nitrogen and phosphorus applications. *South African Journal of Botany* **54**: 558–564.

Larson, D. L., L. Phillips-Mao, G. Quiram, et al. 2011. A framework for sustainable invasive species management: environmental, social, and economic objectives. *Journal of Environmental Management* **92**: 14–22.

Le Maitre, D. C., B. W. van Wilgen, R. A. Chapman, and D. H. McKelly. 1996. Invasive plants and water resources in the Western Caper Province, South Africa: modeling and the consequences of a lack of management. *Journal of Applied Ecology* **33**: 161–172.

Leung, B., D. M. Lodge, D. Finnoff, et al. 2002. An ounce of prevention or a pound of cure: bioeconomic risk analysis of invasive species. *Proceedings of the Royal Society of London, Series B: Biological Sciences* **269**(1508): 2407–2413.

Leung, B., M. R. Springborn, J. A. Turner, and E. G. Brockerhoff. 2014. Pathway-level risk analysis: the net present value of an invasive species policy in the U.S. *Frontiers in Ecology and Evolution* **12**: 273–279.

Liabunya, W. W. 2007. Malawi aquatic weeds management at hydro power plants. *In*: Proceedings of Hydro Sri Lanka, The International Conference on Small Hydropower, 22–24 October 2007, Kandy, Sri Lanka. Available from: http://www.ahec.org.in/links/Kandy_index.html [Accessed January 2016].

Lindbloom, A. J., K. P. Reese, and P. Zager. 2004. Seasonal habitat use and selection of Chukars in west central Idaho. *Western North American Naturalist* **64**: 338–345.

Long, G. E. 1988. The larch casebearer in the intermountain northwest, pp. 233–242. *In*: Berryman, A. A. (ed.). *Dynamics of Forest Insect Populations, Patterns, Causes, Implications*. Plenum Press. New York.

Lonsdale. M. 2010. Pest risk assessment and invasion pathways: invasive weeds. *New Zealand Journal of Forestry Science* **40** (suppl.): S73–S76.

Lovell, S. J. and S. F. Stone. 2005. The economic impacts of aquatic invasive species: a review of the literature. Working paper series, Working paper #05-02. NCEE (National Center for Environmental Economics). U.S. Environmental Protection Agency, Washington, DC. (see http://www.epa. gov/economics). Available from: http://yosemite.epa.gov/ee/ epa/eed.nsf/WPNumber/2005-02?OpenDocument [Accessed January 2016].

Lubulwa, G. and S. McMeniman. 1998. ACIAR-supported biological control projects in the South Pacific (1983-1996): An economic assessment. *Biological Control News and Information* **3**: 91N–98N.

Lyons, D. B. 2014. Introduced pine sawfly (*Diprion similis* [Hartig]) (Hymenoptera: Diprionidae), pp. 115–125. *In*: Van Driesche, R. G. and R. Reardon (eds.). *The Use of Classical Biological Control to Preserve Forests in North America*. FHTET-2013-02. USDA Forest Service, Morgantown. West Virginia, USA. Available from: http://www.fs.fed.us/foresthealth/ technology/pub_titles.shtml [Accessed January 2016].

Martin, T. G. and R. D. van Klinken. 2006. Value for money? Investment in weed management in Australian rangelands. *The Rangeland* **28**: 63–75.

McConnachie, A. J., M. P. de Wit, M. P. Hill, and M. J. Byrne. 2003. Economic evaluation of the successful biological control of *Azolla filiculoides* in South Africa. *Biological Control* **28**: 25–32.

Mackey, A. P. 1997. The biology of Australian weeds. #29: *Acacia nilotica* ssp. *indica* (Benth.) Brenan. *Plant Protection Quarterly* **12**(1): 7–17.

MacQuarrie, C. J. K. 2014. European spruce sawfly (*Gilpinia hercyniae* [Hartig]) (Hymenoptera: Diprionidae), pp. 127–133. *In*: Van Driesche, R. G. and R. Reardon (eds.). *The Use of Classical Biological Control to Preserve Forests in North America*. FHTET-2013-02. USDA Forest Service. Morgantown, West Virginia, USA. Available from: http://www.fs.fed.us/foresthealth/ technology/pub_titles.shtml [Accessed January 2016].

McFadyen, R. C. 2004. *Chromolaena odorata* in East Timor: history, extent, and control, pp. 8–10. *In*: Day, M. D. and R. E. McFadyen (eds.). *Proceedings of the 6th International Workshop on Biological Control and Management of Chromolaena, May 6–9, 2003, Cairns, Australia*. Technical Report # 55. ACIAR (Australian Centre for International Agricultural Research). Canberra, Australia.

McFadyen, R. 2008. Winning the lottery: return on investment from weed biological control programs, pp. 231–233. *In*: Anon. *Proceedings of the 16th Australian Weeds Conference*, Cairns Convention Centre, North Queensland, Australia, 18–22 May, 2008. Queensland Weed Society, Queensland, Australia. Available from: http://econpapers.repec.org/paper/agsaciatr/ 113791.htm [Accessed January 2016].

Mérel, P. R. and C. A. Carter. 2008. A second look at managing import risk from invasive species. *Journal of Environmental Economics and Management* **56**: 286–290.

Millennium Ecosystem Assessment. 2005. *Ecosystems and Human Well-being: Synthesis*. Island Press. Washington, DC.

Mitchell, D. S. 1981. The management of *Salvinia molesta* in Papua New Guinea, pp. 31–34. *In*: Delfosse, E. S. (ed.). *Proceedings of the 5th International Symposium on Biological Control of Weeds*. CSIRO (Commonwealth Scientific and Industrial Research Organization). Melbourne, Australia.

Moorhead, K. K., K. R. Reddy, and D. A. Graetz. 1988. Waterhyacinth productivity and detritus accumulation. *Hydrobiologia* **157**: 179–185.

Moran, V. C., J. H. Hoffmann, and H. G. Zimmermann. 2005. Biological control of invasive alien plants in South Africa: necessity, circumspection, and success. *Frontiers of Ecology and the Environment* **3**: 71–77.

Murbuah, G., I. M. Gren, and B. McKie. 2014. Economics of harmful invasive species: a review. *Diversity* **6**: 500–523.

Myers, J. H., C. Jackson, H. Quinn, et al. 2009. Successful biological control of diffuse knapweed, *Centaurea diffusa*, in British Columbia, Canada. *Biological Control* **50**: 66–72.

Nagler, P. L., E. P. Glenn, K. Didan, and J. Osterberg. 2008. Wide-area estimates of stand structure and water use of *Tamarix* spp. on the lower Colorado River: implications for restoration and water management projects. *Restoration Ecology* **16**: 136–145.

Naylor, R. L. 2000. The economics of alien species invasions, pp. 241–259. *In*: Mooney, H. A. and R. J. Hobbs. (ed.). *Invasive Species in a Changing World*. Island Press. Washington, DC.

Nechols, J. R., L. A. Andres, J. W. Beardsley, et al. 1995. *Biological Control in the Western United States, Accomplishments and Benefits of Regional Project W-84, 1964–1989*. University of California. Oakland, California, USA.

Neuenschwander, P., M. H. Julien, T. D. Center, and M. P. Hill. 2009. *Pistia stratiotes* L. (Araceae), pp. 332–352. *In*: Muniappan, R., G. V. Reddy, and A. Raman (eds.). *Biological Control of Tropical Weeds Using Arthropods*. Cambridge University Press. New York.

Opande, G. O., J. C. Onyango, and S. O. Wagai. 2004. Lake Victoria: the waterhyacinth (*Eichhornia crassipes* [Mart.] Solms), its socio-economic effects, control measures and resurgence in the Winam gulf. *Limnologica* **34**: 105–109.

Paige, A. R. and K. L. Lacey. 2006. Economic impact assessment of Australian weed biological control. Technical Series No. 10. CRC for Australian Weed Management. 150 pp.

Pejchar, L. and H. A. Mooney. 2009. Invasive species, ecosystem services, and human-well being. *Trends in Ecology & Evolution* **24**: 497–504.

Pemberton, R. W. and A. Ferriter. 1998. Old World climbing fern (*Lygodium microphyllum*), a dangerous weed in Florida. *American Fern Journal* **88**: 165–175.

Perrings, C., M. Williamson, E. B. Barbier, et al. 2002. Biological invasion risks and the public good: and economic perspective. *Conservation Ecology* **6**: 1–7.

Petit, J. N., M. S. Hoddle, J. Grandgirard, et al. 2008. Invasion dynamics of the glassy-winged sharpshooter *Homalodisca vitripennis* (Germar) (Hemiptera: Cicadellidae) in French Polynesia. *Biological Invasions* **10**: 955–967.

Pimentel, D. (ed.). 2002. *Biological Invasions: Economic and Environmental Costs of Alien Plant, Animal, and Microbe Species.* CRC Press. Boca Raton, Florida, USA. 384 pp.

Pimentel, D., L. Lach, R. Zuniga, and D. Morrison. 2000. Environmental and economic costs of nonindigenous species in the United States. *Bioscience* **50**: 53–65.

Pimentel, D., R. Zuniga, and D. Morrison. 2005. Update on the environmental and economic costs associated with alien-invasive species in the United States. *Ecological Economics* **52**: 273–288.

Poland, T. M. and D. McCullough. 2006. Emerald ash borer: invasion of the urban forest and the threat to North America's ash resource. *Journal of Forestry* **104**: 118–124.

Pratt, P. D., M. B. Rayamajhi, P. W. Tipping, et al. 2013. Establishment, population increase, spread, and ecological host range of *Lophodiplosis trifida* (Diptera: Cecidomyiidae), a biological control agent of the invasive tree *Melaleuca quinquenervia* (Myrtales: Myrtaceae). *Environmental Entomology* **42**: 925–935.

Pyke, D. A., M. L. Brooks, and C. D'Antonio. 2010. Fire as a restoration tool: a decision framework for predicting the control or enhancement of plants using fire. *Resoration Ecology* **18**: 274-284.

Ricciardi, A., M. E. Palmer, and N. D. Yan. 2011. Should biological invasions be managed as natural disasters? *BioScience* **61**(4): 312–317.

Ryan, R. B. 1990. Evaluation of biological control: introduced parasites of larch casebearer (Lepidoptera: Coleophoridae) in Oregon. *Environmental Entomology* **19**: 1873–1881.

Ryan, R. and R. G. Van Driesche. 2014. Larch casebearer (*Coleophora laricella* [Hübner]) (Lepidoptera: Coleophoridae), pp. 93–100. *In*: Van Driesche, R. G. and R. Reardon (eds.). *The Use of Classical Biological Control to Preserve Forests in North America.* FHTET- 2013-02. USDA Forest Service. Morgantown, West Virginia, USA. Available from: http://www.fs.fed.us/foresthealth/technology/pub_titles.shtml [Accessed January 2016].

Sapir, D. G. and M. F. Lechat. 1986. Reducing the impact of natural disasters: why aren't we better prepared? *Health Policy and Planning* **1**: 118–126.

Schmidt, J. P., M. Springborn, and J. M. Drake. 2012. Bioeconomic forecasting of invasive species by ecological syndrome. *Ecosphere* **3**(5): 1–18. article 46. Available from: http://dx.doi.org/10.1890/ES12-00055.1 [Accessed January 2016].

Scholes, R. J. and R. Biggs. 2005. A biodiversity intactness index. *Nature* **434**: 45–49.

Seawright, E. K., M. E. Rister, R. D. Lacewell, et al. 2009. Economic implications for the biological control of *Arundo donax* in the Rio Grande Basin. *Southwestern Entomologist* **34**: 377–394.

Sharp, R. L., L. R. Larson and G. T. Green. 2011. Factors influencing public preferences for invasive alien species management. *Biological Conservation* **144**: 2097–2104.

Sharratt, B. S. and D. Lauer. 2006. Particulate matter concentration and air quality affected by windblown dust in the Columbia Plateau. *Journal of Environmental Quality* **35**: 2011–2016.

Simpson, R. D. 2010. If invasive species are pollutants, should polluters pay? pp. 83–109. *In*: Perrings, C., H. Mooney and M. Williamson (eds.). *Bioinvasions and Globalization: Ecology, Economics, Management, and Policy.* Oxford University Press. Oxford, UK.

Smith, L., M. Cristofaro, E. de Lillo, et al. 2009. Field assessment of host plant specificity and potential effectiveness of a prospective biological control agent, *Aceria salsolae*, of Russian thistle, *Salsola tragus*. *Biological Control* **48**: 237–243.

Smith, M. C., E. C. Lake, P. D. Pratt, et al. 2014. Current status of the biological control agent *Neomusotima conspurcatalis* (Lepidoptera: Crambidae), on *Lygodium microphyllum* (Polypodiales: Lygodiaceae) in Florida. *Florida Entomologist* **97**: 817–820.

Stock, W. D., K. T. Wienand, and A. C. Baker. 1995. Impacts of invading N_2-fixing *Acacia* species on patterns of nutrient cycling in two Cape ecosystems: evidence from soil incubation studies and ^{15}N natural abundance values. *Oecologia* **101**: 375–382.

Strang, R. M., K. M. Lindsay, and R. S. Price. 1979. Knapweeds: British Columbia's undesirable aliens. *Rangelands* **1**: 141–143.

Thomas, P. A. and P. M. Room. 1986. Successful control of the floating weed *Salvinia molesta* in Papua New Guinea: a useful biological invasion neutralizes a disastrous one. *Environmental Conservation* **13**: 242–248.

Tipping, P. W., M. R. Martin, P. D. Pratt, et al. 2008. Suppression of growth and reproduction of an exotic invasive tree by two introduced insects. *Biological Control* **44**: 235–241.

Tipping, P. W., M. R. Martin, K. R. Nimmo, et al. 2009. Invasion of a West Everglades wetland by *Melaleuca quinquenervia* countered by classical biological control. *Biological Control* **48**: 73–78.

Tipping, P. W., M. R. Martin, R. Pierce, et al. 2012a. Post-biological control invasion trajectory for *Melaleuca quinquenervia* in a seasonally inundated wetland. *Biological Control* **60**: 163–168.

Tipping, P. W., M. R. Martin, R. Pierce, et al. 2012b. Post-biological control invasion trajectory for *Melaleuca quinquenervia* in a seasonally inundated wetland. *Biological Control* **60**: 163–168.

Tisdell, C. A. 1990. Economic impact of biological control of weeds and insects, pp. 301–316. *In*: Mackauer, M., L. E. Ehler, and J. Roland. *Critical Issues in Biological Control.* Intercept. Andover, UK.

Tomley, A. J. and H. C. Evans. 2004. Establishment of, and preliminary impact studies on, the rust *Maravalia cryptostegiae*, of the invasive weed *Cryptostegia grandiflora* in Queensland, Australia. *Plant Pathology* **53**: 475–484.

Tracy, J. L. and C. J. DeLoach. 1999. Suitability of classical biological control for giant reed (*Arundo donax*) in the United

States, pp. 73–109. *In*: Bell, C. E. (ed.). *Proceedings of the Arundo and Saltcedar Workshop, Arundo and saltcedar: The Deadly Duo, 17 June 1998, Ontario, California*. University of California Cooperative Extension Service, Holtville, California. Available from: http://www.cal-ipc.org/symposia/archive/pdf/Arundo_ Saltcedar1998_1-71.pdf [Accessed January 2016].

Turner, C. E. and P. B. McEvoy. 1995. Tansy ragwort, pp. 264–269. *In*: Nechols, J. R., L. A. Andres, J. W. Beardsley, et al. (eds.). *Biological Control in the Western United States: Accomplishments and Benefits of Regional Research Project W-84, 1964–1989*. Publication No. 3361. University of California, Division of Agriculture and Natural Resources, Oakland, California, USA.

Van Driesche, R. G. and R. Reardon (eds.). 2014. *The Use of Classical Biological Control to Preserve Forests in North America*. FHTET- 2013-02. USDA Forest Service. Morgantown, West Virginia, USA. Available from: http://www.fs.fed.us/ foresthealth/technology/pub_titles.shtml [Accessed January 2016].

Van Driesche, R. G., S. Healy, and R. Reardon. 1996. *Biological Control of Arthropod Pests of the Northeastern and North Central Forests in the United States: A Review and Recommendations*. FHTET-96-19, USDA, Forest Service. Morgantown, West Virginia, USA.

Van Driesche, R. G., B. Blossey, M. Hoddle, et al. (eds.). 2002. *Biological Control of Invasive Plants in the Eastern United States*. FHTET-2002-04, USDA Forest Service. Morgantown, West Virginia, USA.

Van Driesche, R. G., R. I. Carruthers, T. Center, et al. 2010. Classical biological control for the protection of natural eco-systems. *Biological Control* **54**: S2–S33.

van Wilgen, B. W., W. J. Bond, and D. M. Richardson. 1992. Ecosystem management, pp. 345–371. *In*: Cowling, R. M. (ed.). *The Ecology of Fynbos: Nutrients, Fire and Diversity*. Oxford University Press. Cape Town, South Africa.

van Wilgen, B. W., M. P. de Wit, H. J. Anderson, et al. 2004. Costs and benefits of biological control of invasive alien plants: case studies from South Africa. *South African Journal of Science* **100**: 113–122.

van Wilgen, B. W., B. Reyers, D. C. Le Maitre, et al. 2008. A biome-scale assessment of the impact of invasive alien plants on ecosystems services in South Africa. *Journal of Environmental Management* **89**: 336–349.

Versfeld, D. B. and B. W. van Wilgen. 1986. Impacts of woody aliens on ecosystem properties, pp. 239–246. *In*: Macdonald, I. A. W., F. J. Kruger, and A. A. Ferrar (eds.). *The Ecology and Control of Biological Invasions in South Africa*. Oxford University Press. Cape Town, South Africa.

Vogler, W. and A. Lindsay. 2002. The impact of the rust fungus *Maravalia crytostegiae* on three rubber vine (*Cryptostegia grandiflora*) populations in tropical Queensland, pp. 180–182.

In: Jacob, H. S., J. Dodd, and J. H. Moore (eds.). *13th Australian Weeds Conference "Threats Now and Forever?" Perth, Western Australia, 8–13 September, 2002*. Plant Protection Society of Western Australia, Victoria Park, Australia.

Wade, D. D. 1981. Some melaleuca-fire relationships, including recommendations for home-site protections. *In*: Geiger, R. K. (ed.). *Proceedings of the Melaleuca Symposium*. Florida Department of Agriculture and Consumer Services, Division of Forestry. Tallahassee, Florida, USA.

Wallner, W. E. 1996. Invasive pests ('biological pollutants') and U.S. forests: whose problem and who pays? *Bulletin of European Plant Protection Organization* **26**: 167–180

Walton, C. 2005. *Reclaiming Lost Provinces: A Century of Weed Biological Control in Queensland*. Queensland Department of Natural Resources and Mines. Brisbane, Australia. 104 pp.

Watson, A. K. and A. J. Renney. 1974. The biology of Canadian weeds. #6: *Centaurea diffusa* and *C. maculosa*. *Canadian Journal of Plant Science* **54**: 687–701.

Webb, F. E. and F. W. Quednau. 1971. *Coleophora laricella* (Hübner), larch casebearer (Lepidoptera: Coleophoridae), pp. 131–136. *In*: Kelleher, J. S. and M. A. Hulme (eds.). *Biological Control Programmes against Insects and Weeds in Canada 1959–1968*. Technical Communication No. 4, Commonwealth Institute of Biological Control. Commonwealth Agricultural Bureaux. Farnham Royal, UK.

Williams, F., R. Eschen, A. Harris, et al. 2010. *The Economic Cost of Invasive Non-Native Species on Great Britain*. CAB International. Wallingford, UK. 198 pp.

Wingfield, M. J., B. Slippers, and B. D. Wingfield. 2010. Novel associations between pathogens, insects, and tree species threaten world forest. *New Zealand Journal of Forestry Science* **4** (suppl.): S95–S103.

Wood, A. R. and M. J. Morris. 2007. Impact of the gall-forming rust fungus *Uromycladium tepperianum* on the invasive tree *Acacia saligna* in South Africa: 15 years of monitoring. *Biological Control* **41**: 68–77.

Yelenik, S. G., W. D. Stock, and D. M. Richardson. 2004. Ecosystem level impacts of invasive *Acacia saligna* in the South African fynbos. *Restoration Ecology* **12**: 44–51.

Young, F. L. 2006. Russian thistle (*Salsola* spp.) biology and management. 145–147. *In*: Preston, C., J. H. Watts, and N. D. Crossman (eds.). *15th Australian Weeds Conference, Papers and Proceedings (Managing Weeds in a Changing Climate), 24–28 September, 2006; Adelaide, South Australia*. Weed Management Society of South Australia, Victoria, Australia.

Zachariades, C., M. Day, R. Muniappan, and G. V. P. Reddy. 2009. *Chromolaena odorata* (L.). King and Robinson (Asteraceae), pp. 130–162. *In*: Muniappan, R., G. V. P. Reddy, and A. Raman (eds.). *Biological Control of Tropical Weeds Using Arthropods*. Cambridge University Press. Cambridge, UK.

CHAPTER 15

The future of biological control: a proposal for fundamental reform

Bernd Blossey

Department of Natural Resources, Cornell University, USA

Introduction

We, the authors, consider biological control the most appropriate and focused stewardship option when targeting widespread introduced species, if carefully and responsibly conducted. Alternative management options can be fraught with problems, unfulfilled promises, and problematic long-term legacies when control measures have to be repeated frequently, or result in undesirable non-target effects. But we have also outlined past and occasionally present problems of conduct and implementation of classical biological control programs. Implementation of rigorous follow-up assessments can help us better understand why some control programs succeed and others fail, and why biological success does not always equate to ecological success (Chapter 8). This final chapter includes a brief status assessment, but its major goal is to chart a way to the future by proposing to make biological control part of holistic invasive species management programs that are scientifically grounded, critically reviewed, better integrated with conservation biology, transparent, and more closely tied to social needs and outcomes. I am not proposing a few small steps. I provide a rigorous re-assessment of the entire enterprise involving both ethical and normative foundations of stewardship, as well as outlining structural problems in scientific, institutional, and social implementation. This chapter and this book hope to inspire a broad discussion and political change that will enable society to harness the power embedded in biological control to benefit species and ecosystems under threat from introduced plants and insects.

Biological control is a collaborative effort involving local, regional, and international partners. Biological control programs involve land managers who deal with problem species in their daily work, scientists and applicators who help land managers in research, development, and implementation of control approaches, ecologists who develop assessment protocols, data analyses and monitoring, regulators in various agencies, and the public. Interactions in these human and ecological networks can be described as diffuse, at best, and dysfunctional in some cases. These problems are, in part, a function of poorly understood or articulated responsibilities, laws, and regulations that differ among countries and jurisdictions (Sheppard et al., 2003) (see also Chapter 10). Despite differences among rules, regulations, and agencies, fundamentally most countries agree on principles of stewardship to safeguard their national and world heritage.

Ethical and normative foundations of stewardship

Throughout human history, the relationship with and responsibilities of humans to other abiotic and biotic members of our communities has been explored in different ways through spirituality, religion, and philosophy (Rolston, 1986; Callicott, 1989; Minteer, 2005; Minteer and Collins, 2008; Pungetti et al., 2012). These ideas are

---

Integrating Biological Control into Conservation Practice, First Edition. Edited by Roy G. Van Driesche, Daniel Simberloff, Bernd Blossey, Charlotte Causton, Mark S. Hoddle, David L. Wagner, Christian O. Marks, Kevin M. Heinz, and Keith D. Warner.
© 2016 John Wiley & Sons, Ltd. Published 2016 by John Wiley & Sons, Ltd.

as old as oral and written histories, and when they are disrespected, societies can be disrupted (Diamond, 2005) and formerly abundant species may vanish (Miller et al., 2005; Ripple and Van Valkenburgh, 2010).

If we agree that normative principles for environmental stewardship exist, albeit in widely different articulations (Weiss, 1989), it then becomes a question of how to enable and enforce their implementation to account for long-term interests of current and future beneficiaries (Hare and Blossey, 2014). But ideas outlining the most suitable approaches to stop devastating anthropogenic transformations of our world's ecosystems (Barnosky et al., 2011) are contested. Should we use economic valuation to base conservation on needs of people, hoping that placing an economic value on ecosystem services will safeguard our environment and protect species (Kareiva and Marvier, 2011)? Or should we appeal to emotional and ethical responses to fairness in how we treat other species (Leopold, 1949; Blossey, 2012)?

Public Trust Thinking (PTT) (Sax, 1970; Weiss, 1989; Blumm and Wood, 2013; Hare and Blossey, 2014; Wood, 2014) can help advance environmental stewardship in policy, law, practice, implementation, and enforcement (Chapter 8). PTT encapsulates environmental ethics that have independently and repeatedly emerged across cultures, and thus are not specific to a region or country but applicable worldwide (Wood, 2014). At its core, PTT recognizes the intimate interdependence of people and ecosystems, and seeks to ensure that benefits we derive from ecosystems are available to everyone, including future generations (Weiss, 1989).

Philosophically, PTT is appealing because it provides a framework for structuring relationships in democratic societies among citizens, their elected governments, and natural resources and the services they provide (Sagarin and Turnipseed, 2012). PTT asserts that elected or appointed trustees hold natural resources in trust in the interests of beneficiaries, all current and future citizens (Turnipseed et al., 2009; Sagarin and Turnipseed, 2012; Blumm and Wood, 2013). Trustees are charged with overseeing trust resources in a manner that ensures long-term viability and does not privilege any individuals, groups, or users. Trustees turn these management obligations over to trust managers (Smith, 2011; Decker et al., 2014). In turn, beneficiaries (the public) are entitled to hold trustees to account if they fail in their obligations (Hare and Blossey, 2014).

I will use PTT principles to offer a critique of the status quo in invasive species management and biological control and then propose fundamental biological control principles as a guide to the future.

Why is biological control a better choice for pest management?

This section briefly outlines alternatives to biological control, such as herbicides and other pesticides (see also Chapter 3). I do not offer a complete assessment; books and reviews have been, and will continue to be, written about these topics. There clearly is a role for herbicides and other pesticides, as well as non-chemical measures, in eradication campaigns (e.g., for mammals on islands [Genovesi, 2011]). However, apart from eradication efforts, the track record of alternatives to biological control in invasive species management is unimpressive. Chemical alternatives, particularly, fail to provide long-lasting benefits, except when early stages of an invasion, individuals, or very small populations are targeted (Keeley, 2006; Pearson and Callaway, 2008; Rinella et al., 2009; Kettenring and Adams, 2011; Louhaichi et al., 2012; Skurski et al., 2013), while their non-target impacts are global and lasting (Muir and de Wit, 2010). Once everything is said and done, once all successes have been tabulated, biological control often emerges as a preferred tool for management of widespread problem species in protected areas.

Some 40 years ago Huffaker and Croft (1976) stated that "the tendency will be toward greater use of science in pest control decision-making, with extensive use of biological monitoring to establish realistic levels of threatened damage to the crop, and greater concern given to possible profit reductions and environmental disturbances of applying an insecticide as well as the possible gain from doing so." It appears hopes for a science-guided and appropriately evaluated pest management approach have not materialized. Twenty years ago the US National Research Council argued for development of oversight procedures for ecologically based pest management (National Research Council, 1996) to replace harmful pesticides. Today, more than 50 years after the publication of *Silent Spring* (Carson, 1962), new pesticides continue to be developed. They are promoted as having reduced toxicity compared to outdated or outlawed older ones. Yet within short periods their devastating ecological and food web effects often become evident (Köhler and Triebskorn, 2013; Chagnon et al., 2015).

Despite widespread use of pesticides, few countries have established wildlife poisoning surveillance programs, and data are either not collected or not made publicly available (Köhler and Triebskorn, 2013). Effects of modern pesticides, particularly insecticides, manifest themselves not through direct toxicity – although evidence for respiratory, cardiovascular, neurological, and immunological toxicity in humans is increasing (Köhler and Triebskorn, 2013) – but rather through indirect effects by disrupting species interactions (Awkerman et al., 2011; Goulson, 2013; van Dijk et al., 2013; Gibbons et al., 2015; Hallmann et al., 2014; Hladik et al., 2014; Main et al., 2014). Even treatments widely considered benign, such as Bti (*Bacillus thuringiensis israelensis*) for mosquito control, when used over large areas, can have devastating consequences rippling through food webs (Poulin, 2012). Thus for all pest control options, whether chemical, biological, mechanical, or physical, we must consider indirect effects mediated through food webs.

Furthermore, pesticides and their metabolites do not remain at the site of application and their impacts reach remote places, for example Antarctica, where seals, whales, penguins, and other species suffer health and reproductive impairments because of pesticide accumulation in their food webs (Muir et al., 1999; Vos et al., 2000; Geisz et al., 2008; Hoferkamp et al., 2010; Muir and de Wit, 2010; Weber et al., 2010; Trumble et al., 2012, 2013). The extent of the contamination can make penguin colonies secondary threatening sources for contaminant accumulation (Roosens et al., 2007); penguins as toxic waste sites – who could have envisioned this scenario? Pesticides and their metabolites, with their threatening impacts, now reach humans through our global food supply (Steuerwald et al., 2000; Schafer and Kegley, 2002).

Many critics of biological control claim that pesticide application, in contrast to biological control, are local, and can be discontinued when negative effects are recognized. The above examples clearly expose this as a myth and wishful thinking. Even 50 years ago, when Rachel Carson finished her book, it was clear that local pesticide applications had accumulating and far-reaching consequences. Yet despite widespread and repeated documentation of negative

effects on humans and wild life, pesticide applications continue nearly unabated (Weber et al., 2010).

Even well intended use of pesticides to protect species considered under threat by plant invaders has resulted in long-term negative consequences for the species needing protection. Except when early stages of an invasion, individuals, or very small populations are targeted, the plant species we want to protect often actually suffer more from pesticide application than if we were to leave them in the presence of introduced plants (Pearson and Callaway, 2008; Rinella et al., 2009; Kettenring and Adams, 2011; Louhaichi et al., 2012; Martin and Blossey, 2013a; Skurski et al., 2013).

This is not to say that we should not target introduced species with control efforts, far from it. Eradication of invasive mammals on islands has had spectacular biological and ecological successes (Genovesi, 2011). But we need to be aware of the impacts introduced species have, as well as the impacts of control methods on targets and species we want to protect. Collectively, we as scientists and managers have largely failed to provide quantitative evidence to the public and those making policy and funding decisions in order to accurately assess success and failures (Chapter 8). Biological control, together with chemical, mechanical, and physical control alternatives, needs to be placed into a holistic risk-assessment framework. Such evaluations would frequently point to biological control as a prime method of choice. To reiterate what I have argued previously: all non-biological management options fail, most of the time, in providing long-term suppression of widespread target species unless treatments are continued in perpetuity. This is occasionally the only option available to safeguard certain conservation-reliant species. But where available, biological control is usually the only method that can provide, at least occasionally, long-lasting and self-sustaining biological success without constant human intervention.

I do not deny, nor want to belittle, any realized, unknown, or potential non-target effects of biological control. But if we consider that through pesticide application we continue to poison our own food and water supply and threaten entire ecosystems and their species near and far, then accepting potential non-target effects of biological control may appear a far superior choice – all things considered. But improvements in the science of biological control, its application, and decision-making processes are urgently needed, as detailed in the following sections.

What is the status quo?

At present, many biological control projects originate as follows: an introduced species has become abundant and widespread, traditional control attempts fail, and the only hope to prevent further (often assumed) negative ecosystem impacts is to develop biological control. Biological control specialists, together with stakeholders, form a working group that raises funds and enlists overseas experts in feasibility evaluations, followed by further in-depth investigations overseas or in quarantine. Fundraising to sustain programs is a constant struggle and delays and interruptions are frequent. After many years (5–20) and if (at least in weed biological control) adequately host-specific agents can be found, they are proposed for introduction, and petitions are reviewed and approved (or rejected) by federal, state, and local entities and landowners before actual releases occur.

Up to this point, stakeholders, often land management agencies, have provided funding. However, soon after agent introduction and initial distribution, funding from these sources is often discontinued (Harris, 1991). Sophisticated and comprehensive evaluations needed to assess biological, ecological, economic, and societal success of a biological control program remain rare, because traditional funding agencies rarely provide sufficient funds for such work, despite frequent calls for long-term assessments (Blossey, 1999; Delfosse, 2005; Downey, 2011). Consequently, opportunities to learn are wasted, the science is not improved, and usually biological control scientists are blamed for lack of follow-through.

A typical charge to biological control is that the ethical and practical responsibilities for outcome assessments rest with those proposing release of biological control agents. But is this really a fair, just, or even appropriate assignment of responsibilities? Can an individual, or even a group of researchers, be responsible for all aspects of a control program? Can they be held accountable for funding, developing the science, and public disclosure of findings over the range of releases (sometimes an entire continent)? When biological control scientists

retire, change their jobs, or die, who is charged with continuing outcome assessments? Can only young scientists develop biological control programs because it may take 20 years from inception to conclusion? Can biological control scientists be held accountable (be fired or sued?) by their employers, or the public, if there is no monitoring or no publication of project outcomes? These questions illustrate, at least to me, that it appears inappropriate to place these responsibilities on research scientists, individually or collectively. This does not eliminate the need for individual transparency and the highest ethical standards in developing and implementing biological control programs by scientists. But what are appropriate responsibilities among various contributors, biological control scientists, stakeholders, and regulators?

Assignment of responsibilities

The following sections outline responsibilities of (1) land managers and land management agencies, (2) biological control scientists, (3) regulatory agencies and reviewers, (4) politicians, and (5) the public at large. Principles of Public Trust Thinking (Hare and Blossey, 2014) will guide the development of ideas regarding these interactions, relationships, and responsibilities. I do not assume that this is more than the beginning of a long discussion that I hope will rejuvenate biological control and lead to stronger relationships with conservation. I have no illusions over the long-term nature of the potential changes that are required. I propose a fundamental overhaul. No single country's current regulatory framework captures all ideas outlined here (Sheppard et al., 2005; Hunt et al., 2008; see also Chapter 11). Laws and enabling legislation to enact these proposed reforms may look different across countries with different judicial systems, but the principles outlined here should hold.

Two important fundamental recommendations make up the core of the proposed changes. The first is establishment of Biological control Review Councils (BRCs) as the core entity at the federal level. This body would resemble the Environmental Protection Authority (EPA, http://biotechlearn.org.nz/themes/bioethics/environmental_risk_management_authority) in New Zealand, but with extended charge and expanded responsibilities. Establishing a BRC requires a political process and passage of a law to establish this entity. The second recommendation is the establishment of funding entities – Biological control Trust Funds (BTFs).

Biocontrol Review Council (BRC)

In 1998 the Hazardous Substances and New Organisms (HSNO) Act came into effect in New Zealand, which established the Environmental Risk Management Authority (ERMA, responsibilities are now part of the Environmental Protection Authority, EPA) (Chapter 11). Many of the functioning and review processes are exemplary and should be consulted for inspiration, but in my view the act fell short in fully accomplishing appropriate oversight and conduct for the full duration of biological control programs. I also consider it problematic that petitioners proposing introduction and field release of biological control agents are not required to be public entities. I propose to charge the BRC with authority to:

1 Approve or reject species proposed by land managers, ecologists or others as *potential* targets for biological control based on (a) the species documented impact, and (b) inability of managers to contain a species' spread or reduce populations by other means.
2 Approve or reject species as targets for full biological control research after an initial feasibility study and public review.
3 Annually rank newly approved species according to severity of their ecological, economic, or social impacts to prioritize species for funding (assuming funding is limited). Funding should be guaranteed for the full duration of a program, unless program reviews indicate otherwise.
4 Issue requests for project proposals, review proposals, and select research teams to conduct full investigations using five-year funding cycles. Funding decisions (how many new programs can be started) are based on availability of resources in biological control trust funds (see below).
5 Review and approve selection of non-target species proposed for host-specificity screening using scientifically valid criteria.
6 Review annual and final reports, and make recommendations for continued funding, or cessation of programs.
7 Review petitions for approval of agents. Issue preliminary summary findings, then invite oral or written public input, followed by a final recommendation.

8 Review and approve or reject petitions for field release submitted by authorized entities (states, provinces, territories, regional consortia of entities, etc.) with authority over management of natural resources in areas under their jurisdiction. Individuals (including biological control scientists, and land managers), companies, and NGOs do not qualify as petitioners. Public entities are responsible and publicly accountable for project implementation and assessment and reporting of outcomes. A monitoring plan (see Chapter 8) is an essential project component, as are assurances of continued funding, to obtain approval for field release of new agents.
9 Conduct regular meetings or symposia for scientists, reviewers, agency personnel, and so on, to advance individual and collective knowledge of the science, of applicable standards, and processes.
10 Conduct timely review of petitions within legally defined time periods that are not to be exceeded.
11 Create a website allowing citizens, land managers, and scientists to assess, follow, and evaluate decisions and processes of the BRC. All meetings and decisions of the BRC are open to the public and all documents are posted on websites for public review and comment.
12 Create flow charts, documents, and outline procedures to help individuals (reviewers of petitions, for example) and entities navigate the BRC process.
13 Establish, as needed, scientific advisory boards for specific tasks or for review purposes to assist permanent members of the BRC and their staff.

The ultimate number of members that constitute a BRC, and their respective areas of expertise, should be discussed and agreed upon in an open democratic process and would differ from country to country. At a minimum, members represent (individually or collectively) fields of conservation biology, ecology, social sciences, ethics, and environmental law. In New Zealand, EPA consists of five members (http://www.epa.govt.nz/EEZ/trans_tasman/decision making-committee/Pages/Committee-members.aspx) appointed by the Minister for the Environment and staff (Hunt et al., 2008) and decisions of the EPA are legally binding.

An important aspect of the EPA is independence from government departments to prevent the process from being captured by special interests or political considerations. ERMA, the precursor to the EPA, was established as an autonomous Crown entity with the HSNO Act specifically prohibiting the Minister of the Environment from giving direction to ERMA relating to any power, duty, or function carried out under the charge of assessment of hazardous substances and new organisms as outline in the Act (Hunt et al., 2008).

Any BRC should have the ability to establish advisory boards, should the need arise owing to special targets or new scientific developments. Council members, reviewers, and staff should regularly organize or attend symposia to keep up to date with the evolving state of the science of biological control. BRC entities should provide, potentially through outside contracts, summary documents outlining what constitutes meaningful evidence, what standards host-specificity testing has to adhere to, what a holistic risk assessment should include, and what components and standard follow-up assessment programs need to include. These documents should be updated every few years to reflect scientific advances. It is particularly important to require reviewers of petitions to adhere to certain standards in what constitutes acceptable or unacceptable risk, particularly when evaluating host-specificity testing procedures (see Chapter 8).

Biological control trust funds

How such funds can be established will differ among countries, but successful examples exist, such as New York States Environmental Protection Fund (http://www.dec.ny.gov/about/92815.html). This fund is financed primarily through a dedicated portion of real estate transfer taxes and has grown from its original annual appropriation of US$31 million in 1994–95. Over the past 20 years, more than US$2.7 billion has been spent on a variety of environmental projects. As a trust fund created in state law, these resources must be kept separate from other state monies. At the US federal level, "Superfund" is the name given to the environmental program established to address abandoned hazardous waste sites (http://www.epa.gov/superfund/policy/cercla.htm). Initially authorized as the Comprehensive Environmental Response, Compensation and Liability Act of 1980, it was reauthorized with additional resources (from US$1.6 to 8.5 billion) by the Superfund Amendments and Reauthorization Act of 1986. This law created a tax on the chemical and petroleum industries and provided broad federal authority to respond directly to releases or threatened releases of hazardous

substances that may endanger public health or the environment. The money is contained in a trust fund and allows the US Environmental Protection Agency to remediate industrial pollution legacy sites.

Other trust funds earmarked for invasive species work exist in various US states (Environmental Law Institute, 2002) and likely in other countries, but usually they are poorly coordinated and lack sound scientifically based decision-making processes. Funding for biological control could come from surcharges on horticultural or ornamental plant sales, the aquarium and pet trade industries, as well as importers of goods, airlines, and shipping companies. These entities create pathways facilitating accidental, and in some cases, purposeful introduction of species. But these organizations and companies do not assume any financial responsibilities for control costs of invasive species (see also Chapter 14). It appears only reasonable and fair to ask for small contributions in the form of taxes or surcharges for control programs rather than outright bans on imports. This should not, however, replace screening and phytosanitary measures to stop initial introductions, which is the best line of defense.

The previous sections established two core ideas for streamlining responsibilities for review and oversight, as well as funding to advance and manage biological control programs. The following sections outline responsibilities of partners involved in biological control programs.

Land managing agencies (public and private)

All entities with responsibilities for protected-area management, including federal agencies and their employees, state or local agencies, NGOs such as land trusts, and land-holding organizations are included in this category. These organizations manage public resources (areas and species) to benefit all present and future generations. Local managers are often the first to recognize newly arriving species and may engage in early eradication efforts. Such approaches should be informed by landscape context and information about the natural and invasion history of a species, including threats identified elsewhere. Land managers, after eradication is deemed infeasible, have the responsibility to determine management needs based on determination of species' impacts. Origin by itself is not a useful indicator to determine the existence of negative impacts (Davis et al., 2011; Cohen et al., 2012; Martin and Blossey, 2013b). Recognition of this fact requires us to assess whether introduced species are indeed threats to conservation interests (MacDougall and Turkington, 2005).

It is the responsibility of land managers to deliver quantified threat assessments to society. In many instances, particularly for small organizations, the expertise to conduct such investigations may not be available and work will need to be contracted out to be accomplished through partnerships with universities and state or federal scientists. Ideally this includes evaluation of population growth rates of species of conservation concern thought to be affected by introduced plants or insects. Land managers can assist in data collection and gain new insights in quantitative assessment protocols. Such assessments are not specific for biological control, but should be required for any control measures, biological, chemical, or mechanical. The public is entitled to this information and land managers are required to assess performance of activities they conduct that are deemed necessary to protect species and their habitats.

Arguments that such investigations take too long, while introduced species continue to spread, are no justification to rush into treatments for well established species. Land managers and society need an assessment of impacts of introduced species, particularly before large-scale (>0.5 ha) chemical or mechanical treatments are attempted. This information on severity of impacts should be collected in the early stages of an invasion before making treatment decisions (eradication attempts are different, but that is rarely the case). Today, biological control is typically considered the last resort when all else has failed. Biological control agents commonly face the enormous task of reducing host abundance over vast areas. The requirement for quantification of effects of chemical and mechanical control, and recognition that success of such approaches is almost always lacking, may shift managers to consider biological control much sooner.

Nevertheless, after presenting evidence of negative impacts, land managers have to wait for feasibility studies, research, and approvals by the BRC. This time should be used to partner with ecologists or biological control scientists to prepare for potential releases of control agents by developing monitoring approaches. After control agent releases, it is the responsibility of land managers and their collaborating partners and scientists to deliver quantitative evidence to entities that are petitioners to the BRC on performance of biological control agents, targeted species, and threatened native or non-target species. Biological control trust fund grants and land management agency funds should jointly fund these evaluations.

Ultimately, agencies and their employees as well as NGOs do "the people's" business and need to be held accountable for their conduct and success or failure. To be held accountable requires that organizations deliver detailed and quantified information about the status of resources they are charged to manage or protect. This is not typically a responsibility organizations readily embrace. Arguments that other activities, such as control treatments (rather than collecting data), are more urgently needed to protect species, habitats, and landscapes are often articulated, but these statements could not be further from the truth. We can immediately see the poor reasoning of this argument if we consider how we select personal investments. We expect and are provided with detailed accounts of asset distribution and performance and we make future investment decisions based on past asset performance, including performance of asset managers. Nothing short of this approach should be required for land managers and their agencies, public or private. It will require a fundamental shift in the way land management and conservation entities operate – but it is a necessary change if we are serious in wanting to stop further erosion of conservation assets and loss of species.

Biological control scientists

Biological control scientists have expert technical knowledge in assessing and evaluating biological control targets and agents. Individuals are often trained in and recruited from specialized disciplines such as entomology, plant pathology, or ecology. But biological control scientists are not elected public officials, and rarely are they experts in additional disciplines such as conservation biology, social sciences, ethics, philosophy, law, risk assessment, or economics. But it appears that proponents and critics alike expect biological control scientists to maneuver biological control programs through federal and state, provincial, or country legislative processes, ethical and economic evaluations, and public opinion (Harris, 1991; Strong and Pemberton,

2000; Louda et al., 2003; Delfosse, 2005; Miller and Aplet, 2005).

Such assignments and acceptance of responsibility are neither justified nor appropriate. While the core expertise and responsibility for ethical conduct of biological control scientists needs to be recognized, biological control scientists should never be applicants submitting petitions for field release of biological control agents or be solely responsible for funding and conducting pre- and post-release evaluations and assessments of program outcomes. Biological control scientists instead are important members of teams informing and conducting such investigations. Only public entities should be petitioners for agent releases; they can be held accountable by voters and they usually continue to exist beyond the lifetime of individual biological control programs and scientists.

Biological control scientists have responsibilities to assess the validity of claims that introduced species are drivers of ecosystem change and threats to native species. Only if biological control scientists are convinced by quantitative evidence that an introduced insect or plant is a severe threat should they actually engage in and help develop biological control programs. After establishment a BRC will actually make these determinations. In the past biological control scientists have been too eager to accept claims of introduced species as causal drivers of change. Garlic mustard, *Alliaria petiolata* (M. Bieb.) Cavara and Grande, claimed as a transformative species, has been the target of biological control development (Blossey et al., 2001). Over a decade later (2014), we now have evidence for lack of impact and abundance declines, and have discovered the importance of earthworms and deer as drivers of ecosystem deterioration and facilitation of plant invasions (Nuzzo et al., 2009; Dávalos et al., 2014; Kalisz et al., 2014; Blossey and others, unpublished data).

Biological control scientists should also accept and embrace their ethical responsibilities and advocate for public review and approval of their activities. Appropriately targeting the right species and assessing outcomes in a sophisticated and accountable way are likely to increase the standing of the discipline and funding to conduct such work. However, releasing biological control agents without public review, regardless of perceived urgency, is unethical. For example, apparently there was no formal public review of the decision to introduce biological control agents

targeting accidentally, or potentially purposefully, introduced herbivores attacking eucalyptus trees in California (Paine et al., 2010, 2011). Ornamental and commercial interests overrode conservation concerns regarding invasiveness of eucalyptus trees. The "willy-nilly" (Strong and Pemberton, 2000) release of entomophagous biological control agents, often en masse and without host-specificity testing (Strong and Pemberton, 2001) for agricultural or forest pests, even if not formally required, is unethical. Similarly, there is no justification for targeting native species with introduced biological control agents (Lockwood, 1993, 1997) despite claims to the contrary (Carruthers and Onsager, 1993).

Until independent and appropriate governance and review structures (such as BRCs) are in place, voluntary host-specificity testing for entomophagous control agents and restraint are the most important principles a biological control scientist should adhere to. Biological control scientists are as accountable as land managers to the public and they should welcome public scrutiny without being defensive. However, biological control scientists should also vigorously develop and defend scientific principles in host-specificity target selection and interpretation of testing results. Attack on a non-target species, even persistent feeding, should be of no concern unless it has negative demographic consequences (Chapter 8). No-choice feeding tests can exclude a large number of species that are outside the fundamental host range of a species, but they fail miserably to assess realized host ranges. This fixation on no-choice tests is, at least in part, "homemade" by biological control scientists, despite significant advances in the science of host-specificity testing procedures (Wapshere, 1974; Sheppard et al., 2005; Barratt et al., 2006; van Klinken and Raghu, 2006), and much better assessment protocols are available for critical test plant species (Briese, 2005). We know too little to make such statements about entomophagous control agents owing to a lack of history of host-specificity screening.

Unfortunately, petitions by weed biological control scientists are often evaluated by biased, secretive, risk-averse regulatory agencies conducting inadequate risk analyses using scientifically questionable evidence. As a consequence, biological control agents remain "on the shelf" while agricultural and conservation entities use inappropriate and ineffective pesticides known to have harmful side effects. It is not in the power of biological control scientists alone to fundamentally change this dysfunctionality, but in concert with other scientists, land managers, and citizens, biological control scientists should promote political enabling of biological control as an integral part of environmental stewardship.

Regulatory and review agencies

Regulatory agencies and their review processes need a fundamental overhaul. I have addressed the required structural changes earlier in this chapter, where the responsibilities of BRCs were outlined. BRCs should not be located in agriculture departments. In the future, regulatory authorities should be independent of special interest groups and political influence and should develop and embrace open, fair, holistic, democratic, and expedited procedures. Regulatory agencies should be critical yet helpful partners in environmental stewardship. In concert with land managers and biological control petitioners, regulatory agencies should accept and embrace their social responsibilities and enforce accountability of their processes and decisions.

Public participation

Concerning biological control, probably the single most important aspect of citizen engagement is the ability to create public pressure to encourage lawmakers at various levels to pass legislation enabling and funding biological control. Without specific and comprehensive legislation, biological control will lack the ability to address important societal mandates. Without enabling legislation, expertise and accountability will continue to erode, or will not be created. At least in the United States, there are no places where this expertise can be gained other than through individual faculty at universities training students. Public pressure and availability of long-term funding are needed to develop this expertise.

In addition, citizens need to recognize and demand accountability by elected officials and management agencies, public or private. Citizens need to be able to review outcomes of management approaches (biological or otherwise) funded with public money for their success in achieving clearly articulated goals. Organizations and agencies may resist these demands; it will be up to citizens, whether they are voters or donors, to encourage accountability. Such success measures are increasingly demanded by large foundations to assess if their investments actually result in benefits, not just expenditures.

Finally, citizens should support establishment of biological control funds (in part through fees on entities benefiting from introduction of non-native species, or entities providing vectors for hitchhiking organisms). In concert with abundant opportunities for citizen participation in an open, informed, and democratic review and approval process by BRCs, this would address issues of fairness around costs and benefits.

Elected officials

Lawmakers at the appropriate, state, federal, or provincial levels need to take leadership and advance legislation to overhaul biological control regulations and funding to improve environmental stewardship. This will not be an easy or convenient process, but is one that is long overdue. Accountability should be a core principle for recipients of public funds, but the right metrics need to be developed through a science-guided approach. Area treated, although easily measured, is not a useful metric. Organizations and entities need to be enabled to provide such assessments; monitoring, or assessment of outcomes, is not a waste of money but an essential component of management.

State, provincial, or territorial governments

This category includes all governmental divisions below the federal (countrywide) level that have independent functioning and budgetary authority. In the United States these are states or territories and in Canada provinces and first nations, with responsibility for natural resource stewardship. In the European Union, member countries and potentially subnational states (Bundesländer, such as in Germany or cantons, as in Switzerland) would be involved. It is not my role to define these responsibilities, but whatever authority administers stewardship and conservation below the federal level needs to create a structure to approve and then administer biological control programs. Responsibilities to ensure follow-up assessments of programs should rest at this level, and it should not be delegated to counties, communities, or municipalities, which usually lack the required qualifications and expertise. Potentially, regional or state councils may form to increase efficiency and decrease costs as determined by state/territorial/provincial governments.

Similar to the federal level, a state-level review and funding mechanism (state council?) is needed to develop oversight and reporting mechanisms. State-level structures should develop their own websites linked to the BRC to increase transparency, openness, and accountability to their citizens about stewardship actions that are taken, or rejected, on their behalf. The reasons for approving or rejecting a biological control option need to be clearly justified in writing and posted on publicly available websites, and decisions need to be made in a timely fashion (three to six months).

State or regional biological control councils have responsibilities including (1) approval of field releases of federally approved biological control agents (this may include a public review of the petition and, very rarely, additional testing under special circumstances, for example cultural or economic interests); (2) evaluation, selection, and funding (after an open bidding process) teams for long-term assessments after biological control agent releases; (3) publication of annual progress reports on their websites, including quantitative findings (see Chapter 8); (4) reviewing of progress and determination of temporal and spatial needs for long-term monitoring within their jurisdiction; and (5) determination of when programs need final evaluations.

The make-up of state/regional councils may take many forms and it is up to the appropriate authorities to develop structures. At a minimum, if, for example, responsibilities are associated with a single office within state government (which should not be the department of agriculture), establishment of an advisory board with sufficient scientific expertise is strongly advised. While led by agency officials, membership in advisory boards should be isolated from political and special interest influences. Members of this council/board should be required to attend regular training and education sessions to retain and update their capabilities as the science develops. The federal BRC may be called upon to assist, when appropriate.

Developing a dedicated funding stream to allow biological control implementation and assessment within the jurisdiction should take the form of trust funds. Such dedicated funding for invasive species work does already exist (Environmental Law Institute, 2002); creating state or regional trust funds should not be a fundamental limitation. Regional councils should operate in similar ways to federal councils, including competitive bidding and selecting appropriate teams to deliver quantitative information to assess biological control programs.

Biological control program flow structure

I have outlined both the fundamental grounding of biological control in environmental stewardship and responsibilities of people and agencies as partners in biological control program development and implementation. The remainder of this chapter charts a hypothetical flow of a program from inception to conclusion, re-iterating major responsibilities.

1 Land managers in public or private entities, in collaboration with ecologists, determine that an introduced species is threatening species (or processes) of conservation significance.

2 Land managers provide quantitative evidence that eradication is no longer possible (ecologically, financially, ethically) and that traditional means (hand pulling, herbicide, or pesticide application in very small areas), or targeting individuals, is unsuccessful in providing control.

3 Land managers petition a federal Biological control Review Council to target a species with biological control.

4 The BRC reviews evidence and a determination is made within three months from submission of evidence. All documents, including decisions and reasons for determinations, are archived and published for public review on a dedicated website. A determination of sufficient evidence allows proposed species to enter a queue for prioritization (see step #5). A determination of insufficient evidence returns the proposal back to step #2, allowing petitioners to assemble additional evidence.

5 After a species enters a prioritization queue, the BRC establishes a public review period. This requires, at a minimum, a widely publicized call for public comments. The process can include, but does not require, public meetings to review or challenge evidence. Written comments are accepted and collated. It is up to the review board to decide how many meetings are needed to complete this stage of the fact-finding period. BRC staff will respond to all concerns and issues raised and will provide responses, which are posted on dedicated websites. Within 12 months after a species advances to a prioritization queue, a determination is made whether it is in the public interest to begin biological control research, or whether other species have higher priority (given usually limited funding).

6 Annually, the BRC ranks new species that have advanced past initial public review. Rankings and reasons for ranking will be made public and posted on the website.

7 Simultaneously with the ranking, the BRC decides how many programs receive initial funding to begin feasibility studies, based on availability of funds. Decisions to conduct feasibility studies do not imply a guarantee that a full program can be funded.

8 The BRC issues a species-specific Request for Proposals and establishes a review board (that includes the original petitioners), which evaluates submitted applications and progress reports. BRC staff assist in logistics and administration. Proposals are for one to three years of funding (in specific justified circumstances, this may be as long as five years). Employees in federal agencies, such as ARS in the United States, are ineligible as Principle Investigators, although they may serve as collaborators, to avoid conflicts of interest. One or occasionally several teams will be selected to perform studies. Timelines from issue of request for proposals to submission will usually be three months, plus maximal an additional three months before funding decisions are made. The BRC and specific boards would have the right to reject all proposals if all are deemed insufficient. A second request for proposals would be issued and would follow procedures as outlined above. If no bidder comes forward, or only insufficient ones, the target species may be considered at a future funding cycle or dropped altogether.

9 After initial feasibility studies are completed, the BRC reviews evidence, and it determines whether the program should be advanced, required to submit additional evidence, or be discontinued. Only in exceptional circumstances would additional funding for this phase be provided. If this need arises, for example to conduct more detailed taxonomic work, supplemental funding requests by the chosen research team can be considered. A major part of feasibility studies is development of preliminary host-specificity testing lists that are reviewed and refined by a specifically assembled review board assembled by the BRC.

10 After review, the BRC issues decisions on which proposed target species should advance to the full research phase. All deliberations are public, except in extraordinary circumstances, and posted on dedicated websites. Availability of resources determines how many targets can be selected annually. Funding decisions will have implications for potentially 20 years. Some programs may cease after additional evidence collection, some programs may never release biological control agents, but funding decisions should ensure availability of funds to study the safety of proposed control agents, develop and implement release procedures, and carry out initial follow-up monitoring (additional resources will need to be provided by states, territories or provinces benefitting from releases, see below).

11 The BRC issues species-specific requests for proposals for each approved target and a review panel ranks proposals. Funding should be provided in five-year increments, with required annual progress reports to release annual increments. After four years, a new request for proposals is issued allowing other teams to bid and the BRC to select the most promising team, which may be the initial team. But a competitive process encourages innovation and expertise building in interdisciplinary teams. Furthermore, a team that excelled in foreign exploration or testing, may not be qualified to advance mass releases or assessments of biological, economic, or social outcomes.

12 After sufficient evidence has been collected, the research team(s) will propose to the BRC to either abandon the target for biological control, eliminate certain potential agents from consideration, or – based on host-specificity testing results and initial risk assessments – propose the release of certain biological control agents. This Petition for Approval for Release should include a detailed follow-up monitoring plan based on initial evidence presented to target a species with biological control (step #1), and information collected during host-specificity testing.

13 The BRC reviews evidence, posts documents on dedicated websites if it is deemed complete, and begins a formal public consultation process that involves open public forums, and input by as many stakeholders as possible. This should be an active and engaged consultation process not merely a note on a website. The BRC conducts a thorough and full risk assessment that will be holistic. Instead of only evaluating potential risks of a biological control agent release, options of doing nothing, continuing with other control methods, and potential risks associated with biological control will be evaluated in a holistic

unified framework. A preliminary finding and recommendations are posted for a minimum of 60 days to allow for public review and comment.

14 The BRC, after reviewing comments, issues a formal recommendation. Whether the BRC ultimately has signatory responsibilities depends on the political structure the BRC is embedded in. Whoever has final signatory authority is obliged to follow BRC recommendations, except when specific circumstances have changed, or new evidence comes to light. If an authority refuses to sign, reasons for refusal need to be clearly articulated and substantiated with evidence that can be challenged by review, including in court.

15 After final signature by the authority, the BRC will request proposals for agent releases. These requests come from states, provinces, or territorial governments with stewardship and natural resource management authority. They cannot come from private citizens, companies, or NGOs but need to be advanced by entities ultimately responsible for funding and overseeing implementation and assessment of biological control programs. It is in these entities where responsibilities for program administration and evaluation reside, not at the level of individual biological control scientists or land managers. An essential part of the application is a detailed release, assessment, and evaluation plan and assurances that funding to carry out the proposed work is available.

16 After approval of release plans, program administration continues at the BRC but at a reduced level. Where requested, the BRC aids entities in developing infrastructure, program execution, or reviews until sufficient local or regional expertise is built. In some instances, such expertise may already exist, in other instances new structure and expertise may need to be developed. The federal BRC website functions as central archive and link to all biological control programs, research teams, and assessments.

Conclusions

I do not usually ask for more bureaucracy, more regulation, and more paperwork. But the status quo in biological control and invasive species management in general is problematic, with poorly defined decision-making authorities, obsolete review procedures, and poorly designed follow-up monitoring after biological control agents are released. These conditions result in little or poor oversight of releases of parasitoids and predators, while in other cases prevent the release of highly specific weed biological control agents. Most land managers and conservationists in NGOs and in state and federal agencies have good intentions, but they have not had a past history or support system to appropriately assess outcomes of their management activities. We need to provide approaches, structures, and funding to enable these assessments, particularly follow-up monitoring. If this requires some additional administrative burden and data collection together with accountability, we should all be the better for it.

Success in conservation and biological control is not measured in days, weeks, or months, but in decades and centuries. Pesticides certainly deliver instant gratification – but we are suffering a continuing hangover. It is time to change.

I have no illusions about the difficulties that lie ahead in making the proposed changes. The recommendations that I make here require fundamental changes in the way we think about our activities and our accountabilities never articulated before. It will require political will, leadership, and willingness to break up entrenched interests and repel meddling by special interests that do not want to be taxed or lose business (such as the horticultural or pet industries and pesticide lobbyists and companies) or even biological control scientists who do not want to be regulated. It will require nothing less than a fundamental change in the culture in many federal and state governments – from the current culture of secrecy and service to special interests to one of accountability, public participation, and democratic and open decision-making processes. This will take time.

Principles of Public Trust Thinking (Hare and Blossey, 2014) are helpful in strategically contemplating roles and responsibilities of land managers, scientists, citizens, and voters. Holding everyone accountable and accepting responsibilities may break inertia and unwillingness to change. Will it deliver better conservation results and fewer non-target disasters? I am convinced it will. Biological control should have a bright future and prosper, if appropriately enabled.

Acknowledgments

I thank Victoria Nuzzo for early reviews of Chapter 15 and Darragh Hare for wide-ranging discussions; suggestions by Richard Casagrande, Kevin Heinz, Hariet Hinz, Mark Hoddle, Christian Marks, Peter McEvoy, Dan Simberloff, David Wagner, and Roy Van Driesche improved earlier versions.

References

Awkerman, J. A., Y. M. R. Marshall, A. B. Williams, et al. 2011. Assessment of indirect pesticide effects on worm-eating warbler populations in a managed forest ecosystem. *Environmental Toxicology and Chemistry* **30**: 1843–1851.

Barnosky, A. D., N. Matzke, S. Tomiya, et al. 2011. Has the Earth's sixth mass extinction already arrived? *Nature* **471**: 51–57.

Barratt, B. I. P., B. Blossey, and H. M. T. Hokkanen. 2006. Post-release evaluation of non-target effects of biological control agents, pp. 166–186. *In:* Bigler, F., D. Babendreier, and U. Kuhlmann (eds.). *Environmental Impact of Invertebrates for Biological Control of Arthropods, Methods and Risk Assessment.* CABI Publishing. Cambridge, Massachusetts, USA.

Blossey, B. 1999. Before, during, and after: the need for long-term monitoring in invasive plant species management. *Biological Invasions* **1**: 301–311.

Blossey, B. 2012. The value of nature. *Frontiers in Ecology and the Environment* **10**: 171.

Blossey, B., V. Nuzzo, H. Hinz, and E. Gerber. 2001. Developing biological control of *Alliaria petiolata* (M. Bieb.) Cavara and Grande (garlic mustard). *Natural Areas Journal* **21**: 357–367.

Blumm, M. C. and M. C. Wood. 2013. *The Public Trust Doctrine in Environmental and Natural Resources Law.* Carolina Academic Press. Durham, North Carolina, USA.

Briese, D. T. 2005. Translating host-specificity test results into the real world: the need to harmonize the yin and yang of current testing procedures. *Biological Control* **35**: 208–214.

Callicott, J. B. 1989. *In Defense of the Land Ethic: Essays in Environmental Philosophy.* State University of New York Press. Albany, New York, USA.

Carruthers, R. I. and J. A. Onsager. 1993. Perspective on the use of exotic natural enemies for biological control of pest grasshoppers (Orthoptera, Acrididae). *Environmental Entomology* **22**: 885–903.

Carson, R. 1962. *Silent Spring.* (40th Anniversary edition, 2002) Houghton Mifflin Company. New York.

Chagnon, M., D. Kreutzweiser, E. D. Mitchell, et al. 2015. Risks of large-scale use of systemic insecticides to ecosystem functioning and services. *Environmental Science of Pollution Research* **22**: 119–134. doi: 10.1007/s11356-11014-13277-x.

Cohen, J. S., J. C. Maerz, and B. Blossey. 2012. Traits, not origin, explain impacts of plants on larval amphibians. *Ecological Appplications* **22**: 218–228.

Dávalos, A., V. Nuzzo, and B. Blossey. 2014. Demographic responses of rare forest plants to multiple stressors: the role of deer, invasive species and nutrients. *Journal of Ecology* **102**: 1222–1233.

Davis, M., M. K. Chew, R. J. Hobbs, et al. 2011. Don't judge species on their origins. *Nature* **474**: 153–154.

Decker, D. J., A. B. Forstchen, J. F. Organ, et al. 2014. Impacts management: an approach to fulfilling public trust responsibilities of wildlife agencies. *Wildlife Society Bulletin* **38**: 2–8.

Delfosse, E. S. 2005. Risk and ethics in biological control. *Biological Control* **35**: 319–329.

Diamond, J. M. 2005. *Collapse: How Societies Choose to Fail or Succeed.* Penguin Books. New York.

Downey, P. O. 2011. Changing of the guard: moving from a war on weeds to an outcome-oriented weed management system. *Plant Protection Quarterly* **26**: 86–91.

Environmental Law Institute. 2002. *Halting the Invasion: State Tools for Invasive Species Management.* Environmental Law Institute. Washington, DC.

Geisz, H. N., R. M. Dickhut, M. A. Cochran, et al. 2008. Melting glaciers: a probable source of DDT to the Antarctic marine ecosystem. *Environmental Science & Technology* **42**: 3958–3962.

Genovesi, P. 2011. Are we turning the tide? Eradications in times of crisis: how the global community is responding to biological invasions, pp. 5–8. *In:* Veitch, C. R., M. N. Clout, and D. R. Towns (eds.). *Island Invasives: Eradication and Management.* IUCN. Gland, Switzerland.

Gibbons, D., C. Morrissey, and P. Mineau. 2015. A review of the direct and indirect effects of neonicotinoids and fipronil on vertebrate wildlife. *Environmental Science and Pollution Research* **22**: 103–118. doi: 10.1007/s11356-014-3180-5.

Goulson, D. 2013. An overview of the environmental risks posed by neonicotinoid insecticides. *Journal of Applied Ecology* **50**: 977–987.

Hallmann, C. A., R. P. B. Foppen, C. A. M. van Turnhout, et al. 2014. Declines in insectivorous birds are associated with high neonicotinoid concentrations. *Nature* **511**: 141–143.

Hare, D. and B. Blossey. 2014. Principles of public trust thinking. *Human Dimension of Wildlife: An International Journal* **19**: 397–406.

Harris, P. 1991. Classical biological control of weeds: its definition, selection of effective agents, and administrative political problems. *The Canadian Entomologist* **123**: 827–849.

Hladik, M. L., D. W. Kolpin, and K. M. Kuivila. 2014. Widespread occurrence of neonicotinoid insecticides in streams in a high corn and soybean producing region, USA. *Environmental Pollution* **193**: 189–196.

Hoferkamp, L., M. H. Hermanson, and D. C. G. Muir. 2010. Current use pesticides in Arctic media; 2000–2007. *Science of the Total Environment* **408**: 2985–2994.

Huffaker, C. B. and B. A. Croft. 1976. Integrated pest management in the U.S.: progress and promise. *Environmental Health Perspectives* **14**: 167–183.

Hunt, E. J., U. Kuhlmann, A. Sheppard, et al. 2008. Review of invertebrate biological control agent regulation in Australia, New Zealand, Canada and the USA: recommendations for a harmonized European system. *Journal of Applied Entomology* **132**: 89–123.

Kalisz, S., R. Spigler, and C. Horvitz. 2014. In a long-term experimental demography study, excluding ungulates reversed invader's explosive population growth rate and restored natives. *Proceedings of the National Academy of Sciences of the United States of America* **111**: 4501–4506.

Kareiva, P. and M. Marvier. 2011. *Conservation Science: Balancing the Needs of People and Nature*. Roberts and Company Publishers. Greenwood Village, Colorado, USA.

Keeley, J. E. 2006. Fire management impacts on invasive plants in the western United States. *Conservation Biology* **20**: 375–384.

Kettenring, K. M. and C. R. Adams. 2011. Lessons learned from invasive plant control experiments: a systematic review and meta-analysis. *Journal of Applied Ecology* **48**: 970–979.

Köhler, H.-R. and R. Triebskorn. 2013. Wildlife ecotoxicology of pesticides: can we track effects to the population level and beyond? *Science* **341**: 759–765.

Leopold, A. 1949. *A Sand County Almanac*. (new edition, 2001: Oxford University Press, New York.)

Lockwood, J. A. 1993. Environmental issues involved in biological-control of rangeland grasshoppers (Orthoptera, Acrididae) with exotic agents. *Environmental Entomology* **22**: 503–518.

Lockwood, J. A. 1997. Competing values and moral imperatives: an overview of ethical issues in biological control. *Agriculture and Human Values* **14**: 205–210.

Louda, S. M., R. W. Pemberton, M. T. Johnson, and P. A. Follett. 2003. Nontarget effects – the Achilles' heel of biological control: retrospective analyses to reduce risk associated with biological control introductions. *Annual Review of Entomology* **48**: 365–396.

Louhaichi, M., M. F. Carpinelli, L. M. Richman, and D. E. Johnson. 2012. Native forb response to sulfometuron methyl on medusahead-invaded rangeland in Eastern Oregon. *Rangeland Journal* **34**: 47–53.

MacDougall, A. S. and R. Turkington. 2005. Are invasive species the drivers or passengers of change in degraded ecosystems. *Ecology* **86**: 42–55.

Main, A. R., J. V. Headley, K. M. Peru, et al. 2014. Widespread use and frequent detection of neonicotinoid insecticides in wetlands of Canada's prairie pothole region. *PLoS One* **9**(3): e92821.

Martin, L. J. and B. Blossey. 2013a. The runaway weed: costs and failures of *Phragmites australis* management in the USA. *Estuaries and Coasts* **36**: 626–632.

Martin, L. J. and B. Blossey. 2013b. Intraspecific variation overrides origin effects in impacts of litter-derived secondary compounds on larval amphibians. *Oecologia* **173**: 449–459.

Miller, G. H., M. L. Fogel, J. W. Magee, et al. 2005. Ecosystem collapse in Pleistocene Australia and a human role in megafaunal extinction. *Science* **309**: 287–290.

Miller, M. L. and G. H. Aplet. 2005. Applying legal sunshine to the hidden regulation of biological control. *Biological Control* **35**: 358–365.

Minteer, B. A. 2005. Environmental philosophy and the public interest: a pragmatic reconciliation. *Environmental Values* **14**: 37–60.

Minteer, B. and J. Collins. 2008. From environmental to ecological ethics: toward a practical ethics for ecologists and conservationists. *Science and Engineering Ethics* **14**: 483–501.

Muir, D., B. Braune, B. DeMarch, et al. 1999. Spatial and temporal trends and effects of contaminants in the Canadian Arctic marine ecosystem: a review. *Science of the Total Environment* **230**: 83–144.

Muir, D. C. G. and C. A. de Wit. 2010. Trends of legacy and new persistent organic pollutants in the circumpolar arctic: overview, conclusions, and recommendations. *Science of the Total Environment* **408**: 3044–3051.

National Research Council. 1996. *Ecologically Based Pest Management*. National Academy Press. Washington, DC.

Nuzzo, V. A., J. C. Maerz, and B. Blossey. 2009. Earthworm invasion as the driving force behind plant invasion and community change in northeastern North American forests. *Conservation Biology* **23**: 966–974.

Paine, T. D., J. C. Millar, and K. M. Daane. 2010. Accumulation of pest insects on eucalyptus in California: random process or smoking gun. *Journal of Economic Entomology* **103**: 1943–1949.

Paine, T. D., C. C. Hanlon, and F. J. Byrne. 2011. Potential risks of systemic imidacloprid to parasitoid natural enemies of a cerambycid attacking eucalyptus. *Biological Control* **56**: 175–178.

Pearson, D. E. and R. M. Callaway. 2008. Weed-biological control insects reduce native-plant recruitment through second-order apparent competition. *Ecological Applications* **18**: 1489–1500.

Poulin, B. 2012. Indirect effects of bioinsecticides on the nontarget fauna: the Camargue experiment calls for future research. *Acta Oecologica* **44**: 28–32.

Pungetti, G., G. Oviedo, and D. Hooke (eds.). 2012. *Sacred Species and Sites: Advances in Biocultural Conservation*. Cambridge University Press. Cambridge, UK.

Rinella, M. J., B. D. Maxwell, P. K. Fay, et al. 2009. Control effort exacerbates invasive-species problem. *Ecological Applications* **19**: 155–162.

Ripple, W. J. and B. Van Valkenburgh. 2010. Linking top-down forces to the Pleistocene megafaunal extinctions. *BioScience* **60**: 516–526.

Rolston, H. 1986. *Philosophy Gone Wild: Essays in Environmental Ethics*. Prometheus Books. Buffalo, New York, USA.

Roosens, L., N. Van Den Brink, M. Riddle, et al. 2007. Penguin colonies as secondary sources of contamination with persistent organic pollutants. *Journal of Environmental Monitoring* **9**: 822–825.

Sagarin, R. D. and M. Turnipseed. 2012. The Public Trust Doctrine: where ecology meets natural resources management. *Annual Review of Environment and Resources* **37**: 473–496.

Sax, J. L. 1970. The Public Trust Doctrine in natural resource law: effective judiial intervention. *Michigan Law Review* **68**: 471–566.

Schafer, K. S. and S. E. Kegley. 2002. Persistent toxic chemicals in the US food supply. *Journal of Epidemiology and Community Health* **56**: 813–817.

Sheppard, A. W., R. L. Hill, R. A. DeClerck-Floate, et al. 2003. A global review of risk-benefit-cost analysis for the introduction of classical biological control agents against weeds: a crisis in the making? *Biological Control News and Information* **24**: 77N–94N.

Sheppard, A. W., R. D. van Klinken, and T. A. Heard. 2005. Scientific advances in the analysis of direct risks of weed biological control agents to nontarget plants. *Biological Control* **35**: 215–226.

Skurski, T. C., B. D. Maxwell, and L. J. Rew. 2013. Ecological tradeoffs in non-native plant management. *Biological Conservation* **159**: 292–302.

Smith, C. A. 2011. The role of state wildlife professionals under the Public Trust Doctrine. *Journal of Wildlife Management* **75**: 1539–1543.

Steuerwald, U., P. Weihe, P. J. Jorgensen, et al. 2000. Maternal seafood diet, methylmercury exposure, and neonatal neurologic function. *Journal of Pediatrics* **136**: 599–605.

Strong, D. R. and J. M. Pemberton. 2000. Biological control of invading species – risk and reform. *Science* **288**: 1969–1970.

Strong, D. R. and R. W. Pemberton. 2001. Food webs, risks of alien enemies and reform of biological control, pp. 57–79. *In*: Wajnberg, E., J. K. Scott, and P. C. Quimby (eds.). *Evaluating Indirect Ecological Effects of Biological Control*. CABI Publishing. Oxford, UK.

Trumble, S. J., E. M. Robinson, S. R. Noren, et al. 2012. Assessment of legacy and emerging persistent organic pollutants in Weddell seal tissue (*Leptonychotes weddellii*) near McMurdo Sound, Antarctica. *Science of the Total Environment* **439**: 275–283.

Trumble, S. J., E. M. Robinson, M. Berman-Kowalewski, et al. 2013. Blue whale earplug reveals lifetime contaminant exposure and hormone profiles. *Proceedings of the National Academy of Sciences of the United States of America* **110**: 16922–16926.

Turnipseed, M., L. B. Crowder, R. D. Sagarin, and S. E. Roady. 2009. Legal bedrock for rebuilding America's ocean ecosystems. *Science* **326**: 183–185.

van Dijk, T. C., M. A. van Staalduinen, and J. P. van der Sluijs. 2013. Macro-invertebrate decline in surface water polluted with imidacloprid. *PLoS One* **8**: e62374.

van Klinken, R. D. and S. Raghu. 2006. A scientific approach to agent selection. *Australian Journal of Entomology* **45**: 253–258.

Vos, J. G., E. Dybing, H. A. Greim, et al. 2000. Health effects of endocrine-disrupting chemicals on wildlife, with special reference to the European situation. *Critical Reviews in Toxicology* **30**: 71–133.

Wapshere, A. J. 1974. A strategy for evaluating the safety of organisms for biological weed control. *Annals of Applied Biology* **77**: 200–211.

Weber, J., C. J. Halsall, D. Muir, et al. 2010. Endosulfan, a global pesticide: a review of its fate in the environment and occurrence in the Arctic. *Science of the Total Environment* **408**: 2966–2984.

Weiss, E. B. 1989. *In Fairness to Future Generations: International Law, Common Patrimony, and Intergenerational Equity*. The United Nations University. Tokyo.

Wood, M. C. 2014. *Nature's Trust: Environmental Law for a New Ecological Age*. Cambridge University Press. New York.

Concluding thoughts on future actions

Roy G. Van Driesche[1], Daniel Simberloff[2], and David L. Wagner[3]

[1] *Department of Environmental Conservation, University of Massachusetts, USA*
[2] *Department of Ecology & Evolutionary Biology, University of Tennessee, USA*
[3] *Department of Ecology & Evolutionary Biology, University of Connecticut, USA*

Through this book, we have sought to stimulate improvements in the way that biological control programs are carried out worldwide and encourage the use of biological control, when appropriate, in wildlands. Much can be done to make modern biological control more scientifically grounded, better reviewed, committed to greater pre- and post-release data collection, and integrated with conservation biology. The integration of biological control into conservation biology, as advocated in this book, is likely to happen slowly based on many events and activities that put the two research disciplines into more frequent, thoughtful contact. Joint efforts should follow. While some countries have already merged these two activities (New Zealand and South Africa are notable examples), for other nations such as the United States much remains to be done. While updating the laws governing biological control in the United States is critically needed (see Chapter 11), this goal has eluded several previous attempts, and legal change may be slow in coming. Until then, we propose that the principles discussed in this book serve as the foundation for training and future work on biological control in wildlands. This movement toward integration can be enhanced by attendance at meetings, involvement in collaborations, and participation in review efforts, beyond each individual's core discipline, allowing opportunities for personal acquaintances to grow and, more importantly, for information and knowledge to be exchanged. Among the useful steps in the short- to mid-term are the following:

- Symposia on biological control within ecology meetings, especially those attended by restoration ecologists and conservation biologists.
- Meetings on how to carry out biological control projects within their ecological context directed at managers of public lands or private preserves.
- Articles on integration of biological control into conservation biology for journals and in-house technical newsletters read by staff of major non-profit ecological organizations such as The Nature Conservancy and land trusts.
- Creation of a website that functions as a public forum on biological control questions posed by ecologists or practitioners to guide selection and development of biological control projects targeting pests of wildlands.
- Development of a review of the magnitude and nature of the non-target effects of biological control agents of insect targets, complementing a recent article reviewing the non-target effects of herbivorous insects and plant pathogens used for weed biological control.
- Development of training materials on integration of biological control into conservation biology that can be made available for use in university classes on ecology or biological control, or in training workshops for land managers.
- Much conversation and good will.

Integrating Biological Control into Conservation Practice, First Edition. Edited by Roy G. Van Driesche, Daniel Simberloff, Bernd Blossey, Charlotte Causton, Mark S. Hoddle, David L. Wagner, Christian O. Marks, Kevin M. Heinz, and Keith D. Warner.
© 2016 John Wiley & Sons, Ltd. Published 2016 by John Wiley & Sons, Ltd.

Index

Note: page numbers followed by f or t refer to Figures or Tables

Integrating Biological Control into Conservation Practice, First Edition. Edited by Roy G. Van Driesche, Daniel Simberloff, Bernd Blossey, Charlotte Causton, Mark S. Hoddle, David L. Wagner, Christian O. Marks, Kevin M. Heinz, and Keith D. Warner.
© 2016 John Wiley & Sons, Ltd. Published 2016 by John Wiley & Sons, Ltd.